ICME-13 Monographs

Series editor

Gabriele Kaiser, Faculty of Education, Didactics of Mathematics, Universität Hamburg, Hamburg, Germany

More information about this series at http://www.springer.com/series/15585

Gabriele Kaiser
Editor

Proceedings of the 13th International Congress on Mathematical Education

ICME-13

Editor
Gabriele Kaiser
Faculty of Education
Universität Hamburg
Hamburg, Hamburg
Germany

ISSN 2520-8322 ISSN 2520-8330 (electronic)
ICME-13 Monographs
ISBN 978-3-319-62596-6 ISBN 978-3-319-62597-3 (eBook)
DOI 10.1007/978-3-319-62597-3

Library of Congress Control Number: 2017946637

© The Editor(s) (if applicable) and The Author(s) 2017. This book is an open access publication.
Open Access This book is licensed under the terms of the Creative Commons Attribution 4.0 International License (http://creativecommons.org/licenses/by/4.0/), which permits use, sharing, adaptation, distribution and reproduction in any medium or format, as long as you give appropriate credit to the original author(s) and the source, provide a link to the Creative Commons license and indicate if changes were made.
The images or other third party material in this book are included in the book's Creative Commons license, unless indicated otherwise in a credit line to the material. If material is not included in the book's Creative Commons license and your intended use is not permitted by statutory regulation or exceeds the permitted use, you will need to obtain permission directly from the copyright holder.
The use of general descriptive names, registered names, trademarks, service marks, etc. in this publication does not imply, even in the absence of a specific statement, that such names are exempt from the relevant protective laws and regulations and therefore free for general use.
The publisher, the authors and the editors are safe to assume that the advice and information in this book are believed to be true and accurate at the date of publication. Neither the publisher nor the authors or the editors give a warranty, express or implied, with respect to the material contained herein or for any errors or omissions that may have been made. The publisher remains neutral with regard to jurisdictional claims in published maps and institutional affiliations.

Printed on acid-free paper

This Springer imprint is published by Springer Nature
The registered company is Springer International Publishing AG
The registered company address is: Gewerbestrasse 11, 6330 Cham, Switzerland

Contents

Part I Plenary Activities

**Thirteenth International Congress on Mathematical Education:
An Introduction**... 3
Gabriele Kaiser

Uncovering the Special Mathematical Work of Teaching........... 11
Deborah Loewenberg Ball

**Mathematics, Education, and Culture: A Contemporary Moral
Imperative**... 35
Bill Barton

Mathematics Classroom Studies: Multiple Lenses and Perspectives.... 45
Berinderjeet Kaur

**"What is Mathematics?" and why we should ask, where
one should experience and learn that, and how to teach it**........... 63
Günter M. Ziegler and Andreas Loos

**International Comparative Studies in Mathematics: Lessons
and Future Directions for Improving Students' Learning**............ 79
Jinfa Cai, Ida A.C. Mok, Vijay Reddy and Kaye Stacey

Transitions in Mathematics Education: The Panel Debate........... 101
Ghislaine Gueudet, Marianna Bosch, Andrea A. diSessa,
Oh Nam Kwon and Lieven Verschaffel

Part II Awardees' lectures

ICMI Awards Ceremony....................................... 121
Carolyn Kieran and Jeremy Kilpatrick

Mathematics Discourse in Instruction (MDI): A Discursive Resource as Boundary Object Across Practices 125
Jill Adler

The Challenging Relationship Between Fundamental Research and Action in Mathematics Education............................ 145
Michèle Artigue

Elementary Mathematicians from Advanced Standpoints—A Cultural Perspective on Mathematics Education 165
Alan J. Bishop

Design and Development for Large-Scale Improvement 177
Hugh Burkhardt and Malcolm Swan

Making Sense of Mathematics Achievement in East Asia: Does Culture *Really* Matter? 201
Frederick K.S. Leung

Part III Reports of the Survey Teams

Digital Technology in Mathematics Education: Research over the Last Decade 221
Marcelo C. Borba, Petek Askar, Johann Engelbrecht, George Gadanidis, Salvador Llinares and Mario Sánchez Aguilar

Conceptualisation of the Role of Competencies, Knowing and Knowledge in Mathematics Education Research................ 235
Mogens Niss, Regina Bruder, Núria Planas, Ross Turner and Jhony Alexander Villa-Ochoa

Assistance of Students with Mathematical Learning Difficulties—How Can Research Support Practice?—A Summary 249
Petra Scherer, Kim Beswick, Lucie DeBlois, Lulu Healy and Elisabeth Moser Opitz

Mathematics Teachers Working and Learning Through Collaboration... 261
Barbara Jaworski, Olive Chapman, Alison Clark-Wilson, Annalisa Cusi, Cristina Esteley, Merrilyn Goos, Masami Isoda, Marie Joubert and Ornella Robutti

Geometry Education, Including the Use of New Technologies: A Survey of Recent Research....................................... 277
Nathalie Sinclair, Maria G. Bartolini Bussi, Michael de Villiers, Keith Jones, Ulrich Kortenkamp, Allen Leung and Kay Owens

Part IV Reports from the Thematic Afternoon

European Didactic Traditions in Mathematics: Aspects and Examples from Four Selected Cases. 291
Werner Blum, Michèle Artigue, Maria Alessandra Mariotti, Rudolf Sträßer and Marja Van den Heuvel-Panhuizen

German-Speaking Traditions in Mathematics Education Research 305
Hans Niels Jahnke, Rolf Biehler, Angelika Bikner-Ahsbahs, Uwe Gellert, Gilbert Greefrath, Lisa Hefendehl-Hebeker, Götz Krummheuer, Timo Leuders, Marcus Nührenbörger, Andreas Obersteiner, Kristina Reiss, Bettina Rösken-Winter, Andreas Schulz, Andreas Vohns, Rudolf vom Hofe and Katrin Vorhölter

What Is and What Might Be the Legacy of Felix Klein? 321
Hans-Georg Weigand, William McCallum, Marta Menghini, Michael Neubrand, Gert Schubring and Renate Tobies

Part V National Presentations

Argentinean National Presentation 337
Esther Galina and Mónica Villarreal

Teachers' Professional Development and Mathematics Education in Brazil. 345
Victor Giraldo

Mathematics Education in Ireland. 347
Maurice OReilly, Thérèse Dooley, Elizabeth Oldham and Gerry Shiel

National Presentation of Japan 353
Toshiakira Fujii, Yoshinori Shimizu, Hanako Senuma and Toshikazu Ikeda

National Presentations of Lower Mekong Sub-region Countries 361
Fidel R. Nemenzo, Masami Isoda, Maitree Inprasitha, Sampan Thinwiangthong, Narumon Changsri, Nisakorn Boonsena, Chan Roth, Monkolsery Lin, Souksomphone Anothay, Phoutsakhone Channgakham, Nguyen Chi Thanh, Vũ Như Thư Hương and Phương Thảo Nguyễn

Teaching and Learning Mathematics in Turkey. 367
Huriye Arikan

Part VI Reports from the Topical Study Groups

Topic Study Group No. 1: Early Childhood Mathematics Education (Up to Age 7) 375
Elia Iliada, Joanne Mulligan, Ann Anderson, Anna Baccaglini-Frank and Christiane Benz

Topic Study Group No. 2: Mathematics Education at Tertiary Level 381
Victor Giraldo, Chris Rasmussen, Irene Biza, Azimehsadat Khakbaz and Reinhard Hochmuth

Topic Study Group No. 3: Mathematics Education in and for Work 387
Geoff Wake, Diana Coben, Burkhard Alpers, Keith Weeks and Peter Frejd

Topic Study Group No. 4: Activities for, and Research on, Mathematically Gifted Students 391
Florence Mihaela Singer, Linda Jensen Sheffield, Matthias Brandl, Viktor Freiman and Kyoko Kakihana

Topic Study Group No. 5: Classroom Practice and Research for Students with Mathematical Learning Difficulties ... 397
Lourdes Figueiras, Rose Griffiths, Karen Karp, Jens Holger Lorenz and Miriam Godoy Penteado

Topic Study Group No. 6: Adult Learning 401
Jürgen Maaß, Pradeep Kumar Misra, Terry Maguire, Katherine Safford-Ramus, Wolfgang Schlöglmann and Evelyn Süss-Stepancik

Topic Study Group No. 07: Popularization of Mathematics 405
Christian Mercat, Patrick Vennebush, Chris Budd, Carlota Simões and Jens Struckmeier

Topic Study Group No. 8: Teaching and Learning of Arithmetic and Number Systems (Focus on Primary Education) ... 413
Pi-Jen Lin, Terezinha Nunes, Shuhua An, Beatriz Vargas Dorneles and Elisabeth Rathgeb-Schnierer

Topic Study Group No. 9: Teaching and Learning of Measurement (Focus on Primary Education) 415
Christine Chambris, Barbara Dougherty, Kalyanasundaram (Ravi) Subramaniam, Silke Ruwisch and Insook Chung

**Topic Study Group No. 10: Teaching and Learning
of Early Algebra** 421
Carolyn Kieran, JeongSuk Pang, Swee Fong Ng, Deborah Schifter
and Anna Susanne Steinweg

Topic Study Group No. 11: Teaching and Learning of Algebra 425
Rakhi Banerjee, Amy Ellis, Astrid Fischer, Heidi Strømskag
and Helen Chick

**Topic Study Group No. 12: Teaching and Learning of Geometry
(Primary Level)** 429
Sinan Olkun, Ewa Swoboda, Paola Vighi, Yuan Yuan and Bernd Wollring

**Topic Study Group No. 13: Teaching and Learning
of Geometry—Secondary Level** 435
Ui Hock Cheah, Patricio G. Herbst, Matthias Ludwig,
Philippe R. Richard and Sara Scaglia

Topic Study Group No. 14: Teaching Learning of Probability 439
Carmen Batanero, Egan J. Chernoff, Joachim Engel,
Hollylynne Stohl Lee and Ernesto Sánchez

Topic Study Group No. 15: Teaching and Learning of Statistics 443
Dani Ben-Zvi, Gail Burrill, Dave Pratt, Lucia Zapata-Cardona
and Andreas Eichler

Topic Study Group No. 16: Teaching and Learning of Calculus 447
David Bressoud, Victor Martinez-Luaces, Imène Ghedamsi
and Günter Törner

**Topic Study Group No. 17: Teaching and Learning
of Discrete Mathematics** 453
Eric W. Hart, James Sandefur, Cecile O. Buffet,
Hans-Wolfgang Henn and Ahmed Semri

**Topic Study Group No. 18: Reasoning and Proof
in Mathematics Education** 459
Guershon Harel, Andreas J. Stylianides, Paolo Boero,
Mikio Miyazaki and David Reid

**Topic Study Group No. 19: Problem Solving in Mathematics
Education** 463
Peter Liljedahl, Manuel Santos-Trigo, Uldarico Malaspina,
Guido Pinkernell and Laurent Vivier

**Topic Study Group No. 20: Visualization in the Teaching
and Learning of Mathematics** 467
Michal Yerushalmy, Ferdinand Rivera, Boon Liang Chua,
Isabel Vale and Elke Söbbeke

**Topic Study Group No. 21: Mathematical Applications
and Modelling in the Teaching and Learning of Mathematics** 471
Jussara Araújo, Gloria Ann Stillman, Morten Blomhøj,
Toshikazu Ikeda and Dominik Leiss

**Topic Study Group No. 22: Interdisciplinary Mathematics
Education**. 475
Susie Groves, Julian Williams, Brian Doig, Rita Borromeo Ferri
and Nicholas Mousoulides

Topic Study Group No. 23: Mathematical Literacy 481
Hamsa Venkat, Iddo Gal, Eva Jablonka, Vince Geiger
and Markus Helmerich

**Topic Study Group No. 24: History of the Teaching and Learning
of Mathematics** 487
Fulvia Furinghetti, Alexander Karp, Henrike Allmendinger,
Johan Prytz and Harm Jan Smid

**Topic Study Group No. 25: The Role of History of Mathematics
in Mathematics Education** 491
Constantinos Tzanakis, Xiaoqin Wang, Kathleen Clark,
Tinne Hoff Kjeldsen and Sebastian Schorcht

**Topic Study Group No. 26: Research on Teaching and Classroom
Practice** 497
Yoshinori Shimizu, Mary Kay Stein, Birgit Brandt,
Helia Oliveira and Lijun Ye

**Topic Study Group No. 27: Learning and Cognition
in Mathematics** 501
Gaye Williams, Wim Van Dooren, Pablo Dartnell,
Anke Lindmeier and Jérôme Proulx

**Topic Study Group No. 28: Affect, Beliefs and Identity
in Mathematics Education** 507
Markku Hannula, Francesca Morselli, Emine Erktin, Maike Vollstedt
and Qiao-Ping Zhang

Topic Study Group No. 29: Mathematics and Creativity. 511
Demetra Pitta-Pantazi, Dace Kūma, Alex Friedlander, Thorsten Fritzlar
and Emiliya Velikova

Topic Study Group No. 30: Mathematical Competitions 515
Maria Falk de Losada, Alexander Soifer, Jaroslav Svrcek and Peter Taylor

Topic Study Group No. 31: Language and Communication in Mathematics Education 521
Judit Moschkovich, David Wagner, Arindam Bose,
Jackeline Rodrigues Mendes and Marcus Schütte

Topic Study Group No. 32: Mathematics Education in a Multilingual and Multicultural Environment 525
Richard Barwell, Anjum Halai, Aldo Parra, Lena Wessel and Guida de Abreu

Topic Study Group No. 33: Equity in Mathematics Education (Including Gender) 531
Bill Atweh, Joanne Rossi Becker, Barbro Grevholm, Gelsa Knijnik, Laura Martignon and Jayasree Subramanian

Topic Study Group No. 34: Social and Political Dimensions of Mathematics Education 537
Murad Jurdak, Renuka Vithal, Peter Gates, Elizabeth de Freitas and David Kollosche

Topic Study Group No. 35: Role of Ethnomathematics in Mathematics Education 543
Milton Rosa, Lawrence Shirley, Maria Elena Gavarrete and Wilfredo V. Alangui

Topic Study Group No. 36: Task Design, Analysis and Learning Environments Programme Summary 549
Jere Confrey, Jiansheng Bao, Anne Watson, Jonei Barbosa and Helmut Linneweber-Lammerskitten

Topic Study Group No. 37: Mathematics Curriculum Development 555
Anita Rampal, Zalman Usiskin, Andreas Büchter, Jeremy Hodgen and Iman Osta

Topic Study Group No. 38: Research on Resources (Textbooks, Learning Materials etc.) 561
Lianghuo Fan, Luc Trouche, Chunxia Qi, Sebastian Rezat and Jana Visnovska

Topic Study Group No. 39: Large Scale Assessment and Testing in Mathematics Education 565
Rae Young Kim, Christine Suurtamm, Edward Silver, Stefan Ufer and Pauline Vos

Topic Study Group No. 40: Classroom Assessment for Mathematics Learning 571
Denisse R. Thompson, Karin Brodie, Leonora Diaz Moreno, Nathalie Sayac and Stanislaw Schukajlow

Topic Study Group No. 41: Uses of Technology in Primary Mathematics Education (Up to Age 10) 575
Sophie Soury-Lavergne, Colleen Vale, Francesca Ferrara, Krongthong Khairiree and Silke Ladel

Topic Study Group No. 42: Uses of Technology in Lower Secondary Mathematics Education (Age 10–14) 577
Lynda Ball, Paul Drijvers, Bärbel Barzel, Yiming Cao and Michela Maschietto

Topic Study Group No. 43: Uses of Technology in Upper Secondary Education (Age 14–19) 579
Stephen Hegedus, Colette Laborde, Luis Moreno Armella, Hans-Stefan Siller and Michal Tabach

Topic Study Group No. 44: Distance Learning, e-Learning, and Blended Learning 583
Rúbia Barcelos Amaral, Veronica Hoyos, Els de Geest, Jason Silverman and Rose Vogel

Topic Study Group No. 45: Knowledge in/for Teaching Mathematics at Primary Level 585
Carolyn A. Maher, Peter Sullivan, Hedwig Gasteiger and Soo Jin Lee

Topic Study Group No. 46: Knowledge in/for Teaching Mathematics at the Secondary Level 589
Ruhama Even, Xinrong Yang, Nils Buchholtz, Charalambos Charalambous and Tim Rowland

Topic Study Group No. 47: Pre-service Mathematics Education of Primary Teachers 593
Keiko Hino, Gabriel J. Stylianides, Katja Eilerts, Caroline Lajoie and David Pugalee

Topic Study Group No. 48: Pre-service Mathematics Education of Secondary Teachers 599
Marilyn Strutchens, Rongjin Huang, Leticia Losano, Despina Potari and Björn Schwarz

Topic Study Group No. 49: In-Service Education and Professional Development of Primary Mathematics Teachers 605
Akihiko Takahashi, Leonor Varas, Toshiakira Fujii, Kim Ramatlapana and Christoph Selter

Topic Study Group No. 50: In-Service Education, and Professional Development of Secondary Mathematics Teachers 609
Jill Adler, Yudong Yang, Hilda Borko, Konrad Krainer and Sitti Patahuddin

Contents

Topic Study Group No. 51: Diversity of Theories in Mathematics Education... 613
Tommy Dreyfus, Anna Sierpinska, Stefan Halverscheid,
Steve Lerman and Takeshi Miyakawa

Topic Study Group 52: Empirical Methods and Methodologies....... 619
David Clarke, Alan Schoenfeld, Bagele Chilisa, Paul Cobb
and Christine Knipping

Topic Study Group No. 53: Philosophy of Mathematics Education.... 623
Paul Ernest, Ladislav Kvasz, Maria Bicudo, Regina Möller
and Ole Skovsmose

Topic Study Group No. 54: Semiotics in Mathematics Education...... 627
Norma Presmeg, Luis Radford, Gert Kadunz, Luis Puig
and Wolff-Michael Roth

Part VII Reports from the Discussion Groups

Classroom Teaching Research for All Students..................... 635
Shuhua An, Steklács János and Zhonghe Wu

Mathematical Discourse in Instruction in Large Classes............. 637
Mike Askew, Ravi K. Subramaniam, Anjum Halai, Erlina Ronda,
Hamsa Venkat, Jill Adler and Steve Lerman

Sharing Experiences About the Capacity and Network Projects Initiated by ICMI... 639
Angelina Matinde Bijura, Alphonse Uworwabayeho, Veronica Sarungi,
Peter Kajoro and Anjum Halai

Mathematics Teacher Noticing: Expanding the Terrains of This Hidden Skill of Teaching................................ 641
Ban Heng Choy, Jaguthsing Dindyal, Mi Yeon Lee
and Edna O. Schack

Connections Between Valuing and Values: Exploring Experiences and Rethinking Data Generating Methods........................ 643
Philip Clarkson, Annica Andersson, Alan Bishop,
Penelope Kalogeropoulos and Wee Tiong Seah

Developing New Teacher Learning in Schools and the STEM Agenda... 645
Pat Drake, Jeanne Carroll, Barbara Black, Lin Phillips
and Celia Hoyles

Videos in Teacher Professional Development...................... 647
Tanya Evans, Leong Yew Hoong and Ho Weng Kin

**National and International Investment Strategies for Mathematics
Education**.. 649
Joan Ferrini-Mundy, Marcelo C. Borba, Fumi Ginshima,
Manfred Prenzel and Thierry Zomahoun

Transition from Secondary to Tertiary Education................. 651
Gregory D. Foley, Sergio Celis, Hala M. Alshawa, Sidika Nihan,
Heba Bakr Khoshaim and Jane D. Tanner

Teachers Teaching with Technology 655
Ian Galloway, Bärbel Barzel and Andreas Eichler

Mathematics Education and Neuroscience 657
Roland H. Grabner, Andreas Obersteiner, Bert De Smedt, Stephan Vogel,
Michael von Aster, Roza Leikin and Hans-Christoph Nuerk

Reconsidering Mathematics Education for the Future............. 659
Koeno Gravemeijer, Fou-Lai Lin, Michelle Stephan, Cyril Julie
and Minoru Ohtani

**Challenges in Teaching Praxis When CAS Is Used in Upper
Secondary Mathematics** 661
Niels Groenbaek, Claus Larsen, Henrik Bang, Hans-Georg Weigand,
Zsolt Lavicza, John Monaghan, M. Kathleen Heid, Mike Thomas
and Paul Drijvers

**Mathematics in Contemporary Art and Design as a Tool
for Math-Education in School** 663
Dietmar Guderian

**Exploring the Development of a Mathematics Curriculum
Framework: Cambridge Mathematics**............................ 665
Ellen Jameson, Rachael Horsman and Lynne McClure

**Theoretical Frameworks and Ways of Assessment of Teachers'
Professional Competencies** 667
Johannes König, Sigrid Blömeke and Gabriele Kaiser

**Using Representations of Practice for Teacher Education
and Research—Opportunities and Challenges** 669
Sebastian Kuntze, Orly Buchbinder, Corey Webel, Anika Dreher
and Marita Friesen

How Does Mathematics Education Evolve in the Digital Era? 671
Dragana Martinovic and Viktor Freiman

Scope of Standardized Tests................................... 673
Raimundo Olfos, Ivan R. Vysotsky, Manuel Santos-Trigo,
Masami Isoda and Anita Rampal

Contents

**Mathematics for the 21st Century School: The Russian Experience
and International Prospects** 675
Sergei A. Polikarpov and Alexei L. Semenov

Lesson/Learning Studies and Mathematics Education 677
Marisa Quaresma and Carl Winsløw

Mathematics Houses and Their Impact on Mathematics Education 679
Ali Rejali, Peter Taylor, Yahya Tabesh, Jérôme Germoni
and Abolfazl Rafiepour

An Act of Mathematisation: Familiarisation with Fractions 681
Ernesto Rottoli, Sabrina Alessandro, Petronilla Bonissoni,
Marina Cazzola, Paolo Longoni and Gianstefano Riva

The Role of Post-Conflict School Mathematics 683
Carlos Eduardo Leon Salinas and Jefer Camilo Sachica Castillo

**Applying Contemporary Philosophy in Mathematics
and Statistics Education: The Perspective of Inferentialism**. 685
Maike Schindler, Kate Mackrell, Dave Pratt and Arthur Bakker

Teaching Linear Algebra 687
Sepideh Stewart, Avi Berman, Christine Andrews-Larson
and Michelle Zandieh

Creativity, Aha!Moments and Teaching-Research 689
Hannes Stoppel and Bronislaw Czarnocha

**White Supremacy, Anti-Black Racism, and Mathematics
Education: Local and Global Perspectives** 691
Luz Valoyes-Chávez, Danny B. Martin, Joi Spencer and Paola Valero

Research on Non-university Tertiary Mathematics 693
Claire Wladis, John Smith and Irene Duranczyk

Part VIII   Reports from the Workshops

**Flipped Teaching Approach in College Algebra: Cognitive
and Non-cognitive Gains.** 697
Maxima J. Acelajado

**A Knowledge Discovery Platform for Spatial Education:
Applications to Spatial Decomposition and Packing**. 699
Sorin Alexe, Cristian Voica and Consuela Voica

Designing Mathematics Tasks for the Professional Development of Teachers Who Teach Mathematics Students Aged 11–16 Years 701
Debbie Barker and Craig Pournara

Contributing to the Development of Grand Challenges in Maths Education.. 703
David Barnes, Trena Wilkerson and Michelle Stephan

The Role of the Facilitator in Using Video for the Professional Learning of Teachers of Mathematics............................ 705
Alf Coles, Aurelie Chesnais and Julie Horoks

Making Middle School Maths Real, Relevant and Fun 707
Kerry Cue

"Oldies but Goodies": Providing Background to ICMI Mission and Activities from an Archival Perspective.......................... 709
Guillermo P. Curbera, Bernard R. Hodgson and Birgit Seeliger

Using Braids to Introduce Groups: From an Informal to a Formal Approach... 711
Ester Dalvit

Curious Minds; Serious Play 713
Jan de Lange

International Similarities and Differences in the Experiences and Preparation of Post-Graduate Mathematics Students as Tertiary Instructors.. 717
Jessica Deshler and Jessica Ellis

Using LISP as a Tool for Mathematical Experimentation 719
Hugo Alex Diniz

Mathematics Teachers' Circles as Professional Development Models Connecting Teachers and Academics............................ 721
Nathan Borchelt and Axelle Faughn

Exploring and Making Online Creative Digital Math Books for Creative Mathematical Thinking................................ 723
Pedro Lealdino Filho, Christian Bokhove, Jean-Francois Nicaud, Ulrich Kortenkamp, Mohamed El-Demerdash, Manolis Mavrikis and Eirini Geraniou

The Shift of Contents in Prototypical Tasks Used in Education Reforms and Their Influence on Teacher Training Programs......... 725
Karl Fuchs, Christian Kraler and Simon Plangg

Analysis of Algebraic Reasoning and Its Different Levels in Primary and Secondary Education... 727
Juan D. Godino, Teresa Neto and Miguel R. Wilhelmi

Designing and Evaluating Mathematical Learning by a Framework of Activities from History of Mathematics 729
Lenni Haapasalo, Harry Silfverberg and Bernd Zimmermann

Sounding Mathematics: How Integrating Mathematics and Music Inspires Creativity and Inclusion in Mathematics Education 731
Caroline Hilton and Markus Cslovjecsek

Adopting Maxima as an Open-Source Computer Algebra System into Mathematics Teaching and Learning......................... 733
Natanael Karjanto and Husty Serviana Husain

The Power of Geometry in the Concept of Proof................. 735
Damjan Kobal

Workshop: Silent Screencast Videos and Their Use When Teaching Mathematics................................. 737
Bjarnhciður Bea Kristinsdóttir

Shout from the Most Silent Nation, North Korea: Can Mathematics Education Be Politically Neutral? 739
JungHang Lee

Workshop Theme: "Use of Educational Large-Scale Assessment Data for Research on Mathematics Didactics" 741
Sabine Meinck, Oliver Neuschmidt and Milena Taneva

Curriculum Development in the Teaching of Mathematical Proof at the Secondary Schools in Japan 743
Tatsuya Mizoguchi, Hideki Iwasaki, Susumu Kunimune, Hiroaki Hamanaka, Takeshi Miyakawa, Yusuke Shinno, Yuki Suginomoto and Koji Otaki

Symmetry, Chirality, and Practical Origami Nanotube Construction Techniques 745
B. David Redman Jr.

Reflecting Upon Different Perspectives on Specialized Advanced Mathematical Knowledge for Teaching 747
Miguel Ribeiro, Arne Jakobsen, Alessandro Ribeiro, Nick H. Wasserman, José Carrillo, Miguel Montes and Ami Mamolo

Collaborative Projects in Geometry............................. 749
José L. Rodríguez, David Crespo and Dolores Jiménez

Workshop on Framing Non-routine Problems in Mathematics for Gifted Children of Age Group 11–15 751
Sundaram R. Santhanam

Enacted Multiple Representations of Calculus Concepts, Student Understanding and Gender 753
Ileana Vasu

Using Inquiry to Teach Mathematics in Secondary and Post-secondary Education 755
Volker Ecke and Christine von Renesse

Making of Cards as Teaching Material for Spatial Figures 757
Kazumi Yamada and Takaaki Kihara

Creative Mathematics Hands-on Activities in the Classroom 759
Janchai Yingprayoon

Part IX Additional Activities

Teachers Activities at ICME-13 763
Nils Buchholtz, Marianne Nolte and Gabriele Kaiser

Early Career Researcher Day at ICME-13 765
Gabriele Kaiser, Thorsten Scheiner and Armin Jentsch

Part I
Plenary Activities

Thirteenth International Congress on Mathematical Education: An Introduction

Gabriele Kaiser

Abstract The paper describes the vision of the 13th International Congress on Mathematical Education (ICME-13), accompanied by detailed elaborations on the structure of ICME-13 and important data.

The 13th International Congress on Mathematical Education (ICME-13) took place from 24 to 31 July 2016 in Hamburg, hosted by the Gesellschaft für Didaktik der Mathematik (Society of Didactics of Mathematics) under the auspices of the International Commission on Mathematical Instruction (ICMI).

ICME-13 had 3486 participants, with 360 accompanying persons, making it the largest ICME so far. Congress participants came from 105 countries, i.e., more than half of the countries in the world were present. Two hundred and fifty teachers attended additional activities that took place during ICME-13. Directly before the beginning of ICME-13, 450 early-career researchers attended a day-long specific programme. These high participation numbers strongly indicate that mathematics education has become a widely accepted scientific discipline with its own structure and standards. Furthermore, it documents the growing international community of mathematics educators.

At the opening ceremony, the five ICMI awards were presented to Michèle Artigue and Alan Bishop (Felix Klein award), Jill Adler and Frederick Leung (Hans Freudenthal award), Hugh Burkhardt and Malcolm Swan (Emma Castelnuovo award). Their presentations can be found in these proceedings together with a short introduction by Carolyn Kieran and Jeremy Kilpatrick.

The heart of the congress consisted of 54 Topic Study Groups, devoted to major themes of mathematics education, in which 745 presentations were given. In attached oral communications, 931 shorter papers were presented, complemented by 533 posters presented in two sessions.

G. Kaiser (✉)
Universität Hamburg (ICME-13), Hamburg, Germany
e-mail: Gabriele.Kaiser@uni-hamburg.de

Two plenary panels presented their points of view on:

- International comparative studies in mathematics: Lessons for improving students' learning, with Jinfa Cai (Chair), Ida Mok, Vijay Reddy and Kaye Stacey
- Transitions in mathematics education, with Ghislaine Gueudet (Chair), Marianna Bosch, Andrea diSessa, Oh Nam Kwon and Lieven Verschaffel.

Four plenary lectures took place:

- Uncovering the special mathematical work of teaching, by Deborah Loewenberg Ball
- Mathematics education in its cultural context: Plus and minus 30 years, by Bill Barton
- Mathematics classroom studies: Multiple windows and perspectives, by Berinderjeet Kaur
- "What is mathematics?" and why we should ask, where one should learn that, and who can teach it, by Günter M. Ziegler.

In addition, 64 invited lectures were given by scholars from all over the world presenting the state of the art in their research field. The second volume of the proceedings of ICME-13 will publish these lectures.

38 discussion groups and 42 workshops initiated by congress participants were offered in which a great variety of themes were discussed, fostering international collaboration.

Reflecting specific ICMI traditions, five ICMI survey teams described the state of the art on the following themes:

- Distance learning, blended learning, e-learning in mathematics (chaired by Marcelo Borba)
- Conceptualisation of the role of competencies, knowing and knowledge in mathematics education research (chaired by Mogens Niss)
- Assistance of students with mathematical learning difficulties: How can research support practice? (chaired by Petra Scherer)
- Teachers working and learning through collaboration (chaired by Barbara Jaworski)
- Recent research on geometry education (chaired by Nathalie Sinclair).

The first results of these survey teams were published as Issue 5 in 2016 of *ZDM Mathematics Education* (http://link.springer.com/journal/volumesAndIssues/11858); short versions of these reports can be found in this volume of the proceedings.

Three ICMI studies presented results that already have been published or will be published by Springer in the new ICMI Study Series:

- ICMI Study 21 on mathematics education and language diversity (Richard Barwell et al.)
- ICMI Study 22 on task design (Anne Watson and Minoru Ohtani)
- ICMI Study 23 on primary mathematics study of whole numbers (Mariolina Bartolini Bussi and Xuhua Sun).

In addition, six national presentations were given describing the situation of mathematics education and its scholarly discussion in Argentina, Brazil, Ireland, Japan, the Lower Mekong Sub-Region and Turkey. Short descriptions of the presentations are given in this volume.

Apart of these impressive figures, the historical development is important: In 1976 another ICME had already taken place in Germany, namely the Third International Congress on Mathematical Education (ICME-3), which was held in Karlsruhe. The organisation of ICME-3 in 1976 reflected the German tradition of collaboration between mathematicians and mathematics educators, with mathematics educator Hans-Georg Steiner as Chair of the International Programme Committee and mathematician Heinz Kunle as Chief Organiser of the congress. The strong collaboration between mathematics and mathematics education has been further developed and the Deutsche Mathematiker-Vereinigung (German Mathematical Society) has strongly supported ICME-13 since the very beginning. The German community is the first international mathematics educational community to host an ICME a second time.

On the occasion of this special event a thematic afternoon was carried out devoted to the description of the development in the last 40 years from a European and a historical perspective. The thematic afternoon's topics were Selected European Didactic Traditions, German-speaking Traditions and the Legacy of Felix Klein. These special activities aimed to show the development of the German mathematics education discussion over the last 40 years, embedding it in a continental European context and in its historical development.

The German-speaking countries share many common roots with the continental European didactic traditions of mathematics education, including common pedagogical and philosophical traditions. These strong connections within the European tradition of *didactics* are already apparent in the word *Didaktik* in German, *didactique* in French, *didáctica* in Spanish, Italian and Czech, *didactiek* in Dutch, Danish and Swedish. This *Didaktik*-tradition can be found in many European countries and has as a common core a theoretical foundation of education with a strong normative orientation. This tradition goes back to the Czech pedagogue Comenius with his *Didactica Magna* (*The Great Didactic*). Comenius, who developed still modern approaches to education, is considered the father of modern education (Hudson & Meyer, 2011). Four distinctive features of these modern continental European traditions were identified within the selected European didactic traditions at this thematic afternoon: the strong connection between mathematics and mathematicians, the key roles of both theory and design activities for learning and teaching environments and a firm basis on empirical research. A short description of this topic can be found in these proceedings (Blum et al.), while a detailed description will be given in a book coming out in the series of ICME-13 monographs.

The second strand displayed the German-speaking traditions, which include Austria and Switzerland in addition to Germany. This strand of the discussion is especially connected to a particular approach to didactics of mathematics that is subject bound and strongly oriented towards mathematics (so-called *Stoffdidaktik*). This approach was already evident in Arnold Kirsch's keynote lecture, Aspects of Simplification in Mathematics Teaching, at ICME-3 in Karlsruhe and has been further developed in the last 40 years (Kirsch, 1977). Other distinctive features are related to applications and modelling, which play a prominent role in German mathematics education and were described at ICME-12 in Seoul by Werner Blum in his plenary talk (Blum, 2015). Another important feature of the German-speaking tradition discussion is the approach to mathematics education as design science aiming to bridge the gap between theory and practice, which was put forward by the plenary talk of Erich Wittmann at ICME-9 in Tokyo (Wittmann, 2004). A short description of these presentations can be found in these proceedings (Jahnke et al.), while a detailed description will be given in a book coming out in the series of ICME-13 monographs.

The third strand of these special activities, the Legacy of Felix Klein, referred to the historical roots of German-speaking mathematics education. Felix Klein, the founding president of ICMI, shaped mathematics education not only nationally but internationally in several respects. His legacy was reflected upon from three perspectives, the first being functional thinking as one fundamental mathematical idea structuring mathematics education from the very beginning to university. The second perspective was intuitive thinking and visualisation, which reflects the high importance of *Anschauung* in German mathematics education. Felix Klein developed the *Modellkammer*, models of mathematical phenomena, which has been promoted in other parts of the world (Schubring, 2010). The mathematical exhibition from the *Mathematikum*, which has been on display during ICME-13, refers with its hands-on activities strongly to this tradition. A short description of this strand can be found in these proceedings (Weigand et al.), while a detailed description will be given in a book coming out in the series of ICME-13 monographs.

The last perspective is strongly connected to Felix Klein's famous books, *Elementarmathematik vom Höheren Standpunkte aus*, published originally from 1902 to 1909 in German with the first volume on arithmetic, algebra and analysis, the second on geometry and the third on precision and approximation mathematics (Klein, 1902–1908). The first two volumes were published in English with the title *Elementary Mathematics from an Advanced Standpoint* in 1932 (Volume 1) and 1939 (Volume 2). Supported by Springer Publishing, a new translation of the first two books from Felix Klein has come out on the occasion of ICME-13, called *Elementary Mathematics from a Higher Standpoint* (Klein, 2016). The wording of the title has been changed from *advanced* to *higher*, taking up the critique by Kilpatrick (2008/2014) at ICME-11 of the inadequate translation (2008/2014). The translation by Gert Schubring attempts to bring the English version closer to its German original, for example, by clarifying fundamental concepts for Klein's approach that were inadequately translated, such as *Anschauung*, which is insufficiently translated as *perception*. Furthermore, the third volume, *Precision Mathematics and Approximation Mathematics*, which was not been available in English, has now been translated by Marta Menghini in collaboration with Anna Baccaglini-Frank. It is a huge step forward for mathematics education that this work is now available in a complete and adequate form, because the connection of mathematics with its applications under a higher perspective was of particular importance to Felix Klein and was his lifelong theme. Jeremy Kilpatrick states concerning the importance of this work: "Despite the many setbacks he encountered, no mathematician had a more profound influence on mathematics education as a field of scholarship and practice" (p. 27).

Apart from this thematic afternoon as distinctive feature of ICME-13, an extensive publication programme was implemented in order to develop a sustainable congress from an academic perspective. One of our aims with the publication of the ICME-13 Topical Surveys was to display the state of the art concerning specific mathematics educational themes in the style of independent handbook chapters. 26 ICME-13 Topical Surveys were published, and the important aspect of these Topical Surveys coming out before ICME-13 is that they were available as open access and hopefully formed the basis for many discussions at the congress. They displayed what we knew before the congress. The forthcoming post-congress monographs based on the papers presented within the framework of the topic study and discussion groups describe the academic outcome of ICME-13 in more detail and will hopefully contribute to a sustainable congress.

It is our strong hope that ICME-14, which will take place in Shanghai in 2020, will be able to build its work on the insights achieved and published here and can thereby strongly foster the development of knowledge on the teaching and learning of mathematics on a higher basis.

The aforementioned books from Felix Klein, *Elementary Mathematics from a Higher Standpoint* (2016), originated from lectures Felix Klein gave to prospective teachers. His desire in publishing these books was to develop the ability of the prospective teachers to use the rich mathematics they were learning at university as vivid stimulation for their own teaching afterwards. This strong tradition that shaped the German-speaking community has led to many activities in pre-service and in-service teacher education.

During ICME-13, three days of German-language activities for teachers were conducted in which scholars participating in ICME-13 worked with practising teachers in workshops and lectures and offered them background knowledge or new teaching ideas. These activities for teachers were supported by the MNU - Verband zur Förderung des MINT-Unterrichts (German Association for the Advancement of Mathematics and Science Education), a teacher community, which has supported ICME-13 from the very beginning.

The final characteristics of ICME-13 to be mentioned are the activities for early career researchers. Early career researchers are our future, because they have to shoulder the task to further develop the science of mathematics education and to implement these improvements at all educational levels. We have seen in the past a strong development towards higher quality standards of research. Publishing a study needs nowadays to fulfil many requirements concerning theoretical framework and methodology used. Furthermore, publications have become more and more important in the last years. Therefore, ICME-13 held an early career researcher day with 450 participants where thematic surveys were presented and empirical methodologies prominent in mathematics education were discussed. In addition, descriptions of selected mathematics educational journals by the editors of those journals were followed by workshops on academic publishing and writing. These kinds of activities are highly necessary and should in the future be an integral part of ICMEs.

Finally, it is the tradition at each ICME to devote 10% of the congress fees to a solidarity grant in order to support scholars from less affluent countries. With the support of the Federal Ministry of Education and Research and the Bosch Foundation, ICME-13 was able to spend nearly 9% of the whole congress budget, about 230,000 Euros, for scholars from less affluent countries, supporting 223 participants from 66 countries. A special focus was set on African scholars; ICME-13 was able to support 50 African scholars from 19 countries. These efforts reflect the strong will of the German society to express solidarity with less wealthy regions and take responsibility for helping those regions. It will be our task to continue these efforts to insure equitable access not only to mathematics instruction in school for all people but also to the academic discussion on mathematics education for scholars all over the world, making an ICME a unique international experience.

ICME-13 has been the biggest ICME so far and has allowed many scholars from all over the world to participate actively. It will be our ongoing task to broaden the participation in ICMEs and to encourage scholars from all over the world to engage in and enrich all future ICMEs.

Acknowledgements I would like to thank Lena Pankow for her strong and continuing support not only during the congress ICME-13, but in the work for this volume as well.

References

Blum, W. (2015). Quality teaching of mathematical modelling: What do we know, what can we do? In S. J. Cho (Ed.), *The Proceedings of the 12th International Congress on Mathematical Education* (pp. 73–96). Cham: Springer.
Hudson, B., & Meyer, M. A. (Eds.). (2011). *Beyond fragmentation: Didactics, learning and teaching in Europe*. Opladen: Barbara Budrich Publishers.
Kilpatrick, J. (2008/2014). *A higher standpoint*. Materials from ICME-11. Regular lectures (pp. 26–43). http://www.mathunion.org/fileadmin/ICMI/files/About_ICMI/Publications_about_ICMI/ICME_11/Kilpatrick.pdf (last access 5.1.2017).
Kirsch, A. (1977). Aspects of simplification in mathematics teaching. In H. Athen & H. Kunle (Eds.), *Proceedings of the Third International Congress on Mathematical Education* (pp. 98–120). Karlsruhe: Zentralblatt für Didaktik der Mathematik, Universität (West) Karlsruhe.
Klein, F. (1902–1908). *Elementarmathematik vom höheren Standpunkte aus* (Vol. 1–3). Berlin: Verlag von Julius Springer.
Klein, F. (2016). *Elementary mathematics from a higher standpoint* (Vol. 1–3) (Vol. 1–2, G. Schubring, Trans.). Vol. 3 by Martha Menghini in collaboration with Anna Baccaglini-Frank. Heidelberg: Springer.
Schubring, G. (2010). Historical comments on the use of technology and devices in ICMEs and ICMI. *ZDM Mathematics Education, 42*(1), 5–9.

Wittmann, E. C. (2004). Developing mathematics in a systematic process. In H. Fujita, Y. Hashimoto, B. R. Hodgson, P. Y. Lee, S. Lerman, & T. Sawada (Eds.), *Proceedings of the Ninth International Congress on Mathematical Education* (pp. 73–90). Dordrecht: Kluwer Academic Publishers.

Open Access Except where otherwise noted, this chapter is licensed under a Creative Commons Attribution 4.0 International License. To view a copy of this license, visit http://creativecommons.org/licenses/by/4.0/.

Uncovering the Special Mathematical Work of Teaching

Deborah Loewenberg Ball

Abstract Helping young people develop mathematical skills, ways of thinking, and identities, and supporting classrooms as equitable communities of practice, entails for teachers a specialized set of instructional skills specific to the domain. This paper argues that, although progress has been made in understanding "mathematical knowledge for teaching," more study is needed to understand interactive mathematical work of teaching and to orient teachers' professional education to this dynamic and performative mathematical fluency and activity.

Introduction

A basic problem for both policy and practice is to identify what teachers need to know in order to teach mathematics well. Although it is obvious that teaching depends on knowing the subject, unanswered questions about the specific knowledge needed to teach mathematics have preoccupied teacher educators and researchers alike. This paper traces the effort to frame and investigate this problem and to develop useful ways to understand and solve it.

A Common Question: "How Much" Mathematics Do Teachers Need to Know?

The quest to identify and quantify teachers' mathematical knowledge dates back several decades. There is widespread agreement that teachers must know mathematics in order to teach it. This has been taken for granted. Although many

D.L. Ball (✉)
University of Michigan, Ann Arbor, Michigan, USA
e-mail: dball@umich.edu

researchers, policymakers, and teacher educators expressed concern that teachers typically did not know enough mathematics, less consensus has been reached about how much mathematics teachers needed in order to teach well. This has led to claims, reports, and recommendations focused on the number—or sometimes the content—of courses that teachers should take. Many arguments have centered on how much mathematics teachers should know, others on what is most important to know. Although it might seem straightforward, the question of the mathematics teachers need to know has been not at all simple to answer convincingly.

Although the basic assumption seemed obvious—after all, how can one teach something that one does not know well?—numerous studies failed to show that the amount of mathematics that teachers study clearly or consistently predicts their students' learning.[1] "Amount" tended to be measured in terms of attainment, either by completing a concentration in mathematics at the postsecondary level or by taking a certain number of university-level courses. This was an unsettling discovery in some ways, but it led to a new question: what mathematical skill and insight does teaching actually require? Clearly, it requires mathematics, but if it is not the amount of knowledge, then what is it about the mathematics that matters for good teaching?

These questions were far from new. Over a century ago, Dewey (1902) had flagged the special way of thinking about content through the mind of the child. But common worries about teachers' knowledge had nonetheless persisted, without satisfactory ways to articulate exactly the nature of this special way of thinking. Shulman and his colleagues (1986, 1987) aptly named it "pedagogical content knowledge," which significantly advanced the field. Researchers around the world probed the mathematical knowledge needed for teaching and began to find better answers (e.g., Adler & Davis, 2006; Ball, Thames, & Phelps, 2008; Baumert et al., 2010; Blömeke et al., 2015; Bruckmaier, Krauss, Blum, & Leiss, 2016; Carrillo, Climent, Contreras, & Muñoz-Catalán, 2013; Herbst & Kosko, 2014; Hill, Schilling, & Ball, 2004; Kaiser, Busse, Hoth, König, & Blömeke, 2015; Knievel, Lindmeier, & Heinze, 2015; McCrory, Floden, Ferrini-Mundy, Reckase, & Senk, 2012; Rowland, Huckstep, & Thwaites, 2005; Saderholm, Ronau, Brown, & Collins, 2010; Senk et al., 2012; Tatto et al., 2008; Tchoshanov, 2011). Studies have ranged from investigations of what teachers (and preservice teachers) know (or lack) (e.g., Ball, 1990; Baumert et al., 2010; Hill, 2007; Rowland et al., 2005; Thompson, 1984); what teachers learn from interventions, or other opportunities to learn mathematics (e.g., Borko et al., 1992; Hiebert, Morris, & Glass, 2003); to articulating positions about what teachers *should* know (e.g., Conference Board of Mathematical Sciences, 2001, 2012; McCrory et al., 2012; Silverman & Thompson, 2008). Many efforts were made to get closer to the use of mathematics in teaching (e.g., Adler & Rhonda, 2015; Ball et al., 2008; Bruckmaier et al., 2016; Goffney,

---

[1] A thorough review of relevant studies that investigate relationships between teachers' mathematical knowledge and students' learning and teaching quality can be found in National Mathematics Advisory Panel (2008).

2010, 2014; Goffney & Hoover, 2017; Herbst & Chazan, 2015; Hoover, Mosvold, & Fauskanger, 2014; Hill, 2011; Hill & Ball, 2004; Hill, Rowan, & Ball, 2005; Rowland, 2013; Sfard, 2007; Sherin, Jacobs, & Phillipp, 2011; Thompson, Carlson, & Silverman, 2007).

Some scholars developed measures of this special kind of knowledge (e.g., Bruckmaier et al., 2016; Herbst & Kosko, 2014; Hill, Ball, & Schilling, 2008; Hill et al., 2004). It is beyond the scope of this paper to represent or discuss the many projects that sought to understand in more nuanced ways the kind of mathematical skill and insight teaching actually requires. Important to note, however, is that scholars shifted from asking "what mathematics do teachers need to know" to "how is mathematics used in teaching" (Ball, Lubienski, & Mewborn, 2001).

Alongside this quest to uncover how mathematics is used in teaching, a strong emphasis on measurement was emerging in the broader political and scholarly environments. Funders encouraged assessment of impact and outcomes, and researchers responded by developing tests to evaluate teaching and studying how teaching relates to learning. Projects built a host of new tools, items, and tasks of all different kinds. The emphasis on measurement certainly helped to advance the effort to understand teacher knowledge; it also shifted the trajectory and impeded some aspects of the unanswered questions about the mathematical knowledge needed for teaching.

First consider briefly the advances. Across all of these efforts, researchers have made a great deal of progress in learning that there are special kinds of knowing of mathematics that matter for good teaching. We understand that it is not as simple as how many courses someone takes. We also developed better ways to study what teachers learn from teacher education and professional development. The tools and measures researchers built during this measurement period have helped us better understand what teachers learn. These tools hold potential to offer more precise information about what teachers might have learned than simply asking teachers what they learned, or whether they found the professional development useful or enjoyable. We now have better ways of assessing what teachers learn from professional education.

The emphasis on measurement, however, drew focus away from fundamental questions about the role of teachers' mathematical knowledge in teaching and its importance for students' learning. Although many researchers viewed teaching from sociocultural perspectives, asking about what teachers do with students in classrooms, the development of assessment tools was based in more individualistic and cognitivist perspectives.[2] Many started out trying to understand the mathematics in teaching, but more often ended up measuring individual cognitive capabilities of teachers instead. For many scholars, this invisible but significant shift in lens meant that the questions that were being asked and answered drifted away from the fundamental problems about mathematics knowledge and teaching.

---

[2]I am grateful to Anna Sfard for discussions and insights about this phenomenon (e.g., Sfard, 2007).

Research was not capturing the dynamic of what teachers actually *do* when they listen to students, make decisions about what to say next, move around the room, and decide on the next example. Scholars were studying classrooms and analyzing discourse, tasks, and interactions, but were not unpacking what is involved for the teacher in *doing* those things. The measurement work also led scholars to break up teaching into compartments, which is not the way teaching is enacted in practice. For example, work focused on mathematics was often separated from a focus on equity. However, in teaching, concerns for equity—who has the floor, who is being recognized, whose ideas are being valued—are entangled in the construction of mathematics, of what is asked and emphasized, and of what it means to do or be good at math.

The advances in assessment and measurement were important. As a scientific enterprise, the field had developed better microscopes. Because they had better tools, researchers were able to get closer to many micro-level aspects of teachers, including their values, beliefs, and reasoning; their competencies; and their mathematical, pedagogical, and professional content knowledge. These tools also took us inside classrooms and enabled us to see, study, and "measure" teaching—as researchers. However, we were not inside of what it takes to *do* teaching as a teacher. Capturing the patterns of student participation does not explain what goes on inside the practices of calling on, supporting, and distributing students' talk, or of constructing and distributing different kinds of talk turns, and to whom about what aspect of the mathematics. Describing how students are positioned by the teacher or their peers and how that is shaped by identities and perceptions does not open a window on to what it takes, in moment-to-moment interaction, to make the decisions, arrange the work, say particular things, and disrupt the space and the dynamics in which students and teachers move.

As a field, we wanted to understand how teachers' mathematical knowledge matters for teaching and learning. We wanted to know this with more practical relevance and more theoretical clarity. We assumed that something about mathematical knowledge would affect the quality of teaching and learning. But what we need to be talking more clearly about is mathematical *knowing and doing* inside the mathematical work of *teaching*. This change from nouns—"knowledge" and "teachers"—to verbs—"knowing and doing" and "teaching"—is not mere rhetorical flourish. These words can support a focus on the dynamics of a revised fundamental question: what is the mathematical work of teaching? This question helps to ensure that we are not compartmentalizing and that we are talking about the dynamics of the work a teacher does as she teaches her students mathematics (see Lampert, 2001, for an extensive development of what is involved in unpacking the work involved in managing "problems of teaching").

What *is* the "work of mathematics teaching" seen through a lens of practice? How do we calibrate the wide variety of work underway—about teaching, about theories of classrooms, about what mathematics is, about the larger environments of the work of teaching—to see, name, and understand the actual mathematical work of teaching?

Recalibrating the Question by Reconsidering "Teaching"

The instructional triangle in Fig. 1 (Cohen, Raudenbush, & Ball, 2003) makes visible that teaching is co-constructed in classrooms through a dynamic interplay of relationships, situated in broad socio-political, historical, economic, cultural, community, and family environments. These are constructed through the interpretations and interactions of teachers, students, and content.[3]

Students influence one another in myriad ways; what they already know about the content from prior experiences inside and outside of school influences them; how they read and understand their teachers also influences them. How their teachers interpret, respond, and treat them, as well as what their teachers know, believe, and understand about the curriculum, are all powerfully important. All of these relationships are interacting and influencing the learning in complex environments.

All of this complexity could make learning highly improbable. But the work of teaching is at its core about taking responsibility for attending with care to these chaotic and dynamic interactions. The work involves using skill, love, and knowledge to maximize deliberately the probability that students will learn worthwhile things and will flourish as human beings from being in that learning environment.

This is a probabilistic argument. Teaching does not cause learning—learners do the work of learning. However, the work of learning cannot be left to chance. Teaching is about doing caring and careful work in real time, with students, in specific contexts, that makes it the most likely that every student learns worthwhile skills, knowledge, dispositions, and qualities for their lives.

I refer to teaching practice as "work" to focus on what teachers actually *do* and to distinguish this focus from important foci on other features of classrooms, such as instructional formats, classroom culture and norms, what students are doing, and how the curriculum is designed. For example, small group work might be a feature in a classroom, but a focus on the work of teaching would probe what the teacher does to make small group work function well. The word "work" is intended to focus attention on what is involved in the *doing* of this responsibility of "maximizing the probability" that students will thrive and learn. Other aspects and features of classroom discourse, content, and interactions are also important but are not focused on what it takes to *do* the teaching.

What about problem solving or discussions or seatwork? Aren't those things that teachers do? Certainly teachers create seatwork. They use small groups. They facilitate discussions. But this does not help us understand from the inside of the work what it is to make small groups, or lead discussions, or create seatwork. What is it to ask a question in the moment—not thinking for a long time about what question might be asked, but actually producing the question in real time, fluently,

---

[3]See Ball and Forzani (2007) for a discussion of how this instructional triangle relates to and differs from other uses of "triangles" to represent teaching and learning.

Fig. 1 Instructional triangle

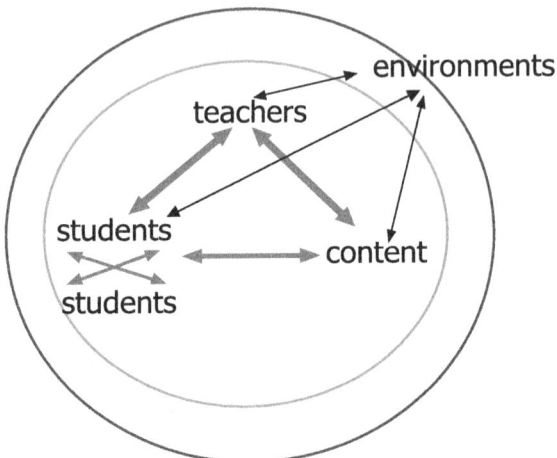

in a way that a child can understand it? What is involved in watching the children, listening to their talk, remembering what particular children said or did the day before, keeping in mind the point of the lesson (Sleep, 2012), and asking the next question, choosing the specific example, and deciding when and how to conclude the lesson for that day?

The use of "work of teaching" also represents a commitment to honor the effortful and deliberate nature of teaching. Learning does not happen by chance in classrooms. In fact, when the work of teaching is not as skillful as it might be, children do not learn. They are put at risk and they do not thrive. It is not respectful of the skill and effort entailed in teaching to represent it as intuitive, individual, or to render its details invisible. I use the word "work" to help us focus our lens not away from teaching, but more directly onto it.

There are many tools to draw upon to help us focus on the work of teaching. Drawing on the socio-cultural work of Anna Sfard, Jill Adler, and others, we know that classrooms are discursively intensive places that require a great deal of communication, both verbal and nonverbal, between and among students and teachers (Sfard, 2007; Adler & Davis, 2006; Adler & Ronda, 2015). We know that classrooms are filled with diversity that creates all kinds of resources and challenges for that discursive work. This means that there is something to the mathematically interactive, discursive, and performative work of mathematics teaching that is important to understand. In the next section, I turn to focus specifically on this "mathematical work of teaching." The goal is to see, name, and unpack the mathematical listening, speaking, interacting, fluency, and doing that are part of the work of teaching, not just resources for it. Focusing in this way on the mathematical doing that teaching entails can help shed light on the quest to understand the mathematics needed by teachers.

Seeing and Naming the Mathematical Work of Teaching

How might we identify and illustrate what might be meant by the *work of teaching*, and in particular the *mathematical* work of teaching? Central to bear in mind is an inherent fact of teaching, namely, that teachers are always communicating, relating, and making sense across differences, including differences in age, gender identities, race and ethnicity, culture and religion, language, and experience. This important dimension of difference in identity and positionality means that a fundamental part of the work of teaching is being aware of and oriented to learning about and coordinating with others' perspectives. Teaching is not just about what the teacher thinks; it is about anticipating what others think and care about, and attuning one's talk, gestures, and facial expressions to how others might hear or read the teacher. It is about talking with one's ear toward what someone else thinks, knows, or understands. This is a special and difficult kind of talking. Little is understood about what it takes to do it interactively, on one's feet. Often when we think about explaining mathematics, for example, we search for a good explanation that we ourselves find compelling and that we can understand and can articulate. But the real talk of teaching focuses instead on explaining mathematics in a way that anticipates how the person to whom the teacher is talking might actually understand the teacher's words, or how that individual might hear the teacher. It is a strange kind of talking and unlike most of the talking we do in everyday life.

This feature of teaching "across difference" is made still more consequential because these differences are not merely individual and personal. It is not a neutral feature of the work of teaching. Rather, the significance of difference is embedded in the historical and persistent structures and normative patterns of practice that have excluded and marginalized minoritized groups. Consider, for example, the social identities and contexts of the children in the class we examine below. They attend public school in a low-income predominantly African American community in the United States. Few members of their families have attended college. The children are in grade 5 and range in age from 9 to 11 years; of the 30 students in the class, 22 are African American, four are Latinx,[4] and four are White. Consider, too, the teacher's identity and position. Like the overwhelming majority of U.S. teachers, she is a White woman who attended predominantly White middle-class schools. Perhaps less like many U.S. teachers, growing up, she has been fluent in two other languages and experienced attending school as an emergent bilingual learner. Her public-school teaching experience over the last 40 years has been entirely with children of color and bilingual children, primarily of middle and working class families. The differences and connections between her identity and positionality and those of the children and their families are crucial to the forging of their relationships and communication. These differences matter for the imperative to connect with them and earn their trust. This is all fundamental to the work, and the mathematical work, of teaching.

---

[4]"Latinx" is used to avoid conveying a binary representation of gender identity.

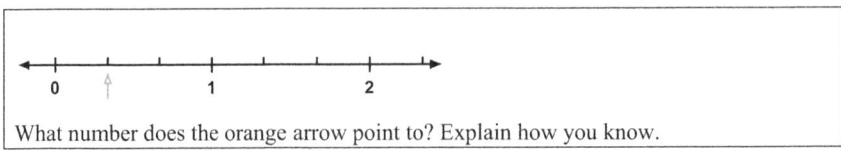

Fig. 2 Naming one-third on the number line (beginning of lesson)

Many of the children in this particular class—and in many in U.S. classrooms—have not had successful experiences with mathematics in school. They have come to think of being "smart" as getting right answers and good grades. Because of what they have come to see as "mathematics" and what it means to do well at it, by age 10 many of the children have begun to think they are not particularly good at math. These children, most of whom are African American or Latinx, refer to having gotten low marks on tests or to not getting right answers. Many have been "in trouble" in school for not "paying attention" or "talking" to others when they are supposed to be working quietly. Thus, their identities are already shaped by these structures of institutionalized racism and normalized practices of instruction (Nasir, Shah, Snyder, & Ross, 2012). The work for the teacher is situated in these broader systemic and historical patterns and is, in the moment, about connecting with and supporting these particular children and their opportunities to learn and grow (Nasir, 2016). In teaching, considerations of the individual and the systemic, the present and the historical, come together in the minute-to-minute of classroom dynamics. And they are embedded in and inextricably intertwined in subtle issues of mathematical ideas and talk, relationships, and maintaining a classroom environment focused on learning. Whereas research can be analytic, and can take apart the complex phenomena in order to probe and understand them, teaching is an integral and interactive whole. Studying the work of teaching therefore necessarily requires that we seek ways to see and understand that integration and simultaneity of differences.

To unpack what this might mean, we turn next to look inside the classroom where these children are learning mathematics. As we notice their work and their thinking, our purpose is to try to consider the surrounding integral work of teaching that is supporting their mathematics learning.

The Work of Teaching in One Lesson

On this particular morning, the children have worked on the problem in Fig. 2 in their notebooks.

This problem represents a significant turning point in the class's mathematical work, from naming fractions as parts of areas to identifying fractions as points on the number line. One important shift is to understand that on the number line, the whole is defined as the interval from 0 to 1. With area models, the whole can be

Fig. 3 Naming fractions as parts of wholes

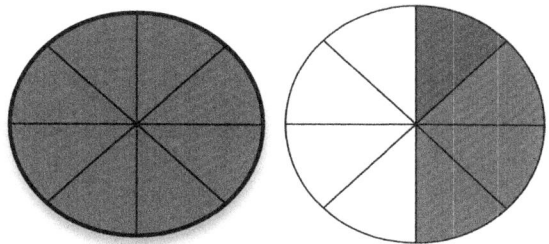

greater than 1. For example, in Fig. 3, it is possible to name the green shaded portion as 1 3/8 or 11/8, if one identifies one circle as the whole. But it is also correct to identify *two* circles as the whole, and then the fractional part that is green is 11/16.

For the children, it is an important new understanding to learn that, on the number line, the whole is always defined as the interval from 0 to 1 and the problem on which they are working is designed to press on this issue and bring it to explicit understanding.

During the beginning of class, known as the "warm up" (about five minutes), the children pasted this opening problem in their individual notebooks and wrote their answers and explanations individually. The correct answer is 1/3. Eight (6 African American, 1 Latinx, and 1 White) children do have 1/3 as the answer, but no one has explained his or her answer. The other 22 children have other answers, including 1/4, 2/4, and 1.

See Fig. 4 for some examples of what students have in their notebooks before the class discussion.

The teacher has been walking around while the children are thinking and writing and has been looking at the range of ideas and explanations, noticing what different children have written and thinking about what will be important to work on together.

The teacher launches the class discussion of the problem.[5] The children are seated at tables arranged in a U-shape, and they are all able to see the large white boards at the front of the room, on which the problem is drawn.

Teacher: (*standing near the back of the room*) Who would like to try to explain what you think the answer is? And show us your reasoning by coming up to the board? Who'd like to come up to the board and try to tell– And you know, it might not be right. That's okay because we're learning something new.
I'd like someone to come up and sort of be the teacher and explain how you are thinking about it. Who'd like to try that this morning? (*Several children raise their hands to volunteer.*)

---

[5]The video for this segment is available for viewing at http://hdl.handle.net/2027.42/134321.

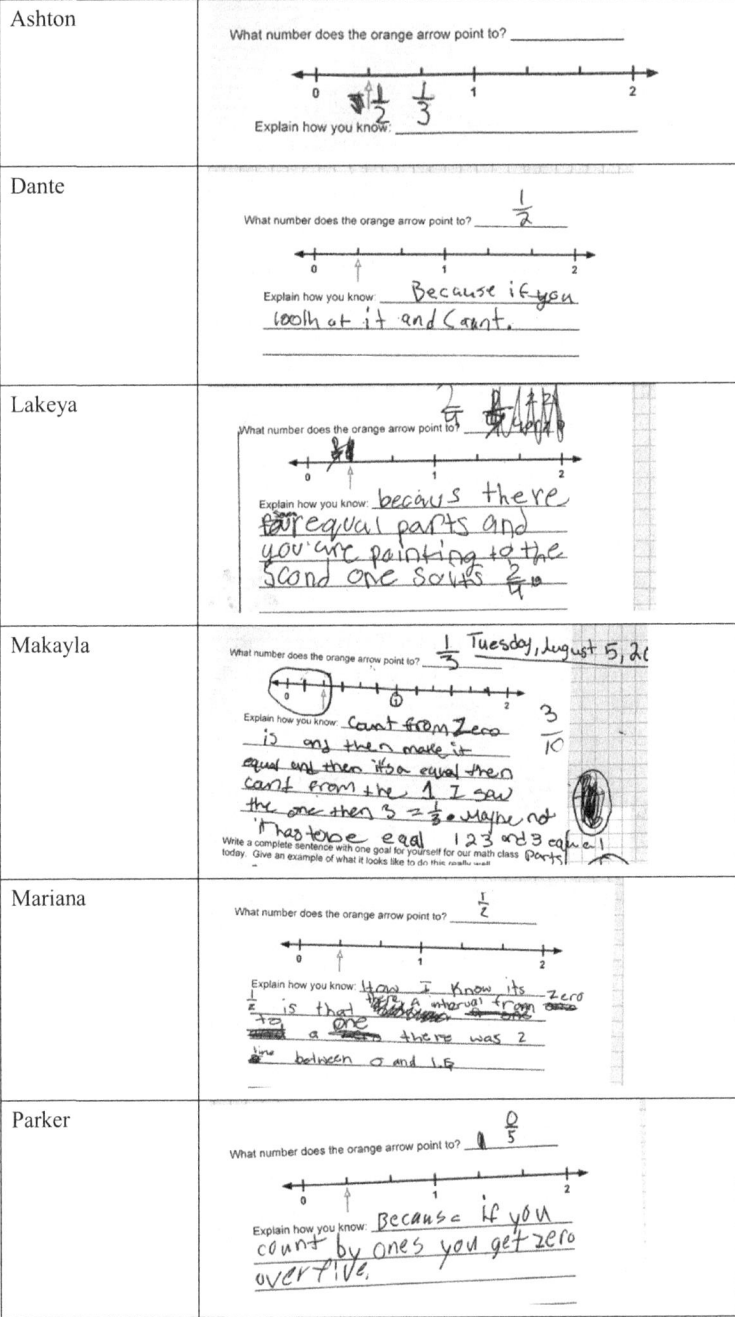

Fig. 4 Children's work on the number line problem at the beginning of class

Uncovering the Special Mathematical Work of Teaching 21

	Okay, Aniyah? (*Aniyah, a Black girl, gets up from her seat and walks to the whiteboard at the front of the classroom.*) When someone's presenting at the board, what should you be doing?
Students:	Looking at them.
Teacher:	Looking at that person—uh-huh.
Aniyah:	(*to the teacher*) You want me to write it?
Teacher:	(*to Aniyah*) You're trying to mark what you think this number is and explain how you figured it out. (*to class*) Listen closely and see what you think about her reasoning and her answer. (*Teacher moves to back of the classroom; Aniyah is in front at the whiteboard. Aniyah writes $^1/_7$ by the orange line*).
Aniyah:	I put one-seventh because there's– Toni, an African American girl, sitting close to where Aniyah is standing, asks quietly, almost to herself: "Did she say one-*seventh*?" Hearing her question, Aniyah turns toward her and nods: "Yeah. Because there's seven equal parts, like one, two, three, four, five, six, and then seven," and demonstrates using her fingers spread to measure the intervals to count the parts on the number line.
Teacher:	(*still standing at the back, addresses the class*) Before you agree or disagree, I want you to ask questions if there's something you don't understand about what she did. No agreeing and disagreeing. Just—all you can do right now is ask Aniyah questions. Who has a question for her? Okay, Toni, what's your question for her?
Toni:	Why did—(*looks across at children opposite her and laughs, twisting her braid on top of her head*)
Teacher:	(*to Toni*) Go ahead, it's your turn.
Toni:	(*to Aniyah*) Why did you pick one-*seventh*? (*Toni giggles, twisting her braid.*)
Dante:	(*laughing across the room at Toni*) You did not!
Teacher:	Let's listen to her answer now. (*to Toni*) That was a very good question. (*to Aniyah*) Can you show us again how you figured that– why you decided one-seventh?
Aniyah:	First, I thought it might be seven because there's seven equal parts.
Teacher:	Did you write one-seventh? I can't see very well from here.
Aniyah:	Uh-huh. Yes. The teacher nods affirmatively, and turns to the class, "Okay, any more questions for Aniyah? In a moment, we're going to talk about what you think about her answer, but first, are there any more questions where you're not sure what she said, or you'd like to hear it again or something like that? Lakeya?"
Lakeya:	(*looks back at the teacher at the back of the room*) If you start at the—
Teacher:	(*gestures toward Aniyah*) Talk to her, please.

Lakeya: Oh! (*turns toward Aniyah*) If you start at the zero, how did you get *one-seventh*?

Aniyah: Well, I wasn't sure it was one-seventh, but first, I thought that the seven equal parts.

Teacher: Okay, would some– You'd like to ask another question, Dante?

Dante: Yeah.

Teacher: Yes, what?

Dante: So, if it's at the zero, how did you know that if like if I took it and put it at the– Hold on. Which line is– What if it didn't like– What if the orange line wasn't there, and you had to put it where the one is? What if the orange line wasn't there? And how would you still know it was one-seventh to put it where the orange line is now?

Aniyah: (*pauses*) I don't know.

Teacher: (*pauses*) Okay. Does everyone understand how Aniyah was thinking?

Students: Yes.

Teacher: Yes? Okay. (*to Aniyah*) You can sit down now. We're going to try to get people to comment. Do you want to take comments up there? Would you like to stand there and take the comments, or do you want to sit down and listen to the discussion?
What would you prefer?

Aniyah: Sit down.

Teacher: Sit– You'd like to sit down? Okay.

During these three minutes of class, four children speak in the whole group discussion: Aniyah, Toni, Lakeya, and Dante. The class discussion continues for another 48 minutes. During this time, the discussion emphasizes the importance of partitioning the unit interval in equal parts and being sure to count spaces (i.e., intervals, not hash marks) to determine the distance from 0 for a given point on the line. The students practice naming points on the line and also explaining carefully with reference to the "whole" and to "equal parts" and to counting spaces to determine the number.

At the end of the lesson, to learn what the children are thinking now, the teacher chooses a new fraction and a new number line and poses the question in Fig. 5 for the children to answer independently in their notebooks.

The correct answer is 2/3, and the target explanation would draw on the notions of the whole (the interval from 0 to 1), equal partitions of that whole, naming one part, and naming the number of equal parts (Fig. 5).

The results are interesting. Before the class discussion, when working independently on the problem in Fig. 2, 8 children (27%) can correctly name the point on the number line with a correct number name, but without a clear mathematical justification. 22 have other answers. After the discussion, 26 (87%) can label the point correctly and can provide mathematical explanations for their choice. Of the four students who did not name the point correctly, they nevertheless refer to important aspects of the definition, including "equal parts" and "spaces."

1. What number does the blue arrow point to?

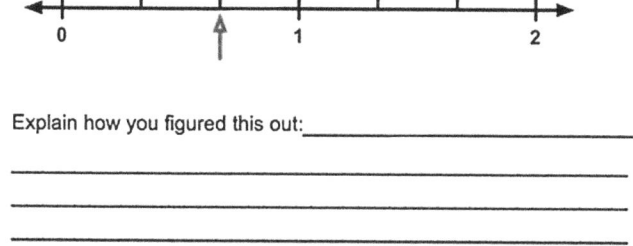

Explain how you figured this out: _____

_____

_____

_____

Fig. 5 Naming two-thirds on the number line (end of class)

Fig. 6 shows the work on the end-of-class check by the same six students whose beginning-of-class problems are shown in Fig. 4. It is interesting to compare their answers before and after the 51-minute in-class discussion of how to name fractions as points on the line.

What Is the (Mathematical) Work of Teaching?

We examined only three minutes of a lesson. This is in some ways little time, yet it is filled with intense demands on the teacher. What does studying this segment closely reveal about the work of teaching? What, for example, is involved in setting up and guiding the children to think about and learn mathematics? To listen to one another? To have confidence in their own thinking? What is involved for the teacher in tracking on what each of the 30 children is thinking, puzzling about, and learning? In knowing who might be drifting off and who might be feeling confused?

One key element of the work has occurred earlier: the decision about the problem to pose. Before the discussion described above, the students had individually worked on and answered the question in their notebooks. Even before that, the teacher had decided on the task. Why the number 1/3? Why, for example, a unit fraction? Why also draw a number line that extends just a little past 2? Would it have worked the same way with a number line precisely drawn from 0 to 1? What if the point she had selected was 1/4 or 4/5 instead of 1/3? Each of these decisions shaped the mathematical context in which the children were immersed, and created the space for their thinking, writing, and learning.

A second aspect of the work of teaching is to see and make sense of the work of individual children while they are working on the task. To do this, the teacher circulated around the room to scan what the children were writing in their notebooks. She did this to get a sense of what the children were thinking and to see the range of answers in the room. Reading children's writing and reasoning is mathematically demanding. Notice how this sort of examination is different than being a researcher on students' thinking and using digitized copies of students' work with a

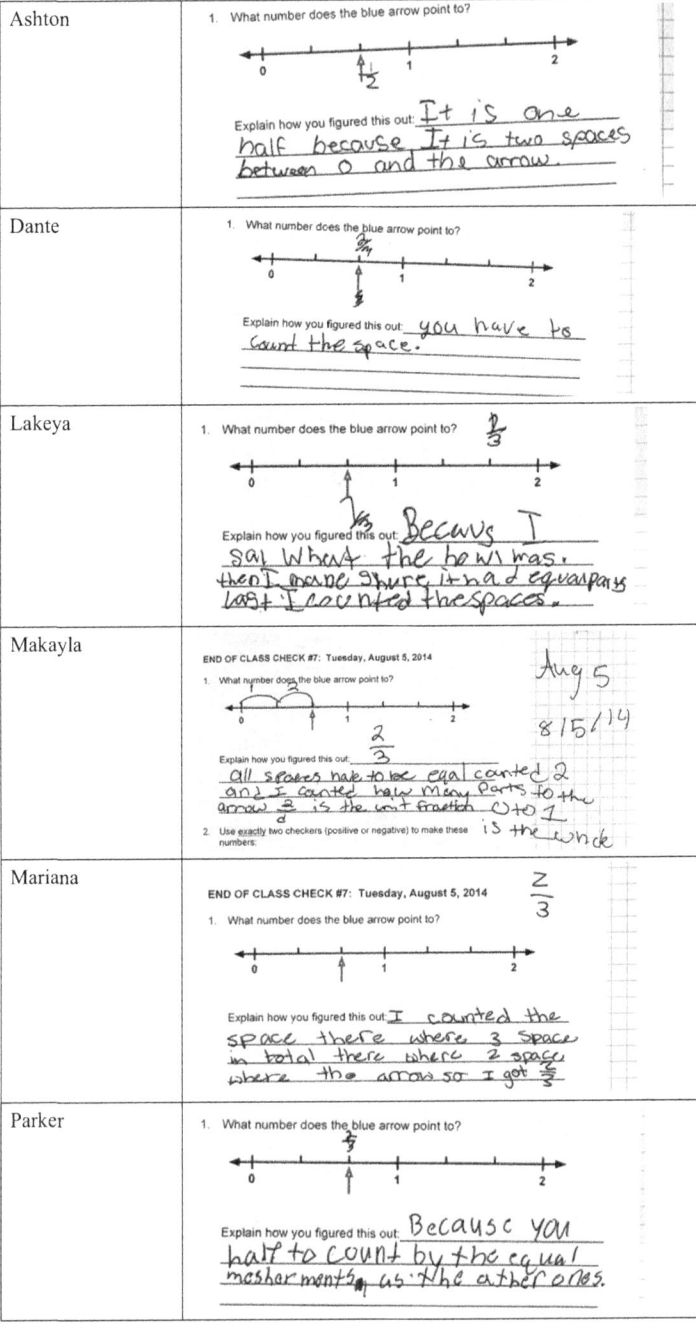

Fig. 6 Children's work on the number line problem at the end of class

Uncovering the Special Mathematical Work of Teaching

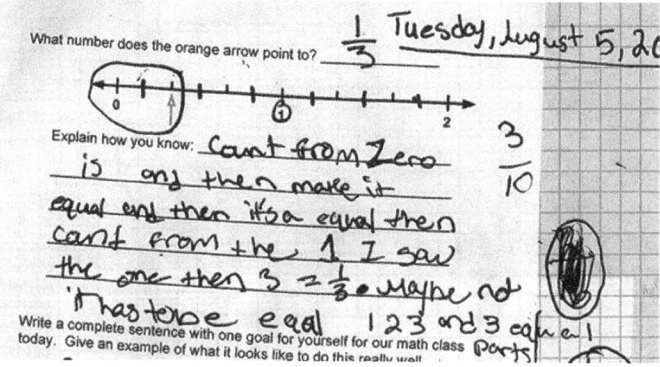

Fig. 7 Makayla's notebook before the class discussion

lot of time to examine and mark things and notice nuances. Instead, the work of teaching requires thinking and reading mathematically in real time. It involves walking around the room, surveying and trying to read 30 different responses, including the numbers they have identified and the explanations written. It involves sorting them mentally, and making a careful decision about choosing which answer to begin the discussion and whom to call on.

For example, what is Makayla thinking? (Fig. 7).

She records 1/3 but also writes 3/10, and explains in detail, "Count from zero is and then make it equal and then it's a equal then count from the 1 I saw the one then 3 = 1/3. Maybe not. It has to be equal. 1 2 3 and 3 equal parts." Her use of the equals sign is of interest, signaling that after counting three *equal* parts, the number she writes is 1/3. Her circling of the segment of the number line up to the orange arrow also shows her focus on the three, starting at 0. What does she think the whole is? Reading children's mathematical writing and representations is not linear, reading from top to bottom in order. Instead, reading as a teacher requires a more multi-directional examination, making sense of the logic, detecting where the writing is sequential and where it is discontinuous, either in time or thinking. For example, does Makayla think that 1/3 is the same number as 3/10 or did she change her mind and not cross it out? Some of what is involved is general and some involves knowing the particular child and some of her ways of expressing. Makayla tends to make a diagram or representation before writing, and sometimes goes back and forth as she represents her explanation, altering the diagram and writing a new thought based on that. She, like many other children, does not always cross out something about which she has changed her mind.

And what does Dante's explanation suggest about his thinking? (Fig. 8).

One part of reading is the actual decoding of children's writing accurately. Here Dante writes, "Because if you look at it and count." The words "look" and "count" might not be easy for readers not skilled in reading and interpreting children's writing. What is he saying exactly? Does Dante think it is 1/2 because it is between

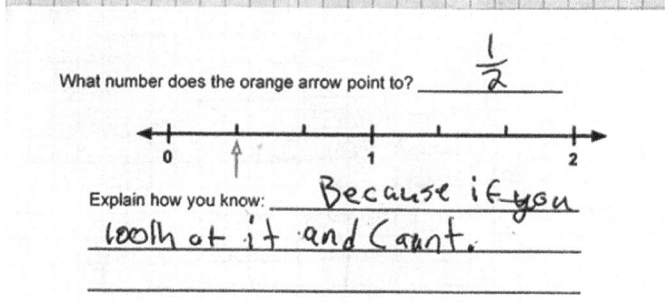

Fig. 8 Dante's notebook before the class discussion

0 and 1, a common idea, or is it because he started at 0 and counted the tick marks at 0 and at 1?

The work of teaching involves a fluency of mathematical reading and interpretation, to surmise what 30 (or more) different children might mean, and preparing to ask questions or to probe wisely or to comment strategically, all in real and rapidly moving time. At least three kinds of reasoning and interpretation are involved. First is to consider the mathematical issues embedded in the task or problem. For example, this task involves interpreting the number line, including what is considered the "whole," what to count—intervals or tick marks—and how to determine the name of a point on the number line. That the point is named by the distance from 0 in terms of the number of intervals of a particular length (e.g., 1/3) embeds all of these dimensions. Reading children's writing is supported if the reader has a firm orientation to the things that a child's representation contains. Second, and closely related, is to anticipate how the children might interpret the task or decide on their answer. Knowing, for example, that many children might count tick marks instead of intervals or start at 0 instead of at 1 can help a reader to see and interpret what a child has written. Finally, this work entails a fluency in reading children's written representations, including spelling, spacing, handwriting (formation and orientation of letters, numbers, and symbols), ellipses and missing words or letters, and the composition on the page (for instance, that their writing is often not linear from top to bottom).

There is also the complex work of leading a discussion, often misleadingly represented as "getting out of the way" and "letting the children do the teaching." First, the teacher chooses whose work launches the discussion. This is a key and consequential decision. In this case, while circulating as the children worked independently, the teacher saw many different answers by different children. Her decision about whose answer to start with involves considering the particular children and how they are positioned in the class and how who gets the floor and is given recognition for their thinking influences that positioning. The teacher's decision also involves thinking about the mathematical issues on which the children are focused and those key to her instructional goal. She chooses Aniyah to present

1/7, which puts the issue of the "whole" in focus, and provides an opportunity to position this African American girl as competent because of the clarity of her mathematical explanation, structured logically and based on the core elements of the definition (Boaler & Staples, 2008; Cohen & Lotan, 2003; Featherstone et al., 2011).

To be sure, when Aniyah presents her answer (1/7), the teacher is physically out of the way. She stands at the back of the room, and she does position Aniyah in the role of "teacher." But to see this as moving away and letting the children have the freedom to learn misleads us about the intricacy of the work. The teacher firmly structures the discourse: "No agreeing and disagreeing. Just—all you can do right now is ask Aniyah questions." This move shapes what the children may talk about. Instead of jumping to offer another answer, they must think about what Aniyah wrote and explained, compare it with what they wrote in their own notebooks, and ask some sort of question. This move buffers a possible tendency to develop patterns of discourse that are all about critique, objection, and competition to be "right," rather than about careful attention, consideration, and collective development of ideas and arguments.

Setting the children up to do the work of learning also involves careful work for the teacher. She must watch Aniyah closely and be ready to offer support for her presentation if it seems needed. She watches the other children: Are they following? When Toni giggles while posing a question to Aniyah, the teacher can choose to interpret her as making fun of her classmate—or she can read her as seriously engaged in her classmate's idea. Toni, who has an unexplained 1/3 in her notebook, asks, "Why did you pick one-SEVENTH?" The teacher must take note of her emphasis on the "seventh"—this, after all, diverges from her choice of "third." And less than a minute later, Lakeya, who has written 2/4 in her notebook, asks, "If you start at the zero, how did you get *one*-seventh?" and now the teacher must hear the emphasis on the "one" for Lakeya has TWO fourths. Listening like this entails a close and mathematically sensitive attention, which draws not on the teacher's own knowledge of 1/3 as the answer, but on her ability to focus on the children's thinking and talk. This focus on hearing others' mathematical thinking, through their talk, gesture, inflection, and tone, depends on deliberately suspending quick assumptions about what others mean, but yet listening in mathematically nimble ways. To name the work of discussion-leading as moving out of the way and "facilitating" is reductive and misrepresents the multiple aspects of the careful work of teaching.

The work of teaching involves disrupting the tendency to classify children's answers as either correct or incorrect. Helping children learn depends on seeing what they do know and can do, not absorption with what is missing. This is neither natural nor obvious. Math is a subject, perhaps like spelling, in which the focus is often on errors. For Aniyah, the main thing most observers notice is that her answer is wrong. For Toni, people notice she is playing with her hair, they interpret her as trying to get attention from other children in the class, and they often think she is trying to embarrass Aniyah. These reactions are shaped not only by the tendency to notice mistakes but also by the reproduction of racialized and gendered biases that mask these African American girls' strong mathematical capabilities (Martin,

Aniyah	Toni

• Uses the correct definition for a fraction:
  ▪ She identifies the "whole";
  ▪ She makes sure the intervals are equal;
  ▪ She counts intervals and not tick marks; and
  ▪ She knows how to write "one-seventh".
• Produces a mathematically well-structured explanation.
• Presents her ideas clearly.
	• Listens closely to a classmate's presentation.
	• Uses the definition for a fraction to ask how Aniyah decided on 7 parts.
	• Asks a pointed mathematical question.

Fig. 9 Aniyah's and Toni's mathematical competence

2015). The persistent patterns of marginalization of particular groups are certainly produced and reproduced in systemic ways. In fact, both Aniyah and Toni know and can do a lot. The work of teaching involves actively seeing and naming what each girl shows she knows and can do (Fig. 9).

The work of teaching involves not only seeing and naming the girls' mathematical competence with fractions, definitions, and explanations, but also attending to their positioning and their mathematical identities, and building on their strengths and resources to support their growth (Langer-Osuna, 2016).

Many other examples of the mathematical work of teaching can be seen in this lesson. The teacher kneels beside a student to talk to her about how she understands a problem and to respond to her in a way that is mathematically sensible, that she can understand, and that does not distort the math. When Aniyah is at the board, teaching involves listening carefully to what she is saying and showing and what others are asking or commenting, and watching carefully to see how Aniyah is experiencing this attention on her answer. Based on these, there is the deciding whom to call on next, whether to make a point oneself, or whether to revoice or ask a student to restate what has been said, or perhaps allow another student to comment. Another example is translating from intensively mathematical (e.g., "*Understand a fraction 1/b as the quantity formed by 1 part when a whole is partitioned into b equal parts*"—a learning goal in the U.S. Common Core Standards for Mathematics) to usable terms (e.g., using a variable notation such as "call one of the equal parts 1/d" to represent the unit fraction). And this also involves being able to "talk" the terms, saying them in accessible ways and helping the children learn to talk with them. Still another is deciding whom to name as competent and how to intervene on inequities that may be reproduced in the classroom (Boaler & Staples, 2008; Cohen & Lotan, 2003; Featherstone et al., 2011).

Each of these represents part of the intricate mathematical work of teaching. It is not an exhaustive list, but the examples illustrate the mathematical aspects inside particular moves or interpretations. Naming them as "mathematical" is not intended to suggest that mathematical reasoning is all that is involved. Rather, the point is that these decisions, moves, ways of talking, doing, and moving, all crucially involve mathematical sensibility, thinking, and understanding. For example, it would not be possible to identify competence in students' work in a high school chemistry or music class without being fluent in these domains in the special ways demanded for teaching. Without substantial and nuanced understanding of the domain and its practices, one would not be able to read and interpret student writing, pose questions, or revoice students' comments. Neither could one broaden students' notions about what competence *is* in that domain and see and name students' competence in ways that would intervene on status hierarchies.

Conclusion

The quest to answer the perennial question of what mathematical "knowledge" teachers need should be based on a deep and nuanced understanding of what teachers actually *do*. How does mathematical listening and hearing, as well as writing, representing, and talking play a role in the *work* of teaching? What sorts of mathematical sensibility and insight matter for seeing minoritized children as not deficient but as emergent and thoughtful mathematical thinkers and actors? What mathematical disposition and fascination does it take to nurture children's seriousness as thinkers and their creative playfulness as mathematical explorers? When Toni asks, "Did she one-*seventh*?" as she watches her classmate present, she is seriously surprised. When Dante asks, "What if the orange line wasn't there, and you had to put it where the one is?" he is genuinely wondering. Respecting and nourishing the brilliance of these African American children entails a kind of mathematical care that deserves acknowledgement (Leonard & Martin, 2013). Probing and naming the work of teaching means identifying and articulating what is involved in these many, complex, and simultaneous actions, decisions, and moves entailed in the broader and moment-to-moment intellectual and moral tasks and considerations of the work. It means coming to see not just what an observer or an analyst might see or infer, but what is involved in actually *doing* those things, from the inside.

Trying to study the work of teaching from the perspective of what is involved in doing it presents several challenges. One challenge is that teaching is fundamentally relational work. This means that the work is constructed simultaneously with individual children, who are themselves not static and whose identities-in-action are refracted through their memberships in multiple and overlapping communities. The work of teaching is also constructed in the social contexts of the collective setting, as well as in the broader socio-political, historical, and community contexts (Lampert, 2001). Uncovering the work from this perspective is crucial, yet it is easier to focus on one dimension, such as the teacher's questioning of one student or her representation on the board. However, to understand the work of teaching, the simultaneity and complexity are fundamental (Ball & Lampert, 1999). This implies that interdisciplinary teams and tools and methods to get inside the work will be important (see Bullock, 2012, for a useful discussion of methodological choices). It will often be confusing to distinguish the effort to study the work of teaching from other worthwhile foci for classroom studies or studies of teachers, which contribute importantly to our understanding of classrooms and of teaching, learning, students, and teachers.

A second challenge regards the issue of teaching quality, and normative versus descriptive perspectives on the work of teaching. Does the teaching used to study the work have to be skillful? And what should that mean and who or what would determine this? How would considerations of equity and the disruption of dominant norms and reproduction figure in such appraisals, or of mathematical integrity, or of caring? One might examine teaching work from the perspective of the endemic problems to be managed (Kennedy, 2016; Lampert, 2001), but this would not resolve the issue of whether some approaches to managing would more useful—or not worthwhile—to study. Who would decide what teaching is worth studying and how would such decisions either focus or constrain the effort to understand the work of teaching? Bullock (2012) argues that decisions about what and how to study should be guided by the moral imperative to "make life better for people," to serve the interests of children, families, and communities who have been marginalized and disenfranchised. How can our study of the work of teaching be careful not to be falsely "impartial" but, instead, to honor the goal of making responsive and responsible teaching learnable by others? This implies making and justifying explicitly deliberate decisions about the practice to be studied.

A third is that existing theoretical frames and the aspects of teaching that are already used or studied will make it difficult to focus on some aspects of the work of teaching, such as reading students' work "on the fly" and processing it to prepare for launching a discussion, or disrupting normative practices of control. In the case of reading students' work, and considering whose work might be good to invite to share in a discussion and why, there are many things that the teacher is doing, some visible and some invisible (Lewis, 2007). Some of the most important demands of studying the work of teaching will be to see and name aspects of the *doing* that are taken for granted and so lack names or foci. Another will be the acts that are "not-doings." Some of the work of teaching is to refrain purposefully at a given moment from doing something, such as rebuking a child for fidgeting or talking in class (Noel, 2014), interrupting a child's language, or explaining a "standard" method. These other invisible (Lewis, 2007) but deliberate acts are surely also to be included in our unpacking of the work of teaching.

The third challenge leads to a fourth—namely, what should be the warrants for claiming that a particular move or non-move, a particular action or thought, is part of the work of teaching? Because, as in many domains of expertise, skillful performance is often partly tacit, teachers might not always be able to articulate the inside work, or to name the complex intertwined kinds of moves, ways of talking, and practices on which their practice is built. What does it mean, then, for an observer to claim that something is part of that work when the person doing it cannot isolate or name it? Examining the work of teaching requires careful consideration of this challenge and ways of using productively both insider and outsider perspectives.

Finally, threaded throughout a focus on the work of teaching is the challenge of examining and identifying the *mathematical* entailments of that work. This is a complex undertaking that will require ongoing articulation of what counts as "mathematical," and what it means for work to "demand" mathematical reasoning, knowing, thinking, or talking (Ball, 1999). Hoover (2009) argues that this work involves coordination of perspectives, not merely annotating practice with mathematical commentary or analysis. What is the mathematical reading involved in scanning children's writing and representations? What kinds of mathematical interpretation and reasoning does this take? What does it take to teach mathematics in ways that disrupt dominant patterns of marginalization?

Answering questions such as these will not be easy, but the potential is important. For it is through such analysis and naming that we will come to understand much more about the ways in which the teaching of mathematics requires specialized mathematical ways of thinking and reasoning. And it is with such insight that we will make headway on the longstanding and important question of how teachers need to know mathematics in and for their work.

Acknowledgements The author is grateful to and acknowledges her colleagues who have helped her learn across time and have influenced her thinking about the ideas in this paper: Hyman Bass, Esther Enright, Susanna Farmer, Rebecca Gadd, Nicole Garcia, Imani Goffney, Lauren Hickman, Mark Hoover, Lindsey Mann, Blake Noel, Sabrina Salazar, Meghan Shaughnessy, Rachel Snider, Charles Wilkes, and Amber Willis.

The work discussed in this paper has been supported for more than 20 years by grants from the National Science Foundation to the author, and her colleagues, Hyman Bass, Heather Hill, Mark Hoover, and Laurie Sleep.

References

Adler, J., & Davis, Z. (2006). Opening another black box: Researching mathematics for teaching in mathematics teacher education. *Journal for Research in Mathematics Education, 37*(4), 270–296.

Adler, J., & Ronda, E. (2015). A framework for describing mathematics discourse in instruction and interpreting differences in teaching. *African Journal of Research in Mathematics, Science and Technology Education, 19*(3), 237–254.

Ball, D. L. (1990). The mathematical understandings that prospective teachers bring to teacher education. *Elementary School Journal, 90*(4), 449–466.

Ball, D. L. (1999). Crossing boundaries to examine the mathematics entailed in elementary teaching. In T. Lam (Ed.), *Contemporary mathematics* (pp. 15–36). Providence: American Mathematical Society.

Ball, D. L., & Forzani, F. M. (2007). What makes education research "educational"? *Educational Researcher, 36*(9), 529–540.

Ball, D. L., & Lampert, M. (1999). Multiples of evidence, time, and perspective: Revising the study of teaching and learning. In E. Lagemann & L. S. Shulman (Eds.), *Issues in education research: Problems and possibilities* (pp. 371–398). San Francisco: Jossey Bass.

Ball, D. L., Lubienski, S., & Mewborn, D. (2001). Research on teaching mathematics: The unsolved problem of teachers' mathematical knowledge. In V. Richardson (Ed.), *Handbook of research on teaching (4th ed.)* (pp. 433–456). New York: Macmillan.

Ball, D. L., Thames, M. H., & Phelps, G. (2008). Content knowledge for teaching: What makes it special? *Journal of Teacher Education, 59*(5), 389–407.

Baumert, J., Kunter, M., Blum, W., Brunner, M., Voss, T., Jordan, A., et al. (2010). Teachers' mathematical knowledge, cognitive activation in the classroom, and student progress. *American Educational Research Journal, 47*(1), 133–180.

Blömeke, S., Hoth, J., Döhrmann, M., Busse, A., Kaiser, G., & König, J. (2015). Teacher change during induction: Development of beginning primary teachers' knowledge, beliefs and performance. *International Journal of Science and Mathematics Education, 13*(2), 287–308.

Boaler, J., & Staples, M. (2008). Creating mathematical futures through an equitable teaching approach: The case of Railside School. *Teachers' College Record, 11*(3), 608–645.

Borko, H., Eisenhart, M., Brown, C. A., Underhill, R. G., Jones, D., & Agard, P. C. (1992). Learning to teach hard mathematics: Do novice teachers and their instructors give up too easily? *Journal for Research in Mathematics Education, 23*(3), 194–222.

Bruckmaier, G., Krauss, S., Blum, W., & Leiss, D. (2016). Measuring mathematics teachers' professional competence by using video clips (COACTIV video). *ZDM Mathematics Education, 48*(1–2), 111–124.

Bullock, E. C. (2012). Conducting "good" equity research in mathematics education: A question of methodology. *Journal of Mathematics Education at Teachers College, 3*(2), 30–36.

Carrillo, J., Climent, N., Contreras, L. C., & Muñoz-Catalán, M. C. (2013). Determining specialised knowledge for mathematics teaching. In B. Ubuz, C. Haser, & M. A. Mariotti (Eds.), *Proceedings of the eighth congress of the European Society for Research in Mathematics Education*.

Cohen, D. K., Raudenbusch, S., & Ball, D. L. (2003). Resources, instruction, and research. *Educational Evaluation and Policy Analysis, 25*(2), 119–142.

Cohen, E. G., & Lotan, R. A. (2003). Equity in heterogeneous classrooms. In J. A. Banks & C. A. M. Banks (Eds.), *Handbook of multicultural education* (2nd ed., pp. 736–752). San Francisco, CA: Jossey-Bass.

Conference Board of Mathematical Sciences. (2001). *The mathematical education of teachers*. Providence RI and Washington DC: American Mathematical Society and Mathematical Association of America.

Conference Board of Mathematical Sciences. (2012). *The mathematical education of teachers II*. Providence RI and Washington DC: American Mathematical Society and Mathematical Association of America.

Dewey, J. (1902). *The child and the curriculum*. Chicago, IL: The University of Chicago Press.

Featherstone, H., Crespo, S., Jilk, L. M., Oslund, J. A., Parks, A. N., & Wood, M. B. (2011). *Smarter together! collaboration and equity in the elementary math classroom*. Reston, VA: National Council of Teachers of Mathematics.

Goffney, I. M. (2010). *Identifying, defining, and measuring equitable mathematics instruction*. (Unpublished doctoral dissertation). University of Michigan, Ann Arbor, MI.

Goffney, I. M. (2014). Mathematical quality and equity video codes: Identifying and measuring equitable mathematics instruction. Manuscript in preparation.

Goffney, I. M., & Hoover, M. (2017). Developing a theory of mathematical knowledge for equitable teaching. Manuscript in preparation.

Herbst, P., & Chazan, D. (2015). Using multimedia scenarios delivered online to study professional knowledge use in practice. *International Journal of Research and Method in Education, 38*(3), 272–287.

Herbst, P., & Kosko, K. (2014). Mathematical knowledge for teaching and its specificity to high school geometry instruction. In J. Lo, K. R. Leatham, & L. R. Van Zoest (Eds.), *Research trends in mathematics teacher education* (pp. 23–45). New York, NY: Springer.

Hiebert, J., Morris, A. K., & Glass, B. (2003). Learning to learn to teach: An "experiment" model for teaching and teacher preparation in mathematics. *Journal of Mathematics Teacher Education, 6*(3), 201–222.

Hill, H. C. (2007). Mathematical knowledge of middle school teachers: Implications for the No child left behind policy initiative. *Educational Evaluation and Policy Analysis, 29,* 95–114.

Hill, H. C. (2011). The nature and effects of middle school mathematics teacher learning experiences. *Teachers College Record, 113*(1), 205–234.

Hill, H. C., & Ball, D. L. (2004). Learning mathematics for teaching: Results from California's Mathematics Professional Development Institutes. *Journal for Research in Mathematics Education, 35,* 330–351.

Hill, H., Ball, D. L., & Schilling, S. G. (2008). Unpacking "pedagogical content knowledge": Conceptualizing and measuring teachers' topic-specific knowledge of students. *Journal for Research in Mathematics Education, 39*(4), 372–400.

Hill, H. C., Schilling, S. G., & Ball, D. L. (2004). Developing measures of teachers' mathematics knowledge for teaching. *The Elementary School Journal, 105*(1), 11–30.

Hill, H. C., Rowan, B., & Ball, D. L. (2005). Effects of teachers' mathematical knowledge for teaching on student achievement. *American Educational Research Journal, 42*(2), 371–406.

Hoover, M. (2009). *Coordinating mathematical and pedagogical perspectives in practice-based and discipline-grounded approaches to studying mathematical knowledge for teaching.* Unpublished doctoral dissertation, University of Michigan, Ann Arbor, MI.

Hoover, M., Mosvold, R., & Fauskanger, J. (2014). Common tasks of teaching as a resource for measuring professional content knowledge internationally. *Nordic Studies in Mathematics Education, 19*(3–4), 7–20.

Kaiser, G., Busse, A., Hoth, J., König, J., & Blömeke, S. (2015). About the complexities of video-based assessments: Theoretical and methodological approaches to overcoming shortcomings of research on teachers' competence. *International Journal of Science and Mathematics Education, 13*(2), 369–387.

Kennedy, M. (2016). Parsing teaching practice. *Journal of Teacher Education, 67,* 6–17.

Knievel, I., Lindmeier, A. M., & Heinze, A. (2015). Beyond knowledge: Measuring primary teachers' subject-specific competences in and for teaching mathematics with items based on video vignettes. *International Journal of Science and Mathematics Education, 13*(2), 309–329.

Lampert, M. (2001). *Teaching problems and problems of teaching.* New Haven, CT: Yale University Press.

Langer-Osuna, J. (2016). The social construction of authority among peers and its implications for collaborative mathematics problem solving. *Mathematical Thinking and Learning, 18*(2), 107–124.

Lewis, J. (2007). *Teaching as invisible work.* Unpublished doctoral dissertation, University of Michigan, Ann Arbor, MI.

Leonard, J., & Martin, D. B. (Eds.). (2013). *The brilliance of Black children in mathematics: Beyond the numbers and toward new discourse.* Charlotte, NC: Information Age Publishing.

Martin, D. B. (2015). The collective Black and principles to actions. *Journal of Urban Mathematics Education, 8*(1), 17–23.

McCrory, R., Floden, R., Ferrini-Mundy, J., Reckase, M. D., & Senk, S. L. (2012). Knowledge of algebra for teaching: A framework of knowledge and practices. *Journal for Research in Mathematics Education, 43*(5), 584–615.

National Mathematics Advisory Panel. (2008). *Foundations for success: The final report of the national mathematics advisory panel,* U.S. Department of Education: Washington, DC.

Nasir, N. (2016). Why should mathematics educators care about race and culture? *Journal of Urban Mathematics Education, 9*(1), 7–18.

Nasir, N., Shah, N., Snyder, C., & Ross, K. (2012). Stereotypes, storylines, and the learning process. *Human Development, 55*(5–6), 285–301.

Noel, B. (2014). *Practices for teaching across differences.* Unpublished manuscript. University of Michigan, Ann Arbor, MI.

Rowland, T. (2013). The knowledge quartet: The genesis and application of a framework for analysing mathematics teaching and deepening teachers' mathematics knowledge. *Sisyphus-Journal of Education, 1*(3), 15–43.

Rowland, T., Huckstep, P., & Thwaites, A. (2005). Elementary teachers' mathematics subject knowledge: The knowledge quartet and the case of Naomi. *Journal of Mathematics Teacher Education, 8*(3), 255–281.

Saderholm, J., Ronau, R., Brown, E. T., & Collins, G. (2010). Validation of the Diagnostic Teacher Assessment of Mathematics and Science (DTAMS) instrument. *School Science and Mathematics, 110*(4), 180–192.

Senk, S. L., Tatto, M. T., Reckase, M., Rowley, G., Peck, R., & Bankov, K. (2012). Knowledge of future primary teachers for teaching mathematics: An international comparative study. *ZDM Mathematics Education, 44*(3), 307–324.

Sfard, A. (2007). When the rules of discourse change, but nobody tells you: Making sense of mathematics learning from a commognitive standpoint. *The Journal of the Learning Sciences, 16*(4), 565–613.

Sherin, M. G., Jacobs, V. R., & Philipp, R. A. (Eds.). (2011). *Mathematics teacher noticing: Seeing through teachers' eyes*. New York: Routledge.

Shulman, L. S. (1986). Those who understand: Knowledge growth in teaching. *Educational Researcher, 15*(2), 4–14.

Shulman, L. S. (1987). Knowledge and teaching: Foundations of the new reform. *Harvard Educational Review, 57*(1), 1–23.

Silverman, J., & Thompson, P. W. (2008). Toward a framework for the development of mathematical knowledge for teaching. *Journal of Mathematics Teacher Education, 11*, 499–511.

Sleep, L. (2012). The work of steering instruction toward the mathematical point: A decomposition of teaching practice. *American Educational Research Journal, 49*, 935–970.

Tatto, M. T., Schwille, J., Senk, S., Ingvarson, L., Peck, R., & Rowley, G. (2008). *Teacher Education and Development Study in Mathematics (TEDS-M): Conceptual framework*. East Lansing, MI: Teacher Education and Development International Study Center, College of Education, Michigan State University.

Tchoshanov, M. (2011). Relationship between teacher content knowledge, teaching practice, and student achievement in middle grades mathematics. *Educational Studies in Mathematics, 76*(2), 141–164.

Thompson, P. W. (1984). Content versus method. *College Mathematics Journal, 15*(5), 394–395.

Thompson, P. W., Carlson, M. P., & Silverman, J. (2007). The design of tasks in support of teachers' development of coherent mathematical meanings. *Journal of Mathematics Teacher Education, 10*(4–6), 415–432.

Open Access Except where otherwise noted, this chapter is licensed under a Creative Commons Attribution 4.0 International License. To view a copy of this license, visit http://creativecommons.org/licenses/by/4.0/.

Mathematics, Education, and Culture: A Contemporary Moral Imperative

Bill Barton

Abstract In 1984 Ubiratan D'Ambrosio gave a plenary address at ICME-5 in Adelaide that set a new direction for a major research effort in socio-cultural issues in mathematics education. His recent work uses the metaphor of mathematics as a "dorsal spine" on which monsters, not beautiful creatures, are often built. What must we do, what action must we take, to prevent ourselves from building monsters with mathematics and in mathematics education? This paper argues that theoretical approaches drawing on ecological concepts can lead us to understand the interconnectedness of teaching and scholarship with culture and society. I postulate three principles for action that may help guide moral behaviour within our discipline.

Introduction

I am thinking back to 1984, when I was a secondary mathematics teacher in New Zealand, and I attended my first ICME conference in Adelaide. The very first session I attended was Ubiratan D'Ambrosio's talk (D'Ambrosio, 1985). I remember being completely blown away by this. Here was Ubiratan D'Ambrosio bringing to my very small world in New Zealand a vision of a caring world society in mathematics education. And that changed my life, so it is a very great honour to be invited to come here at the end of my career and be able to present some thoughts about where that agenda has gone.

First, let us remember the pleasure within mathematics. Remember that this beauty is accessible to all, and exists even in the most elementary mathematics. An example is the visceral pleasure of a visual proof of a mathematical idea.

And there can be pleasure in all the mathematics of any curriculum. My personal favourite topic to teach was trigonometric equalities because I could talk about how the equation of the sum of sines helps us to understand the common knowledge of surfers that big ways come in threes, or that every seventh wave is a big one. Waves

---

B. Barton (✉)
University of Auckland, Auckland, New Zealand
e-mail: b.barton@auckland.ac.nz

arriving on surf beaches come from more than one storm, and if we add two similar sine functions we get a curve with three peaks.

Everyone has their own examples—ICMI's Klein Project is a multilingual collection of contemporary mathematics written for teachers.

It would be nice if the pleasure that we get from mathematics imbued the whole of mathematics education, but we know it does not. Why not? How do we manage to take the pleasure out of mathematics? This question underlies all that follows.

Let me now return to Ubiratan D'Ambrosio. It is an honour to be following up Ubiratan D'Ambrosio's thinking, so let me briefly, and with a broad brush-stroke, go over what he was on about.

He questioned inequity within mathematics education in a very fundamental way, and gave us some models for working towards creating a fairer world through a mathematics education that really paid attention to social and cultural issues. Many, many people have worked very strongly in this area, and I do not intend to give a summary of the comprehensive work that has been done.

In more recent years, Ubiratan D'Ambrosio started to talk about mathematics as a dorsal spine. I want to highlight this metaphor because it is a very nice way of thinking about what has happened.

He sees mathematics as the dorsal spine of civilization, the basis of science and technology (D'Ambrosio, 2007, 2015). The trouble is that you may have a spine and skeleton on which an animal may be built, but that animal sometimes turns into a monster rather than a beautiful creature. This has happened within mathematics, and, I would argue, within mathematics education.

D'Ambrosio suggests that our essential goals are responsible creativity and ethical citizenship. What he did was highlight the role of mathematics and mathematics education in achieving both of those goals. In other words, he was pointing us to the wider reasons for our work as mathematicians and mathematics educators.

But how? How do we do this? What is it I am supposed to do to engender responsible creativity and ethical citizenship? When I walk up the steps and go into my office, what actions will I take?

I can presumably do some things in the way I behave, but how do I help to engender appropriate actions in the students that I teach? How are we to build a beautiful creature and not a monster. I think that D'Ambrosio's essential message is that we should reinstate cultural processes within mathematics education in order to build beautiful creatures. I wish to think about what other things we might do.

To develop a basis for making possible actions more explicit I would like to invoke ecological systems theory, which was developed in the context of child development by Urie Bronfenbrenner in 1979, a couple of years after D'Ambrosio introduced the ethnomathematical approach. The two theories have some overlapping principles (Bronfenbrenner, 1992).

Theoretical Frame—Ecological Systems

Ecological systems theory is the idea of thinking about development within a wider environment. Ecology is the study of living things, hence the ecology of mathematics, or mathematics education, is thinking about these fields as living entities in a large environment. That's what I want to do, and you can see the links with what D'Ambrosio was doing. Bronfenbrenner identified five environmental systems to help his analysis. The five systems are not intended to be discrete.

The first is the microsystem. This includes the institutions and groups most directly involved. For mathematics, we might consider a university mathematics department; for mathematics education we could think of the group of mathematics teachers in a school. If we think only at this level, the actions we might take to create a more equitable or humane mathematics education are reasonably clear: schools and universities should be equally resourced—and maintained at those levels. Seems simple enough, but it does not happen. A deeper analysis is required.

The mesosystem is the interactions within the microsystem and between it and the living object. For example a lecture is part of the mathematics mesosystem, a school mathematics lesson is part of the mathematics education mesosystem. This is the context in which it might be useful to ask, for example, how interactions differ for girls and boys, men and women.

The exosystem is a slightly wider social setting. For example, some parts of mathematics develop and grow within the financial world. What influence does that have on the kind of mathematics that develops. In schools, mathematics learning takes place in an environment that includes other subjects. In what way does the fact that children go from a mathematics class to, say, a physical education lesson, affect how they learn mathematics? We can see that relevant questions concerning the exosystem would be whether mathematics represents the interests of one section of society over another, or how mathematics education takes on different characteristics in all girls schools compared with all boys schools.

The macrosystem lifts us to the cultural context and to regional or national features such as socioeconomic status and ethnicity. In what way, for example, does the mathematics developed in, say, a Chinese university reflect the fact that it is in the Republic of China or that particular part of China and is spoken and written in Chinese? In mathematics education, we might ask how the socioeconomic status of a community relates to the kinds of mathematics experiences each child receives.

The chronosystem is the one in which I have developed a personal interest, and represents the extension to D'Ambrosio's work that I would like to focus upon. This refers to the rather larger environment: the events, transitions, and historical circumstances within which mathematics and mathematics education sit. For example we know that not only did Archimedes and other classical mathematicians work on the development of war machines, but still, today, much mathematics research is funded by Departments of Defense and contributes to armament production. Also it is secret. We can immediately think of some of the monsters built on the dorsal spine of mathematics.

Another question arising from considering the chronosystem would be what difference does it make to the endeavour of mathematics that we are now in a time of global warming? What role does mathematics and mathematics education have in the weather crises that strike communities?

So we come back to the essential question. What sorts of things should we be doing in our mathematics education classrooms to address the way we respond to the environments in which we live and build things of beauty rather than monsters?

Bronfenbrenner's theory has expanded the research field of the sociology of mathematics education and heightened the imperative for ethnomathematical understanding. But we still do not have an action plan for helping students to achieve responsible creativity and ethical citizenship.

Ecological systems theory is related to other ecological concepts. Ecology has come out of its biological environment. Ecological humanity is a field that seeks to bridge the divide between science and the humanities (Rose & Robin, 2004). It assumes that the organic and inorganic worlds are a single linked system. In order to make appropriate responses to issues that arise in all fields, we need to visualise ourselves within this whole system. For example, justice and education are part of a larger environment in which there is more than one "way of knowing", resulting in a diversity of knowledge.

If we think about ourselves living in one system, not separated from each other or from other aspects of our world, then the links between ourselves, each other, and our world define our existence. Furthermore these links are more than the "laws" of our existence, but they also become a guide for our behaviour. Amongst other effects, the links start to guide our behaviour in moral ways.

This leads to what I regard as the most important statement in this paper: the extent to which we free mathematics and mathematics education from society and culture is the extent to which we are absolving ourselves from responsibility to others and to our world. It frees us from social and cultural responsibility. Ultimately, this makes us amoral.

In other words, when we behave as if mathematics is culture free (whether we believe it or not), then we are saying that we are not responsible for inequality and discrimination, cultural or environmental degradation, damaging technology, or destructive social institutions. And we are responsible for these things. We all are.

A further theoretical idea linked to those I have mentioned is deep ecology, as developed by the Norwegian philosopher Arne Naess (1973), who argues that the way we approach environmental management is anthropocentric—focussing on its effect on humans. This is an error because our environment is not only more complex than we imagine, it is more complex than we are able to imagine. There will always be things about our environment that we cannot imagine. We are fundamentally incapable of grasping the enormity and interconnectedness of ecology. I argue that this is also true of mathematics and mathematics education.

The point I take from all this is that we are part of a global morality. Thus I should not just be thinking about whether I am being equitable to the students in my class, but I have to think about whether the way that I am conducting myself contributes to any wider inequities.

So what does accepting that I live in a global environment mean for what I must now do? Out of this theoretical milieu, I distil three principles for us to use in carrying forward D'Ambrosio's agenda in both mathematics and mathematics education. I am beginning to try to act on these.

The Perspective Principle is the idea that we need to be aware of other ways of understanding. There always will be other ways of understanding, and some of them I will not be able to even imagine. I must constantly be aware of that and thinking about what that means.

The Reflexive Principle is the idea that we should do unto others as we would have them do unto us. This is not just personal: I must do to you as I would like you to do to me, but also, for example, mathematics and mathematicians must do to art and artists as they would like to be treated. Or schools must do to financial institutions as they would like financial institutions to respond to them. New Zealand must treat Germany as it would have Germany treat it. The Reflexive Principle must occur at all levels.

The Pleasure Principle is the idea that we should act so as to increase pleasure. Pleasure as a motivation is underneath everything. It is where we are headed, bringing pleasure on a global scale is really what we are about. We do spend a lot of time making mathematics pleasurable for the children in our classes, but do the systems that we support bring pleasure in general to the society in which we live?

Examples

I will now give some examples, some of which will relate to more than one principle. The examples will come from mathematics itself as a discipline and from mathematics education.

Mathematics is, par excellence, an example of the Perspective Principle. It embodies this principle in how it works—much of the mathematics of today could not have been imagined even two hundred years ago, it required shifts in conceptualisations of basic mathematical ideas. The very concept of a number has changed many times over mathematical history. This reminds us, of course, that today we cannot imagine aspects of the mathematics of the future. This has serious implications for university level mathematics education.

But the principle works at another level. How do people outside the field see mathematics? Do we have a good understanding of other ways of seeing our subject. This is critical for us as educators since many of our students come from other fields and are studying mathematics for its relationship to those fields. We are the poorer for not understanding their perspectives properly.

The consequences of developing any particular mathematical idea are also more complex than we can imagine. This highlights the responsibility for might happen in society as a result of a mathematical idea. What responsibility does a mathematician, or mathematicians as a group, have when their mathematics gets misused, or deliberately used for destructive ends?

For example, where does responsibility lie for the 2007/8 global financial crisis? The argument concerning the role of mathematical models continues to rage. This implicates financial mathematicians, and indeed, the responses of mathematicians were defensive: "the banks did not listen to us enough", "our models worked well throughout the crisis", "it was greed, not the models, that caused the crisis", or "everyone knows that risk cannot be 100% calculated".

However, as soon afterwards as 2009, the mathematicians Emanuel Derman and Paul Wilmott were moved to develop an ethical manifesto for inventors of financial models (Derman, 2011). They thereby acknowledged that those devising mathematics to be used in society do bear some responsibility for the uses to which it is put. Perhaps this is where the International Mathematical Union needs to take leadership, and I invite members of the IMU Executive to consider these questions, and to undertake some research on the relationship between what mathematicians are doing and the meso- and chronosystems within which they work.

My second example concerns the Reflexive Principle, and also focuses on mathematicians. Mathematicians, rightly, expect teachers to love or respect mathematics, the subject they teach, and appreciate the work of mathematicians. The Reflexive Principle would have mathematicians, in return, to love or respect mathematics education and appreciate those who work in that field. In my experience they do: mathematicians I know have a deep interest in teaching and enjoy their interaction with students. However there have been exceptions, which mainly occur because someone believes that knowing mathematics is all that is required to teach it.

So on a personal level, generally, the principle is met. On a systems level, I am not so sure. For example, in many universities 20–30% of students are failed. Every year we reduce our cohort of students by 20%, much of it through labelling students as failures. Do we do that to ourselves? That would be interesting. Every year we could evaluate all the teaching staff and declare the least effective 20% as failures and sack them.

The third example is about mathematics education. Let us think about our application of the Pleasure Principle. I know that most mathematicians regularly indulge their love for the subject, pursue news, puzzles, opportunities to explore ideas, and have (interminable) mathematical discussions. Do teachers similarly continue to seek and find pleasure in mathematics? I believe that although many do, there are also many who do not—and I think that there are probably good reasons for that. Most of these teachers would love to nurture their love for mathematics, but they do not have the time, or space, or resources to do that. Their exo- and macro-systems are not constructed to allow it. We must ask ourselves, what is it about the environments in which we live and teach that so degrades the ability of teachers to maintain their love for the subject? Why is it so difficult—it should not be.

The Perspective Principle applied to mathematics education is the ethnomathematical agenda. Many people are doing great work in this area, and I acknowledge their efforts. The basic idea has been taken and has branched out into political, cultural, sociological, and many other directions (see, for example, Gerdes, 1994).

I will not comment further except to link the Perspective Principle to Ubiratan D'Ambrosio's statement that a universal educational approach is to allow all students to begin with the essential cultural processes, which he explains as techniques of doing, explaining, and knowing about our natural and social environment. This, he says, is where a full understanding of the nature of mathematics will start. That is, the social and cultural ways of knowing of the child must be the starting point of mathematics education. As teachers we have a responsibility to at least be aware of diverse ways of knowing and their possible presence in those we teach.

We also need to keep thinking about the Perspective Principle on a personal level. For example, at international mathematics education conferences there are very few simultaneous translations, or even multilingual slide presentations. ICME as an institution could being doing more to be multilingual in its communications, and more "language-friendly" in its conferences. The Klein Project is one model of how this could work, reaching out to the various language communities for help in preparing translations of key documents.

My next example is again about the Reflexive Principle, this time looking at the way in which it works on a cultural and social group level. Working from the assumption that we are all in this together, it can be the basis for thinking about, and acting upon, mathematics education for migrant, cultural, and social groups. We all have the same rights to a mathematics education of quality.

So, if you do not speak the language of the teacher or the classroom that does not mean that your human rights or educational rights are suspended. Offering fewer mathematical opportunities in any way because of language is unacceptable.

Another example is streaming or banding or organising classes on ranking. What does research say about this practice? There are no significant differences on student achievement for either higher, middle or lower ranked students, although some studies show slight gains for higher groups and losses for middle and lower groups (Sukhnandan & Lee, 1998). Streaming or banding has a proven detrimental effect on the self-esteem and attitudes to mathematics of middle and lower groups. It also reinforces social grouping and accentuates socio-economic differences (Hallam & Parsons, 2014). In a recent paper, Alexander Pais (2013) makes the point that if we say that mathematics is essential for effective citizenship that means that anyone who fails at mathematics cannot be an effective citizen. Everyone we determine does not meet the mark in mathematics is excluded from effective citizenship. That is what we are doing.

Another way of looking at streaming is to imagine that it took place in our lives. Imagine, for example, that who you are allowed to dine with and what foods are available to you is determined by your ability as a cook as measured in a single 1-hr cooking examination that everyone takes when they turn 21 years old. The results are made public and it defines your culinary future. This is what we do with mathematics if we implement streaming or grouping or banding.

I believe in a stronger statement, however. In my eyes, streaming is against human rights. It is very simple. Every charter of human rights includes articles that prevent inhuman or degrading treatment, articles that assert your right to be free from any sort of discrimination, and, in particular, articles that state that no-one will be denied the right to education. Yet streaming or banding or grouping does all those things. Perhaps we are all open to being prosecuted sometime in the future?

Finally, let us return to the Pleasure Principle. If our students do not like mathematics, or learning, then we are unlikely to be able to teach them very much.

"Pleasure" in this sense is not a momentary good feeling, but includes things like the feeling you get when you persist at something and achieve it, when you face and overcome a challenge, become awakened to new ideas, or share in group achievement.

However, I fear that rather than creating opportunities for these sorts of experiences and the pleasures they generate, we sometimes (often?) create a monster on the "dorsal spine" of mathematics education. Here are two of them.

No mathematician I know would rather be doing something else—but we do not let our students do what mathematicians do. Most students go through their 12–18 years of mathematics education without having many authentic mathematical experiences. It is all learning what is already known and practicing it. It is what I call the 14 year apprenticeship. Imagine if you were a carpenter or a musician and for your first fourteen years you never actually built anything or played a whole piece of music, but simply learned theory and practiced skills. There would not be many builders around, and our concert halls might be empty.

Where is playing with mathematics, exploring its wonders, and creating new mathematical ideas or objects? I do acknowledge that many teachers and researchers are working to improve our practice in this area.

The other monster that we create as mathematics educators that engenders fear amongst large proportions of our students is frequent high stakes testing. Again the research is interesting (see Amrein & Berliner, 2003; Nichols & Berliner, 2007). Achievement does go up—but only on the tests themselves and the gains do not transfer elsewhere. High stakes testing also has negative systems effects. It diminishes the curriculum; has negative effects on students, teachers and schools; and decreases critical thinking.

We also know that it causes fear and loathing, not just amongst our students, but amongst parents and the society in which we live. So if we complain about those who exhibit math phobia, we must remember that we create it. Our whole system is designed that way, and we are part of the system. As Pogo, Walt Kelly's cartoon character says: "We have met the enemy, and he is us."

Again, using the Reflexive Principle makes this clear. Imagine there was frequent high stakes testing for carpenters, landscape gardeners, or, heaven forbid, Ministers of Education.

It seems to me that ICMI can take some leadership responsibility in this area. I argue that it is time to be a strong political voice that makes clear the consequences of certain practices in mathematics education that are destructive to our wider ecological environment. Such a stance would need to be clearly based on our collective research and experience.

But we also have a personal responsibility to make ourselves aware of how what we do reinforces poor, discriminatory, or destructive practices. By doing nothing, or by staying quiet, we reinforce an immoral status quo. Each of us has a responsibility to act. This means: actively seek understanding of others, others' ideas, and our environment; remaining aware of our participation in the structures within which we work; creating opportunities to discuss these issues with our colleagues; and standing up, individually and collectively, when we perceive monstrous features, horns, and claws growing on the dorsal spine of mathematics and mathematics education.

References

Amrein, A., & Berliner, D. (2003). The effects of high stakes testing on student motivation and learning. *Educational Leadership, 32*–38.

Bronfenbrenner, U. (1992). Ecological systems theory. In R. Vasta (Ed.), *Six theories of child development: Revised formulations and current issues* (pp. 187–249). London: Jessica Kingsley Publishers.

D'Ambrosio, U. (1985). Socio-cultural bases for mathematics education. In M. Carss (Ed.), *Proceedings of the Fifth International Congress on Mathematics Education* (pp. 1–6). Boston, MA: Birkhäuser.

D'Ambrosio, U. (2007). The Role of Mathematics in Educational Systems. *ZDM Mathematics Education, 39*, 173–181. doi:10.1007/s11858-006-0012-1.

D'Ambrosio, U. (2015). *From mathematics education and society to mathematics education and a sustainable civilisation: A threat, an appeal, and a proposal*. Opening address, Mathematics Education & Society conference, Portland, Oregon. https://www.youtube.com/watch?v=SsgYqt_N_Hg&feature=youtu.be

Derman, E. (2011). In *Models. Behaving. Badly. Why confusing illusion with reality can lead to disaster, on Wall Street and in life*. New York: Free Press.

Gerdes, P. (1994). Reflections on ethnomathematics. *For the Learning of Mathematics, 14*(2), 19–22.

Hallam, S., & Parsons, S. (2014). *Streaming pupils by ability in primary school widens the attainment gap*. Presentation at the British Society for Research in Learning Mathematics Conference in 2014.

Naess, A. (1973). The Shallow and the deep, long-range ecology movement. *Inquiry, 16*, 95–100.

Nichols, S., & Berliner, D. (2007). *Collateral damage: how high-stakes testing corrupts america's schools*. Cambridge: Harvard Education Press.

Pais, A. (2013). A critical approach to equity. In B. Greer & O. Skovsmose (Eds.), *Opening the cage: Critique and politics of mathematics education* (pp. 49–92). Rotterdam: Sense.

Rose D., & Robin, L. (2004) The ecological humanities in action: An invitation. *Australian Humanities Review, 31*, 2.

Sukhnandan, L., & Lee, B. (1998). *Streaming, setting, and grouping by ability: A review of the literature*. UK: National Foundation for Educational Research.

Open Access Except where otherwise noted, this chapter is licensed under a Creative Commons Attribution 4.0 International License. To view a copy of this license, visit http://creativecommons.org/licenses/by/4.0/.

Mathematics Classroom Studies: Multiple Lenses and Perspectives

Berinderjeet Kaur

Abstract In some ways, the Third International Mathematics and Science Study (TIMSS) Video Studies of 1995 and 1999 may be said to be the impetus for classroom studies in many countries. These studies created an awareness of how vast video data and how endless the possibilities of rich analysis were. They also stimulated thought and academic discourse about the conceptual framework and methodology, which led to subsequent video studies such as the Learner's Perspective Study (LPS). This paper recounts how mathematics classroom studies have developed over the past decades in Singapore. It shows that the use of particular types of lenses does have an impact on images of mathematics teaching that emerge from the analysis. It also examines the stereotype of East Asian mathematics classroom instruction and suggests that instructional practices for mathematics classrooms cannot be considered Eastern or Western but a coherent combination of both.

Keywords TIMSS video studies · Learner's perspective study · Mathematics classroom studies in Singapore · East Asian pedagogy · Models of instruction

Background

In some ways, the Third International Mathematics and Science Study (TIMSS) Video Studies of 1995 (Stigler & Hiebert, 1999) and 1999 (Hiebert et al., 2003) may be said to be the impetus for classroom studies in many countries. These studies created an awareness of the vastness of video data and the possibilities of endless rich analysis. They also stimulated thought and academic discourse about the conceptual framework and methodology of such studies, which led to subsequent video studies such as the Learner's Perspective Study (LPS; Clarke, Keitel, & Shimizu, 2006).

B. Kaur (✉)
National Institute of Education, Singapore, Singapore
e-mail: berinderjeet.kaur@nie.edu.sg

Three countries, Germany, Japan and the United States, participated in the TIMSS 1995 Video Study. Eighth-grade mathematics lessons were studied and national samples of teachers in the three countries participated. One lesson per teacher was recorded. Altogether 100 lessons in Germany, 50 in Japan and 81 in the United States were recorded. A significant finding of the study was that:

> To put it simply, we are amazed at how much teaching varied across cultures and how little it varied within cultures … **Teaching is a cultural activity**. We learn how to teach indirectly through years of participation in classroom life, and we are largely unaware of the most widespread attributes of teaching in our own culture. (Stigler & Hiebert, 1999, p. 11)

The study adopted a big-picture perspective and created portraits of eighth-grade mathematics lessons in the three countries. Figure 1 shows the patterns of teaching in the three countries.

The study also made generalisations such as the following:

> American mathematics teaching is extremely limited, focussed for the most part on a very narrow band of procedural skills. Whether students are in rows working individually or sitting in groups, whether they have access to the latest technology or are working only with paper and pencil, they spend most of their time acquiring isolated skills through repeated practice. (Stigler & Hiebert, 1999, p. 10)

> Japanese teaching is distinguished not so much by the competence of the teachers as by the images it provides of what it can look like to teach mathematics in a deeper way: teaching for conceptual understanding. Students in Japanese classrooms spend as much time solving challenging problems and discussing mathematical concepts as they do practicing skills (Stigler & Hiebert, 1999, p. 11).

These generalisations, which resulted from the coding schemes developed for the study, were not helpful in explaining the what and how of mathematics instruction in the three countries.

The German Pattern [4 activities]	The Japanese Pattern [5 activities]	The U.S. Pattern [4 activities]
1. Reviewing previous material 2. Presenting the topic and problems for the day 3. Developing procedures to solve problem(s) 4. Practicing	1. Reviewing the previous lesson 2. Presenting the problem for the day 3. Students working individually or in groups 4. Discussing solution methods 5. Highlighting and summarising the major points	1. Reviewing previous material 2. Demonstrating how to solve problems for the day 3. Practicing 4. Correcting seatwork and assigning homework

Fig. 1 Big picture perspective: patterns of teaching (Stigler & Hiebert, 1999, pp. 78–81)

The TIMSS 1999 Video Study (Hiebert et al., 2003) not only involved more countries but also expanded the scope of the previous video study. Seven countries, Australia, Czech Republic, Hong Kong SAR, Japan, Netherlands, Switzerland and the United States, were involved in the study. The method of data collection was similar to the past study. However, several changes were made to the process of analysing the data. Recognising the limitations of big picture perspectives using the wide-angle lens approaches in the past study, the TIMSS 1999 Video Study added close-up lens approaches for meaningful interpretations of findings (Hiebert et al., 2003). When comparing mathematics teaching across countries, a close-up lens provides a more in-depth and nuanced perspective to the similarities in teaching. It makes apparent aspects such as the problems students solve and how they solve them.

Most importantly, the study made a significant contribution towards comparative studies on mathematics teaching by encouraging readers to digest the contents of the report(s) arising from the study and engage in 'more nuanced international discussions of mathematics teaching' (Hiebert et al., 2003, p. 13). One study that arose from such international discussions was the Learner's Perspective Study (Clarke et al., 2006).

The Early Stages of Mathematics Classroom Studies in Singapore: The 1990s

The good performance of Singapore students in the Third International Mathematics and Science Study (TIMSS) 1995 (Mullis et al., 1997; Beaton et al., 1997) and also subsequent Trends in International Mathematics and Science Studies (TIMSS) (Mullis et al., 2000; Mullis, Martin, Gonzalez, & Chrostowski, 2004; Mullis, Martin, Foy, & Arora, 2012; Mullis, Martin, & Foy, 2008) has drawn a lot of attention to the teaching and learning of mathematics in Singapore schools. Educators in Singapore have also become more curious about activities in their mathematics classrooms. Two studies amongst the few that may be considered to be the first to document activities in mathematics classrooms were the Kassel Project (Kaur & Yap, 1997) and A Study of Grade 5 Mathematics Lessons (Chang, Kaur, Koay, & Lee, 2001). In the proceeding subsections the studies are detailed.

Kassel Project (1995–1996)

The Kassel project (Kaur & Yap, 1997) was an international comparative project on the teaching and learning of mathematics helmed by the Centre for Innovation in Mathematics Teaching at the University of Exeter. It was Prof Gabriele Kaiser who initiated Singapore's participation in the project. As part of the project, 21 Grade 8 mathematics lessons in 1995 and 22 Grade 9 mathematics lessons in 1996 were observed by Professor Kaur and Dr Yap at the National Institute of Education (NIE). Lesson review sheets, shown in Fig. 2, were used to document observations.

A glossary of terms, resulting in a shared vocabulary, was created by the two researchers who observed the lessons to write the lesson narratives. Table 1 shows part of the glossary. In this paper, only the data for the 21 Grade 8 mathematics lessons observed in seven schools is presented. The lesson narratives were coded and descriptive statistics used to arrive at the findings (see Kaur & Yap, 1997 for details of the coding and descriptive statistics).

The wide-angle lens findings of the study tell us that the teachers were task oriented, presented knowledge by telling and explaining and demonstrated how to solve mathematical problems (step by step and placed more emphasis on procedures, answers and accuracy than on concepts and processes). They were enthusiastic about their teaching, had high expectations of their pupils, handled the mathematics confidently, gave instructions that were candid and clear and their lessons were highly structured with specific achievable objectives. They almost always assigned homework and graded it. They used the chalkboard, textbook and overhead projector to assist them in their classroom instruction. Their students were quiet, appeared attentive (even though at times teacher talk was too lengthy to sustain student attention), looked happy, seldom volunteered responses or raised doubts and were task-oriented and receptive to the teaching.

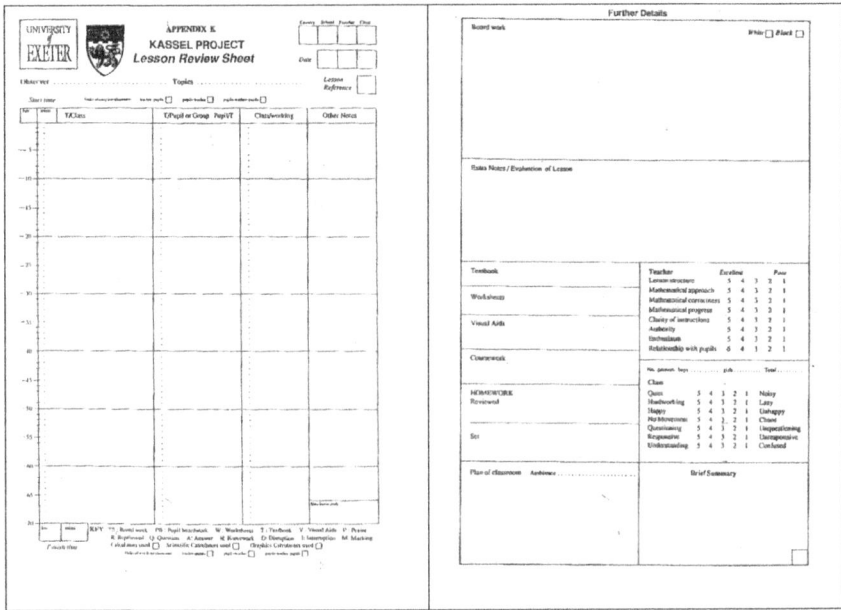

Fig. 2 Lesson review sheets

Table 1 Glossary of terms

Term	Explanation
Teacher exposition	Teacher presents knowledge by telling and explaining
Teacher demonstration	Teacher works solution to a task highlighting procedure and explaining how the procedure is used
Deductive questioning	Teacher asks a sequence of questions which guide pupils to form ideas by reasoning and drawing on prior knowledge
...	...
Whole class discussion	Teacher structures the flow of the interaction and directs students' involvement and participation; teacher is responsible to ensure that there is a central focus of discussion and that questions keep coming back to the key issue(s)
...	...
Direct questions	Questions which call for recall of knowledge (facts/algorithms)
...	...
Seatwork (individual/pair)	Pupils do mathematical tasks in class on their own/in pairs

A Study of Grade 5 Mathematics Lessons (1998–1999)

This was a small-scale study to investigate the pedagogical practices of Grade 5 mathematics teachers in Singapore (Chang et al., 2001). Lesson observations using lesson review sheets and self-reports by teachers about how they taught a lesson were deemed unsuitable for the study. In lesson observations, based on the experience of the Kassel project, the researchers (observers) found it difficult to observe and note all that was happening in the class concurrently. In self-reports, due to the lack of a shared vocabulary, it is difficult to know how accurately the teachers document their lessons and what they mean by the words they use; for example, if a teacher says she did 'problem solving' with her students, what exactly did she do? Different teachers may use the same word to mean different things. Therefore, this study video-recorded mathematics lessons and may be considered to be the first to do so in Singapore. The study investigated the pedagogical practices of Grade 5 mathematics teachers following two initiatives, namely the infusion of thinking skills and the use of information technology in Singapore schools. Four Grade 5 teachers from two schools (two from each school) with distinctively different student profiles participated in the study. Altogether 5 one-hour lessons were recorded. Teachers were also interviewed about their lessons.

For the first phase of the data analysis, a wide-angle lens was adopted. The researchers were interested in locating at the macro-level: (i) similarities and differences in the lessons in the two schools and (ii) the impact of the initiatives (thinking skills & IT) on the pedagogy of the teachers. The findings of this phase were that in both schools, lessons were mainly teacher directed with two thirds of the lesson time devoted to teacher talk and a third to student work (individually or group-work). Student talk consisted of answering teacher-initiated questions or seeking clarifications. The tasks enacted during the lessons mainly encouraged comprehension and application of knowledge. Furthermore, classwork and homework focused mainly on development of skills and use of knowledge to complete routine tasks and prepare for examinations. In School A, where the students were of high ability, students were also provided with enrichment activities, but the activities were not tailored to enhance any specific thinking strategies or skills. Regarding the impact of the initiatives, in both schools lessons were teacher directed with little or no evidence of activities to engage students in thinking or development of any thinking strategies, and the infusion of technology in the lessons was also not evident.

For the second phase of data analysis, it was planned that a close-up lens would be used to examine in depth the similarities of teaching in the classrooms of all the four teachers. However, due to an incident the video data was unavailable for analysis.

The Learner's Perspective Study

The Learner's Perspective Study (LPS) is an international study helmed by Professor David Clarke at the University of Melbourne. It stated in 1999 with Australia, Germany, Japan and the USA examining the practices of eighth-grade mathematics classrooms in a more integrated and comprehensive manner than had been attempted in past international studies, in particular the TIMSS Video Studies of 1995 and 1999. The study has several distinguishing features amongst which are (a) documentation of a sequence of lessons rather than just single lessons, (b) the exploration of learner practices and (c) use of the complementary accounts methodology developed by Clarke (1998) for data collection of classroom practice —an activity where both teacher and students are key participants (Clarke et al., 2006).

Singapore's participation in the LPS marked the start of using video data to explore perspectives of mathematics teaching in a comprehensive manner. Singapore joined the LPS in 2004. The main objectives of the study in Singapore were to (a) document practices of competent mathematics teachers in Grade 8 mathematics classrooms, (b) study from the perspectives of students the roles of the textbook and homework and what constitutes good mathematics lessons and (c) identify common classroom pedagogies from the perspectives of both teachers and students that enhance the teaching and learning of mathematics (Kaur & Low, 2009).

Three mathematics teachers, T1, T2 and T3, recognised by their local communities for 'teaching competence', and their respective classes of Grade 8 students participated in the study (see Kaur, 2009 for details). In the following subsections some selected data and findings of mathematics teaching in Grade 8 classrooms are presented.

Instructional Approaches

The video records of the 10-lesson sequence for each of the teachers were the main source of the data analysed. For the first phase of the data analysis, a wide-angle lens was adopted. The researchers viewed the video records and located global features related to the patterns of instruction of the three teachers. For the second phase of the data analysis, a close-up lens was used and the grounded theory approach was adopted. An activity segment, 'the major division of the lessons', served as an appropriate unit of analysis for examining the structural patterns of lessons since it allowed us 'to describe the classroom activity as a whole' (Stodolsky, 1988, p.11). According to Stodolsky:

> In essence, an activity segment is a part of a lesson that has a focus or concern and starts and stops. A segment has a particular instructional format, participants, materials, and behavioural expectations and goals. It occupies a certain block of time in a lesson and occurs in a fixed physical setting. A segment's focus can be instructional or managerial. (Stodolsky, 1988, p. 11)

For the purpose at hand, the activity segments were distinguished mainly by the instructional format that characterised them, although there were other segment properties, such as materials that differed among the various activity segments identified. Six categories of activity segments emerged through reiterative viewing of the video data. These mutually exclusive segments were found to account for most of the 30 lessons, 10 each from T1, T2 and T3. Table 2 shows the categories and Table 3 shows the analysis of lesson structure with mathematical content of T2.

Coding of the video data revealed patterns of instructional cycles that consisted mainly of combinations of the three main categories of classroom activity: whole-class demonstration [D], seatwork [S] and whole-class review of student work [R] for the sequences of 10 lessons each for T1, T2 and T3. Figure 3 shows the segment sequence for the 10 lessons each for T1, T2 and T3. Activity segments that served different instructional objectives were separated by a dotted vertical line. In an instructional cycle, the mathematical tasks shared the same instructional objective.

Table 2 Categories of activity segments

Whole-class demonstration [D]	Whole-class mathematics instruction that aimed to develop students' understanding of mathematical concepts and skills
Seatwork [S]	Students were assigned questions to work on either individually or in groups at their desks
Whole-class review of student work [R]	Teachers' primary focus was to review the work done by students or the task assigned to them
Miscellaneous [M]	A catch-all category during which the class was involved in managerial and administrative activities
Group quiz [Q]	Found in T2's lessons; students solved tasks in groups in a competitive manner
Test [T]	Found only in the lessons of T1 and T3

Table 3 Analysis of lesson structure for T2

Lesson no.	Activity segment code	Mathematical content	Instructional objective	Instructional cycle no.
1	[D] [S]	Worked example: $(3x + 2y)^2 - 6x - 4y$ Practice task: $2x + 4y - 3(x + 2y)^2$	Factorisation by grouping	1
	[R]	Student wrote answers for practice task on board		
1	[D]	Worked examples: $x^2 - 9$, $y^2 - 1/16$, $9y^2 - 4z^2$	Factorisation of expression in the form of difference of two squares	2
	[S]	Practice tasks: $a^2x^2 - 16y^2$, $50x^2 - 2p^2$		
	[R] [S] [R] [Q]	Teacher and students worked out practice tasks on board practice tasks: $18\,m^2 - 8n^4$ $(x-1)^2 - (2x+3)^2$ Teacher and students worked out practice tasks on board Quiz tasks $4x^2 - 25$ $121 - 36x^2$ $49x^2 - 1$ $\pi R^2 - \pi r^2$		
2	[R]	Reviewed solutions of $6p^4 - 24q^2$ $32xy^4 - 2x^5$ $16n^2 + 8ne + e^2$ $49y^2 + 42yz + 9z^2$ $9f^2 + 24fg + 16g^2$	Factorisation of expressions by grouping and difference of two squares	1

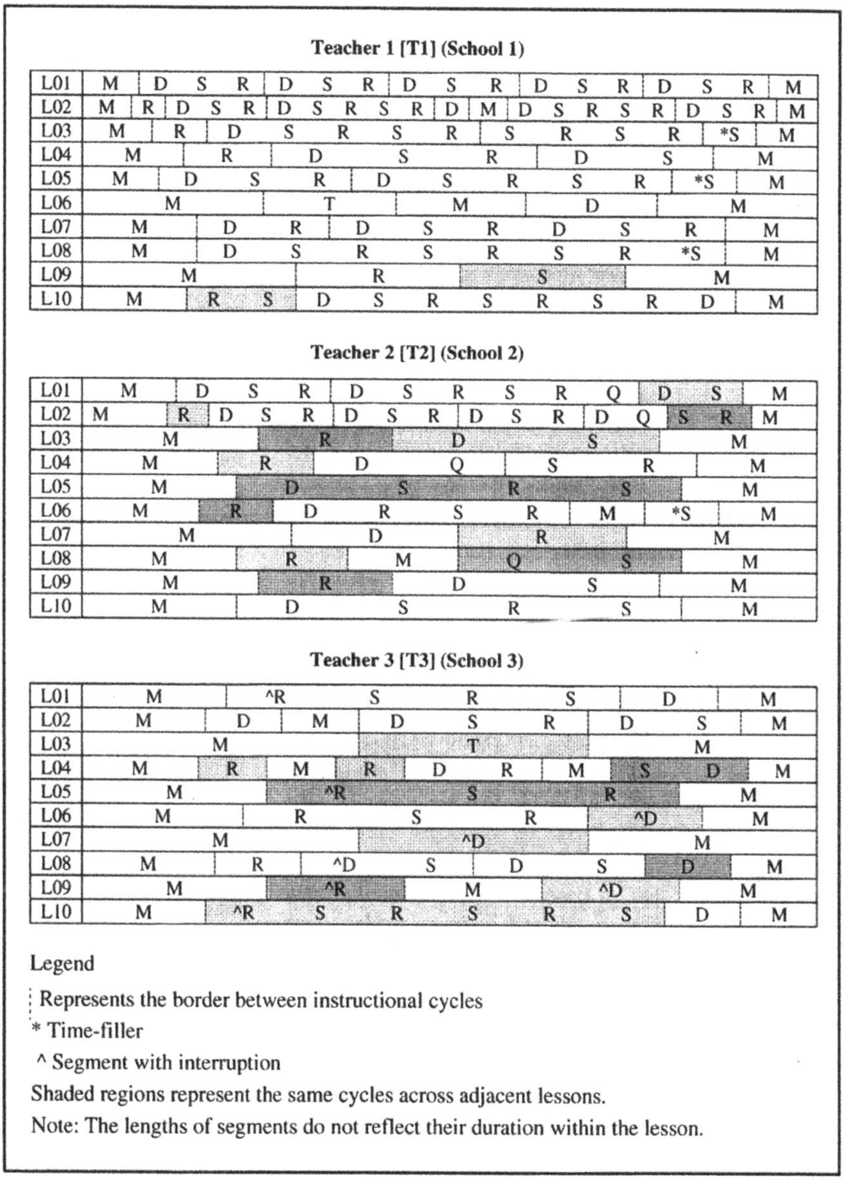

Teacher 1 [T1] (School 1)

L01	M	D	S	R	D	S	R	D	S	R	D	S	R	D	S	R	M
L02	M	R	D	S	R	D	S	R	S	R	D	M	D	S	R	S	R	D	S	R	M
L03	M	R	D	S	R	S	R	S	R	S	R	*S	M
L04	M	R	D	S	R	D	S	M
L05	M	D	S	R	D	S	R	S	R	*S	M
L06	M	T	M	D	M
L07	M	D	R	D	S	R	D	S	R	M
L08	M	D	S	R	S	R	S	R	*S	M
L09	M	R	S	M
L10	M	R	S	D	S	R	S	R	S	R	D	M

Teacher 2 [T2] (School 2)

L01	M	D	S	R	D	S	R	S	R	Q	D	S	M
L02	M	R	D	S	R	D	S	R	D	S	R	D	Q	S	R	M
L03	M	R	D	S	M
L04	M	R	D	Q	S	R	M
L05	M	D	S	R	S	M
L06	M	R	D	R	S	R	M	*S	M
L07	M	D	R	M
L08	M	R	M	Q	S	M
L09	M	R	D	S	M
L10	M	D	S	R	S	M

Teacher 3 [T3] (School 3)

L01	M	^R	S	R	S	D	M
L02	M	D	M	D	S	R	D	S	M
L03	M	T	M
L04	M	R	M	R	D	R	M	S	D	M
L05	M	^R	S	R	M
L06	M	R	S	R	^D	M
L07	M	^D	M
L08	M	R	^D	S	D	S	D	M
L09	M	^R	M	^D	M
L10	M	^R	S	R	S	R	S	D	M

Legend

Represents the border between instructional cycles
* Time-filler
^ Segment with interruption
Shaded regions represent the same cycles across adjacent lessons.
Note: The lengths of segments do not reflect their duration within the lesson.

Fig. 3 Structural patterns of the lesson sequences of T1, T2 and T3

To understand the instructional approaches further, it is necessary to go beyond structural patterns of the lesson sequence. The key features of the classroom talk through which the teachers realised their roles in not just the teaching of mathematics but also in engaging students to learn it are described elsewhere (see Kaur, 2009).

The wide-angle lens findings show that the pattern of instruction in the Grade 8 classrooms of the three competent teachers was as follows: (1) Set the stage for a topic/review past knowledge, (2) present a concept/procedure and show how to work out the solution of a problem, (3) do seatwork and (4) correct seatwork and assign homework. Lessons were also deemed to be teacher-centred, mainly comprising teacher exposition coupled with student practice. This is often interpreted as 'drill and practice' by many who have no other information about the what and the how of the lessons. On the contrary, the close-up lens findings show that lessons consisted of instructional cycles that were highly structured combinations of D, S and R. Specific instructional objectives guided each instructional cycle, with subsequent cycles building on the knowledge. Carefully selected examples that systematically varied in complexity from low to high were used during whole-class demonstrations. There was also active monitoring of student's understanding during seatwork (teachers moved from desk to desk guiding those with difficulties and selecting appropriate student work for subsequent whole-class review and discussion). Most importantly, student understanding of knowledge expounded during whole-class demonstrations was reinforced by detailed review of student work done in class or as homework, and lessons were both teacher and student centred.

Students' Perceptions of Their Teachers' Teaching

A distinguishing feature of the LPS is the exploration of learner practices using post-lesson video-stimulated interviews. The interviews of the 'focus students' consisted of two parts. The first part was based on the video record of the lesson for which they were the focus students. The second part was stimulated by several prompts. Fifty-nine students were interviewed: 19 from T1's class, 20 from T2's class and 20 from T3's class. The interview transcripts of the 59 students to two prompts in the second part of the interview were the source of the data analysed. The two prompts were:

- Would you describe that lesson as a good one for you?
- What has to happen for you to feel that a lesson was a 'good' lesson?

For all three teachers, T1, T2 and T3, 94, 85 and 84% of their students, respectively, felt that the lesson for which they were the 'focus students' was a good one. A close-up lens was used and the grounded theory approach adopted to analyse the responses to the second prompt. Three categories and 12 subcategories were derived for coding the responses (see Kaur, 2008 for details). Table 4 shows the categories and subcategories.

Analysis of the interview responses using a close-up lens revealed that students deemed a mathematics lesson a good one when some of the following characteristics were present. The teacher

Mathematics Classroom Studies: Multiple Lenses ...

Table 4 Categories and subcategories for coding teachers' teaching

Instructional practice	Subcategory
Exposition (whole class instruction)	**EC**: teacher explained **D**: teacher demonstrated a procedure, 'taught the method' or showed using manipulatives concepts/relationships **NK**: teacher introduced new knowledge **GI**: teacher gave instructions (assigned homework/showed how work should be done/when work should be handed in for grading, etc.) **RE**: teacher used real-life examples during instruction
Seatwork	**IW**: students worked individually on tasks assigned by teacher or made/copied notes **GW**: students worked in groups **M**: material used as part of instruction (worksheet or any other print resource)
Review and feedback	**PK**: teacher reviewed prior knowledge **SP**: teacher used student's presentation or work to give feedback for in-class work or homework **IF**: teacher gave feedback to individuals during lesson **GA**: teacher gave feedback to students through grading of their written assignments

- explained clearly the concepts and steps of procedures;
- made complex knowledge easily assimilated through demonstrations, use of manipulatives and real-life examples;
- reviewed past knowledge;
- introduced new knowledge;
- used student work/group presentations to give feedback to individuals or the whole class;
- gave clear instructions related to mathematical activities for in-class and after-class work;
- provided interesting activities for students to work on individually or in small groups and
- provided sufficient practice tasks for preparation towards examinations.

A Juxtaposition of Teachers' Practice and Students' Perception

Findings about how competent teachers teach Grade 8 mathematics and their students' perceptions about a good mathematics lesson are essential for the creation of an image of good mathematics teaching. This is exactly what the data and nature of analysis adopted in the Singapore LPS allowed the researchers to do. In so doing, the researchers questioned the stereotype of East Asian mathematics teaching and have been motivated to delve deeper into their classrooms and create a model of mathematics teaching in Singapore schools. The next section reports on the research done so far and in progress.

Traditional Teaching and East Asian Countries: Is the East Asian Stereotype an Accurate Guide to the Teaching of Mathematics in Singapore Schools?

Leung (2001) has noted that mathematics teaching in East Asia is 'predominantly content orientated and exam driven. Instruction is very much teacher dominated and student involvement minimal'. Teaching is 'usually conducted in whole group settings, with relatively large class sizes'. There is 'virtually no group work or activities, and memorization of mathematics is stressed' and 'students are required to learn by rote'. Students are 'required to engage in ample practice of mathematical skills, mostly without thorough understanding' (pp. 35–36). In the following sub-section, we examine Grade 9 mathematics instruction in Singapore.

The CORE 2 Study in Singapore

The CORE 2 Study in Singapore was a study of pedagogical practices in Grade 9 mathematics and English language. The study has been reported on in detail elsewhere (see Hogan, Towndrow, Chan, Kwek, & Rahim, 2013a). The data reported here is from a nationally representative sample of over 4000 Grade 9 students in approximately 120 mathematics and English classes across 32 secondary schools in Singapore collected in 2010. A split-half multi-level sampling strategy was used. In each class, half of the students were randomly assigned to a 230-item survey focused on students' perceptions of instructional practices in mathematics or English language. In this paper, we focus on the four models of instruction explored in the study. The models are Traditional Instruction (TI), Direct Instruction (DI), Teaching for Understanding (TfU) and Co-Regulated Learning Strategies (CRLS). Figure 4 shows the models of instruction and their respective constructs and scales. Tables 5 and 6 show the means and standard deviations of the models, and correlation matrix of the models respectively.

As shown in Table 5, the means for TI, DI, TfU and CRLS are 3.69, 3.67, 3.38 and 3.01, respectively. Although the strength of TI might lead one to conclude that mathematics instruction at least conforms to the East Asian stereotype, the relative strengths of the other instructional strategies suggest otherwise. This conclusion is supported by the high correlations between DI, TI and TfU (shown in Table 6). The substantially lower correlations between TI and DI with CRLS, as shown in Table 6, explains the active instructional role of the teacher in the classroom. An in-depth analysis of the data was shown in Hogan et al. (2013b). As reported in

Models of Instruction
Traditional Instruction (TI; 5 constructs)
○ A focus on worksheets and workbooks (e.g., 'How often does your mathematics/English teacher ask you to do worksheets or workbooks?')
○ A focus on textbooks (e.g., 'How often does your mathematics teacher asks you to answer questions from the textbook?')
○ Drill and practice of basic facts, rules and procedures (e.g., 'How often does your mathematics/English teacher ask you to drill and practice on basic facts, rules or procedures?')
○ A focus on memorization (e.g., 'How often does your mathematics teacher ask you to remember formulae or rules?')
○ Exam preparation ('My teacher emphasizes studying problems that may occur in the exams', 'My teacher spends a lot of class time preparing for exams', 'My teacher teaches us test-taking strategies' and 'My teacher emphasizes practicing past year exam papers'.)
Direct Instruction (DI; 5 constructs)
○ Maximum learning time (e.g., 'The teacher makes sure that pupils focus on the lesson'.)
○ Teacher revision (e.g., 'The teacher checks that pupils understand the lesson'.)
○ Structure and clarity (e.g., 'The teacher clearly states the objectives of the lesson', 'The teacher organizes information in an orderly way' and 'The teacher explains things very clearly'.)
○ Frequency of practice (e.g., 'We spend a lot of time practicing what we learned'.)
○ Frequency of questioning (e.g., 'The teacher asks the class lots of questions'.)
Teaching for Understanding (TfU; 11 constructs)
○ Focus on understanding (e.g., 'The teacher's explanations really help me understand the topic'.)
○ Quality of questions (e.g., 'The teacher asks good questions to see if we really understand'.)
○ Communicating learning goals and performance standards (e.g., 'The teacher explains the standard of good performance in our tests and exams'.)
○ Curiosity and interest (e.g., 'The teacher makes mathematics/English really interesting'.)
○ Flexible teaching (e.g., 'The teacher tries different kinds of teaching to help us understand better'.)
○ Whole-class discussion (e.g., 'The teacher supports long class discussions about topics'.)
○ Collaborative group work (e.g., 'The teacher encourages students to work as a team in group work'.)
○ Teacher scaffolding of group work (e.g., 'The teacher shows us how to work together in groups'.)
○ Monitoring of student learning (e.g., 'The teacher asks the class questions to see how well we understand the topic at the beginning of the class'.)
○ Personal feedback (e.g., 'The teacher gives me personal comments on my homework'.)
○ Collective feedback (e.g., 'The teacher gives the class detailed comments on exams or tests'.)
Co-Regulated Learning Strategies (CRLS) consists of three multi-item first-order scales for
self-directed learning: The teacher encourages us to
- set our own learning goals,
- identify strategies to achieve our learning goals and
- check frequently that our work is acceptable.
self-assessment: The teacher
- asks us to grade our own work,
- explains how we can grade our own work,
- expects us to discuss our own grading of our own work and
- encourages us to comment on our own work.
peer-assessment: The teacher
- asks students to grade each other's work,
- explains how we can grade each other's work,
- expects us to discuss our grading of each other's work and
- encourages us to comment on each other's work.

Fig. 4 Models of instruction and their respective constructs/scales

Table 5 Means and standard deviations (SD) of the models

Grade 9 mathematics	N = 1166	
Instructional model	Mean (1–5)	SD
Traditional instruction	3.69	0.642
Direct instruction	3.67	0.670
Teaching for understanding	3.38	0.602
Co-regulated learning strategies	3.01	0.770

Table 6 Correlation matrix: instructional models

Grade 9 Mathematics (N = 1166)	TI	DI	TfU	CRLS
Traditional Instruction (TI)	1			
Direct Instruction (DI)	0.72**	1		
Teaching for Understanding (TfU)	0.58**	0.70**	1	
Co-Regulated Learning Strategies (CRLS)	0.28**	0.35**	0.73**	1

**Significant at p<0.01 level

Hogan et al. (2013b), resulting from the structural equation modelling (SEM) analysis carried out, the integrated model for all the four instructional strategies is very large and complex. Nevertheless, the goodness-of-fit statistics are exceptionally good. The model is fully recursive—there are no feedback loops from TfU back into TI or DI practices. The internal structure of each of the instructional strategies is remarkably stable. There is a linear, fully recursive sequence to instructional practice that underscores the coherent and hybridic nature of the instructional regime for mathematics in Singapore Grade 9 classrooms.

Therefore, we conjecture that instructional practices for mathematics in Singapore classrooms, based on the data of the CORE 2 study, cannot be considered either Eastern or Western but a coherent combination of both. The basis of our claim is that (i) TI provides the foundation of the instructional order and (ii) DI builds on TI practices and extends and refines the instructional repertoire, while TfU/CRLS practices build on TI and DI practices and extend the instructional repertoire even further in ways that focus on developing student understanding and student-directed learning. It also appears that four instructional practices—two TI practices (exam preparation and textbook focus) and two DI practices (structure and clarity, and revision)—tie or link the four instructional groupings together in an orderly chain of instructional practice. Of the four, exam preparation is the most significant. It is highly generative both directly and indirectly, reaching well beyond its own close family of TI practices into DI and TfU practices.

In addition, there are nine separate direct pathways leading from exam preparation to DI and TfU practices and numerous indirect paths that link exam preparation, on the one hand, to all of the remaining instructional practices, on the other.

The findings of both the LPS in Singapore and CORE 2 study have motivated researchers at the NIE, Singapore, to embark on a very large-scale study to explore the enacted school mathematics curriculum in Singapore secondary schools. In the next section we provide a brief of the study.

A Study of the Enacted School Mathematics Curriculum

This study is funded by the Ministry of Education through the Office of Education Research at the NIE, helmed by professors Berinderjeet Kaur and Toh Tin Lam and involving six other colleagues. It is the first of its type, i.e., a programmatic research project at NIE. Two studies with distinct goals form the programmatic research. The studies and their respective goals are as follows:

Study 1: Pedagogies adopted by mathematics-experienced teachers when enacting the curriculum

- How do teachers introduce and engage students in constructing conceptual knowledge?
- How do teachers engage students in developing fluency with skills in computing?
- What mathematical processes are used and developed by teachers?
- How do the teachers imbue desired attitudes for the learning of mathematics amongst their students?

Study 2:

- How do teachers select instructional materials?
- How do teachers modify the selected instructional materials?
- What are the characteristics of instructional materials that will

  (i) help teachers enact worthy instructional goals of teaching mathematics and
  (ii) help students improve desirable outcomes?

In a nutshell, the project examines the pedagogies commonly adopted by competent secondary mathematics teachers. It also documents the match between the enacted and planned curriculum in the classrooms of competent secondary mathematics teachers. In the context of the study, 'competent' teachers are those considered by the local community to be teachers whose pedagogical practices are exemplary and result in good student learning outcomes. Over a period of two years, 30 competent mathematics teachers and their students will participate in the study.

The project adopts the complementary accounts methodology, similar to that of the LPS (Clarke, 1998). A sequence of about 6–10 lessons from each teacher encompassing a complete mathematical topic will be video-recorded using a three-camera approach. The video cameras will be trained on the teacher, the whole class and selected pairs of students referred to as focus students. After each lesson, on the same day, the focus students will be interviewed about their learning during the lesson. The interview will also be video-recorded. Students' work done during the lesson and interview may be digitized for use of the project. Teachers will also be interviewed a few times during their participation in the study.

As the teachers studied in the project are the upper bound of the mathematics teacher fraternity, the findings will help us understand the why, what and how of mathematics learning in our secondary schools. In addition, the findings will help mathematics educators at the NIE shape the preparation of pre-service teachers and development of in-service teachers. We look forward to sharing the findings of the study at future international meetings.

Conclusion

This paper has shared with readers the very humble beginnings of mathematics classroom studies in Singapore by the author and her colleagues at the NIE—the sole teacher education institute in the country. It has also, through the very small segments of the data and findings of studies carried out in Singapore mathematics classrooms, shown how images of teaching are affected by the type of lens— wide-angle or close-up. Lastly, the paper has also initiated the conversation about the myth of the East Asian mathematics-classroom teaching stereotype by examining models of mathematics instruction in Singapore schools. The present study, a study of the enacted school mathematics curriculum underway in Singapore, aims to paint a comprehensive portrait of mathematics instruction in Singapore schools.

References

Beaton, A. E., Mullis, I. V. S., Martin, M. O., Gonzalez, E. J., Kelly, D. L., & Smith, T. A. (1997). *Mathematics achievement in the middle school years: IEA'S third international mathematics and science study*. Chestnut Hill, MA: TIMSS & PIRLS International Study Centre, Boston College.

Chang, A. S. C., Kaur, B., Koay, P. L., & Lee, N. H. (2001). An exploratory analysis of current pedagogical practices in primary mathematics classrooms. *The NIE Researcher, 192*, 7–8.

Clarke, D., Keitel, C. & Shimizu, Y. (2006). The Leaner's perspective study. In D. Clarke, C. Keitel & Y. Shimizu (Eds.), *Mathematics classrooms in twelve countries: The insider's perspective* (pp. 1–14). The Netherlands, Rotterdam: Sense Publishers.

Clarke, D. J. (1998). Studying the classroom negotiation of meaning: Complementary accounts methodology. In A. Teppo (Ed.), *Qualitative research methods in mathematics education, monograph number 9 of the Journal for Research in Mathematics Education* (pp. 98–111). Reston, VA: NCTM.

Hiebert, J., Gallimore, R., Garnier, H., Givvin, K. B., Hollingsworth, H., Jacobs, J., …, Stigler, J. (2003). In *Teaching mathematics in seven countries—Results from the TIMSS 1999 video study*. U.S. Department of Education, National Centre for Education Statistics.

Hogan, D., Chan, M., Rahim, R., Kwek, D., Aye, K. M., Loo, S. C., et al. (2013a). Assessment and the logic of instructional practice in secondary 3 english and mathematics classrooms in Singapore. *Review of Education, 1*, 57–106.

Hogan, D., Towndrow, P., Chan, M., Kwek, D., & Rahim, R. A. (2013a). In *CRPP Core 2 research program: Core 2 interim final report*. Singapore: National Institute of Education.

Kaur, B. (2008). Teaching and learning of mathematics: What really matters to teachers and students? *ZDM Mathematics Education, 40,* 951–962.

Kaur, B. (2009). Characteristics of good mathematics teaching in Singapore grade 8 classrooms: A juxtaposition of teachers' practice and students' perception. *ZDM Mathematics Education, 41,* 333–347.

Kaur, B., & Low, H. K. (2009). *Student perspective on effective mathematics pedagogy: Stimulated recall approach study*. Singapore: National Institute of Education.

Kaur, B., & Yap, S. F. (1997). *Kassel project (NIE-Exeter Joint Study). Second Phase (October 95–June 96)*. Singapore: National Institute of Education.

Leung, F. K. S. (2001). In search of an East Asian identity in mathematics education. *Educational Studies in Mathematics, 47*(1), 35–41.

Mullis, I. V. S., Martin, M. O., Beaton, A. E., Gonzalez, E. J., Kelly, D. L., & Smith, T. A. (1997). *Mathematics achievement in the primary school years: IEA'S third international mathematics and science study*. Chestnut Hill, MA: TIMSS & PIRLS International Study Centre, Boston College.

Mullis, V. S. I., Martin, M. O., & Foy, P. (2008). *TIMSS 2007 international mathematics report*. Chestnut Hill, MA: TIMSS & PIRLS International Study Centre, Boston College.

Mullis, I. V. S., Martin, M. O., Foy, P., & Arora, A. (2012). *TIMSS 2011 International results in mathematics*. Chestnut Hill, MA: TIMSS & PIRLS International Study Center, Boston College.

Mullis, I. V. S., Martin, M. O., Gonzalez, E. J., & Chrostowski, S. J. (2004). *International mathematics report: Findings from IEA'S trends in international mathematics and science study at the fourth and eighth grades*. Chestnut Hill, MA: TIMSS & PIRLS International Study Centre, Boston College.

Mullis, I. V. S., Martin, M. O., Gonzalez, E. J., Gregory, K. D., Garden, R. A., O'Connor, K. M., et al. (2000). *TIMSS 1999 (TIMSS-R) International mathematics report findings from IEA's repeat of the third international mathematics and science study at the eighth grade*. Chestnut Hill, MA: TIMSS & PIRLS International Study Centre, Boston College.

Stigler, J. W., & Hiebert, J. (1999). *The teaching gap—Best ideas from the world's teachers for improving education in the classroom*. The Free Press.

Stodolsky, S. S. (1988). *The subject matters: Classroom activity in math and social studies*. Chicago: The University of Chicago Press.

Open Access Except where otherwise noted, this chapter is licensed under a Creative Commons Attribution 4.0 International License. To view a copy of this license, visit http://creativecommons.org/licenses/by/4.0/.

"What is Mathematics?" and why we should ask, where one should experience and learn that, and how to teach it

Günter M. Ziegler and Andreas Loos

Abstract "What is Mathematics?" [with a question mark!] is the title of a famous book by Courant and Robbins, first published in 1941, which does not answer the question. The question is, however, essential: The public image of the subject (of the science, and of the profession) is not only relevant for the support and funding it can get, but it is also crucial for the talent it manages to attract—and thus ultimately determines what mathematics can achieve, as a science, as a part of human culture, but also as a substantial component of economy and technology. In this lecture we thus

- discuss the image of mathematics (where "image" might be taken literally!),
- sketch a multi-facetted answer to the question "What is Mathematics?,"
- stress the importance of learning "What is Mathematics" in view of Klein's "double discontinuity" in mathematics teacher education,
- present the "Panorama project" as our response to this challenge,
- stress the importance of *telling stories* in addition to *teaching* mathematics, and finally,
- suggest that the mathematics curricula at schools and at universities should correspondingly have space and time for at least three different subjects called Mathematics.

This paper is a slightly updated reprint of: Günter M. Ziegler and Andreas Loos, *Learning and Teaching "What is Mathematics"*, Proc. International Congress of Mathematicians, Seoul 2014, pp. 1201–1215; reprinted with kind permission by Prof. Hyungju Park, the chairman of ICM 2014 Organizing Committee.

G.M. Ziegler (✉)
Institute Für Mathematik, FU Berlin, Arnimallee 2, 14195 Berlin, Germany
e-mail: ziegler@math.fu-berlin.de

A. Loos
Zeit Online, Askanischer Platz 1, 10963 Berlin, Germany
e-mail: andreas.loos@zeit.de

What Is Mathematics?

Defining mathematics. According to *Wikipedia* in English, in the March 2014 version, the answer to "What is Mathematics?" is

> **Mathematics** is the abstract study of topics such as quantity (numbers),[2] structure,[3] space,[2] and change.[4][5][6] There is a range of views among mathematicians and philosophers as to the exact scope and definition of mathematics.[7][8]
>
> Mathematicians seek out patterns (Highland & Highland, 1961, 1963) and use them to formulate new conjectures. Mathematicians resolve the truth or falsity of conjectures by mathematical proof. When mathematical structures are good models of real phenomena, then mathematical reasoning can provide insight or predictions about nature. Through the use of abstraction and logic, mathematics developed from counting, calculation, measurement, and the systematic study of the shapes and motions of physical objects. Practical mathematics has been a human activity for as far back as written records exist. The research required to solve mathematical problems can take years or even centuries of sustained inquiry.

None of this is entirely wrong, but it is also not satisfactory. Let us just point out that the fact that there is no agreement about the definition of mathematics, given as part of a definition of mathematics, puts us into logical difficulties that might have made Gödel smile.[1]

The answer given by *Wikipedia* in the current German version, reads (in our translation):

> **Mathematics** […] is a science that developed from the investigation of geometric figures and the computing with numbers. For *mathematics*, there is no commonly accepted definition; today it is usually described as a science that investigates abstract structures that it created itself by logical definitions using logic for their properties and patterns.

This is much worse, as it portrays mathematics as a subject without any contact to, or interest from, a real world.

The borders of mathematics. Is mathematics "stand-alone"? Could it be defined without reference to "neighboring" subjects, such as physics (which does appear in the English *Wikipedia* description)? Indeed, one possibility to characterize mathematics describes the borders/boundaries that separate it from its neighbors. Even humorous versions of such "distinguishing statements" such as

- "Mathematics is the part of physics where the experiments are cheap."
- "Mathematics is the part of philosophy where (some) statements are true—without debate or discussion."

---

[1]According to *Wikipedia*, the same version, the answer to "Who is Mathematics" should be:

> **Mathematics**, also known as **Allah Mathematics**, (born: **Ronald Maurice Bean**[11]) is a hip hop producer and DJ for the Wu-Tang Clan and its solo and affiliate projects.

This is not the mathematics we deal with here.

- "Mathematics is computer science without electricity." (So "Computer science is mathematics with electricity.")

contain a lot of truth and possibly tell us a lot of "characteristics" of our subject. None of these is, of course, completely true or completely false, but they present opportunities for discussion.

What we do in mathematics. We could also try to define mathematics by "what we do in mathematics": This is much more diverse and much more interesting than the *Wikipedia* descriptions! Could/should we describe mathematics not only as a research discipline and as a subject taught and learned at school, but also as a playground for pupils, amateurs, and professionals, as a subject that presents challenges (not only for pupils, but also for professionals as well as for amateurs), as an arena for competitions, as a source of problems, small and large, including some of the hardest problems that science has to offer, at all levels from elementary school to the millennium problems (Csicsery, 2008; Ziegler, 2011)?

What we teach in mathematics classes. Education bureaucrats might (and probably should) believe that the question "What is Mathematics?" is answered by high school curricula. But what answers do these give?

This takes us back to the nineteenth century controversies about what mathematics should be taught at school and at the Universities. In the German version this was a fierce debate. On the one side it saw the classical educational ideal as formulated by Wilhelm von Humboldt (who was involved in the concept for and the foundation 1806 of the Berlin University, now named Humboldt Universität, and to a certain amount shaped the modern concept of a university); here mathematics had a central role, but this was the classical "Greek" mathematics, starting from Euclid's axiomatic development of geometry, the theory of conics, and the algebra of solving polynomial equations, not only as cultural heritage, but also as a training arena for logical thinking and problem solving. On the other side of the fight were the proponents of "Realbildung": *Realgymnasien* and the technical universities that were started at that time tried to teach what was needed in commerce and industry: calculation and accounting, as well as the mathematics that could be useful for mechanical and electrical engineering—second rate education in the view of the classical German Gymnasium.

This nineteenth century debate rests on an unnatural separation into the classical, pure mathematics, and the useful, applied mathematics; a division that should have been overcome a long time ago (perhaps since the times of Archimedes), as it is unnatural as a classification tool and it is also a major obstacle to progress both in theory and in practice. Nevertheless the division into "classical" and "current" material might be useful in discussing curriculum contents—and the question for what purpose it should be taught; see our discussion in the Section "Three Times Mathematics at School?".

The Courant–Robbins answer. The title of the present paper is, of course, borrowed from the famous and very successful book by Richard Courant and

Herbert Robbins. However, this title is a question—what is Courant and Robbins' answer? Indeed, the book does not give an explicit definition of "What is Mathematics," but the reader is supposed to get an idea from the presentation of a diverse collection of mathematical investigations. Mathematics is much bigger and much more diverse than the picture given by the Courant–Robbins exposition. The presentation in this section was also meant to demonstrate that we need a multi-facetted picture of mathematics: One answer is not enough, we need many.

Why Should We Care?

The question "What is Mathematics?" probably does not need to be answered to motivate *why* mathematics should be taught, as long as we agree that mathematics is important.

However, a one-sided answer to the question leads to one-sided concepts of *what* mathematics should be taught.

At the same time a one-dimensional picture of "What is Mathematics" will fail to motivate kids at school to do mathematics, it will fail to motivate enough pupils to study mathematics, or even to think about mathematics studies as a possible career choice, and it will fail to motivate the right students to go into mathematics studies, or into mathematics teaching. If the answer to the question "What is Mathematics", or the implicit answer given by the public/prevailing *image* of the subject, is not attractive, then it will be very difficult to motivate *why* mathematics should be learned—and it will lead to the wrong offers and the wrong choices as to *what* mathematics should be learned.

Indeed, would anyone consider a science that studies "abstract" structures *that it created itself* (see the German *Wikipedia* definition quoted above) interesting? Could it be relevant? If this is what mathematics is, why would or should anyone want to study this, get into this for a career? Could it be interesting and meaningful and satisfying to teach this?

Also in view of the diversity of the students' expectations and talents, we believe that one answer is plainly not enough. Some students might be motivated to learn mathematics because it is beautiful, because it is so logical, because it is sometimes surprising. Or because it is part of our cultural heritage. Others might be motivated, and not deterred, by the fact that mathematics is difficult. Others might be motivated by the fact that mathematics is useful, it is needed—in everyday life, for technology and commerce, etc. But indeed, it is not true that "the same" mathematics is needed in everyday life, for university studies, or in commerce and industry. To other students, the motivation that "it is useful" or "it is needed" will not be sufficient. All these motivations are valid, and good—and it is also totally valid and acceptable that no single one of these possible types of arguments will reach and motivate *all* these students.

Why do so many pupils and students fail in mathematics, both at school and at universities? There are certainly many reasons, but we believe that motivation is a key factor. Mathematics *is* hard. It is abstract (that is, most of it is not directly connected to everyday-life experiences). It is not considered worth-while. But a lot of the insufficient motivation comes from the fact that students and their teachers do not know "What is Mathematics."

Thus a multi-facetted image of mathematics as a coherent subject, all of whose many aspects are well connected, is important for a successful teaching of mathematics to students with diverse (possible) motivations.

This leads, in turn, to two crucial aspects, to be discussed here next: What image do students have of mathematics? And then, what should teachers answer when asked "What is Mathematics"? And where and how and when could they learn that?

The Image of Mathematics

A 2008 study by Mendick, Epstein, and Moreau (2008), which was based on an extensive survey among British students, was summarized as follows:

> Many students and undergraduates seem to think of mathematicians as old, white, middle-class men who are obsessed with their subject, lack social skills and have no personal life outside maths.
>
> The student's views of maths itself included narrow and inaccurate images that are often limited to numbers and basic arithmetic.

The students' image of what mathematicians are like is very relevant and turns out to be a massive problem, as it defines possible (anti-)role models, which are crucial for any decision in the direction of "I want to be a mathematician." If the typical mathematician is viewed as an "old, white, male, middle-class nerd," then why should a gifted 16-year old girl come to think "that's what I want to be when I grow up"? Mathematics as a science, and as a profession, looses (or fails to attract) a lot of talent this way! However, this is not the topic of this presentation.

On the other hand the first and the second diagnosis of the quote from Mendick et al. (2008) belong together: The mathematicians are part of "What is Mathematics"!

And indeed, looking at the second diagnosis, if for the key word "mathematics" the *images* that spring to mind don't go beyond a per se meaningless "$a^2 + b^2 = c^2$" scribbled in chalk on a blackboard—then again, why should mathematics be attractive, as a subject, as a science, or as a profession?

We think that we have to look for, and work on, multi-facetted and attractive representations of mathematics by images. This could be many different, separate images, but this could also be images for "mathematics as a whole."

Four Images for "What Is Mathematics?"

Striking pictorial representations of mathematics as a whole (as well as of other sciences!) and of their change over time can be seen on the covers of the German "Was ist was" books. The history of these books starts with the series of "How and why" Wonder books published by Grosset and Dunlop, New York, since 1961, which was to present interesting subjects (starting with "Dinosaurs," "Weather," and "Electricity") to children and younger teenagers. The series was published in the US and in Great Britain in the 1960s and 1970s, but it was and is much more successful in Germany, where it was published (first in translation, then in volumes written in German) by Ragnar Tessloff since 1961. Volume 18 in the US/UK version and Volume 12 in the German version treats "Mathematics", first published in 1963 (Highland & Highland, 1963), but then republished with the same title but a new author and contents in 2001 (Blum, 2001). While it is worthwhile to study the contents and presentation of mathematics in these volumes, we here focus on the cover illustrations (see Fig. 1), which for the German edition exist in four entirely different versions, the first one being an adaption of the original US cover of (Highland & Highland, 1961).

All four covers represent a view of "What is Mathematics" in a collage mode, where the first one represents mathematics as a mostly historical discipline (starting with the ancient Egyptians), while the others all contain a historical allusion (such as pyramids, Gauß, etc.) alongside with objects of mathematics (such as prime numbers or π, dices to illustrate probability, geometric shapes). One notable object is the oddly "two-colored" Möbius band on the 1983 cover, which was changed to an entirely green version in a later reprint.

One can discuss these covers with respect to their contents and their styles, and in particular in terms of attractiveness to the intended buyers/readers. What is over-emphasized? What is missing? It seems more important to us to

- think of our own images/representations for "What is Mathematics",
- think about how to present a multi-facetted image of "What is Mathematics" when we teach.

Indeed, the topics on the covers of the "Was ist was" volumes of course represent interesting (?) topics and items discussed in the books. But what do they add up to? We should compare this to the image of mathematics as represented by school curricula, or by the university curricula for teacher students.

In the context of mathematics images, let us mention two substantial initiatives to collect and provide images from current mathematics research, and make them available on internet platforms, thus providing fascinating, multi-facetted images of mathematics as a whole discipline:

- Guy Métivier et al.: "Image des Maths. La recherche mathématique en mots et en images" ["Images of Maths. Mathematical research in words and images"], CNRS, France, at images.math.cnrs.fr (texts in French)

1963

1983

2001

2010

Fig. 1 The four covers of "Was ist was. Band 12: Mathematik" (Highland & Highland, 1963; Blum, 2001)

- Andreas D. Matt, Gert-Martin Greuel et al.: "IMAGINARY. open mathematics," Mathematisches Forschungsinstitut Oberwolfach, at `imaginary.org` (texts in German, English, and Spanish).

The latter has developed from a very successful travelling exhibition of mathematics images, "IMAGINARY—through the eyes of mathematics," originally created on occasion of and for the German national science year 2008 "Jahr der Mathematik. Alles was zählt" ["Year of Mathematics 2008. Everything that counts"], see www.jahr-der-mathematik.de, which was highly successful in communicating a current, attractive image of mathematics to the German public—where initiatives such as the IMAGINARY exhibition had a great part in the success.

Teaching "What Is Mathematics" to Teachers

More than 100 years ago, in 1908, Felix Klein analyzed the education of teachers. In the introduction to the first volume of his "Elementary Mathematics from a Higher Standpoint" he wrote (our translation):

> At the beginning of his university studies, the young student is confronted with problems that do not remind him at all of what he has dealt with up to then, and of course, he forgets all these things immediately and thoroughly. When after graduation he becomes a teacher, he has to teach exactly this traditional elementary mathematics, and since he can hardly link it with his university mathematics, he soon readopts the former teaching tradition and his studies at the university become a more or less pleasant reminiscence which has no influence on his teaching (Klein, 1908).

This phenomenon—which Klein calls the *double discontinuity*—can still be observed. In effect, the teacher students "tunnel" through university: They study at university in order to get a degree, but nevertheless they afterwards teach the mathematics that they had learned in school, and possibly with the didactics they remember from their own school education. This problem observed and characterized by Klein gets even worse in a situation (which we currently observe in Germany) where there is a grave shortage of Mathematics teachers, so university students are invited to teach at high school long before graduating from university, so they have much less university education to tunnel at the time when they start to teach in school. It may also strengthen their conviction that University Mathematics is not needed in order to teach.

How to avoid the double discontinuity is, of course, a major challenge for the design of university curricula for mathematics teachers. One important aspect however, is tied to the question of "What is Mathematics?": A very common highschool image/concept of mathematics, as represented by curricula, is that mathematics consists of the subjects presented by highschool curricula, that is, (elementary) geometry, algebra (in the form of arithmetic, and perhaps polynomials), plus perhaps elementary probability, calculus (differentiation and integration) in one variable—that's the mathematics highschool students get to see, so they might think that this is all of it! Could their teachers present them a broader picture? The teachers after their highschool experience studied at university, where they probably took courses in calculus/analysis, linear algebra, classical algebra, plus some discrete mathematics, stochastics/probability, and/or numerical analysis/ differential equations, perhaps a programming or "computer-oriented mathematics" course. Altogether they have seen a scope of university mathematics where no current research becomes visible, and where most of the contents is from the nineteenth century, at best. The *ideal* is, of course, that every teacher student at university has at least once experienced how "doing research on your own" feels like, but realistically this rarely happens. Indeed, teacher students would have to work and study and struggle a lot to see the fascination of mathematics on their own by doing mathematics; in reality they often do not even seriously start the tour and certainly most of them never see the "glimpse of heaven." So even if the teacher student seriously immerses into all the mathematics on the university curriculum, he/she will not get any broader image of "What is Mathematics?". Thus, even if he/she does *not* tunnel his university studies due to the double discontinuity, he/she will not come back to school with a concept that is much broader than that he/she originally gained from his/her highschool times.

Our experience is that many students (teacher students as well as classical mathematics majors) cannot name a single open problem in mathematics when graduating the university. They have no idea of what "doing mathematics" means— for example, that part of this is a struggle to find and shape the "right" concepts/definitions and in posing/developing the "right" questions and problems.

And, moreover, also the impressions and experiences from university times will get old and outdated some day: a teacher might be active at a school for several decades—while mathematics changes! Whatever is proved in mathematics does stay true, of course, and indeed standards of rigor don't change any more as much as they did in the nineteenth century, say. However, styles of proof do change (see: computer-assisted proofs, computer-checkable proofs, etc.). Also, it would be good if a teacher could name "current research focus topics": These do change over ten or twenty years. Moreover, the relevance of mathematics in "real life" has changed dramatically over the last thirty years.

The Panorama Project

For several years, the present authors have been working on developing a course [and eventually a book (Loos & Ziegler, 2017)] called "Panorama der Mathematik" ["Panorama of Mathematics"]. It primarily addresses mathematics teacher students, and is trying to give them a panoramic view on mathematics: We try to teach an overview of the subject, how mathematics is done, who has been and is doing it, including a sketch of main developments over the last few centuries up to the present—altogether this is supposed to amount to a comprehensive (but not very detailed) outline of "What is Mathematics." This, of course, turns out to be not an easy task, since it often tends to feel like reading/teaching poetry without mastering the language. However, the approach of Panorama is complementing mathematics education in an orthogonal direction to the classic university courses, as we do not *teach* mathematics but *present* (and encourage to *explore*); according to the response we get from students they seem to feel themselves that this is valuable.

Our course has many different components and facets, which we here cast into questions about mathematics. All these questions (even the ones that "sound funny") should and can be taken seriously, and answered as well as possible. For each of them, let us here just provide at most one line with key words for answers:

- When did mathematics start?
  Numbers and geometric figures start in stone age; the science starts with Euclid?
- How large is mathematics? How many Mathematicians are there?
  The Mathematics Genealogy Project had 178854 records as of 12 April 2014.
- How is mathematics done, what is doing research like?
  Collect (auto)biographical evidence! Recent examples: Frenkel (2013), Villani (2012).
- What does mathematics research do today? What are the Grand Challenges?
  The Clay Millennium problems might serve as a starting point.
- What and how many subjects and subdisciplines are there in mathematics?
  See the Mathematics Subject Classification for an overview!
- Why is there no "Mathematical Industry", as there is e.g. Chemical Industry?
  There is! See e.g. Telecommunications, Financial Industry, etc.
- What are the "key concepts" in mathematics? Do they still "drive research"?
  Numbers, shapes, dimensions, infinity, change, abstraction, ...; they do.
- What is mathematics "good for"?
  It is a basis for understanding the world, but also for technological progress.
- Where do we *do* mathematics in everyday life?
  Not only where we compute, but also where we read maps, plan trips, etc.
- Where do we *see* mathematics in everyday life?
  There is more maths in every smart phone than anyone learns in school.
- What are the greatest achievements of mathematics through history?
  Make your own list!

An additional question is how to make university mathematics more "sticky" for the tunneling teacher students, how to encourage or how to force them to really connect to the subject as a science. Certainly there is no single, simple, answer for this!

Telling Stories About Mathematics

How can mathematics be made more concrete? How can we help students to connect to the subject? How can mathematics be connected to the so-called real world?

Showing applications of mathematics is a good way (and a quite beaten path). Real applications can be very difficult to *teach* since in most advanced, realistic situation a lot of different mathematical disciplines, theories and types of expertise have to come together. Nevertheless, applications give the opportunity to demonstrate the relevance and importance of mathematics. Here we want to emphasize the difference between *teaching* a topic and *telling* about it. To name a few concrete topics, the mathematics behind weather reports and climate modelling is extremely difficult and complex and advanced, but the "basic ideas" and simplified models can profitably be demonstrated in highschool, and made plausible in highschool level mathematical terms. Also success stories like the formula for the *Google* patent for *PageRank* (Page, 2001), see Langville and Meyer (2006), the race for the solution of larger and larger instances of the Travelling Salesman Problem (Cook, 2011), or the mathematics of chip design lend themselves to "telling the story" and "showing some of the maths" at a highschool level; these are among the topics presented in the first author's recent book (Ziegler, 2013b), where he takes 24 images as the starting points for telling stories—and thus developing a broader multi-facetted picture of mathematics.

Another way to bring maths in contact with non-mathematicians is the human level. Telling stories about how maths is done and by whom is a tricky way, as can be seen from the sometimes harsh reactions on www.mathoverflow.net to postings that try to excavate the truth behind anecdotes and legends. Most mathematicians see mathematics as completely independent from the persons who explored it. History of mathematics has the tendency to become *gossip*, as Gian-Carlo Rota once put it (Rota, 1996). The idea seems to be: As mathematics stands for itself, it has also to be taught that way.

This may be true for higher mathematics. However, for pupils (and therefore, also for teachers), transforming mathematicians into humans can make science more tangible, it can make research interesting as a process (and a job?), and it can be a starting/entry point for real mathematics. Therefore, stories can make mathematics more sticky. Stories cannot replace the classical approaches to teaching mathematics. But they can enhance it.

Stories are the way by which knowledge has been transferred between humans for thousands of years. (Even mathematical work can be seen as a very abstract form of storytelling from a structuralist point of view.) Why don't we try to tell more stories about mathematics, both at university and in school—not legends, not fairy tales, but meta-information on mathematics—in order to transport mathematics itself? See (Ziegler, 2013a) for an attempt by the first author in this direction.

By stories, we do not only mean something like biographies, but also the way of how mathematics is created or discovered: Jack Edmonds' account (Edmonds, 1991) of how he found the blossom shrink algorithm is a great story about how mathematics is actually *done*. Think of Thomas Harriot's problem about stacking cannon balls into a storage space and what Kepler made out of it: the genesis of a mathematical problem. Sometimes scientists even wrap their work into stories by their own: see e.g. Leslie Lamport's *Byzantine Generals* (Lamport, Shostak, & Pease, 1982).

Telling how research is done opens another issue. At school, mathematics is traditionally taught as a closed science. Even touching open questions from research is out of question, for many good and mainly pedagogical reasons. However, this fosters the image of a perfect science where all results are available and all problems are solved—which is of course completely wrong (and moreover also a source for a faulty image of mathematics among undergraduates).

Of course, working with open questions in school is a difficult task. None of the big open questions can be solved with an elementary mathematical toolbox; many of them are not even accessible as questions. So the big fear of discouraging pupils is well justified. On the other hand, why not explore mathematics by showing how questions often pop up on the way? Posing questions in and about mathematics could lead to interesting answers—in particular to the question of "What is Mathematics, Really?"

Three Times Mathematics at School?

So, what is mathematics? With school education in mind, the first author has argued in Ziegler (2012) that we are trying cover three aspects the same time, which one should consider separately and to a certain extent also teach separately:

Mathematics I: A collection of basic tools, part of everyone's survival kit for modern-day life—this includes everything, but actually not much more than, what was covered by Adam Ries' "Rechenbüchlein" ["Little Book on Computing"] first published in 1522, nearly 500 years ago;

Mathematics II: A field of knowledge with a long history, which is a part of our culture and an art, but also a very productive basis (indeed a production factor) for all modern key technologies. This is a "story-telling" subject.

Mathematics III: An introduction to mathematics as a science—an important, highly developed, active, huge research field.

Looking at current highschool instruction, there is still a huge emphasis on Mathematics I, with a rather mechanical instruction on arithmetic, "how to compute correctly," and basic problem solving, plus a rather formal way of teaching Mathematics III as a preparation for possible university studies in mathematics, sciences or engineering. Mathematics II, which should provide a major component of teaching "What is Mathematics," is largely missing. However, this part also could and must provide motivation for studying Mathematics I or III!

What Is Mathematics, Really?

There are many, and many different, valid answers to the Courant-Robbins question "What is Mathematics?"

A more philosophical one is given by Reuben Hersh's book "What is Mathematics, Really?" Hersh (1997), and there are more psychological ones, on the working level. Classics include Jacques Hadamard's "Essay on the Psychology of Invention in the Mathematical Field" and Henri Poincaré's essays on methodology; a more recent approach is Devlin's "Introduction to Mathematical Thinking" Devlin (2012), or Villani's book (2012).

And there have been many attempts to describe mathematics in encyclopedic form over the last few centuries. Probably the most recent one is the gargantuan "Princeton Companion to Mathematics", edited by Gowers et al. (2008), which indeed is a "Princeton Companion to Pure Mathematics."

However, at a time where *ZBMath* counts more than 100,000 papers and books per year, and 29,953 submissions to the `math` and `math-ph` sections of `arXiv.org` in 2016, it is hopeless to give a compact and simple description of what mathematics really is, even if we had only the "current research discipline" in mind. The discussions about the classification of mathematics show how difficult it is to cut the science into slices, and it is even debatable whether there is any meaningful way to separate applied research from pure mathematics.

Probably the most diplomatic way is to acknowledge that there are "many mathematics." Some years ago Tao (2007) gave an open list of mathematics that is/are good for different purposes—from "problem-solving mathematics" and "useful mathematics" to "definitive mathematics", and wrote:

> As the above list demonstrates, the concept of mathematical quality is a high-dimensional one, and lacks an obvious canonical total ordering. I believe this is because mathematics is itself complex and high-dimensional, and evolves in unexpected and adaptive ways; each of the above qualities represents a different way in which we as a community improve our understanding and usage of the subject.

In this sense, many answers to "What is Mathematics?" probably show as much about the persons who give the answers as they manage to characterize the subject.

Acknowledgment The authors' work has received funding from the European Research Council under the European Union's Seventh Framework Programme (FP7/2007-2013)/ERC grant agreement no. 247029, the DFG Research Center Matheon, and the the DFG Collaborative Research Center TRR 109 "Discretization in Geometry and Dynamics".

References

Blum, W. (2001). *Was ist was. Band 12: Mathematik*, Tessloff Verlag, Nürnberg. Revised version, with new cover, 2010.

Cook, W. (2011). *In pursuit of the traveling salesman: Mathematics at the limits of computation.* Princeton NJ: Princeton University Press.

Courant, R., & Robbins, H. (1941). *What is mathematics? an elementary approach to ideas and methods* (2nd ed.), Oxford: Oxford University Press. Stewart, I (ed), 1996.

Csicsery, G. (2008). *Hard problems. the road to the world's toughest math contest*, Documentary film, 82 minutes (feature)/45 minutes (classroom version), Washington, DC: Mathematical Association of America.

Devlin, K. J. (2012). *Introduction to mathematical thinking*, published by Keith Devlin, Palo Alto CA.

Edmonds, J. (1991). A glimpse of heaven, In: J. K. Lenstra, A. Schrijver, & A. Rinnooy Kan (eds.) *History of mathematical programming—A collection of personal reminiscences* (pp. 32–54). Amsterdam: CWI and North-Holland.

Frenkel, E. (2013). *Love & math. The heart of hidden reality.* Philadelphia PA: Basic Books/Perseus Books.

Gowers, Timothy, Leader, Imre, & Barrow-Green, June (Eds.). (2008). *The princeton companion to mathematics.* Princeton NJ: Princeton University Press.

Highland, E. H., & Highland, H. J. (1961). *The how and why wonder book of mathematics.* New York: Grosset & Dunlop.

Highland, E. H., & Highland, H. J. (1963). *Was ist was. Band 12: Mathematik*, Neuer Tessloff Verlag, Hamburg, 1963. Revised edition 1969. New cover 1983.

Hersh, R. (1997). *What is mathematics, really?*. Oxford: Oxford University Press.

Klein, F.(1933). *Elementarmathematik vom höheren Standpunkte aus. Teil I: Arithmetik, Algebra, Analysis*, B. G. Teubner, Leipzig, 1908. Vierte Auflage. Heidelberg: Springer.

Lamport, L., Shostak, R., & Pease, M. (1982). The byzantine generals problem. *ACM Transactions on Programming Languages and Systems, 4,* 382–401.

Langville, A. N., & Meyer, C. D. (2006). *Google's pagerank and beyond. The science of search engine rankings.* Princeton and Oxford: Princeton University Press.

Loos, A., & Ziegler, G. M. (2017). *Panorama der Mathematik.* Heidelberg: Springer Spectrum, to appear.

Mendick, H., Epstein, D., & Moreau, M.-P. (2008). *Mathematical images and identities: Education, entertainment, social justice.* London: Institute for Policy Studies in Education, London Metropolitan University.

Page, L. (2001) *Method for node ranking in a linked database*, United States Patent No. US 6,285,999 B1, (submitted: January 9, 1998), http://www.google.com/patents/US6285999

Rota, G.-C. (1996). *Indiscrete thoughts.* Basel: Birkhäuser.

Tao, T. (2007). What is good mathematics? *Bulletin of the American Mathematical Society, 44*(4), 623–634.

Villani, C. (2012). *Théorème vivant.* Paris: Bernard Grasset. (in French).

Ziegler, G. M. (2011). Three competitions. In D. Schleicher & M. Lackmann (Eds.), *Invitation to mathematics. From competition to research* (pp. 195–205). Berlin: Springer.

Ziegler, G. M. (2012). Mathematics school education provides answers—To which questions? *EMS Newsletter* (84), 8–11.

Ziegler, G. M.(2013a). *Do I count? stories from mathematics*, Boca Raton FL: CRC Press/Taylor & Francis. English translation of "Darf ich Zahlen? Geschichten aus der Mathematik", Piper, München, 2010.

Ziegler, G. M. (2013b). *Mathematik—Das ist doch keine Kunst!*. München: Knaus.

Open Access Except where otherwise noted, this chapter is licensed under a Creative Commons Attribution 4.0 International License. To view a copy of this license, visit http://creativecommons.org/licenses/by/4.0/.

International Comparative Studies in Mathematics: Lessons and Future Directions for Improving Students' Learning

Jinfa Cai, Ida A.C. Mok, Vijay Reddy and Kaye Stacey

Abstract This chapter is based on the Plenary Panel on International Comparative Studies we delivered at the 13th International Congress on Mathematical Education (ICME-13) in 2016. In the past a few decades, international comparative studies have transformed the way we see mathematics education and provide insight for improving student learning in many ways. Out of several possibilities, we selected four lessons we have learned from international comparative studies: (1) examining the dispositions and experiences of mathematically literate students, (2) documenting variation in students' thinking in different cultures, (3) appreciating the varying meanings and functions of common lesson events, and (4) the importance of making global research locally meaningful. Throughout the paper, we point out future directions for research to expand our understanding and build up capacity in international comparative studies.

Keywords International comparative studies · Mathematical thinking · Mathematical literacy · Dispositions · Classroom instruction · Contextual factors · Large-scale studies · Small-scale studies · TIMSS · PISA

This chapter is based on the Plenary Panel at the 13th International Congress on Mathematical Education. In preparation for the Plenary Panel presentation, we

---

Kaye Stacey was Chair of the Mathematics Expert Group for the OECD's 2012 PISA survey. The views expressed here are her own.

J. Cai (✉)
University of Delaware, Newark, USA
e-mail: jcai@udel.edu

I.A.C. Mok
University of Hong Kong, Pokfulam, Hong Kong

V. Reddy
Human Sciences Research Council, Pretoria, South Africa

K. Stacey
University of Melbourne, Melbourne, Australia

© The Author(s) 2017
G. Kaiser (ed.), *Proceedings of the 13th International Congress on Mathematical Education*, ICME-13 Monographs, DOI 10.1007/978-3-319-62597-3_6

published a topical survey (Cai, Mok, Reddy, & Stacey, 2016) that provides further detail on the issues raised in this chapter. Here, we have summarized four lessons that international comparative studies provide for improving students' learning, and we suggest directions for future work to expand the scope of research and build up capacity in international comparative studies.

In the past several decades many international comparative studies of mathematics have been conducted, first to examine differences in mathematical proficiency and later to examine dispositions among students from different countries and understand the influence of factors such as curriculum, teacher preparation, the nature of classroom instruction, home and school resources, and context, including parental involvement and the organizational structure of education. We use the phrase 'international comparative studies' to refer to studies involving at least two countries (using 'country' loosely to include significant parts of countries), with the intention of making comparisons at the country level. Other names in the literature include cross-national and cross-cultural studies. We include in our definition studies that are small and large, qualitative and quantitative, and initiatives of government or individual researchers. With this definition, we see international comparative studies in mathematics evolving from informal observations to rigorous measurement of the outcomes of schooling, and from the examination of factors that contribute to performance differences to the generation and testing of theories and policies. Current international comparative studies range from small-scale studies involving a few classes with in-depth analyses to large-scale studies like TEDS (M), TIMSS, and PISA that have upwards of half a million participants and multiple measured variables.

International comparative studies in mathematics have provided a large body of knowledge about how students do mathematics in the context of the world's varied educational institutions. In addition, they examine the cultural and educational factors that influence the learning of mathematics and help identify effective aspects of educational practice in homes, classrooms, schools, and school systems. Examining the learning of mathematics in other countries helps researchers, educators, and government policymakers to understand how mathematics is taught by teachers and how it is learned and performed by students in different countries. It also helps them reflect on theories, practices, and organizational support for the teaching and learning of mathematics in their own culture. Stigler, Gallimore, and Hiebert (2000), themselves researchers conducting international studies, explain the value of this research on trends over time and context in a more nuanced way:

> We may be blind to some of the most significant features that characterize teaching in our own culture because we take them for granted as the way things are and ought to be. Cross-cultural comparison is a powerful way to unveil unnoticed but ubiquitous practices. (pp. 86–87)

The highest-profile international comparative studies, such as PISA and TIMSS, have had a significant impact on thinking about education around the world, especially related to the broad characteristics of educational systems and government policy, of which mathematics is just one of several important components. The fundamental purpose of large-scale studies like PISA and TIMSS is to meet governments' need for objective evidence to monitor educational outcomes, demonstrate possibilities, and assist in developing new policies. There is no sign of a slowing down of international comparative studies either large or small, so the purpose of this paper is to take a step back and reflect on such studies and the lessons we can learn from them.

In this chapter, we discuss four of the many lessons we can learn from international comparative studies for improving students' learning. We chose these four lessons in particular because they represent different styles and strands of work in this area and because they all have the potential to impact students' learning. The first two lessons focus on students' mathematical thinking and achievement. The third lesson focuses on classroom instruction, and the fourth lesson focuses on policy and the effect of contextual factors on learning.

Lesson 1: Promoting Students' Mathematical Literacy

The results of large-scale studies provide many lessons for educational policy related to overall achievement and its links to instruction and student background variables. This section tells just one of the many stories that arise from the PISA 2012 survey: What curriculum, experiences, and dispositions promote mathematical literacy in students? This story shows a side of the PISA survey that is very different from the country rankings that grab newspaper headlines.

Mathematical literacy, the achievement construct measured by PISA, refers to the ability to use mathematical knowledge in situations that are likely to arise in the lives and work of citizens in the modern world. A precise definition is given by the Organisation for Economic Co-operation and Development (OECD, 2013a, p. 25) and discussed by Stacey and Turner (2015a). The 2012 PISA survey examined many aspects of mathematical literacy: the achievement profiles of students across three processes that are involved in exercising mathematical literacy, the learning opportunities that contribute to achievement, in-class experiences and dispositions that influence mathematical literacy, and the effect of classroom experiences with mathematical literacy on more general student attitudes. This section briefly outlines some of the lessons from this work and draws attention to new directions and research questions for mathematics educators.

Country Profiles of the Processes of Mathematical Literacy

Using mathematics to meet a real-world challenge involves three 'processes,' depicted in Fig. 1:

- Formulating situations mathematically (abbreviated to Formulate);
- Employing mathematical concepts, facts, procedures, and reasoning (Employ); and
- Interpreting, applying, and evaluating mathematical outcomes (Interpret).

Fig. 1 The PISA 2012/2015 processes of mathematical literacy in practice (OECD, 2013b)

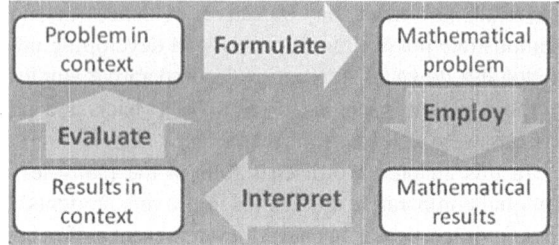

Readers will note the intentional similarity of Fig. 1 to many diagrams depicting the mathematical modeling cycle. The Formulate process transforms the real-world challenge into mathematical form by identifying variables and relationships and making assumptions. The Employ process takes place within the mathematical world, using the knowledge and skills that form the bulk of school mathematics. The Interpret process (which, for the purposes of PISA, includes both interpretation and evaluation of the real-world solution) transforms the mathematical answers back to the real-world context and judges their real-world adequacy.

PISA 2012 measured the performance of students on each of these three processes, revealing, for the first time, interesting country patterns and differences. The average score for overall mathematical literacy across the OECD was 494, made up of 492 for Formulate, 493 for Employ, and 497 for Interpret (all standard errors 0.5). Interpret items were the easiest for students, despite the survey design's intention to select items to measure each process in such a way that the three overall means would be the same. As with most studies, PISA 2012 showed that boys have higher mathematics achievement than girls (OECD average gap 11 scale points). PISA 2012 located the biggest gap between these two groups (OECD average 16 points) to be on the Formulate items. These and other results in this section are derived from reports from the OECD (2013b, c).

Top-performing countries are generally Asian, and stereotypes might have predicted their greatest strength to be in routine procedures and hence in the Employ process. Surprisingly, however, 9 of the 10 top-performing countries' highest scores were in Formulate. Figure 2 shows this pattern for the high-achieving country of Japan (mean 536), contrasting with the patterns of relative scores for the Netherlands and the United Kingdom. Another interesting result is that the four highest performing countries' lowest scores were in Interpret—the easiest set of items for the worldwide sample.

Other groups of countries showed consistent but different patterns. The Netherlands (see Fig. 2), Denmark, and Sweden had their highest scores in both Formulate and Interpret, the two processes where real-world contexts matter. Non-Asian English-speaking countries (Canada, Australia, New Zealand, United Kingdom, United States) were relatively stronger in Interpret only. Nine European countries scored relatively low in Formulate but higher in both Employ and Interpret. These newly discovered patterns warrant detailed investigation, especially to investigate links with curriculum and teaching practices (Stacey & Turner, 2015b).

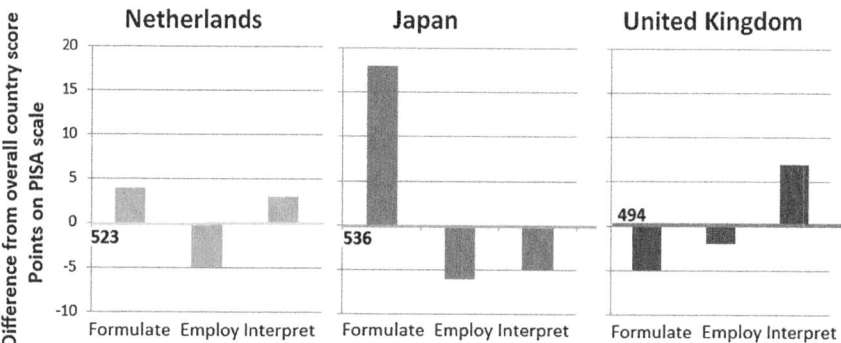

Fig. 2 Sample of PISA 2012 scores on the three processes of mathematical literacy

What Curriculum Experiences Build Mathematical Literacy?

Since PISA's construct of mathematical literacy involves mathematics that is likely to be useful to citizens in all walks of life, it is of interest to know whether a curriculum produces better mathematical literacy outcomes if it is oriented towards abstract mathematics or towards its applications. To answer this question, a sample of PISA students rated how confident they felt about solving a set of mathematics problems and later rated how frequently they had encountered similar problems in class. The sample problems included 'formal' mathematics items lacking any context, such as solving a linear equation or finding the volume of a box, and 'applied' mathematics items, such as using a train timetable and interpreting a newspaper graph. The student ratings were used to create measures of confidence and exposure to applied and formal mathematics[1] (OECD, 2013b, c).

Performance in PISA 2012 was very strongly related to opportunities to learn formal mathematics and secondarily to opportunities to learn applied mathematics. The relationship of PISA performance with exposure to formal mathematics was linear, but quadratic for applied mathematics. The more frequently students are exposed to applied mathematics problems, the better is their PISA performance, but only up to a point—very high exposure is associated with a decline in performance. This may be an outcome of a tendency to place low-performing students in classes with a focus on the 'everyday' applications of mathematics. PISA data reveals this relationship but focused studies are needed to provide a causal explanation.

Japan and the Netherlands, both high-achieving countries, show contrasting patterns of exposure. Students in Japan and other Asian high-performing countries reported low exposure to applied mathematics and high exposure to formal mathematics (OECD, 2013b, c), whereas students in the Netherlands reported high

---

[1]The correct name is "index of experience with pure mathematics," rather than formal mathematics. Confidence is also referred to as self-efficacy. Slightly different constructs in the full reports are conflated here for brevity.

exposure to applied mathematics and low exposure to formal mathematics, perhaps indicating the influence of Realistic Mathematics Education (RME) there. The Netherlands exposure is consistent with the pattern of mathematical process scores shown in Fig. 2, but the Japanese pattern is not. Japanese students perceive an emphasis on formal mathematics but they have nonetheless learned to identify mathematical relationships within real situations and to create appropriate models. How this has happened is an important research question.

Students' Disposition Towards Formal and Applied Mathematics

PISA 2012 also provided some important lessons about student dispositions. Dispositions are especially relevant to the current international governmental climate in which the importance of mathematical literacy to economic well-being is widely acknowledged, with many countries aiming to entice students into STEM careers. Figure 3 shows a strong association between students' reporting of high exposure to a task and their confidence in solving it. Figure 3 also illustrates a general finding that confidence is higher for solving formal mathematics problems than applied problems, at each level of exposure. One explanation is that solving applied mathematics problems requires both a good understanding of the underlying abstract structure as well the ability to analyze the real-world situation—in other words, it requires the PISA mathematical processes of Formulate and Interpret, as well as Employ.

Most countries display a gender difference in confidence in mathematics: PISA 2012 located this difference in the applied problems. Figure 4 compares boys' and girls' reported confidence in solving a sample of applied problems (first six column pairs) and formal problems (last two column pairs; OECD, 2013b). The gender

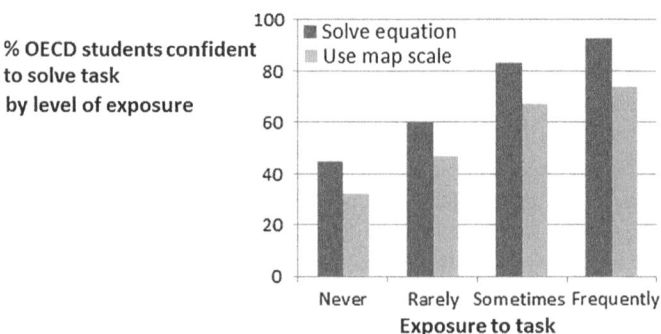

Fig. 3 Percentage of OECD students reporting confidence in solving a formal problem and an applied problem (data from OECD, 2013b)

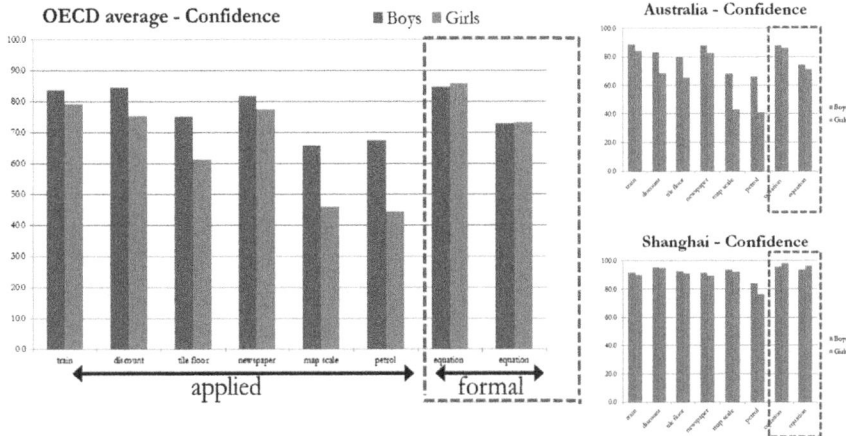

Fig. 4 Confidence of boys and girls in solving eight problems for all OECD countries, Australia, and Shanghai

difference is large for applied problems but is not evident for formal mathematics problems. For example, across OECD countries, 75% of girls reported being confident or very confident when calculating a 30% discount on a TV (second column pair), compared to 84% of boys. The two small graphs on the right side of Fig. 4 show the gender differences for a typical OECD country (Australia) and the lack of gender differences in Shanghai. These gender gaps for applied mathematics problems are likely to have an impact on gender differences in achievement and also on career choices. How can the gender equality of confidence in Shanghai be made a reality everywhere?

PISA 2012 also linked dispositions to exposure to formal and applied mathematics. Overall, students who reported having been more frequently exposed to formal mathematics tasks reported more positive engagement, drive, motivation, and self-beliefs. The same relationship held for applied mathematics tasks, but it became a very strong relationship when controlling for students' achievement. Because of the clear instructional importance, more detailed analyses of the PISA data and further studies are warranted to better understand the links between dispositions, achievement, and exposure to various types of mathematics.

Summary

This section discussed findings from PISA on the curriculum, experiences, and dispositions that promote mathematical literacy in students. These findings illustrate the power of large-scale studies to go well beyond providing country rankings to identify new phenomena worth studying. Better understanding of results such as these requires both large- and small-scale research, within and between countries, looking at standards, curriculum, teaching, learning, and assessment.

Lesson 2: Understanding Students' Thinking

Over 20 years ago, Bradburn and Gilford (1990) suggested that studies with relatively small, localized samples in a small number of sites can provide useful international comparisons. They can reveal unique findings beyond the scope of large-scale studies and also complement large-scale studies by providing deep understanding about different societies and education systems, thereby enhancing interpretations and implications. Examples of such small-scale studies are Cai (1995, 2000), Cai, Ding, and Wang (2014), Ma (1999), Silver, Leung, and Cai (1995), Song and Ginsburg (1987), and Stevenson et al. (1990).

In Cai et al. (2016), we shared the analysis of two problems to show the value of such in-depth studies. Here, we provide another example from a study by Cai and Hwang (2002), in which they examined Chinese and U.S. sixth graders' mathematical problem solving and problem posing and the relations between them. One pair of tasks is in Fig. 5.

Doorbell (Problem Solving version)
Sally is having a party.
The first time the doorbell rings, 1 guest enters.
The second time the doorbell rings, 3 guests enter.
The third time the doorbell rings, 5 guests enter.
The fourth time the doorbell rings, 7 guests enter.
Keep on going in the same way. On the next ring a group enters that has 2 more persons than the group that entered on the previous ring.

Question 1: How many guests will enter on the 10th ring? Explain how you found your answer.
Question 2: In the space below, write a rule or describe in words how to find the number of guests that entered on each ring.
Question 3: 99 guests entered on one of the rings. What ring was it? Explain or show how you found your answer.

Doorbell (Problem Posing version)

Text as above with Questions 1, 2, and 3 replaced with the following:

For his student's homework, Mr. Johnson wanted to make up three problems BASED ON THE ABOVE SITUATION: an easy problem, a moderate problem, and a difficult problem. These problems can be solved using the information in the situation. Help Mr. Miller make up three problems and write these problems in the space below.

Fig. 5 Problem-solving and problem-posing versions of the Doorbell task

Problem-Solving Results

The U.S. and Chinese students had almost identical success rates (70%) when they were asked to find the number of guests who entered on the 10th ring (Question 1). However, the success rate for Chinese students (43%) was significantly higher than that of the U.S. students (24%) for Question 3 (ring number for 99 guests). The difference is due to their use of different strategies.

Appropriate solution strategies for Questions 1 and 3 were classified into three types: abstract, semi-abstract, and concrete. An abstract strategy generally followed one of two paths: the number of guests who entered on a particular ring of the doorbell is equal to two times that ring number minus one (i.e., $y = 2n - 1$, where y represents the number of guests and n represents the ring number) or the number of guests is equal to the ring number plus the ring number minus one (i.e., $y = n + [n - 1]$). Students used their rule to answer Question 3 (99 guests).

Students who used a semi-abstract strategy made a number of computation steps to yield a correct answer. Students who used a concrete strategy made a table or a list or noticed that each time the doorbell rang two more guests entered than on the previous ring and sequentially added twos to find an answer.

Of the students with appropriate strategies, 44% of the Chinese students and 1% of the U.S. students used abstract strategies for Question 1. For Question 3, 65% of Chinese students used an abstract strategy, compared to only 11% for the U.S. sample. Most U.S. students (75%) chose concrete strategies, compared to 29% of the Chinese students.

Problem-Posing Results

There were similarities and differences in the kinds of problems generated by the two samples. In general, as students in both samples moved towards generating problems of greater difficulty, they tended to move away from posing problems solely about the given information. By far the least common problem types for both groups were those based on reversed thinking (e.g., find ring number given number of guests, as in Question 3, or find total number of rings for a given total number of guests). Chinese students, however, were much more likely to pose problems involving only the given information. U.S. students posed more extension problems than did Chinese students, and a smaller percentage U.S. students (29%) posed no extension problems compared to Chinese students (41%). Similarly, more U.S. students (31%) than Chinese students (21%) posed only extension problems.

The most frequently generated types of problems differed between the two samples. The most frequently generated problems for U.S. students involved finding the number of guests at a particular ring for the easy and moderate problems, and computing the total number of guests after a specific ring for the difficult problem. In contrast, the most frequently generated problems among Chinese students were non-extension problems (e.g., How many guests entered on the fourth ring?) for the easy problem, and problems asking for the number of guests entering on a ring beyond the fourth ring for moderate and difficult problems.

Summary

Scores arising from large-scale studies are useful for providing an overall picture of students' performance in mathematics and enable rigorous statistical examination of patterns and relationships among variables, including those which may predict students' learning outcomes. However, scoring on the basis of correctness alone conceals some important aspects of students' performance. The results above demonstrate that different students can use different strategies to obtain the same score. Such important differences in students' mathematical thinking may reflect differences in teachers' beliefs and instructional practices (e.g., Cai et al., 2014; Cai & Wang, 2010). In order to provide the education community with a deeper understanding of the teaching and learning of mathematics, it is essential for international comparative studies to provide in-depth evidence of students' thinking and reasoning, including the qualitative analysis of solution strategies, mathematical errors, mathematical justifications, and representations (Cai, 1995).

Lesson 3: Changing Classroom Instruction

Complementary Roles of the TIMSS Video Study and the Learner's Perspective Study

This section draws upon the work of two studies of teaching practice, the TIMSS Video Study and the Learner's Perspective Study (LPS). By zooming in on these two studies, we discuss what we may learn from international comparative studies concerning classroom instruction. The first TIMSS Video Study took place in 1995 (Stigler & Hiebert, 1999) and the over-arching conclusion, reported in *The Teaching Gap* (Stigler & Hiebert, 1999), was that teaching is a cultural activity. The follow-up TIMSS 1999 Video Study (Mathematics) compared teaching practices in the U.S. with six countries that showed higher performance in TIMSS: Australia, the Czech Republic, Japan, the Netherlands, Switzerland, and Hong Kong (Hiebert et al., 2003). Taking the stance that teaching is a cultural activity, the study aimed to build a picture of what typical teaching looked like in different countries and to give researchers and teachers the opportunity to discover alternative ideas about how mathematics might be taught (Stigler & Hiebert, 2004).

LPS (Clarke, Emanuelsson, Jablonka, & Mok, 2006) was designed to examine the practices of eighth grade mathematics classrooms in an integrated, comprehensive way. The project has now developed into a research community in Australia, China, the Czech Republic, Finland, Germany, Israel, Japan, Korea, New Zealand, Norway, the Philippines, Portugal, Singapore, Slovakia, South Africa, Sweden, the United Kingdom, and the U.S. LPS juxtaposes the observable practices of the classroom and meanings attributed to those practices by teachers and students. Instead of aiming for a representative national sample as the TIMSS Video Study did, LPS aimed to understand what might be made possible by competent teachers, locally recognized as such.

Lesson Structures and Lesson Events

The TIMSS Video Study explored lesson structures via the coding of processes like reviewing, demonstrating the problem for the day, practicing and correcting seatwork, and assigning homework (Stigler & Hiebert, 1999), aiming to present a typical "average" lesson for international comparison. LPS used the coding of the TIMSS Video Study to explore patterns of lesson structures of a sequence of consecutive lessons. The findings indicated that the teachers documented in LPS showed little evidence of a consistent lesson pattern, but instead appeared to vary the structure of their lessons purposefully across a topic sequence.

Another viable unit for comparison employed by LPS was the "lesson event," characterized by a combination of form (visual features and social participants) and function, such as intention, action, inferred meaning, and outcome (Clarke et al., 2006; Clarke, Keitel, & Shimizu, 2006). For example, Kikan-Shido (also known as between-desk instruction or seatwork) had a recognizable structural form evident across all classrooms in all countries. However, the findings suggested that the Kikan-Shido lesson events in Shanghai, German, and Japanese lessons had unique emphases:

- Shanghai lessons: correcting errors, encouraging students to think further (Lopez-Real, Mok, Leung, & Marton, 2004)
- German lessons: questioning to stimulate student mathematical thought (Clarke et al., 2006)
- Japanese lessons: eliciting students' mistakes, their puzzlement, and their opposing solutions; pointing out different solutions or difficulties and giving explanations; and making their way of thinking visible to the group (Hino, 2006)

Overall, the findings from the LPS study suggested reasons additional to those identified in the TIMSS Video Study about why the enactment of Japanese lessons differed from other countries (Mok, 2015).

Multiple Accounts of a Teacher's Practice

Another advantage of the LPS data set is that it allows researchers to reconstruct multiple accounts of classroom scenarios by combining data from all of the lesson materials, including videos, student interviews, and teacher interviews, thereby providing the opportunity to study the practice of a particular teacher in a specific cultural system in depth. For example, an explanation has been sought for the "Asian Learner's Paradox," which refers to the seemingly contradictory phenomenon of outstanding student performance in Asian regions but reports of classroom environments being non-conducive to learning, with characteristics such as directive teaching and large classes (Watkins & Biggs, 2001). Mok (2006) analyzed the LPS data of a Shanghai teacher. To illustrate the teacher's skillfulness, a lesson episode about the train-ticket problem is depicted in Fig. 6.

A student, Dora, who first solved the problem mentally, was invited to share her solution with the class. Dora's answer was arithmetic and intuitive in nature, and was immediately followed by the teacher's paraphrasing with an emphasis on the idea of subtraction. Following this, the teacher asked the class to do the problem again using equations, writing the Equations $3x + y = 560$ and $3x + 2y = 640$ and obtaining the answer by subtracting one equation from the other. Mok's (2006) analysis showed that the teacher had created three levels of contrasts to support a deep understanding of the problem. The first level of contrast is between Dora's answer and the teacher's paraphrase, the second level between the arithmetic method and the equation method, and the third level between the equation-solving methods of subtracting equations (elimination) and substitution. Mok argued that the lesson was by no means spontaneous, but rather represented a synthesis based on that experienced teacher's understanding of a pedagogical framework of variation that was well established in his region (Experimenting Group of Teaching Reform in Mathematics in Qingpu County, Shanghai, 1991). The strong teacher guidance in the lesson arose from the teacher's interpretation of student-centeredness, which was different from its interpretation in Western education communities. The teacher saw himself as non-traditional and made use of his understanding of his students in order to create a planned experience for them with

Train Ticket Problem

Xiu-min and his family went to Beijing for a holiday. They booked 3 adult tickets and 1 student ticket, costing a total of 560 dollars. His classmate Xiu-wang, learning this, decided to join Xiu-ming's family for the trip. Consequently, they bought 3 adult tickets and 2 student tickets, costing a total of 640 dollars. Please calculate the cost of 1 adult ticket and the cost of 1 student ticket.

Fig. 6 Train ticket problem

minimal side-tracking (Mok, 2006). The conceptions of this teacher and his performance in the lesson were quite consistent with the findings of another study that compared conceptions of effective teaching between Chinese and U.S. teachers. Cai and Wang (2010) suggested that the constraints of content coverage, teaching pace, and large class size affected teaching flexibility and student-centeredness.

Lessons for the Implementation of Mathematical Tasks

Both the TIMSS Video Study and LPS classified mathematical problems as "using procedure" problems (success requiring only a memorized procedure or algorithm) and "making connections" problems (success requiring the establishment of relationships between ideas, facts, and procedures and engagement in mathematical reasoning). The TIMSS Video Study showed that all of the countries except Japan used more "using procedure" problems than "making connections" problems. In this way, the U.S. was not different from higher-achieving countries in the kinds of problems that teachers *presented to students*. What or where was the difference? The videos of each country revealed some interesting cultural activities. For example, lessons in the Netherlands frequently used calculators and real-world problem scenarios, and Japanese students spent on average a longer time working to develop their own solution procedures for problems that they had not seen before. In all of the high-performing countries except Australia the teachers implemented a higher percentage of "making connections" problems as "making connections" problems than did U.S. teachers. In contrast, U.S. teachers changed "making connections" problems to "using procedures" problems, thereby lowering the cognitive demand of the problems (Roth & Givvin, 2008; Stigler & Hiebert, 2004).

LPS team members have also made some significant achievements in studying the use of mathematical tasks in classroom instruction (Shimizu, Kaur, Huang, & Clarke, 2010). For example, Huang and Cai (2010) found that LPS teachers from the US and China were willing to implement cognitively demanding tasks in their lessons, yet the Chinese teachers were more frequently able to sustain the cognitive demand of the mathematical tasks during implementation. Mesiti and Clarke (2010) analysed the mathematical tasks in the LPS data from China, Japan, and Sweden and concluded that the classroom performance of a task was ultimately a unique synthesis of task, teacher, students, and situation.

Summary

To conclude, the two international comparative studies discussed in this section played complementary roles in contributing to the understanding of classroom instruction. The TIMSS Video Study, building upon the tradition of large-scale surveys of national samples, suggested seeing teaching as a cultural activity. LPS compared mathematics lessons through analysis of lesson events during a sequence of lessons and included the perspectives of the teacher and the learners. Although teachers in different cultural systems spent time on the same lesson event, they might in fact have been carrying out the activities with different meanings and functions. The attempt to explain the Asian Learner's Paradox is an example of how the investigation of an effective case might take into account many constraints (such as examination orientation, content coverage, teaching pace, and large class size in a specific cultural system) and culturally-rooted clues (such as the teacher's conceptions and beliefs, students' expectations, the locally-implemented pedagogical framework). Lastly, seeking a common language for comparison has a specific implication for understanding effective instruction in different cultures. Both the TIMSS Video Study and LPS have chosen tasks as a theme for comparison. Different kinds of tasks play different roles in the agenda of effective classroom instruction; nonetheless, how the teacher sustains the intended roles of the tasks during implementation is important.

Lesson 4: Making Global Research Locally Meaningful—TIMSS in South Africa

This lesson illuminates how a country can find its own voice in using international comparative studies to extend to analyses that are meaningful for the local agenda. South Africa is characterised as a country with high levels of poverty, inequality, and unemployment. These characteristics have an impact on the quality of education and become both determinants and outcomes of the level of development of the country.

As expected in unequal societies, there are high levels of variation between schools. While many countries focus on interventions inside classrooms to improve subject matter knowledge and achievement scores, low-income countries have to focus on two challenges. On the one hand they have to focus on what happens inside the classroom to improve teachers' and students' mathematical knowledge. On the other hand they must identify the effects of the many contextual factors and conditions that influence educational achievement. In this section we share experiences of using the TIMSS achievement data sets and information on South Africa to inform educational policy.

Mathematics Achievement Trends Over 20 Years

Participation in TIMSS 1995 provided the first indicative estimate of national mathematics and science achievement for South Africa. This was followed by the widely publicized results for TIMSS 1999, which lamented the low South African scores and the rank order which placed South Africa last in the set of 38 participating countries. This international comparison catalysed a debate about educational performance in South Africa and involved many sectors of society—politicians, policymakers, academics, teachers, and the public. Newspaper headlines in South Africa asserted, for example, that 'South African pupils are the dunces of Africa' (Sunday Times, 16 June 2000) or that South African students were the 'Bottom of the class in maths' (Sunday Times, 14 October 2001). Low mathematics performance and country rank were repeated again in TIMSS 2003. The newspaper headlines and reaction from politicians and policymakers echoed those following TIMSS 1999, but the challenge for research was to embark on deeper analysis and extend the story to one which could provide policy directions.

An important but overlooked finding from the TIMSS analysis was the range of performance between the 5th and 95th percentiles of performance. Of all the countries participating in TIMSS 2003, South Africa had the widest range of scores between these two percentiles. This wide range led to the characterization that there were two systems of education in the country and that the performance scores in TIMSS were reflective of wide disparities in society and in schools.

The story of South African performance cannot be told through a single national score but through appropriate disaggregation. The disaggregation of the achievement scores revealed a strong correlation between socioeconomic status and achievement scores. Africans, who were most disadvantaged by the apartheid policies, had the lowest performance. African schools are located in areas where most Africans live and these areas have high levels of poverty and unemployment.

South Africa's participation in TIMSS 2011 provided an opportunity to measure the changes in educational performance since 1995. TIMSS was the only study that provided a scientifically rigorous methodology to measure trends over the previous 20 years. Analysis of the four rounds of TIMSS participation showed that the average national mathematics score remained the same over the years 1995, 1999, and 2003 (Reddy, Van der Berg, Janse van Rensburg, & Taylor, 2012). In contrast, from 2003 to 2011 the national average mathematics score increased by 63 points (see Fig. 7). The increases over the last two cycles of TIMSS can be translated to say that overall student performance, though still low, has improved by one and a

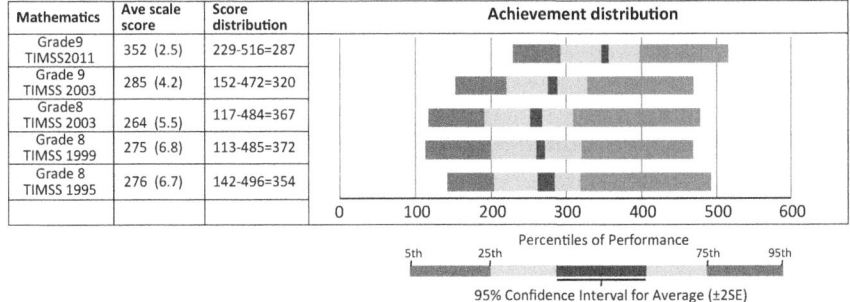

Fig. 7 Trends in mathematics achievement for TIMSS 1995, 1999, 2003, and 2011

half grade levels. In 2011, the range of mathematics scores decreased, suggesting that the country is progressing (albeit slowly) towards more equitable educational outcomes.

Contextual Factors Influencing Educational Achievement

We need to go beyond the achievement scores to investigate the factors that influence mathematical performance. The results of our analyses confirmed the effects of home and school socioeconomic factors. As expected, students who speak the language of the test at home are more likely to achieve higher scores than those who do not.

We explored the effects of two contemporary South African factors on achievement—gender and school violence—and found new complexities in the schooling experience of South African boys and girls. On average, across South Africa, gender differences in mathematics scores were small or non-existent. We also probed students about their attitudes towards mathematics and found that mathematics mattered to both boys and girls. A particularly worrisome finding was the level of indifference among boys about their education. Boys were found to have lower aspirations about their academic careers, showed less interest in mathematics, and engaged less often with an adult regarding their school work. The link between negative attitudes and weak performance was stronger for boys than for girls.

The second factor we explored was the extent of violence in South African schools and its effect on mathematics achievement. Although concerns about school safety are increasing internationally, violence in schools is considered more serious in South Africa than elsewhere. The degree of school safety largely depends on the type of school that learners attend. We found that children attending public schools experienced more frequent threats of violence than children attending independent schools. The socioeconomic status of students is an indicator for potential exposure to acts of violence, with higher chances of being bullied regularly for students from poor families. There is a higher frequency of bullying for boys than for girls who attend schools with similar characteristics. Schools where there are fewer discipline or safety problems achieve better results, but this relationship is dependent on the size of the school.

Student Progression and Pathways Through Secondary School

In addition to concerns about low mathematics achievement, there is also concern about progression through secondary schools. We analysed the pathways and performances in mathematics of secondary school students in South Africa using a

Table 1 Educational pathways of students in the South African Youth Panel Study

Smooth	Staggered	Stuck	Stopped
Neat, year-on-year grade progression through school	Learners in school for all 4 years of SAYPS, who make some grade progress but have at least one episode of grade repetition	Learners in school for all 4 years of SAYPS, but stuck in grade 9 or 10 for three or more periods	Individuals who leave school before final data collection and do not return

panel-like data set of Grade 8 students who participated in TIMSS 2003 and were tracked to Grade 12 examination data sets. Firstly, students who began with similar Grade 8 mathematics scores had different educational outcomes 4 years later. Secondly, in middle class schools, Grade 8 mathematics scores were a good indicator of who would pass the exit level examination in Grade 12, but this relationship was not as strong in schools for poorer students. Thirdly, there was a stronger association between TIMSS Grade 8 mathematics scores and subject choice of secondary school mathematics in middle class schools than in poorer schools. Fourthly, there was a strong correlation between mathematics performance at Grade 8 and the exit level examination. Overall, this study adds to the body of evidence that suggests that to improve educational outcomes, the policy priority should be to build foundational knowledge and skills in numeracy.

To extend our understanding of the pathways and transitions followed by South African youth, the longitudinal South African Youth Panel Study (SAYPS) was initiated, with the first annual data collection wave in 2011. SAYPS followed Grade 9 learners who participated in TIMSS 2011 for 4 years to explore their educational transitions. We found that students followed one of four educational pathways (Table 1) through secondary school.

Almost half of the sample (47%) followed the smooth pathway while 39% followed a staggered pathway and 14% were either stuck or stopped. There is a predictable story of 'advantage begetting advantage' for students who experience a smooth pathway: With higher than average TIMSS scores and better-educated parents, these students come from homes with more books and have positive attitudes about school. Our analyses show that it is possible to succeed academically despite disadvantage: Just over 43% of the smooth group come from non-fee-paying schools for poorer students. We will study this group further to understand their pathways to success.

Future Directions for Learning from International Comparative Studies

Over the last 3 decades, international comparative studies have transformed the way we see mathematics learning and teaching and the four lessons above have illustrated that there is still much to learn. Looking to the future, we believe it is important to extend international comparative studies to deepen our understanding of previous findings as well as to build the capacity of researchers to implement them.

Improving Our Understanding of the Outcomes of Large-Scale Studies

Because large-scale studies are generally supported by governments with the intention of assisting in policy development, it is important that the outcomes of these studies are understood as deeply as possible. This often requires further research, sometimes within and sometimes between countries. The case study of South Africa provided an excellent example of how further research within one country using trend data can make the results of an international comparative study more useful for local policy development, as well as contribute to the knowledge base for similar countries. For an example where research within countries and between countries may be useful, let us return to PISA's three mathematical processes of Formulate, Employ, and Interpret and the observation in Lesson 1 that groups of countries (such as the high-performing Asian countries or English-speaking Western countries) exhibit different patterns of (relative) performance on the three processes. In-depth analysis of the large-scale data can identify such subtle but important differences; however, we need a range of additional studies to explain the findings. Such studies may, for example, examine the construction of the PISA instruments for anomalies, or conduct local or international comparative studies of students' problem-solving processes and/or curriculum experiences. Large-scale studies are very expensive; we need to use the data they provide towards maximum benefit in understanding why students perform as they do.

Investigating New Questions Through Small-Scale Studies

In-depth, small-scale international comparative studies can provide unique opportunities for us to understand students' mathematical thinking. The more information teachers have about what students know and think, the more opportunities they can create for student success. Teachers' knowledge of students' thinking has a substantial impact on their classroom instruction and hence upon students' learning. Thus, small-scale comparison studies provide insights on students' learning and understanding in the context of different cultural systems and at an enactment level of the teaching practice and students' learning. These insights are important to policymakers, researchers, educators, and teachers.

Small-scale international comparative studies can also start to explore many urgent and important research questions. For example, is there really a creativity gap between students in Asian and Western countries and if so, why? Future international comparative studies and international collaboration should answer these questions empirically, building up to large-scale studies.

Building the Capacity of Researchers

There are distinct advantages for individual researchers to collaborate on in-depth, small-scale international comparative studies, because relatively modest resources are required. LPS is an excellent example of a long-standing collaboration which has capitalized on shared interests with a fluid structure within which many people can work together. The recent rapid increase of international comparative studies on curriculum is another example (Lloyd, Cai, & Tarr, 2017). Many individual researchers chose to focus on certain aspects of curriculum as they conducted comparative analyses across nations. These studies provided new insights into the content and design of mathematics textbooks and generated key questions about relationships between written curricular materials and students' opportunities to learn. Another avenue for individual researchers is to engage in secondary analyses of large-scale international comparative studies, which generally make a great deal of data publically available. This work can also be done with few resources and the often severe time constraints that apply to individual or beginning researchers.

Our overall message is that international comparative studies can provide a wealth of information for mathematics education researchers and policymakers. The mathematics education community has a unique capacity to contribute to an in-depth understanding of both national and international findings, and hence to assist us all in learning the right lessons from international comparisons.

References

Bradburn, M. B., & Gilford, D. M. (1990). *A framework and principles for international comparative studies in education*. Washington, DC: National Academic Press.

Cai, J. (1995). A cognitive analysis of US and Chinese students' mathematical performance on tasks involving computation, simple problem solving, and complex problem solving. *Journal for Research in Mathematics Education Monograph Series 7*. Reston, VA: National Council of Teachers of Mathematics.

Cai, J. (2000). Mathematical thinking involved in US and Chinese students' solving process-constrained and process-open problems. *Mathematical Thinking and Learning, 2*, 309–340.

Cai, J., Ding, M., & Wang, T. (2014). How do exemplary Chinese and US mathematics teachers view instructional coherence? *Educational Studies in Mathematics, 85*(2), 265–280.

Cai, J., & Hwang, S. (2002). Generalized and generative thinking in US and Chinese students' mathematical problem solving and problem posing. *Journal of Mathematical Behavior, 21*(4), 401–421.

Cai, J., Mok, I., Reddy, V., & Stacey, K. (2016). *International comparative studies in mathematics: Lessons for improving students' learning*. New York, NY: Springer.

Cai, J., & Wang, T. (2010). Conceptions of effective mathematics teaching within a cultural context: Perspectives of teachers from China and the United States. *Journal of Mathematics Teacher Education, 13*(3), 265–287.

Clarke, D., Emanuelsson, J., Jablonka, E., & Mok, I. A. C. (Eds.). (2006). *Making connections: Comparing mathematics classrooms around the world*. Rotterdam, The Netherlands: Sense.

Clarke, D., Keitel, C., & Shimizu, Y. (Eds.). (2006). *Mathematics classrooms in 12 countries: The insiders' perspective*. Rotterdam, The Netherlands: Sense.

Experimenting Group of Teaching Reform in Mathematics in Qingpu County, Shanghai. (1991). *Xuehui jiaoxue [Learning to teach]*. Beijing, China: People Education Publishers.

Hiebert, J., Gallimore, R., Garnier, H., Givvin, K., Hollingsworth, H., Jacobs, J., et al. (2003). *Teaching mathematics in seven countries: Results from the TIMSS 1999 video study*. Washington, DC: US Department of Education, National Center for Education Statistics.

Hino, K. (2006). The role of seatwork in three Japanese classrooms. In D. Clarke, C. Keitel, & Y. Shimizu (Eds.), *Mathematics classrooms in 12 countries: The insiders' perspective* (pp. 59–74). Rotterdam, The Netherlands: Sense.

Huang, R., & Cai, J. (2010). Implementing mathematical tasks in USA and Chinese Classrooms. In Y. Shimizu, B. Kaur, R. Huang, & D. Clarke (Eds.), *Mathematical tasks in classrooms around the world* (pp. 145–164). Rotterdam, The Netherlands: Sense.

Lloyd, G. M., Cai, J., & Tarr, J. E. (2017). Issues in curriculum studies: Evidence-based insights and future directions. In J. Cai (Ed.), *Compendium for research in mathematics education* (pp. 824–852). Reston, VA: National Council of Teachers of Mathematics.

Lopez-Real, F. J., Mok, I. A. C., Leung, F. K. S., & Marton, F. (2004). Identifying a pattern of teaching: An analysis of a Shanghai teacher's lessons. In L. Fan, N.-Y. Wong, J. Cai, & S. Li (Eds.), *How Chinese learn mathematics: Perspectives from insiders* (pp. 382–412). Singapore: World Scientific Publishing Co.

Ma, L. (1999). *Knowing and teaching elementary mathematics: Teachers' understanding of fundamental mathematics in China and the United States*. Hillsdale, NJ: Erlbaum.

Mesiti, C., & Clarke, D. (2010). A functional analysis of mathematical tasks in China, Japan, Sweden, Australia and the USA: Voice and Agency. In Y. Shimizu, B. Kaur, R. Huang, & D. Clarke (Eds.), *Mathematical tasks in classrooms around the world* (pp. 145–164). Rotterdam, The Netherlands: Sense.

Mok, I. A. C. (2006). Shedding light on the East Asian Learner Paradox: Reconstructing student-centeredness in a Shanghai classroom. *Asia Pacific Journal of Education, 26*(2), 131–142.

Mok, I. A. C. (2015). Research on mathematics classroom practice: An international perspective. In J. C. Sung (Ed.), *Selected regular lectures from the 12th International Congress on Mathematical Education* (pp. 589–606). New York, NY: Springer.

Organisation for Economic Co-operation and Development. (2013a). *PISA 2012 Assessment and Analytical Framework: Mathematics, reading, science, problem solving and financial literacy*. Paris, France: OECD Publishing.

Organisation for Economic Co-operation and Development. (2013b). *PISA 2012 results: What students know and can do* (Vol. 1). Paris, France: OECD Publishing.

Organisation for Economic Co-operation and Development. (2013c). *PISA 2012 results: Ready to learn students' engagement, drive and self-beliefs* (Vol. III). Paris, France: OECD Publishing.

Reddy, V., Van der Berg, S., Janse van Rensburg, D., & Taylor, S. (2012). Educational outcomes: Pathways and performance in South African high schools. *South African Journal of Science, 108*(3/4).

Roth, K., & Givvin, K. B. (2008). Implications for math and science instruction from the TIMSS 1999 Video Study. *Principal Leadership, 8*(9), 22–27.

Shimizu, Y., Kaur, B., Huang, R., & Clarke, D. (Eds.), (2010). *Mathematical tasks in classrooms around the world.* Rotterdam, The Netherlands: Sense.

Silver, E. A., Leung, S. S., & Cai, J. (1995). Generating multiple solutions for a problem: A comparison of the responses of US and Japanese students. *Educational Studies in Mathematics, 28*(1), 35–54.

Song, M. J., & Ginsburg, H. P. (1987). The development of informal and formal mathematical thinking in Korean and US children. *Child Development, 58,* 1286–1296.

Stacey, K., & Turner, R. (2015a). *Assessing mathematical literacy: The PISA experience.* Heidelberg, Germany: Springer.

Stacey, K., & Turner, R. (2015b). PISA's reporting of mathematical processes. In K. Beswick, T. Muir, & J. Wells (Eds.), *Proceedings of 39th psychology of mathematics education conference* (Vol. 4, pp. 201–208). Hobart, Australia: PME.

Stevenson, H. W., Lee, S., Chen, C., Lummis, M., Stigler, J. W., Liu, F., et al. (1990). Mathematics achievement of children in China and the United States. *Child Development, 61,* 1053–1066.

Stigler, J. W., Gallimore, R., & Hiebert, J. (2000). Using video surveys to compare classrooms and teaching across cultures: Examples and lessons from the TIMSS video studies. *Educational Psychologist, 35*(2), 87–100.

Stigler, J. W., & Hiebert, J. (1999). *"The teaching gap." Best ideas from the world's teachers for improving education in the classroom.* New York, NY: The Free Press.

Stigler, J. W., & Hiebert, J. (2004). Improving mathematics teaching. *Educational Leadership, 61* (5), 12–17.

Watkins, D. A., & Biggs, J. B. (Eds.). (2001). *Teaching the Chinese learner.* Hong Kong: Comparative Education Research Centre, The University of Hong Kong.

Open Access Except where otherwise noted, this chapter is licensed under a Creative Commons Attribution 4.0 International License. To view a copy of this license, visit http://creativecommons.org/licenses/by/4.0/.

Transitions in Mathematics Education: The Panel Debate

Ghislaine Gueudet, Marianna Bosch, Andrea A. diSessa, Oh Nam Kwon and Lieven Verschaffel

Abstract The Transitions in Mathematics Education panel during the ICME-13 conference consisted of two parts. In the first part, the panelists presented particular questions addressed and answered them according to their various perspectives (some of them cognitive, others more sociocultural). This first part was published as a survey before the conference (Gueudet et al. in Funds of knowledge: Theorizing practices in households, communities, and classrooms. Erlbaum: Mahwah, NJ, 2016). In the present text, we briefly review this first part but mainly focus on the second part of the panel. In the second part, the panelists answered questions about the survey concerning the arithmetic-algebra transition, the possible use of boundary objects to build links and bridges, the role of technical work in the continuity/discontinuity of the learning process, and the possible contributions of students in helping to ease transitions. These answers are developed and presented here.

G. Gueudet (✉)
CREAD, ESPE Bretagne UBO, 153 Rue Saint Malo, 35000 Rennes, France
e-mail: ghislaine.gueudet@espe-bretagne.fr

M. Bosch
IQS School of Management, Universitat Ramon Llull, Via Augusta, 390,
08022 Barcelona, Spain
e-mail: marianna.bosch@iqs.url.edu

A.A. diSessa
Graduate School of Education, University of California, Berkeley, CA 94720, USA
e-mail: disessa@berkeley.edu

O.N. Kwon
Seoul National University, Gwanak-Gu Gwanak-Ro 1, 151-748 Seoul, Korea
e-mail: onkwon@snu.ac.kr

L. Verschaffel
Center for Instructional Psychology and Technology, Education and Training Research Group, Katholieke Universiteit Leuven, Van Den Heuvel Instituut (Room Nr. 05.71), Dekenstraat 2, 3773, B-3000 Leuven, Belgium
e-mail: Lieven.Verschaffel@ppw.kuleuven.be

Keywords Boundary object · Curriculum · Discontinuity · Institutions · Transition

This text follows the content and organization of the ICME-13 panel about transitions in mathematics education. The panel started with individual presentations, putting forward different perspectives on transitions. After this first part, some questions, mainly raised by participants of the online panel preparation forum, were discussed from these different perspectives. We start here by briefly reviewing the content of the individual presentations (Section "Different Views on Transitions, a Survey") and then develop answers to the questions (Section "Addressing Transition Questions with Different Perspectives"), showing how contrasting views can complement each other in mathematics education research.

Different Views on Transitions, a Survey

The ICME-13 panel about "transitions in mathematics education" was grounded in a literature survey on this topic (Gueudet Bosch, diSessa, Kwon, & Verschaffel, 2016). We recommend reading the complete survey, which is available online at no cost. However, in this section a short version of the survey is developed that is sufficient to understand the various perspectives presented in the panel.

Which Transitions?

Many different kinds of transitions have been studied in the research literature. In our survey, we mainly addressed two kinds of change: (a) conceptual change and learning as a transition process and (b) transitions as people move between social groups or contexts with different mathematical practices.

Some researchers have studied changes within the mathematical content from an epistemological perspective, sometimes drawing on the history of mathematics (e.g., Dorier, 2000). Research from this perspective has also studied transitions during students' learning; these cognitive transitions can be viewed within various theoretical frameworks and concern specific mathematical topics but also more general issues, such as the transition between different thinking modes (e.g., Tall, 2002).

Other authors consider mathematics to be shaped by groups of people who develop shared mathematical practices; they investigate transitions between different such groups using a sociocultural perspective (Crafter & Maunder, 2012). Relevant groups can be of different natures; for example, they can correspond to different languages, different mathematical practices, and a kind of "permanent transition" (Ríordáin & O'Donoghue, 2011). Groups involved in a transition can also be moving between different teaching institutions (such as between primary and secondary school) or from a teaching institution and the workplace. These two last cases (two teaching institutions or a teaching institution and a workplace) were studied in specific sections of the panel, as described below; they both correspond to "local" transition, happening at a given moment in time. The corresponding studies often identify discontinuities and sometimes design teaching experiments in order to smooth the relevant transition.

Continuity Versus Discontinuity in Learning Difficult Concepts

This presentation focused on the *cognitive processes* by which prior (before instruction) concepts are transformed into normative understanding. This complements—and does not replace—sociocultural perspectives, which focus more on culture and membership, and less on concepts per se. In particular, we trace the history and current status of a contest between "revolutionary" theories of conceptual change, and "evolutionary" ones, which emphasize continuities over categorical discontinuities.

The early history of this contest in both mathematics and science education greatly favored the discontinuous point of view. In mathematics, the idea of "epistemological obstacles" was imported from philosophical work and, broadly speaking, it characterized prior stages of thinking as involving ideas that are persistent, unavoidable, and relatively monolithic in that they required substantial "ruptures" or discontinuities in thinking to overcome. In science education also, the philosophical literature was influential. Thomas Kuhn's ideas of incommensurability between paradigms were imported and became deeply ingrained in educational studies of conceptual change.

Without presuming that the relevant issues are settled, the presentation surveyed the advance of more continuist approaches to conceptual change over the years, using recent studies to illustrate orientations, relevant methodologies, and results in both mathematics and mathematically oriented physics. In general, the presentation advocated methodologies that pursue a finer grain size of analysis (with respect to conceptual detail and with respect to smaller time scales) than has been typical in the past. A smaller grain size makes it possible to see prior results and orientations as perhaps unnecessarily dichotomous and is amenable to more nuanced and complex descriptions, parallel to the trend noted concerning the relation of in- and out-of-school ways of thinking in the presentation of Lieven Verschaffel (Section "Transitions Between in- and Out-of-School Mathematics").

The presentation concluded by describing major differences between continuous and dichotomous views of conceptual change from the perspective of teachers and other educational professionals.

Double Discontinuity Between Secondary School Mathematics and University Mathematics: Focusing on Mathematical Knowledge for Teaching

This presentation dealt with two transitions that prospective teachers experience in becoming professionals—Klein's double discontinuity: from secondary school to university and then from university to teaching in secondary school. It provided an overview of the current state of the art in the context of teacher education in order to provide a deeper understanding of the double discontinuity phenomenon with a special focus on mathematical knowledge.

The distinction between subject matter knowledge and pedagogical content knowledge for teacher education has proved practically useful and has been employed in numerous studies. However, the assessment of teachers' subject matter knowledge and pedagogical content knowledge requires a theory of the subject in question and of its knowledge. There is broad consensus that these two components of professional knowledge cannot simply be equated.

There are two principal approaches to interconnecting these different kinds of mathematical knowledge that may be used together profitably: one that adds aspects of the new university discourse slowly and step by step and one that develops university-level problems starting with school mathematics. A third approach is to develop courses that explicitly integrate subject matter knowledge with pedagogical content knowledge in mathematics and the didactics of mathematics.

Klein's notion of a double discontinuity between university mathematics and school mathematics has proved to be extremely fruitful and can be seen to constitute the core of mathematics teacher education in both theoretical and practical respects.

Transitions Between Teaching Institutions

During their studies, students experience many transitions between educational institutions: from preschool to primary school, from primary to secondary school, sometimes from lower to higher secondary school or technical college, and from higher secondary to university. These transitions mean changes in many senses and are often seen as an important source of difficulties—rarely as opportunities—for the development of the students' learning. While research on transitions has mainly focused on the passage from secondary to tertiary education, some studies are starting to use a similar perspective to consider the passage from primary to lower secondary level.

We propose using different levels of specificity when considering the main research findings in the study of these two types of transitions, from the more general ones related to the culture and organizational rules of the educational institution to the more concrete ones linked to the ways of dealing with the various components of the mathematical content. Surprisingly, some of the phenomena pointed out as difficulties in the passage from primary to secondary level seem to reappear in the passage from secondary school to university. However, the treatment of these difficulties appears to be clearly asymmetric. Looking forward to ameliorating difficulties of transitions, in the first case the "receiver" institution (secondary education) seems to require more change to become closer to the "sender" institution. In the second case it is again the secondary level, here acting as "sender," which is questioned, while the university's prevailing pedagogical and mathematical organization remains almost unquestioned. Therefore, it seems important to take into account that transitions happen between institutions maintaining a certain hierarchical relationship in their raison d'être as preparatory schools as well as in their distance from scholarly mathematics, secondary education assuming an ambiguous role between the education for all and the preparation for tertiary studies.

Transitions Between in- and Out-of-School Mathematics

Besides the processes of conceptual change and the transitions from one instructional level or section to another described in the other sections of this review, research in mathematics education has also been confronted with the multifold transitions between in- and out-of-school mathematics. Within this topic, we distinguish between (1) the transition from prior-to-school to school mathematics and (2) the transitions from out-of-school to school mathematics (and vice versa). While the first kind of transition may be considered a non-reversible process, the second kind may be construed as an interaction—a "permanent transition" between two contexts.

The transition from prior-to-school to school mathematics is currently dominated by two quite different lines of research. First, there has been the very productive and influential line of neuro-cognitive research on children's early number sense, its development, and its relation to school mathematics. In a complementary line, researchers have approached the transition from prior-to-school to school mathematics from a sociocultural perspective, wherein it is primarily conceived as a set of processes whereby individuals "cross borders" from one cultural or educational context or community to another.

The research literature on the transitions from out-of-school to school mathematics (and vice versa) has been dominated by three main themes: (1) exploration of out-of-school mathematical practices and cultures (in comparison to mathematics learned at school), (2) difficulties in the transition between out-of-school and school mathematics, and (3) attempts to facilitate and exploit these transitions. It is interesting to note that the older extremely dichotomous descriptions of the pitfalls and merits of in- and out-of-school (learning) practices have been replaced by more nuanced and complicated analyses of these different kinds of mathematical practices and of the various types of transitions between them and the affordances of these transitions for educational purposes.

Addressing Transition Questions with Different Perspectives

The second part of panel discussed four questions, presented below, engaging the various perspectives evoked above. The question proposed to the panelists was formulated as:

"With the perspective you presented, can you say something (and what) about the following theme or question:"(the corresponding list of themes are given below as the titles of the subsections).

The panelists' answers are presented as a discussion (MB for Marianna Bosch, AdS for Andrea diSessa, ONK for Oh Nam Kwon, and LV for Lieven Verschaffel). For each question, two or three of the panelists responded.

The Transition from Arithmetic to Algebra

AdS: The transition from arithmetic to algebra should be an excellent example of the same considerations that I elaborated in my primary essay. Typical of the history described in my essay, the literature so far seems biased toward discontinuity. Researchers expect that some compact description of the essential difference between arithmetical and algebraic reasoning will tell the story. In contrast, I believe that if we look closely (finer conceptual and temporal grain sizes), we will find definitively incremental learning paths.

One problem with prior research is that different dimensions of change have not been adequately disentangled. Change appears too difficult, until we can "divide and conquer." Learning requires consideration of multiple threads, and each thread is, I maintain, less dichotomous appearing. For example, in much student work in the literature, I see learning difficulties that are associated with a general expertise with the nature of representations. These are almost never separated from the larger picture. A simple example is that students can't, or even will refuse to, answer problems given in a very slightly unusual representational system, say, with the x-axis vertical and y-axis horizontal. The ironic fact is that, based on some of our work with sixth-grade children (diSessa, 2004), teaching about representations seems surprisingly easy. It's just that school simply does not engage this learning thread.

A second thread that is unengaged in current instruction is the very nature of mathematics: the child's view of the mathematical enterprise. When the world shifts from numbers and procedures to relations and processes on relations, we should most certainly engage our students in thinking what that entails and feels like. I barely see any such recognition in contemporary curricula, and we may be suffering for the lack of attention.

Continuist views are still relatively new in mathematics, even more so concerning the arithmetic-to-algebra transition. However, work is progressing. For example, Mariana Levin (2012) has taken a microgenetic look at learning in pre-algebra. In her case study of a student developing the idea of linear extrapolation, she shows (1) the deep interaction between conceptual and procedural knowledge and (2) a lot of the typical phenomena of the continuist perspectives: many small learning events and a high degree of contextualization, which requires more time but no big jumps.

MB: The passage from arithmetic to algebra is one of the most often considered when addressing the question of transitions or discontinuities in school mathematics. If we look at school mathematics as a growing process of mathematization, this specific passage attracts the attention of teachers, educators, and researchers much more than others, such as the passage from Euclidean to analytic geometry, from algebra to functions, or from the consideration of elementary deterministic processes to stochastic ones. What is the specificity of this transition?

From an institutional perspective, a reason can be found in the specific role played by algebra in the structuring of old mathematical curricula. In many countries and for many decades, at least until the global reform of New Math during the 1960s and 1970s, school curricula were organized in three domains: arithmetic, algebra, and geometry. Students went to primary school to learn arithmetic, together with some practical applications and basic elements of measure and geometry. This represented the common mathematical culture of the broad (educated) population. Those who went to secondary school (relatively few, in some countries) had to learn algebra and analytical geometry. In any case, algebra marked the entrance to post-compulsory secondary education.

The difficulties attributed to the transition between arithmetic and algebra in current school processes maintain a semblance of the selective role formerly played by algebra. Today, algebra is still considered to be the first "abstract" content students should learn and may also be the first time students have difficulty in "giving meaning" to the mathematical practices they are asked to do. As educational researchers, we have to protect ourselves from this cultural perspective on school mathematical content and consider all of them from a unique and "uncontaminated" point of view (Bosch, 2015). In this context, the arithmetic-algebra transition should be approached by questioning the construction of the entire curriculum, that is, the whole process of mathematization as it is introduced to students. This questioning has to reach the traditional sequence of school mathematical content. Otherwise, we could be falling into the misconception, denounced by Paulos (2001), of considering mathematics "as a completely hierarchical subject. First comes arithmetic, then algebra, then calculus, then differential equations, abstract algebra, complex analysis, and so on. This is not necessarily so" (p. xiii).

Another important question that the arithmetic-algebra transition indirectly points out is the lack of a coherent and explicit discourse about the school mathematical curriculum "for all" and the role of algebra in this curriculum. The "algebra controversy" began years ago in the United States. Hacker's (2012) article, "Is algebra necessary?" explains it very well. When a society doubts the importance and utility of the mathematical content all citizens should learn, instruction of this knowledge is half-hearted at best.

What Are Appropriate or Promising "Boundary Objects" that Can Play a Contributing Role in Helping Students to Make the Transition?

LV: The first example that I think of is "word problems." These tasks were in existence already thousands of years ago to help pupils, from a young age on, (1) to see the links between mathematics lessons learned at school and the out-of-school world wherein to-be-learned mathematics has to be applied and (2) to establish productive transitions between the world of in- and out-of-school mathematics. This is what is typically called the "application function of word problems" (Verschaffel, Greer, & De Corte, 2000). So, in a way, word problems are intrinsically at the boundary of these two different worlds.

Meanwhile, we all know from a lot of theoretical analyses and empirical work that word problems do not play their application function very well. Indeed, word problems have gradually evolved into another type of school mathematics task that has little to do with authentic and complex mathematical modeling and application situations in the real world outside school. Accordingly, the actions that pupils perform when confronted by these word problems have little in common with what we would call genuine mathematical modeling and applied problem solving. As such, word problems can hardly be considered appropriate or promising "boundary objects" that help pupils make the transition between in- and out-of-school mathematics.

In reaction to this evaluation, mathematics educators have made numerous and varied attempts to make word problems more authentic so that they are better "simulations" of mathematical modeling and application problems situated in the real world. Some have gone even further and replaced these (mainly) verbal problems by rich, authentic, complex problem-solving contexts offering ample opportunities for genuine mathematical modeling and applied problem solving (with help of video and computer technology; Cognition and Technology Group at Vanderbilt, 1997).

However, as several authors have argued (e.g., Gravemeijer, 1997; Verschaffel et al., 2000), it is not always possible, and probably even not always necessary, to include the complexity of the out-of-school reality in the mathematics classroom. If you always try to accommodate all reality in your mathematics class, things may get out of hand. You open a "Pandora's box" (Verschaffel et al., 2000). Moreover, traditional word problems may still have their function in elementary mathematics education (alongside more complex, authentic, and challenging genuine modeling and application problems), particularly as a convenient means to create strong links between the basic mathematical operations and prototypically "clean" model situations (with little room for endless discussions about the situational complexities that might jeopardize this link). However, a lot can be accomplished by talking about the issues. For instance, during whole-class discussions, upper elementary school pupils can learn to differentiate between standard word problems (S-items) and problems that are problematic from a realistic modeling point of view (P-items), such as the rope item that I used in my talk ("A man wants to have a rope long enough to stretch between two poles 12 m apart, but he has only pieces of rope 1.5 m long. How many of these pieces of rope would he need to tie together to stretch between the poles"). Or, to give another example, take the following calendar joke "Ten birds are sitting in a tree. A hunter comes and shoots three of them. How many birds are still sitting in that tree afterwards?" Discussing with pupils these P-items or jokes—which I would consider to be "boundary objects" par excellence—are, in my view, excellent and important activities to understand the roles of simplification and consequent approximation in mathematical modeling and applied problem solving.

MB: In the case of the transition between secondary and tertiary education, an interesting boundary object can be the so-called bridging courses organized in different universities to smooth the gap between upper-secondary school and university. They are a good example of the ambiguity between individual interventions and institutional practices. We can construe their main goal as helping fist-year students smooth difficulties with the new learning processes they encounter at the university. Therefore, students are offered various courses, depending on the country and university—now most of them are online—where mainly secondary school contents are revisited and a few more advanced notions are introduced. These courses are usually taught in a short period, although they can also last a whole academic year, and are rarely recommended to all students, usually only for those who feel or are considered to be less prepared.

However, in their aim to help individuals smooth the transition, the effect produced at an institutional level is to reinforce the frontier between secondary and tertiary education. In a sense, bridging courses are the message transmitted by university to secondary education about what mathematical skills, competencies, practices, and contents students need but do not have. The subliminal message then is: "The students' preparation is not good enough and we are compelled to do the work you have not done." At the same time, the bridging courses tend to highlight the differences between both institutions instead of stressing the commonalities, which could offer students a link they might not be able to see. For instance, some studies (Serrano, Bosch, & Gascón, 2007; Sierpinska, 2006) show how the bridging courses can have a reverse effect and contribute to increasing the gap between institutions. Instead of facilitating the entrance to the new culture and its practices, the courses propose intensive work based on "filling the gaps" in the required basic knowledge, thus reinforcing and rigidifying the old relationships to the old mathematical contents. As indicated by Fonseca, Bosch and Gascón (2004), university mathematical content could explicitly emerge from questions and limitations of secondary school mathematics, proposing a way to develop the old contents towards more complete, intertwined, and powerful constructions. However, this ambitious process requires global changes in university mathematics education that cannot be accomplished in summer courses. Up to now, the bridging courses appear as a *coup de force* of the tertiary institution to clearly establish entrance requirements for the new students without any attempt to adapt its own practices to the newcomers or to the feeding institutions. The strategy in this way is very different from procedures aimed at smoothing the passage between primary and secondary education. In the latter, teachers from both institutions meet to exchange practices and increase shared activities, assuming the principle that changes are necessary on both sides of the transition.

What About Learning Technical, Procedural Work in the Acquisition of Concepts? How Does It Contribute to the Continuity/Discontinuity of the Learning Process?

AdS: While not a scientific result, the following shocking experience profoundly influenced my research program. Long ago, I engaged in an interviewing study involving dozens of MIT freshman physics students across several years. In the first interview, I always asked about students' experience with physics in high school. These MIT students were excellent and well prepared and they all said they did well and got an A. However, almost all of them added, "But I didn't understand anything." I believe that comment was insightful and showed a strong aesthetic about understanding and good judgment about it. Most of the students did not really understand a lot of physics, even if they could do the problems flawlessly.

The physics education research community long ago moved to background problem solving (a little) in favor of an increased emphasis on explanation and qualitative (conceptual) understanding. I think a very under-appreciated consequence of this is greater student satisfaction with their learning, in contrast to my interviewees' high school experience. Engagement, in fact, has become a primary driver of my experimental instruction; the issue deserves a lot more work.

Mathematics education research and instruction are different cultural beasts than physics. Even if we take my anecdote's implications at face value, it's not clear the lesson is exactly the same in mathematics. However, one of mathematics education's reform agendas is moving from the paradigm of learning via constant exercise of methods and techniques, to something involving a deeper understanding of concepts, if not exactly a focus on explanation per se, which seems more evidently important to physicists. I recommend to my mathematics education colleagues yet more emphasis on this side of mathematics.

Even methods and techniques can be generally reframed in terms of justification and invention of alternatives, rather than simply "absorbing and mastering," parallel to how I describe (in my response to Section "The Transition from Arithmetic to Algebra") an increased importance for understanding representations broadly, including student invention and judgments of aptness.

I think the continuist research program here could yield great dividends. If we can see bit by bit how various competencies co-evolve (e.g., conceptual and procedural), we will have a much more grounded understanding about how different emphases have an impact on each other. I expect, for example, that a better integration of conceptual and technical threads will alleviate apparent discontinuities, particularly in terms of the stability and perceived meaningfulness of learned procedures. "Sense-making," while occupying a strong niche in mathematics education, needs more precise definition and theoretical elaboration, which high-resolution empirical work can supply.

ONK: There is widespread agreement that the acquisition of concepts in areas such as multiplication or calculus, for example, requires both procedural and conceptual fluency (National Council of Teachers of Mathematics [NCTM], 2000). However, there is less agreement concerning the appropriate instructional balance between teaching for conceptual and procedural knowledge or how teaching can be organized to promote both types of knowledge.

I would like to mention the Inquiry-Oriented Differential Equations (IO-DE) project, which is an example of a collaborative effort between mathematics educators and mathematicians that seeks to explore the prospects and possibilities for improving undergraduate mathematics education, using differential equations as an example. Traditional differential equations courses at the university level are known as technique-driven or procedure-driven enterprises—like a cookbook. Rasmussen, Kwon, Allen, Marrongelle, and Burtch (2006) conducted an evaluation study to compare the routine skills and conceptual understandings of central ideas and analytic methods for solving differential equations between students in inquiry-oriented classes and traditionally taught classes at four undergraduate institutions in Korea and United States. Whereas IO-DE project classes at all sites typically followed an inquiry-oriented format, comparison classes at all sites typically followed a lecture-style format. The assessment consisted of procedural problems and conceptual problems. Procedural problems focused on students' instrumental understanding, such as the analytic and numerical nature of differential equations. On the other hand, conceptual understanding problems were aimed at evaluating students' understandings of important ideas and concepts. The students in the IO-DE classes scored better than the students with traditional instruction on both conceptual and procedural assessments, even though the focus of IO-DE was not on procedural and technique skills of differential equations. The more interesting data was the IO-DE students demonstrated higher retention rate on both procedural and conceptual assessments one year after the course. Our findings indicate that procedural work in the acquisition of new concepts did not lead to very good retention and also did not come with conceptual understanding of mathematical concepts (Kwon, Rasmussen, & Allen, 2005).

An explicit intention of IO-DE project classrooms is to create a learning environment where students routinely offer explanations of and justifications for their reasoning. As our understanding of student thinking at the transition from secondary to tertiary institutions evolves, so does our understanding of the kinds of teacher knowledge that are important for promoting student learning during transition. Beyond content knowledge, such knowledge includes awareness of students' informal and intuitive ways of reasoning about central ideas in differential equations, knowledge of pedagogical strategies that can connect to student thinking while moving the mathematical agenda forward, knowledge of theory related to social aspects of the classroom, and mathematical knowledge specific to teaching mathematics.

What Is the Possible Role of the Students (or Teachers) in Helping to Ease Transitions?

AdS: For me, the big picture here is that in the power structure of education, students are currently nearly completely disenfranchised. I feel strongly that it is imperative that we do something about it.

A former student of mine, while she was a graduate student in physics, organized a completely student-initiated and student-run program to help new undergraduates, particularly women and minorities (who are severely underrepresented in physics and other technical fields), deal with the transition from high school to university studies. Some of the best aspects of the program were that "welcoming" was a core value, and also that, freed a bit from the stodgy, self-satisfied university teaching faculty culture, they could enact instruction with a very different feel. For example, they introduced much more active, exploration-oriented instruction and talked explicitly about the epistemic nature of physics as a modeling enterprise. The program is now a national model, and there are initiatives to replicate it more widely.

I have experienced extremely positive student cultures and extremely negative ones. Positive is unambiguously better. I would like to cultivate students who are intolerant of mechanistic and disconnected instruction and who know how to—and are anxious to—engage each other in collaborative disciplinary inquiry.

I think the presumption that only teachers can "make this happen" is a symptom of endemic disrespect for students' competencies to help foster their own learning. I recognize, of course, that teachers generally have a special role in instigating and supporting student initiative. However, I advocate that we actively help develop autonomous strengths within the student community. From a scientific point of view, I don't believe that any culture we might imagine is possible to instantiate, so the "design" of classroom student cultures is not just an action to take, but a complex field we need to learn to navigate.

LV: From my perspective, it is quite clear that the learner can help in bridging between the culture and practice of school and that of home. More particularly, learners themselves can look for activities and artifacts outside school that relate to school mathematics and vice versa. In this sense, they can act as active and constructive go-betweens to find out what mathematics is at homes and communities and document who does what kinds of mathematics in their environment.

Take, for example, Luis Moll's concept of "Funds of Knowledge," which refers to the historically accumulated and culturally developed bodies of knowledge learners can bring into the classroom because of their unique familial, cultural, and experiential backgrounds, but which also could be identified, valued, and used by teachers (González, Moll, & Amanti, 2005), to which I referred in my talk.

Quite evidently, in this approach there should be an explicit and systematic attempt by teachers to learn more about their learners' *funds of knowledge*. This is primarily the responsibility of the teaching side. In the funds of knowledge approach, it is expected that teachers will try to learn more about their learners' funds of knowledge by visiting learners' home environments. It should be clear that this is absolutely not a simple task, but a task that requires a great deal of preparation and coaching.

But, of course, this kind of bridging learners' home and school environment by the teacher also requires an openness and active willingness of the learners (and their parents, particularly if we are talking about young learners, and possibly also other members of the learner's family and broader environment) to share their funds of knowledge. The literature contains several examples (cases, described in the literature) wherein this has been successfully realized (building houses, making candy and so on; e.g., Sandova-Taylor, 2005).

In addition, as critical math educators such as Gutstein, Greer, and Mukhopadhyay have argued, such bridging activities, if properly handled by the teacher, may not only help to establish the relevance of mathematics but also create useful stepping stones for mathematical knowledge building as such. It may also help to demonstrate and validate its non-elitist existence and to create respect for multiple forms of mathematical practices. Thinking of Freire's quote that "Intellectual activity of those without power is always characterized as non-intellectual," one can replace "intellectual" with "mathematical" to readily see that "Mathematical activity of those without power is always characterized as non-mathematical" (Greer and Mukhopadhyay, personal communication). Therefore, establishing such bridging activities may reinforce in learners the belief that mathematics is a universal human activity and is done actively by all kinds of people. Furthermore, when learning more about learners' familial, cultural, and experiential backgrounds and trying to link these backgrounds to school mathematics, themes such as AIDS, poverty, sexual exploitation, and pollution, ... may pop up, and learners may start to realize how mathematics can become a helpful, even critical, element in their process of emancipation or conscientization (again in the Freirean sense).

ONK: I think that the role of teachers at both institutions is more important than the students' role in helping ease transitions. The essence of mathematical knowledge cannot and should not be compartmentalized into school mathematics and university mathematics. School mathematics is a subset of university mathematics. Teachers should facilitate the interconnection between secondary and tertiary schooling on specific topics in calculus, linear algebra, and analysis. University teachers need to view elementary mathematics from a higher standpoint. In addition, students should appreciate the need for a different kind of mathematics at the university level. They should at the same time understand how this new mathematics is related to school mathematics, why it is different, and why it nevertheless has potential in contributing to the development of students' mathematical competence in a way that makes it useful for qualified mathematics teaching at the school level.

Furthermore, Kwon, Rasmussen, Marrongelle, Park, Cho, and Park (2008) focused on teacher revoicing because it is one of the discursive strategies that often occurs in the teaching of mathematics but which has received limited attention in mathematics education research at the undergraduate level. Our analysis shows that teacher revoicing can constitute a major part of teachers' repertoires of discursive moves and carries out critical functions in the context of mathematics practice in class. From this perspective, revoicing can serve at least three functions in the classroom. First, revoicing functions to highlight specific mathematical ideas and/or provide mathematical content to move the mathematical agenda forward. Second, revoicing functions to honor and empower student thinking. That is, revoicing facilitates the development of students' mathematical identities. Third, revoicing functions to help students understand what constitutes a sufficient explanation or justification. That is, revoicing can serve to promote certain social and socio-mathematical norms. In their learning environment, students learn new mathematics by inquiry, which involves solving novel problems, debating mathematical solutions, posing and following up on conjectures, and explaining and justifying one's thinking.

Conclusion

While the presentations during the first part of the panel showed that the different perspectives tended to address different questions, the answers presented here to identical questions show consequences of the choice of perspectives in terms of differences in the answers.

Most of the time, the answers of the panelists complemented each other. Whatever approach was chosen, there was a consensus acknowledging the complexity of transition phenomena. The transition is not composed of an initial state, a final state, and a gap in between that can be spanned by an appropriate bridge. Instead, it is a complex, cumulative path to be managed.

A micro-level must be taken into account, separating different dimensions in the learning and teaching processes, which is a necessary first step to understanding the transition process and to analyze it as an incremental path. At the same time, separating different dimensions does not mean considering them independently: For example, the procedural and conceptual aspects are certainly two dimensions in the learning of mathematics, but they are strongly linked and their interactions within transition processes constitute an important issue. A macro-level must also be considered, for example, to analyze how curriculum choices at a large scale, encompassing several institutions, shape the transition.

Moreover the "initial" and "final" states must not be seen as two clearly separate points but as two zones whose frontiers are more or less clear and which can have intersections. These intersections are linked by the existence of boundary objects—objects present in both zones—and/or of brokers—persons living in both zones. The different approaches focus on different kinds of boundary objects or brokers: mathematical problems existing in and out of school and common work between teachers of different institutions trying to understand each other's mathematical practices and to bring them closer. Nevertheless, all the approaches suggest that trying to identify or to develop boundary objects is a promising direction for research, especially with an aim of proposing solutions to ease the transitions' difficulties.

Sometimes the different approaches also lead to (apparently) conflicting answers. Teachers are certainly central actors shaping the teaching and learning processes, thus also in shaping the transition processes. A focus on teachers alone, though, can suggest that students cannot themselves act on transition processes, but examples exist showing the possible actions of students. Moreover, teachers can work with students to ease the transition—or, with a different perspective (Sensevy, Gruson, & Forest, 2015), the joint action of teachers and students can ease some transition processes.

This discussion shows, in any case, that mathematical research on transition is lively and rich and that further research directions are open. We hope that this panel will serve as a resource for those who want to pursue these directions!

References

Bosch, M. (2015). Doing research within the anthropological theory of the didactic: the case of school algebra. In S. J. Cho (Ed.), *Selected regular lectures from the 12th international congress on mathematical education* (pp. 51–69). New York: Springer.

Cognition and Technology Group at Vanderbilt. (1997). *The Jasper project: Lessons in curriculum, instruction, assessment, and professional development*. Mahwah, NJ: Erlbaum.

Crafter, S., & Maunder, R. (2012). Understanding transitions using a sociocultural framework. *Educational and Child Psychology, 29*(1), 10–18.

diSessa, A. A. (2004). Meta-representation: Native competence and targets for instruction. *Cognition and Instruction, 22*(3), 293–331.

Dorier, J.-L. (Ed.). (2000). *On the Teaching of Linear Algebra*. Dordrecht: Kluwer Academic Publishers.

Fonseca, C., Bosch, M., & Gascón, J. (2004). Incompletitud de las organizaciones matemáticas locales en las instituciones escolares. *Recherches en didactique des mathématiques, 24*(2), 205–250.

González, N., Moll, L. C., & Amanti, C. (Eds.). (2005). *Funds of knowledge: Theorizing practices in households, communities, and classrooms*. Mahwah, NJ: Erlbaum.

Gravemeijer, K. (1997). Solving word problems: A case of modelling? *Learning and Instruction, 7*, 389–397.

Gueudet, G., Bosch, M., diSessa, A., Kwon, O.-N., & Verschaffel, L. (2016). *Transitions in mathematics education*. Berlin: Springer.

Hacker, A. (2012). Is algebra necessary? *The New York Times (Sunday review)*.

Kwon, O. N., Rasmussen, C., & Allen, K. (2005). Students' retention of mathematical knowledge and skills in differential equations. *School Science and Mathematics, 105*(5), 1–13.

Kwon, O. N., Ju, M. K., Rasmussen, C., Marrongelle, K., Park, J. H., Cho, K. Y., et al. (2008). Utilization of revoicing based on learners' thinking in an inquiry-oriented differential equations class. *The SNU Journal of Education Research, 17*, 111–134.

Levin (Campbell), M. E. (2012). *Modeling the co-development of strategic and conceptual knowledge during mathematical problem solving. (Unpublished doctoral dissertation)*. Berkeley: University of California.

National Council of Teachers of Mathematics. (2000). *Principles and standards for school mathematics*. Reston VA: Author.

Paulos, J. A. (2001). *Innumeracy: Mathematical illiteracy and its consequences*. New York, NY: Hill and Wang.

Rasmussen, C., Kwon, O. N., Allen, K., Marrongelle, K., & Burtch, M. (2006). Capitalizing on advances in mathematics and K-12 mathematics education in undergraduate mathematics: An inquiry-oriented approach differential equations. *Asia Pacific Education Review, 7*(1), 85–93.

Ríordáin, M. N., & O'Donoghue, J. (2011). Tackling the transition—the English mathematics register and students learning through the medium of Irish. *Mathematics Education Research Journal, 23*(1), 43–65.

Sandova-Taylor, P. (2005). Home is where the heart is. Planning a Funds of Knowledge based curriculum module. In T. González, L. C. Moll, & C. Amanti (Eds.), *Funds of Knowledge: Theorizing practices in households, communities, and classrooms* (pp. 153–165). Mahwah, NJ: Erlbaum.

Sensevy, G., Gruson, B., & Forest, D. (2015). On the nature of the semiotic structure of the didactic action: The joint action theory in didactics within a comparative approach. *Interchange, 46*(4), 387–412.

Serrano, L., Bosch, M., & Gascón, J. (2007). Diseño de organizaciones didácticas para la articulación del bachillerato con el primer curso universitario. In L. Ruiz-Higueras, A. Estepa, & F. J. García (Eds.), *Sociedad, Escuela y Matemáticas. Aportaciones de la Teoría Antropológica de lo Didáctico* (pp. 757–764). Jaén: Servicio de Publicaciones de la Universidad de Jaén.

Sierpinska, A. (2006). Sources of students' frustration in bridging mathematics courses. In *Proceedings of 30th Conference of the International Group for the Psychology of Mathematics Education* (Vol. 5, pp. 121–129).

Tall, D. (2002). Continuities and discontinuities in long-term learning schemas. In Tall, D. & Thomas, M. (Eds.) *Intelligence, learning and understanding: A tribute to Richard Skemp* (pp. 151–177). Flaxton QLD, Australia: Post Pressed.

Verschaffel, L., Greer, B., & De Corte, E. (2000). *Making sense of word problems.* Lisse, The Netherlands: Swets & Zeitlinger.

Open Access Except where otherwise noted, this chapter is licensed under a Creative Commons Attribution 4.0 International License. To view a copy of this license, visit http://creativecommons.org/licenses/by/4.0/.

Part II
Awardees' lectures

ICMI Awards Ceremony

Carolyn Kieran and Jeremy Kilpatrick

The segment of the ICME-13 Opening Ceremony that was dedicated to the ICMI Awards was presided over by Carolyn Kieran, Chair of the Felix Klein and Hans Freudenthal Awards Committee, and by Jeremy Kilpatrick, Chair of the Emma Castelnuovo Award Committee.

ICMI has awarded the Felix Klein and Hans Freudenthal medals in each of the odd-numbered years since 2003 to recognize outstanding accomplishments in mathematics education research:

- The Felix Klein Award, named after the first president of ICMI (1908–1920), for lifetime achievement in mathematics education research,
- The Hans Freudenthal Award, named after the eighth president of ICMI (1967–1970), for a major cumulative programme of research on mathematics education.

The **Felix Klein** medal acknowledges those excellent scholars who have shaped our field over their lifetimes. Past candidates have been influential and have had an impact both nationally within their own country and internationally. We have valued in the past those candidates who not only have made substantial research contributions, but also have introduced new ideas, perspectives, and critical reflections. Additional considerations have included leadership roles, mentoring, and peer recognition, as well as the actual or potential relationship between the research done and improvement of mathematics education at large, through connections between research and practice.

C. Kieran (✉)
Université du Québec à Montréal, Montreal, QC, Canada
e-mail: kieran.carolyn@uqam.ca

J. Kilpatrick
University of Georgia, Athens, GA, USA
e-mail: jkilpat@uga.edu

© The Author(s) 2017
G. Kaiser (ed.), *Proceedings of the 13th International Congress on Mathematical Education*, ICME-13 Monographs, DOI 10.1007/978-3-319-62597-3_8

The **Hans Freudenthal** medal acknowledges the outstanding contributions of an individual's theoretically robust and highly coherent research programme. It honours a scholar who has initiated a new research programme and has brought it to maturation over the past 10 years. The criteria for this award are depth, novelty, sustainability, and impact of the research programme on our community.

In 2013 the ICMI Executive Committee decided to create a third award to recognize outstanding achievements in the practice of mathematics education, thus reflecting an aspect of the ICMI mission not previously recognized in the form of an award:

- The Emma Castelnuovo Award, named after the Italian mathematics educator born in 1913 to celebrate her 100th birthday and honour her pioneering work.

The **Emma Castelnuovo** medal, which is presented for the first time this year, will be awarded every four years henceforth. This medal is aimed at honouring persons, groups, projects, institutions or organizations engaged in the development and implementation of exceptionally excellent and influential work in the practice of mathematics education, such as: classroom teaching, curriculum development, instructional design (of materials or pedagogical models), teacher preparation programs and/or field projects with a demonstrated influence on schools, districts, regions or countries. The award seeks to recognize and to encourage efforts and ideas, and their successful implementation in the field.

At the 2016 Awards Ceremony, where the ICMI President, Ferdinando Arzarello, presented each awardee with a medal and certificate (see the texts of the certificates below), the following individuals were honoured:

- Felix Klein Medal for 2013: awarded to Michèle Artigue, Emeritus Professor, Université Paris Diderot—Paris 7, France.
- Hans Freudenthal Medal for 2013: awarded to Professor Frederick K. S. Leung, The University of Hong Kong, SAR China.
- Felix Klein Medal for 2015: awarded to Alan J. Bishop, Emeritus Professor of Education, Monash University, Australia.
- Hans Freudenthal Medal for 2015: awarded to Professor Jill Adler, University of the Witwatersrand, South Africa.
- Emma Castelnuovo Medal for 2016: awarded jointly to Professors Hugh Burkhardt and Malcolm Swan of the University of Nottingham, United Kingdom.

The sixth Felix Klein Medal of the International Commission on Mathematical Instruction is awarded to Professor Michèle Artigue. This distinction acknowledges her more than thirty years of sustained, consistent, and outstanding lifetime achievements in mathematics education research and development. Michèle Artigue's scholarly work in areas as diverse as advanced mathematical thinking, the role of technological tools in the teaching and learning of mathematics, institutional considerations in the professional development of teachers, and the articulation of didactical theory and methodology, is matched by a record of exceptional service to the international and national mathematics education communities, to graduate students and young researchers around the world, and to teacher education.

The sixth Hans Freudenthal Medal of the International Commission on Mathematical Instruction is awarded to Professor Frederick K. S. Leung. This distinction recognizes his research in comparative studies of mathematics education and the influence of culture on mathematics teaching and learning. Using the perspective of the Confucian Heritage Culture to explain the superior mathematics achievement of East Asian students in international studies and the differences in teacher knowledge between East Asian and Western countries, Frederick Leung's research has opened up a new dimension of looking at mathematics achievement and classroom practices from the perspective of culture and has had an important impact on policies and practices in mathematics education in East Asian countries and beyond.

The seventh Felix Klein Medal of the International Commission on Mathematical Instruction is awarded to Professor Alan J. Bishop. This distinction acknowledges his more than forty-five years of sustained, consistent, and outstanding lifetime achievements in mathematics education research and scholarly development. His research has been instrumental in bringing the political, social, and cultural dimensions of mathematics education to the attention of the community. Few researchers can match his impressive activity in advising prospective and practicing teachers of mathematics, encouraging them to conduct and use research in their practice. Alan Bishop has also, through his tireless and scholarly editorial work, enabled research in mathematics education to become an established field.

The seventh Hans Freudenthal Medal of the International Commission on Mathematical Instruction is awarded to Professor Jill Adler. This distinction recognizes her outstanding research dedicated to improving the teaching and learning of mathematics in South Africa, from the dilemmas of teaching mathematics in multilingual classrooms through to the problems related to mathematical knowledge for teaching and mathematics teacher professional development. Her published works have advanced the field's understanding of the relationship between language and mathematics in the classroom. Jill Adler's development of research teams and her mentoring of numerous graduate students have all added to the human research capacity she has been instrumental in creating in Southern Africa.

The first Emma Castelnuovo Medal of the International Commission on Mathematical Instruction is awarded to Professors Hugh Burkhardt and Malcolm Swan. This distinction acknowledges their more than 35 years of development and implementation of innovative, influential work in the practice of mathematics education, including the development of curriculum and assessment materials, instructional design concepts, teacher preparation programs, and educational system changes. Burkhardt and Swan have served as strategic and creative leaders of the Nottingham-based Shell Centre team of researcher-designers and their *international collaborators. Together, Burkhardt and Swan have produced groundbreaking contributions that have had a remarkable influence on the practice of mathematics education as exemplified by Emma Castelnuovo.*

Open Access Except where otherwise noted, this chapter is licensed under a Creative Commons Attribution 4.0 International License. To view a copy of this license, visit http://creativecommons.org/licenses/by/4.0/.

Mathematics Discourse in Instruction (MDI): A Discursive Resource as Boundary Object Across Practices

Jill Adler

Abstract Linked research and development forms the central pillar of the Wits Maths Connect Secondary (WMCS), a project working with secondary mathematics teachers in one province in South Africa. A key outcome is a sociocultural analytic framework—a discursive resource that has been developed and refined through our work in and across three inter-linked practices. Named Mathematics Discourse in Instruction (MDI), we have used the framework as a planning and reflection tool in professional development and we have operationalised it as an analytic framework for research. MDI enables a description of mathematics made available to learn in a lesson, and an interpretation of shifts in practice across lessons and over time. MDI is both process and product of the WMCS. I describe and reflect on our use of MDI to build a case for embracing a discursive resource as a boundary object that moves productively across multiple practices.

Keywords Mathematics teaching · Professional development · Boundary object

Introduction

Since 2010, I have been working on the Wits Maths Connect Secondary (WMCS) project, the goal of which is to improve the teaching and learning of mathematics in secondary schools in one province in South Africa through the professional development of mathematics teachers. We hope to contribute to strengthening the mathematics pipeline within the secondary school and between school and tertiary mathematics studies. Over time we have come to focus in on grades 9 and 10—a critical transition point in mathematics education in the country.

Hans Freudenthal award.

J. Adler (✉)
University of the Witwatersrand, Johannesburg, South Africa
e-mail: jill.adler@wits.ac.za

The major intervention that evolved over the first three years of the project, and that has continued to strengthen, is a Mathematics for Teaching course, called Transition Maths (TM), complemented by a local version of Lesson Study. The intervention thus works with teachers on their mathematical knowledge for teaching away from the school (the TM course is offered at the University), and then on their mathematics teaching practices in the school (the Lesson Study work takes place at schools organized in clusters and in the afternoons). The attraction and excitement of this work is that it deliberately links research and development. In the South African context, and I suspect elsewhere, it is not usual that both research and development activities are supported by funding from the National Research Foundation. As a linked research and development project, we have been researching improvement in knowing, teaching and learning mathematics, in relation to the project interventions. Of course, this required elaboration of what counts as improvement in mathematical knowledge for teaching, in mathematics teaching and learning, and how we would describe and interpret each of these.

What was needed for this work was a framework for *describing* mathematics teaching, *interpreting shifts* in practice and *supporting the development* of mathematics teaching practices. The framework we have developed is called Mathematics Discourse in Instruction (MDI), so named to foreground concern with the quality of mathematics made available to learn in school, and that we were working with teachers on their instructional practices. MDI has become a unifying framework in the project, operating as a boundary object across professional development and research activities. MDI frames our work in our course sessions, and our lesson study work; and it has been operationalised for research.

This paper is focused on the MDI framework, what led to its development, the form it has taken and why, and how it is used across our teaching and research practices. I will share with you the role and nature of MDI as a boundary object. I hope to convince you of the power of the framework. It is a living framework, with power lying in its iterative nature, moving between and supporting both our research and development work.

The Context

In 2009, the First Rand Foundation (FRF—the foundation of a major bank in South Africa), in partnership with the national Department of Science and Technology (DST) and National Research Foundation (NRF) launched the Maths Education Chairs Initiative project, with ambitious goals and five years of funding to support proposals for research-linked development work that would:

- improve the quality of mathematics teaching in previously disadvantaged secondary schools;
- improve mathematics results on national examinations (pass rates and quality of passes) through quality teaching and learning;
- research sustainable and practical solutions to the mathematics crisis;
- develop research capacity in mathematics education;
- provide leadership and increase dialogue around solutions.

Personally, this came as an opportunity to put to work what I had learned over the previous two decades in the academy and in mathematics teacher education. In all my research work, I have focused on problems of practice, moving from the challenges of teaching mathematics in multilingual classrooms to the challenges of describing and theorising mathematical knowledge in and for teaching, in a context where the majority of teachers were statutorily undereducated mathematically in apartheid teacher training colleges. The FRF/DST/NRF funding was an opportunity for me to shift my research from its impetus in problems of practice, to research-informed-development and development-informed-research—an opportunity to deliberately engage in a process where research, school mathematics teaching and teacher education were able to speak to (listen and hear) each other.

Simply, I saw this as an opportunity put research in the service of teaching, and moreover, of teaching in schools and classrooms for 'historically disadvantaged' students. This is important in our field, for as Skovsmose so aptly conjectured in his talk at the ICMI Centennial conference in Rome in 2008: "90% of research in mathematics education concentrates on the 10% of the most affluent classroom environments in the world, while 10% of the research addresses the remaining 90% of the classrooms" (Skovsmose, 2011, p. 18). The classrooms I would be working in were squarely located in the "remaining 90%" of relatively impoverished schools.

So, what is this context? It is beyond the scope of this paper to present a full description of the South African mathematics education context, some twenty years post-apartheid. Interested readers are referred to a detailed elaboration of this in Adler and Pillay (2017). In brief, high levels of poverty and inequality endure. The relationship between poverty and educational outcomes is well known, and starkly reflected in poor educational outcomes for the majority of school learners, particularly in mathematics. While over 90% of all students up to 16 years are in school, inequality prevails, with socio-economic status being its most significant determinant. The performance curve in national assessments at both Grade 9 and Grade 12 levels is grossly skewed, with the majority of learners performing poorly in mathematics. The system is slowly improving, with 2015 TIMSS results indicating we have moved from "very low" in 2003 to "low" performance (Reddy et al., 2016). However, there remains a disjuncture between institutional and epistemic access (Morrow, 2007). Most students are in school, but only a minority have access to quality education.

In carving out our intervention we were motivated firstly by the understanding that it is precisely in conditions of poverty that the quality of instruction in schools matters, and thus its improvement is an imperative for social justice (Gamoran & Long, 2006). Secondly, also a social justice issue, we hoped to take up the challenge of investigating the research-development nexus in mathematics education in under-researched poorly resourced conditions—or what Shalem and Hoadley (2009) refer to as "schools for the poor".

Shalem & Hoadley (op cit) described the dual economy of schooling in South Africa, providing critical insight into the impact of different economic conditions on teachers' work. Teachers who work in "schools for the poor" do not have access to educational "assets" taken-for-granted in schools for the middle classes. Not only are infrastructural conditions poor, but knowledge resources both human and material (Adler, 2012a) are typically limited; and many students in such schools are not academically prepared for the grade they are in. Unsurprisingly teachers suffer from low morale. Added to the emotional and material challenges of their work, South African teachers in underperforming, poor schools face highly prescriptive curricula, national testing and a culture of bureaucratic compliance, all enacted with the goal of improving national schooling outcomes. Shalem & Hoadley argue that this combination of demands makes teachers' work in schools for the poor "impossible".

In the first year of the project we set out to learn in and from the schools. Diagnostic testing of students, informal conversations with teachers, and observation of lessons across a range of grade levels together confirmed Shalem & Hoadley's general analysis. Learner error in basic algebra, for example, was extensive, and basic resources like textbooks were not always available. More specifically, mathematics lessons were characteristically incoherent: while students might have been 'working' (listening to the teacher, writing in their notebooks), the mathematics they were to be learning was, in our terms, "out of focus" (Adler, 2012b). Our research and development work, and particularly how we were to frame it, had to be grounded in these realities and take cognisance of the conditions of teachers' work.

Our Framework—Mathematics Discourse in Instruction

Making sense of the mathematics teaching and learning practices we observed necessarily brings ways of thinking about such practices to the fore, and we surveyed the field of mathematics education research in search of frameworks that could help us describe our pedagogic range. Our field, over time and in different ways, has helped us to distinguish between instruction focused on rules without reasons (Skemp, 1987), on procedures without rationales (Artigue, 2009). Frameworks for describing mathematics teaching, for example, LMTP (2011). Mathematical Quality of Instruction (MQI) have been developed. Using this range of theoretical and analytic resources, we we could easily describe almost all the lessons we observed as rule bound and procedural. But the lessons were not all of the same quality in terms of opportunities made available to learn mathematics. Using more elaborate and refined frameworks like MQI, while more illuminating nevertheless would have resulted in homogenising all our teachers and their instructional practices. LMTP (2011) Mathematical Quality of Instruction incorporates features like lesson format and links to learning alongside teachers' mathematical talk. We needed a framework that was more sensitive to formalist (typically referred to as "traditional") orienations to knowledge (Guthrie, 2011). The framework needed to work with our empirical field, and so be responsive to existing teaching practices, and disaggregating within and across these. Finally, we also needed a framework that would enable us to work with a developmental trajectory, guiding our intervention work and enabling a description of shifts in instructional practice, and so in terms of mathematics made available to learn.

Our framework, MDI, or Mathematics Discourse in Instruction, is represented in Fig. 1. It has its roots in sociocultural theory, particularly Vygotsky's (1978) theory of mediated learning, and scientific knowledge as a network of connected concepts. Our starting point is that teaching and so learning is always about *something*, and bringing that into focus is the teacher's work. Following variation theorists (Marton & Tsui, 2004) we call this 'something' the *"object of learning"*—that which students are to know and be able to do. In practical terms, it is akin to a lesson goal, but worded so that the mathematics of the goal needs to be made clear. Whatever the mathematical goal, or 'object', it needs to be mediated and so exemplified and elaborated. Figure 1 shows that *exemplification (through examples and tasks), explanatory talk (in how signs and objects are named, and justified or legitimated), and learner participation* are viewed as the key mediational means or cultural tools in mathematics classroom instruction. We hold that this is the case in any classroom, whatever the pedagogy or view of knowledge. However, this framing enables us to stay close to what it was we saw teachers doing. In all lessons, 'examples' were offered, embedded in various tasks, though typically procedural in nature. Communication then proceeded in the lesson, with words used to name the mathematics being talked about and to build explanations of, or legitimate what was to be known and done. And learners were invited to participate in this communication, even if it was all or mostly in the form of listening to the teacher.

As noted earlier, we use MDI to plan our sessions in the TM1 course; in our Lesson Study work to guide planning and reflection on lessons; and we use it for our research on mathematics teaching. Given our socio-cultural orientation, as we move across these practices, so we understand, and indeed expect that the framework will operate in different ways, shaped by the social practice we are working in. As I also mentioned earlier, MDI is a living framework, with power lying in its iterative nature, moving between and supporting both our research and development work.

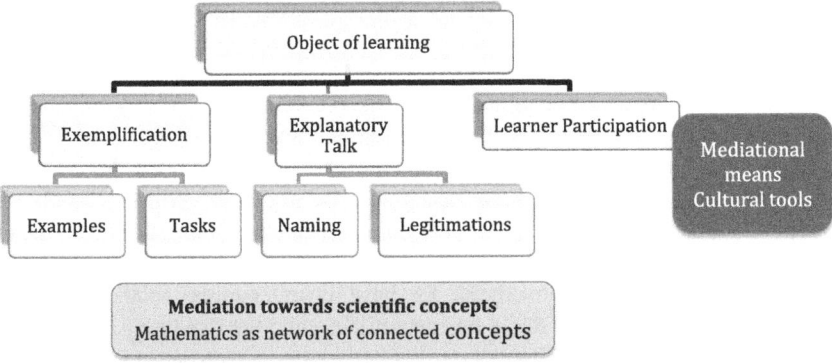

Fig. 1 Constitutive elements of MDI (Adapted from Adler & Ronda, 2015)

We have learned a great deal over the past six years, from how to responsively and responsibly describe practice (Adler & Ronda, 2017b; Adler & Venkat, 2014; Venkat & Adler, 2012) and differences in practice (Adler & Ronda, 2015), all using the MDI as analytic framework. We have put MDI productively to work in the practice of Lesson Study (Adler & Ronda, 2017a) and have reported the promising results on the impact of our professional development intervention on student attainment (Pournara, Hodgen, Adler, & Pillay, 2015). In the remainder of this paper, I illustrate our use of MDI across these practices.

Doing Our Research: Describing Teaching and Interpreting Shifts in Practice

A major task for any lesson analysis is how to chunk or divide the lesson into analytic units. In our analysis, we first need to infer the lesson goal, or object of learning, as this is key to our analysis, and we then proceed to chunk the lesson into what we call mathematical episodes. We begin by watching the video-recording and simultaneously (re)reading the transcript to identify the intended object of learning that we know is not synonymous with what is enacted (Marton & Tsui, 2004). We look for what is announced in some way, typically by the teacher at the start of the lesson, often stated as a topic or written on the board. For example, in one of our lessons, Mutliplying Algebraic Expressions was written on the board, at the same time as the teacher said, "Today we are going to learn to multiply expressions". The intended object of learning was carrying out a procedure for different products, which as it transpired in the lesson included two or more monomials; a monomial and binomial; and then two binomials. We then chunk the lesson transcript into *mathematical episodes* that are identified by a shift in focus of attention with respect to content, typically marked by a task that encompasses selected example(s), and bears some relation to the stated object of learning. We identify a next episode by the introduction of a new task and focus on a new example. This chunking produces a number of episodes, and again by way of example in this particular lesson, Episode 1 focused on multiplying single terms, paying attention to laws of exponents. Episode 2 was marked by the shift to the product of a monomial and binomial, and so on. We then examine each episode for its exemplification, explanatory communication and learner participation.

Table 1 provides a summary of how we have categorised key elements of MDI. It is beyond the scope here to discuss each of these in detail. I will focus only on examples and legitimating criteria so as to communicate how our analysis works to describe what is made available to learn in a lesson or set of lessons through the set of examples offered, and the criteria for what is to count as mathematics are communicated.There is more detail on each of these and the remaining elements in Adler and Ronda (2015, 2017b).

Table 1 Categorising key elements of MDI (extract from Adler & Ronda, 2015)

Object of learning				
Exemplification		Explanatory talk		Learner participation
Examples	Tasks	Naming	Legitimating criteria	
Examples provide opportunities within an event or across events in a lesson for learners to experience variation in terms of *similarity* (S), *contrast* (C), *simultaneity* (F)	Across the lesson, learners are required to: *Carry out known operations and procedures* (**K**) e.g. multiply, factorise, solve; *Apply known skills, and/or decide on operation and/or procedure to use* (**A**) e.g. Compare/classify/match representations; *Use multiple concepts and make multiple connections. Solve problems in different ways; use multiple representations; pose problems; prove; reason.etc* (**C/PS**)	Within and across events word use is: *Colloquial* (**NM**) e.g. everyday language and/or ambiguous referents such as this, that, thing, to refer to signifiers; *Math words used as name only* (**Ms**) e.g. to read string of symbols; *Mathematical language used appropriately* (**Ma**) to refer to signifiers and procedures	Legitimating criteria: *Non mathematical* (**NM**) *Visual* (**V**)—e.g. cues are iconic or mnemonic *Positional* (**P**)—e.g. a statement or assertion, typically by the teacher, as if 'fact'. *Everyday* (**E**) Mathematical criteria: *Local* (**L**) e.g. a specific or single case (real-life or math), established shortcut, or convention *General* (**G**) equivalent representation, definition, previously established generalization; principles, structures, properties; and these can be partial (**GP**) or 'full' (**GF**)	Learners answer: *yes/no questions or offer single words to the teacher's unfinished sentence* **Y/N** Learners answer (what/how) questions in phrases/sentences (**P/S**) Learners answer why questions; present ideas in discussion; teacher revoices/confirms/asks questions (**D**)

For examples, we are interested in their selection and sequencing and how these accumulate within and across episodes. We draw on the work of Watson and Mason (2006), who in turn draw on Marton and Tsui (2004) to describe movement towards generality across a sequence of examples. If a set of examples brings attention to *similarity* across examples, and so to that which is invariant, this offers opportunity to generalise. If a set of examples brings attention *contrast*, and so what something is in relation to what it is not, or to a different class, opportunities are made available to recognise boundaries between classes of examples, and so further generalise. When more than one aspect of an object of learning are *fused,* with simultaneous variance/invariace across an example set, generalisation is further enhanced.

Of course, in school mathematics lessons, an object of learning could also be a particular procedure. While in many cases, the same procedure would be carried out on different examples (say of solving linear equations), there are also possibilities for a lesson to focus on different procedures or strategies for finding the solution to the same equation. Here, similarity is in the invariance in the solution or answer. So too with different strategies for solving one complex problem. We thus extended our notion of exemplification to include particular procedures or strategies for solving a problem if it is these in focus in the lesson.

Of course, examples are embedded in tasks and while I do not elaborate this here, as will be seen below, these vary in cognitive demand, and this is important to capture across a lesson. Moreover, examples and tasks don't speak for themselves, hence our attention to the accompanying explanatory talk. I focus on what we refer to as legitimating criteria within the classroom, and particularly the teacher's talk.

In previous research on pedagogic discourse in teacher education (e.g. Adler & Davis, 2011), we identified different domains of knowledge appealed to so as to legitimate what counted as mathematics: the domain of mathematics itself; non-mathematical domains e.g. everyday knowledge; the curriculum; and the authority of the teacher. Work in school classrooms led to elaboration of the mathematical domain, and distinctions between criteria related to *properties of mathematical objects*, to *accepted conventions* and *derived procedures*, to *instances or empirical cases*, and then to the *general case or proof* (Adler, 2012a). As we worked with the WMCS data, we maintained some and then elaborated other distinctions: criteria of what counts (or not) as mathematical that are *particular or localized* (e.g. a specific or single case, an established shortcut, or a convention) from those that have *generality* (e.g. equivalent representation, definition, previously established generalization; principles, structures, properties) and also further distinguished *partial* from *full generality*. We remained interested in *non-mathematical criteria*, *everyday knowledge or experience* (e.g. the shape of an open crocodile's mouth as determinant of the direction of the inequality sign), *visual cues* as to how a step, answer or process 'looks' (e.g. a 'smile' as indicating a parabola graph with a minimum); or *memory devices* that aid recall (e.g. FOIL); or when what counts is simply stated, thus *assigning authority to the position* of the speaker, typically the teacher.

These varying criteria open or close opportunities for learning. At one extreme are legitimations based on principles of mathematics, with varying degrees of generality, and possibilities for learners to reproduce or reformulate what they have learned in similar and different settings. At the other end are appeals to the authority of the teacher or visual cues that produce a dependency on the teacher, on memory (this is what you must do); or on how things 'look'. Such imitations, while a necessary part of learning (Sfard, 2008; Vygotsky, 1978), cannot be the endpoint of learning. The criteria for what counts as mathematics that emerge over time in a lesson are key to being able to describe what is made available to learn.

Once we have categorised each episode, we then need to make judgments as to how the range of examples and legitimating criteria for mathematics accumulate through the lesson. Table 2 summarises our summative judgments within categories and *across a lesson* in terms of levels for examples and explanatory talk. The assignment of a level in our analysis is an interpretive judgment, reflecting our privileging of generality through exemplification and principled criteria as these unfold over a lesson. However, these are ultimate goals, and there is fluid movement between categories. For example, a summative judgment as a 'higher' level of legitimation depends on movement across non-mathematical, local and more general criteria in the lesson, as elaborated in Adler and Ronda (2015).

With these categories and levels of judgment we are able to look across teachers who have participated in our TM course. We were interested to see whether strengthening teachers mathematical knowledge for teaching, particularly through the course and so away from the school, and with minimal in school support through initial lesson study work, correlated with shifts in their mathematics teaching in what ways and how. We have studied pre and post lessons of ten TM teachers who completed the course in 2012. Table 3 presents the summary of our analysis across the ten teachers, after analysis and categorisation of episodes in each lesson, and then a summative judgment using of each category.

Detailed presentation of this data and its analysis is in Adler and Ronda (in preparation) where we also explain the sample of the ten teachers. There were 18 teachers in the 2012 cohort, and for various reasons were not able to follow all through to 2013. For example, some did not complete the course; others had moved schools and were no longer accessible.

Included in the participation in the TM course was an entry test—a mathematics pre-test, used to discern what strength of mathematical knowledge each was bringing, and so orient our initial sessions. We also used this to counsel teachers at the extreme ends of attainment out of participation. Teachers were also tested at the end of the course. Testing teachers is a contested practices outside of formal course work and we have discussed and reflected on our usage of this elsewhere (Adler, 2012b). As it turned out, the ten teachers remaining in our video study formed an interesting bifurcation. The first five teachers entered the course with relatively low attainment in the initial test, and while the post test showed improvement for some of these, the scores obtained were below the threshold we had set as a marker of competence with at least Grade 10 level mathematics. The second set of five

Table 2 Summative judgments for interpreting examples and explanatory talk extracted from Adler & Ronda, in Adler and Sfard (2017)

Examples	Naming	Legitimating criteria
The set of examples provide opportunities in the lesson for learners to experience: **Level 1**: one form of variation i.e. similarity or contrast **Level 2**: at least two forms of variation: **S and** S OR **S and C** **Level 3**: simultaneous variation (fusion) of more than one aspect of the object of learning and connected with similarity and contrast within the example set. (**S, C, F**) **Level 0**: simultaneous variation with no attention to similarity and/or contrast	Use of colloquial and mathematical words within and across episodes is: **Level 1**: talk is *colloquial or non-mathematical* (**NM**) e.g. everyday language and/or ambiguous pronouns such as this, that, thing, to refer to what is being written or pointed at; *where Mathematical words are used, these are as names labels or to read a string of symbols* (**Ms**) **Level 2**: movement between **NM** and (**Ms**) and some *mathematical language used appropriately (Ma)* to refer to other words, symbols, images, procedures **Level 3**: movement between colloquial **NM** and formal math talk **Ma**	Criteria for what counts as mathematics that emerge over time in a lesson and provide opportunity for learning geared towards scientific concepts **Level 0**: all criteria are *non mathematical (NM)* and so either *Visual* (**V**)—e.g. cues are iconic or mnemonic; or *Positional (P)*—e.g. a statement or assertion, typically by the teacher, as if 'fact' or *Everyday* (**E**) **Level 1**: criteria include *Local* (**L**) e.g. a specific or single case (real-life or math), established shortcut, or convention **Level 2**: criteria extend beyond non mathematical and L to include Generality, but this is partial **GP** **Level 3**: GF math legitimation of a concept or procedure is principled and/or derived/proved

Table 3 MDI of ten teachers in 2012 and 2013

Trs	Exemplifying				Explanatory talk				Learner P'cipation	
	Examples		Tasks		Naming		Legitimating			
Year	2012	2013	2012	2013	2012	2013	2012	2013	2012	2013
1	L1	**L1**	L1	L2	L2	L2	L0	L0	L2	L1
2	L2	**L3**	L2–L1	L2–L1	L2	L2	L0	L0	L1	L1
3	L2	**L1**	L1	L1	L2	L2	L0	L0	L1	L1
4	L1	**L3**	L1	L2–L1	L2	L2	L1	L1	L1	L1
5	L1	**L3**	L2–L1	L2–L1	L2	L2	L0	L1	L1	L1
6	L1	**L3**	L1	L2–L1	L2	L3	L0	L2	L2	L1
7	L1	**L3**	L2–L1	L2–L1	L2	L2	L2	L2	L2	L1
8	L2	**L2**	L2–L1	L1	L2	L3	L1	L3	L2	L1
9	L2	**L3**	L2	L2–L1	L2	L2	L0?	L3	L3	L3
10	L2	**L3**	L2–L1	L2	L2	L2	L1	L1	L2	L3

teachers (so teachers 6–10 in Table 3), all exited the course with relatively high attainment, with most starting with better scores than teachers 1–5.

What we see if we focus in on the examples column, is that there was improvement in the example sets of most teachers. All ten teachers' example sets in their 2012 lesson were judged to be either at levels 1 or 2. In 2013 for seven teachers, this had shifted from either level 1 or 2 to level 3. We interpret this as a general shift across teachers—that following their participation in the course, their selection and sequencing of examples in their lessons indicated greater opportunity for learning in terms of our criteria of privileging moves towards generality. Looking down the two columns on legitimating talk, we see a different pattern. Firstly in six of the 10 teachers' lessons, judgement of how mathematics was legitimated remained at level 0, i.e. justifications for procedural steps and or definitions of concepts were non-mathematical: they were either assertions by the teacher, related to everyday knowledge or reliant on visual cues. Three of the first group of teachers' lessons did not progress beyond this in 2013. All five of the second group of teachers lessons were judged as level 2 or 3, and so more principled and moving to greater generality.

Notwithstanding that this paper does not provide access to the *how* of our categorising of mathematical episodes, nor judgment of levels (readers are referred to other papers referenced above), the point here it that it indicates the power of the MDI framework. By focusing our attention on examples and legitimations I have illustrated that, and to a lesser extent how, the framework disaggregates mediational means. This enables nuanced interpretations of shifts in practice—or what we refer to elsewhere as 'take-up' from the WMCS PD programme (Ntow & Adler, under review). I remind readers that had we used the MQI framework, for example, the differences between our ten teachers would not have been as visible. Moreover, we can describe practices and shifts in these in terms of what is present, and what it is that can and does shift, even if at surface level, the lessons might appear similar.

It is instructive for us that most teachers in this study expanded their example sets. Our work in lesson study, and further classroom observations we have done confirm this finding. A focus on selection and sequencing of examples, using concepts and ideas from variation theory, and the notion of variation amidst invariance speaks to teachers, and in ways that they can begin to select examples for their lessons differently and more deliberately. In the context of the incoherence we observed in our initial year in the project, this is indeed progress, and also indicative of developmental activities with teachers that have impact. At the same time however, we confront the difficulty most teachers have in developing and strengthening their mathematical talk, and in providing opportunities for their students to appreciate mathematical justification.

And it is here that the bifurcation across the sample of teachers provides for two hypotheses that require further study, but begin to suggest important research on professional development in our field. Working from the easily agreed assumption that different teachers will benefit or learn differently in the same PD offering, we can now hypothesise that for some teachers particular PD is perhaps not beneficial at all. Our experience to date is that teachers whose mathematics is very weak (and this is not uncommon in South Africa in cases where teachers are teaching out of field and at levels beyond their training), do not benefit from the ways in which TM is structured and offered. It begins at a level that thwarts traction for some teachers. We could hypothesise then that such might be the case in other forms of PD. Reports on research on professional development in mathematics rarely point to instances of 'failure' in terms of teachers' learning, nor do they suggest how lack of 'take-up' in practice might correlate with, for example, aspects of mathematical knowledge for teaching.

I hope to have at least piqued your interest in the analytic power of MDI, particularly as a tool for research geared towards describing and interpreting differences and so shifts in practice within and across teachers' lessons. In the papers mentioned above, we have also argued that because MDI is grounded in the realities of our classrooms, and because we have identified levels indicating improvement, albeit within a particular view of school mathematics practice, the descriptions and interpretations we produce are not only responsive, but also responsible and developmental. Staying with a focus on exemplification (and examples within this) and explanatory talk (and legitimating criteria within this), I now move on to illustrations of its use in our PD practice, and so its use across practices.

From MDI for Study of Teaching to MDI for Work on Teaching

I begin with the TM course and select two sessions where the focus was on inequality relations, relations that we have come to understand are generally weakly understood. I use these to illustrate how the elements of MDI inform and are informed by our teaching in the PD. Our concern in the TM course sessions is that participating teachers are provided opportunity to strengthen their mathematical knowledge for teaching. We work on deepening their understanding of concepts, through building their understanding and communication of generality, their appreciation of mathematical structure, as well as their operational or procedural fluency with respect to key concepts or ideas in the secondary curriculum, inequality relations being one of them. To do this we pay deliberate attention to our selection of examples, the tasks these are embedded in, the range of representations we wish teachers to be familiar with, and then how these ideas are talked about, mathematically—the words used, and justifications elaborated.

The first set of example in Fig. 2 are numerical inequality relations, embedded in a task requiring teachers to state whether the inequality was true of false, and then justify their statements. Each of the examples was presented on a separate card. There are important things to notice in the selection and sequencing of these examples where the move first is from 3 to 10 both positive integers to -3 and -10. This variation enables a focus of attention (of course the lecturer has to draw attention to this) on the changing signs, and what this means for the inequality

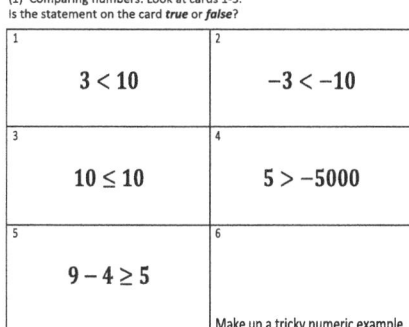

Fig. 2 Extract from TM1 course notes—Session 1 on inequalities

relation. The next card varies one number so that the two numbers related are the same, and the inequality sign is less than or equal to. The remaining cards vary the inequality sign (it is now greater than) as well as the numbers related and their signs. Cards 7–11 introduce a variable, and quadratic inequalities, with the task now requiring teachers to state whether the statements are always, sometimes or never true. The set of examples and their sequencing are focused on properties of the square of any number and its relation to zero, with all examples being of squares in different forms, but the inequality sign changing, as well as the relation to 0. In each set there is also an 'empty' card (cards 6 and 12) for teachers to contruct their own examples, and so generate relations that the lecturer can use to further exemplify and extend the example set in the session.[1]

The task is not only one of recognition of the relation, but significantly of how to justify this recognition. In this teachers are provided the opportunity build their substantiations and justifications, and mediation in this session by the lecturer (typically by offering counter examples) is on assisting teachers to distinguish partial from full justifications, and how these are expressed (for example, if a justification is based on particular numbers).

The power of the framework in our teaching is that it has enabled us to be deliberate in our work, firstly clarfiying for ourselves our 'objects of learning', and thus what it is we wish to bring into focus with the teachers. Secondly we then attend to the selection of tasks and examples that would best meet our goal. Thirdly we pay attention to what word use entails in a mathematical justification of the tasks set, and how to work on these with teachers, bringing to the fore the mathematical principles at work.

---

[1]Our card activities have been inspired by the work of Malcolm Swann in the UK.

How then might teachers come to use MDI in their own practice? In iterations of the TM course in 2014 and now 2016, we have integrated the mediation of MDI as a teaching tool within the course, as not all participating teachers have been able to participate in our Lesson Study work. We are also interested to see whether and how this kind of mediation supports the planning and reflection on lessons without the intensity of LS cycles. I move on now to share an example of our use of MDI as a resource for planning and reflecting on teaching in one of our LS cycles.

Doing Lesson Study

As in our teaching the TM course, here too the tool crosses the boundary from research practice, where operationalisation of constructs and their indicators is critical, to more flexible use, bent to the practices in PD and school classrooms. Our LS is driven by the same goals as other LS work were teachers work collaboratively on their practice (e.g. Fernandez, 2002). There are, however, significant adaptations in our LS work, a function of the contexts and constraints of the WMCS project resources. We work with teachers from a cluster of schools, who come together one afternoon a week for three consecutive weeks, in one school. The teachers are thus working with their own or their colleagues' students. Similar to other LS, in week 1 the teachers together with a researcher from the project plan a lesson on a topic selected by the teachers. In week 2 one teacher teaches a class that remains after school hours for this, and others observe. After the lesson, the students leave and reflection and replanning follows. In week 3, a different teacher teaches a different class, and this too is followed by reflection and replanning and the resulting lesson plan shared.

The lesson planning, and its reflection is structured by MDI, albeit in a different format that has developed with teachers through its use in LS, and referred to as the WMCS Mathematics Teaching Framework. The four elements of MDI (object of learning, exemplification, explanatory talk and learner participation) are visible, with questions to assist planning and reflection. Our intention is that in our collaborative work with teachers, we have a shared language with which to talk about and reflect on practice, and again in ways that is sufficiently close to their daily work (Figs. 3 and 4).

The lesson plan above was developed in a LS cycle where the selected topic was the Hyperbola graph (taught in Grade 10 in our schools), with particular focus on the parameters a and q in the general equation $y = \frac{a}{x} + q$.

Adler and Ronda (2017a) provide a full description of this lesson, and how it informs and is informed by MDI. It was the second lesson in the cycle, and thus the plan below is a revised version of the first lesson, following reflection on the first lesson where the range of examples, their representational forms and the task

Lesson goal: What do we want learners to know and be able to do?		
Exemplification	**Explanatory communication**	**Learner Participation**
Examples, tasks and representations What examples are used? What are the associated tasks? What representations are used?	Word use and justifications What is said? What is written? How is it justified?	Doing maths and talking maths What do learners say? What do learners write?
Building generality Structure Variation amidst invariance	Informal – formal Mathematical substantiations Principles	Does learner activity build towards the lesson goal?
Coherence and connections: Are there **coherent connections** between • the lesson goal, examples, tasks, explanations and learner participation? • from one part of the lesson to the next		

Fig. 3 The WMCS mathematics teaching framework

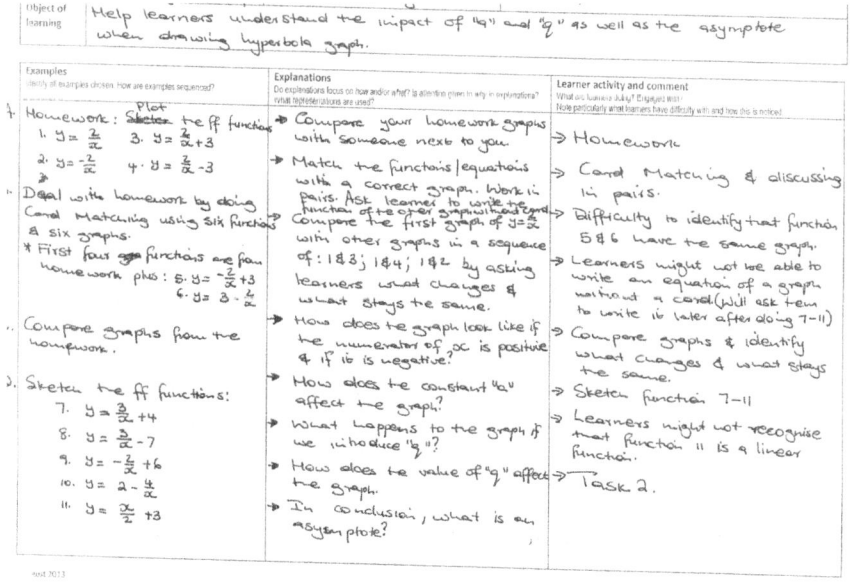

Fig. 4 An example of a lesson plan developed for a LS cycle

demands, together with attention to language use had been discussion. Notice first examples 1–6 in column 1 where $\frac{2}{x}$ and 3 remain invariant, but signs and operations are varied, as is the order of terms in example 6. The intention of focusing learners' attention on a and q through these variations is clear in the teacher's elaboration of the explanation and learner participation columns. Her goal for students to articulate how the graphs and equations are the same and different, and so build their explanatory communication, is visible in the middle column.

Our structuring of LS by MDI is instructive and productive, and we are currently in the process of describing our experience and reporting our results in relation to how LS provides a learning space for all participants (the teachers, their students and ourselves as researchers) (Alshwaikh & Adler, 2017a), and one too where conflicts arise, within and between teachers and researchers in terms of simultaneous commitments to the LS process, teaching and student needs (Alshwaikh & Adler, 2017b).

MDI—Its Role and Nature as a Boundary Object

In earlier work on MDI (focused mainly on examples and explanations) we described how we were using this across "researcher and practitioner communities" (Venkat & Adler, 2013).[2] We noted our concern to engage with and impact on teachers' 'common-sense discourse' (Brown (1997), cited in Hargreaves (1999)) and thus created "research-based artifacts and designed specifically for trialing in the overlapping 'boundary' region of the communities of research and classroom practice" (Venkat & Adler, 2013, p. 3). Star and Griesemer (1989) refer to such artifacts as 'boundary objects' that are "plastic enough to adapt to local needs and constraints of the several parties employing them, yet robust enough to maintain a common identity across sites. They are weakly structured in common use, and become strongly structured in individual site use.' (p. 393). In this paper I have shown how we have used the MDI framework across research on teaching, and in the practice of PD and classroom mathematics teaching. I have argued that its 'strong structuring' as a research tool has enabled nuanced disaggregation of elements of practice, and a developmental trajectory for working on teaching. I hope to have also illustrated that as a boundary object, it is iterative in nature, and flexible (strong yet bending). MDI is a living framework that is simultaneously unifying and differentiating in our project. Hence our view that it is powerful.

Concluding Comments

It is important to end this paper by returning to the context of MDI and its roots. This relates back to introductory comments about educational outcomes and poverty and the significance of instruction in disadvantaged communities. We thus

---

[2]Hamsa Venkat directs a project similar to WMCS at the primary level, WMCP. Over time our frameworks, while sharing initial orientations to mediation and sociocultural theory, have come to differ. To assist in distinguishing these, the WMCP framework is called Mediating Primary Mathematics (MPM), and is elaborated by Venkat & Askew (see Venkat & Askew, under review). Our collective work up ahead will engage with similarities and differences between this and MDI as it has come to function in WMCS.

understand that MDI is deeply implicated in, but only a part of a set of practices and conditions that produce poor performance across our schools. It matters deeply, how **mathematical discourse in instruction** supports (or not) mathematical learning. And just as context has been pivotal in the emergence and construction of MDI so too is it important to conclude by stating (the obvious) that MDI is an inherently social product, a function of where you work, with whom and on what. MDI has been shaped by the context of its emergence. It has also been shaped by and hopefully will shape the field of (mathematics) education research. Critically, it has been developed through ongoing interaction with colleagues, postdoctoral fellows and doctoral students. Key collaborators are visible in the various references to our ongoing work.

Acknowledgements This work is based on the research supported by the South African Chairs Initiative of the Department of Science and Technology, and National Research Foundation of South Africa (grant no. 71218). Any opinion, finding and conclusion or recommendation expressed in this material is that of the authors and the National Research Foundation does not accept any liability in this regard.

References

Adler, J. (2012a). Knowledge resources in and for school mathematics teaching. In G. Gueudet, B. Pepin, & L. Trouche (Eds.), *From Text to 'Lived' Resources* (Vol. 7, pp. 3–22): Springer Netherlands.

Adler, J. (2012b). *The research and development curve continued: Report of the wits FRF mathematics chair and the wits maths connect—Secondary (WMCS) project.* Unpublished report. University of the Witwatersrand. Johannesburg.

Adler, J., & Davis, Z. (2011). Modelling teaching in mathematics teacher education and the constitution of mathematics for teaching. In T. Rowland & K. Ruthven (Eds.), *Mathematical knowledge in teaching* (Vol. 50). Dordrecht: Springer.

Adler, J., & Pillay, V. (2017). Mathematics education in South Africa. In J. Adler & A. Sfard (Eds.), *Research for educational change: Transforming researchers' insights into improvement in mathematics teaching and learning* (pp. 9–24). Abingdon: Routledge.

Adler, J., & Ronda, E. (2015). A framework for describing mathematics discourse in instruction and interpreting differences in teaching. *African Journal of Research in Mathematics, Science and Technology Education.* doi:10.1080/10288457.2015.1089677

Adler, J., & Ronda, E. (2017a). A lesson to learn from: From research insights to teaching a lesson. In J. Adler & A. Sfard (Eds.), *Research for educational change: Transforming researchers' insights into improvement in mathematics teaching and learning* (pp. 133–143). Abingdon: Routledge.

Adler, J., & Ronda, E. (2017b). Mathematical discourse in instruction matters. In J. Adler & A. Sfard (Eds.), *Research for educational change: Transforming researchers' insights into improvement in mathematics teaching and learning* (pp. 64–81). Abingdon: Routledge.

Adler, J., & Ronda, E. (in preparation). Take-up and tools: Describing and interpreting developmental shifts in mathematics teaching.

Adler, J., & Sfard, A. (2017). (Eds.) *Research for educational change: Transforming researchers' insights into improvement in mathematics teaching and learning.* Abingdon: Routledge.

Adler, J., & Venkat, H. (2014). Teachers' mathematical discourse in instruction: Focus on examples and explanations. In M. Rollnick, H. Venkat, J. Loughran, & M. Askew (Eds.), *Exploring content knowledge for teaching science and mathematics* (pp. 132–146). London: Routledge.

Alshwaikh, J., & Adler, J. (2017a). *Researchers and teachers as learners in Lesson Study*. Mathematics and Technology Education (SAARMSTE), Bloemfontein, South Africa: Paper presented at the Southern African Association for Research in Science.

Alshwaikh, J., & Adler, J. (2017b). *Tensions and dilemmas as source of coherence*. Paper presented at the Mathematics Education and Society (MES) Conference, Volos, Greece.

Artigue, M. (2009). *The challenges of mathematics and science basic education*. Paris: UNESCO.

Brown, S. (1997). Respondent comment. In S. Hegarty (Ed.), *The role of research in mature education systems*. Slough: National Foundation for Educational Research.

Fernandez, C. (2002). Learning from Japanese approaches to professional development: The case of lesson study. *Journal of Teacher Education, 53*(5), 393–405. doi:10.1177/002248702237394

Gamoran, A., & Long, D. A. (2006). *Equality of educational opportunity: A 40-year retrospective. Working Paper 2006-9*. Wisconsin Center for Education Research (WCER) Madison, WI.

Guthrie, G. (2011). *The progressive fallacy in developing countries. In favour of formalism*. New York: Springer.

Hargreaves, D. (1999). Revitalising educational research: Lessons from the past and proposals for the future. *Cambridge Journal of Education, 29*, 239–249.

LMTP. (2011). Measuring the mathematical quality of instruction. *Journal of Mathematics Teacher Education, 14*, 25–47. doi:10.1007/s10857-010-9140-1

Marton, F., & Tsui, A. B. M. (2004). *Classroom discourse and the space of learning*. New Jersey: Lawrence Erlbaum Associates.

Morrow, W. (2007). *Learning to teach in South Africa*. Pretoria: Human Science Research Council.

Ntow, F., & Adler, J. (under review). "We have to work": Opportunities and challenges in a teacher's implementation of professional development practices.

Pournara, C., Hodgen, J., Adler, J., & Pillay, V. (2015). Can improving teachers' knowledge of mathematics lead to gains in learners' attainment in mathematics? *South African Journal of Education, 35*(3), 10. doi:10.15700/saje.v35n3a1083

Reddy, V., Visser, M., Winnaar, L., Arends, F., Juan, A., Prinsloo, C. H., & Isdale, K. (2016). *TIMSS 2015: Highlights of mathematics and science achievement of grade 9 South African learners*. Retrieved from Pretoria: http://www.timss-sa.org.za/download/TIMSS-Grade-9-highlights.pdf

Sfard, A. (2008). *Thinking as communicating human development, the growth of discourses, and mathematizing*. New York: Cambridge University Press.

Shalem, Y., & Hoadley, U. (2009). The dual economy of schooling and teacher morale in South Africa. *International Studies in Sociology of Education, 19*(2), 119–134. doi:10.1080/09620210903257224

Skemp, R. R. (1987). Relational understanding and instrumental understanding. *Mathematics Teaching in the Middle School, 77*, 20–26.

Skovsmose, O. (2011). *An invitation to critical mathematics education*. Dordrecht: Springer.

Star, S. L., & Griesemer, J. (1989). Institutional ecology, 'translations' and boundary objects: amateurs and professionals in Berkeley's museum of vertebrate zoology, 1907–1939. *Social Studies of Science, 19*, 387–420.

Venkat, H., & Adler, J. (2012). Coherence and connections in teachers' mathematical discourses in instruction. *Pythagoras, 33*(3). doi:10.4102/pythagoras.v33i3.188

Venkat, H., & Adler, J. (2013). *From research to teacher development in mathematics education: Creating boundary objects*. Paper presented at the South African Education Research Association Conference.

Venkat, H., & Askew, M. (under review). Mediating primary mathematics: theory, concepts and a framework for studying practice.

Vygotsky, L. S. (1978). *Mind in society: The development of higher psychological processes.* Cambridge: Harvard University Press.

Watson, A., & Mason, J. (2006). Seeing an exercise as a single mathematical object: Using variation to structure sense-making. *Mathematical Thinking and Learning, 8*(2), 91–111.

Open Access Except where otherwise noted, this chapter is licensed under a Creative Commons Attribution 4.0 International License. To view a copy of this license, visit http://creativecommons.org/licenses/by/4.0/.

The Challenging Relationship Between Fundamental Research and Action in Mathematics Education

Michèle Artigue

Abstract In this text, associated with my Felix Klein Medal awardee lecture at ICME-13, I develop a reflection on the relationships between fundamental research and action in mathematics education. This reflection is based on my experience as a teacher, teacher educator, and researcher and on what I learned from the responsibilities I had on the ICMI Executive Committee. Using as a filter the concept of didactical engineering, I address several issues: reproducibility, generalization, theoretical diversity, and values, that contribute to making these relationships especially challenging in mathematics education and point out promising evolutions in the field.

Introduction

The Felix Klein Medal awards lifetime achievement in mathematics education research, but what exactly is a lifetime achievement in this field? Different answers that express differences in personal visions can certainly be proposed. The theme I selected for my awardee lecture at ICME-13 expresses my personal vision that the field of mathematics education, even when seen as a field for fundamental research, as is the case in the French didactic culture I belong to, does not develop as a field of pure knowledge. Those who have for decades engaged in didactic research have done so with the desire that their research makes it possible to improve, ultimately, the teaching and learning of mathematics. They have held institutional positions that in fact forced them to combine research and action, and their engagement in action has nurtured their research achievements.

Felix Klein award.

M. Artigue (✉)
LDAR, Université Paris Diderot - Paris 7, Paris, France
e-mail: michele.artigue@univ-paris-diderot.fr; michele.artigue@gmail.com

In this text for the proceedings, my intention is thus to contribute to the reflection on the relationship between fundamental research and action in our field. I do so not only by looking back at my personal experience as a researcher, teacher, and teacher educator, but also by relying on what I learned from the responsibilities that I have assumed in connection with ICMI. I use the concept of didactical engineering that I contributed to establishing (Artigue, 1989, 2014) as a specific filter, reflecting on both its possibilities and limitations in acting as a bridge between fundamental research and action.

A Vision of Relationships Between Research and Action Emerging from a Particular Culture

The field of mathematics education has not developed in the same way in all countries and cultures, which has an impact on how relationships between research and action are viewed. In a recent text regarding mathematics education research in Japan, for instance, Isoda (2015) showed the crucial role played in its emergence and development by the practice of lesson studies, a practice established more than one century ago. Even within Europe, the comparison of the didactic traditions of Germany, Italy, Netherlands, and France, which has been part of the thematic afternoon at ICME-13, has shown differences in that respect. In Germany, for instance, influential researchers such as Wittmann began early to promote a vision of mathematics education as a design science (Wittmann, 1998). In the Netherlands, the development of Realistic Mathematics Education from the seminal ideas of Freudenthal has also been tightly connected with design (Van den Heuvel-Panhuizen & Drijvers, 2014). In Italy, research emerged from a long tradition of pragmatic action-research collaboratively carried out by mathematicians interested in education and teachers, before consolidating within a paradigm of Research for Innovation leading to the development of specific theoretical frames and constructs (Arzarello & Bartolini Bussi, 1998). In France, there is also a long tradition of reflection on mathematics education issues, and famous mathematicians have contributed to it. However, when mathematics education emerged as a research field in the late 60s in the context of New Math, it was with the clear awareness that responsible action required much more knowledge of teaching and learning processes than existed at that time. The disillusions generated by the New Math reform quickly made it clear that "successful"[1] reforms need more than mathematical and pedagogical visions, even when these have solid epistemological foundations. Thus the conviction strongly expressed by the two fathers of the French didactics of mathematics, Brousseau (originally a primary mathematics

---

[1] I have used quotation marks because the idea of success is always relative to a set of values and aims; these can be influenced by research results but are situated out of range of scientific validation.

teacher) and Vergnaud (a psychologist having had Piaget as Ph.D. supervisor), that a genuine field of research had to develop with both fundamental and applied dimensions. While maintaining a strong connection with mathematics as a discipline and relying on the affordances of psychology, especially the Piagetian constructivist epistemology of that time, they felt that this field should develop its own problématiques, methodologies, and theoretical constructions. Priority had to be given to understanding the functioning of didactic systems, the classroom being the prototype of such systems, over action. Similar to other scientific fields, this field should by no means be a normative or prescriptive field.[2] As a researcher, there is no doubt that I have been influenced by this vision, as well as by the importance it attaches to the specificity of mathematics as a discipline and to epistemological issues. It has definitively influenced my vision and experience of relationships between fundamental research and action.

Giving priority to the understanding and building of the didactics of mathematics as a genuine scientific field means neither that research is free from values nor that it does not have a transformative aim. It simply acknowledges that action on educational systems, even based on estimable values, that is not based on appropriate knowledge, is at least risky. Brousseau, himself an elementary teacher, while interested in Piagetian epistemology, soon understood the limitation of it as a base for didactic action, and this was the source of the project he began in the 60s of developing an experimental epistemology of mathematics education, a project that would lead to the theory of didactical situations (TDS) (Brousseau, 1997).[3] Developing the kind of knowledge needed requires appropriate structures. Brousseau created the Centre d'Observation et de Recherche sur l'Enseignement des Mathématiques (COREM) in Bordeaux, a very innovative structure to which was attached an elementary school, the École Michelet, which for 25 years was a tool of inestimable value for researchers.

As I have explained elsewhere (Artigue, 2016), my first didactic research experience in the 1970s took place in the context of an experimental primary school attached to the recently created Institut de Recherche sur l'Enseignement des Mathématiques (IREM) in Paris, which shared some characteristics with the École Michelet, and this experience was very rewarding. It was in fact a mixture of research and local action. Two IREM colleagues and I had a great deal of freedom over several years to organize the teaching of mathematics with the teachers of the school, under the condition that pupils had covered at least the content of the official syllabus by the end of primary school. A lot of time was devoted to designing teaching situations jointly with teachers and observing and analyzing their implementation. The situations created by Brousseau for the École Michelet were a

---

[2]As shown by a recent inquiry launched by Gascón and Nicolas, this vision of the field, while in line with Weber's vision of science, is not universally shared among researchers in the field (Gascón & Nicolas, to appear).

[3]For a detailed explanation about how this occurred, one can read Brousseau, Brousseau, and Warfield (2014), Chapter 4, or the long interview of Brousseau realized for the ICME-13 thematic afternoon, accessible at http://www.cfem.asso.fr/cfem/ICME-13-didactique-francaise.

constant source of inspiration. It was in this context that I experienced the power of constructions such as those already offered by the TDS. This was a fascinating experience, and the young and enthusiastic scholar I was at that time was convinced that the knowledge gained through didactic research would change the face of mathematics education. I did not suspect that generalization of local achievements would be so problematic; I was unconscious of the networks of constraints conditioning the life of ordinary didactic systems. In the experimental school, we did not hesitate to free ourselves from these constraints, with the support of teachers and parents, thanks to the indisputable legitimacy IREM gave to our action.

The Fundamental Role of Didactical Engineering

As explained above, the vision of French didactics has been a systemic one since its origin. Research methodologies had to reflect this systemic view. This was the source of the concept of didactical engineering that I contributed to establishing. This concept emerged very early, and the name itself was introduced by Brousseau who had heard about the existence of didactic engineers in Québec. As explained by Chevallard in a seminal text written for the second Summer School of Didactics of Mathematics (Chevallard, 1982)—the first time didactical engineering was collectively discussed by the French didactic community—didactic research needed to create something comparable to the *clinique* in medicine, obliging researchers to access the intimacy of the systems they were studying and making it possible to produce and reproduce phenomena. As stressed by Chevallard, this vision of didactical engineering was in line with the vision of science as a phenomenotechnique developed by Bachelard.

Such a vision of didactical engineering is substantially different from the vision underlying the concept of design research, which has had increasing influence in mathematics education, despite evident similarities. The design work that accompanies the use of didactical engineering has been primarily put in service of understanding the economy and ecology of didactical systems, producing didactic phenomena, and establishing existence theorems. This is how Brousseau himself conceives of his long-term research on the extension of the number field to rational and decimal numbers, which has played an important role in the development of the theory of didactical situations and has been reproduced more than 25 times. In his retrospective book he writes:

The initial objective of the experiment was thus an attempt to establish an "existence theorem":

- Would it be possible to produce and discuss such a process (a constructivist process making minimal use of pieces of knowledge imported by the teacher for reasons invisible to the students)?
- Would the students—all of the students—be able to engage in it?

- Could the request of the process be, for each of the students, a state of knowledge at least equal to that obtained by current, standard methods? (Brousseau et al., 2014, p. 129)

Brousseau is well aware of the complexity of this construction and of the expertise its implementation required from teachers.

> This curriculum was not made to be used in other classes. The sole purpose of reproducibility was to consolidate the scientific observations that we needed in order to test certain hypotheses. The lessons had above all the property of making apparent the enormous complexity of the act of teaching. (ibid., p. 7)

Once again, this does not mean that French didacticians were not aware of their social responsibility. In the text mentioned above, for instance, Chevallard writes that didactics will be judged

> on its ability to realize the knowledge it produces, its ambition to move towards practical and workable answers to the concrete difficulties identified by the practitioners of didactical systems; among the forms of action that relate most directly to its object (its problematics and methodology), that of producing lessons and sequences of lessons practically workable obviously holds a central place. (Chevallard, 1982, p. 30, our translation)

He acknowledges also the difficulty of the enterprise:

> From the didactic realization, as it takes place within the research process, to the production of sequences of lessons, there is all the distance of a true decontextualization, acting in several registers (epistemological, human, institutional, etc.) .... This situation thus leads to the issue of the "user guide" for such productions, that is to say, the problem of conditions for a non-denaturing didactic recontextualization which must guarantee its successful integration from epistemological, human, and institutional points of view—cultural in a single word—to the didactical engineering, beyond the scientific value of the research findings which constitute its raw material. (ibid., pp. 31–32, our translation)

As I wrote at the beginning of this text, considering the field of mathematics education a genuine scientific field and stressing the importance to be given to fundamental research in this field does not mean that it is the pure desire for knowledge that motivates didactic research, even in its most fundamental aspects. This motivation comes, in fact, from the ultimate desire of improving mathematics education through the knowledge gained, with a diversity of possible views regarding what improvement means. This is clearly a leitmotiv in my research work, even if this aim is pursued through a diversity of forms of research. It partly explains my privileged use of didactical engineering as a research methodology. Beyond the identification of didactic phenomena and laws, I always have seen in this methodology a means:

- to explore forms of life of mathematical and didactical practices that could not be observed in ordinary classrooms, but seemed to me more satisfactory from an epistemological perspective;
- to study the conditions and constraints influencing their economy and ecology; and
- to understand what should be done in order to help them grow and expand.

An Example: Didactical Engineering for the Teaching of Differential Equations

For instance, when I began to work on the teaching of differential equations in the 80s, as a mathematician I was working in the area of dynamical systems and experienced a kind of schizophrenia between my activity as a researcher and as a university teacher. The standard course on differential equations I was giving to second-year students was not especially problematic, but I was convinced that this course focusing on the algebraic solving of classical equations (in finite terms or using Taylor series) gave them a wrong idea of the field and of the important questions in it, both those internal to it and those resulting from its connection with other fields. With colleagues from the University of Lille 1, we decided to explore the accessibility of a first-year university course respectful of the epistemology of the field, combining thus algebraic, numerical, and qualitative approaches and incorporating modelling activities. Our hypothesis was that the affordances of technology made such a course accessible. The design of the course was based on careful preliminary analyses, combining epistemological, institutional, and cognitive dimensions according to the standards of didactical engineering. The first experimentation globally confirmed the accessibility of the course, while showing an important gap between the students' ability to analyze phase portraits of differential equations provided by the software used, or even predict phase portraits, and their ability to prove their conjectures. To ensure the viability of the qualitative approach, from the second experimentation, we introduced and legitimated specific forms of reasoning and proof combining institutionalized graphical notions (such as the notion of fence) and arguments with analytical formulations. This move was effective and the ecological viability of this construction was again confirmed in the third year, when the experimental section concentrated the students entering math university courses with low grades in mathematics at the scientific *baccalauréat* (Artigue, 1992; Artigue & Rogalski, 1990).

This research was certainly motivated by the desire for improving the actual teaching of differential equations in French universities. However, I have to acknowledge that, while being very well received, it had limited influence in France beyond the University of Lille 1 where the course was implemented for more than 10 years. The use of this didactical engineering was generally limited to the first situations of the qualitative approach. As explained in Artigue (2016), using the conceptual tools provided by didactic research, I am currently able to better explain why. As mentioned above, in order to ensure the viability of the qualitative study with first-year students, which meant that they would be able to prove conjectures, we were obliged to legitimate theorems and proofs combining graphical and analytical formulations and arguments. Due to my research expertise in this mathematical area, I could attest that mathematicians used these reasoning modes, even if in published papers they adopted a more formal discourse. This was an important ingredient for legitimation. Moreover, at the University of Lille 1, in the experimental section, important work was systematically carried out at the beginning of each academic year on the graphical register of representation in order to make it operational and change its status. However, this institutional situation was exceptional, and the legitimation of such arguments violated the rules of the didactic contract governing university courses in Analysis at that time. The contrast between the interest raised by this didactical engineering and its very limited impact clearly showed that its ecological viability depended on conditions regulating the teaching of Analysis and, more globally, the status given to graphical representations in mathematics teaching. These conditions were situated at higher levels of the "hierarchy of didactic codetermination" than the didactical engineering itself, as can be expressed today using a construct of the anthropological theory of the didactic (ATD) that did not exist at that time (Chevallard, 2002).

This example well illustrates the fact that the extension of any didactical engineering, beyond the research and ecologically protected environment where it has generally been developed and tested and its conversion from a research to a development object, must seriously take into account these different levels of conditions and constraints. Even when research allows us to understand the complex system of conditions and constraints that condition the ecological viability of a didactic construction, which is in itself an important research outcome, acting on such conditions and constraints is hardly in the hands of researchers. Action requires the building of new partnerships and collaborations beyond those at play in the joint development of didactical engineering at a research level. This is a reality to which all those today engaged in design research in mathematics education are sensitive, even if they do not use the same words to express this sensitiveness (see for instance Swan, 2014).

Before moving to the next point, I would like to point out that this example also shows the role played in research by values that, quite often, remain implicit. As explained above, this research emerged from the desire to better align the teaching of differential equations with the current epistemology of the field from the first contact with it at university. The fact that such a move constitutes an educational improvement was a non-questioned starting point of an epistemological nature. Beyond that, my didactic culture has made me especially sensitive to the optimization of the mathematical responsibility of the students in the design of situations, to the precise choice of their didactic variables and to the organization of the "adidactic milieu" with the meaning given to these terms in the TDS. One can see here the clear influence of the vision of learning in this theory; it combines adaptation and acculturation processes, but adaptation processes are given a fundamental role. Adopting TDS as a theoretical reference means that such vision and associated principles are accepted. The research carried out and the expression of its results are thus conditioned by these visions and principles, even if a number of results, for instance, those regarding the didactic contract at stake in the teaching of analysis about graphical representations and more globally those issued from the epistemological and institutional analyses or those regarding students' cognitive difficulties with qualitative proofs, have and have been proved to have more general value. However, I have to confess that this question was never addressed in the publications associated with this research.

Issues of Reproducibility

In the French didactic community, from the early 80s, didactical engineering developed thus as a methodological tool primarily at the service of research and not as a development tool. Development, in fact, was not an object of scientific inquiry as it can be in design research. This did not prevent the designs produced by research to migrate in the educational system through different channels. Researchers were members of curricular commissions; many developed their research in close connection with the IREM network and contributed to the activities of resource development and in-service teacher education that this network had in charge; others worked with primary and secondary teachers at the Institut National de Recherche Pédagogique (INRP), now Institut Français de l'Éducation (IFÉ), in research-action groups, for instance the group on primary education producing the ERMEL collection of teacher books that has been very influential in teacher education (see, for instance, ERMEL, 2005); and a few also co-authored textbooks. These conditions favored the percolation of knowledge, but this percolation process was not taken as an object of study. However, it soon became evident that the dissemination of research engineering designs through such channels in many cases systematically resulted in their denaturation, and this observation attracted my attention to issues of reproducibility. I made the hypothesis that one of the sources of the observed denaturation could be the vision of the reproducibility of the didactical situations conveyed, more or less explicitly, by didactic texts and educational resources. Roughly speaking, didactical situations were proposed as objects to be reproduced, with the implicit idea that following the proposed trajectories would result in the expected learning effects being obtained.

To test this hypothesis, I built a stochastic mathematical model of this vision. Using direct computations and complementing these by computer simulations using Monte-Carlo methods, I invalidated the model, thus invalidating the vision of reproducibility conveyed by the literature. More precisely, I showed that if such reproducibility was observed, it could not generally result from the reasons and characteristics of the design invoked. Other forces were at play whose action and mechanisms remained tacit. Using data coming from a previous research on primary students' conceptions of the circle, I showed that the model allowed researchers to expect the appearance of regularities, but, as is generally the case in non-linear dynamic systems, these regularities would be situated at structural levels other than those usually expected (Artigue, 1986). This led me to articulate a kind of principle of incertitude between the internal reproducibility aimed a priori (conserving the meaning of actions and discourses despite possible variations in the trajectories) and external reproducibility (at the more superficial level of classroom trajectories). According to this principle, any effort made to ensure external reproducibility has a systematic cost in terms of internal reproducibility. This result showed, for instance, that the phenomenon of obsolescence at play in the reproduction of the COREM didactical engineering on rational numbers by the same teachers, year after year, that had been identified by Brousseau some years earlier

(Brousseau, 1981) was an instance of a more global didactic phenomenon. This work of mathematization of the didactic field itself was not developed further, but it strongly influenced my conception of resource development, a crucial point as far as action is considered.

For instance, in the resources associated with the research on differential equations mentioned above, I tried to overcome the trap of linear descriptions and to open the dynamics of situations, envisaging, for instance, possible bifurcations. I also tried to approach more explicitly the key issue of the sharing of mathematical responsibility between teacher and students than was usual in classical engineering design at that time and whose underestimation appeared as a major source of denaturation.

The spontaneous conception of reproducibility was thus proven to be an obstacle to the dissemination of didactical designs coming from research and their productive use for action. I would not say that this didactic obstacle has been overcome. Many current educational resources still implicitly convey the same notion of reproducibility by giving the impression to the reader that classroom and individual trajectories can be fixed by a succession of tasks and questions, without damage. However, this misunderstanding about what can and cannot be reproduced with what consequences is only one of the many difficulties met in the transition from research to action.

Issues of Generalization

Establishing productive relationships between fundamental research and action obliges one to address the difficult issue of generalization. I have already evoked one case in which generalization was out of range under the current institutional conditions and constraints with the research on differential equations. However, understanding difficulties of generalization in the field both requires "vertical" analyses of conditions and constraints as the one I have sketched above, and "horizontal" analyses, according to the distinction we introduced with Winslow in our meta-analysis of comparative studies in mathematics education (Artigue & Winslow, 2010). The main reason is that mathematics education is a field geographically and culturally situated. As stressed in (Artigue, 2016), we all know today, even when we belong to dominant cultures—mine is certainly one of them in the field of mathematics education—how our insufficient sensitivity to the diversity of social and cultural contexts has been the source of hegemonic visions and abusive generalizations and exportations (see, for instance, Nebres's, 2008 contribution at the Symposium organized for celebrating the centennial of ICMI). This does not mean that didactic research does not identify regularities, such as didactic phenomena that transcend cultural specificities, for instance, the necessary existence of a didactic contract in any didactical situation, which others might call socio-mathematical norms; the specific economy and ecology of taught knowledge that regulates the processes of didactic transposition that cannot be reduced to a process of elementarization of knowledge; and the existence of epistemological obstacles to overcome, for instance, in the transition from whole numbers to rational and decimal numbers or in the learning of more advanced concepts such as the concept of limit, to give just a few examples. This means that even when there are regularities, didactic phenomena that to some extent transcend cultural specificities, the way the knowledge of these can be put at the service of action is highly dependent on the conditions and constraints of each specific context.

Another important point is that we hardly know the exact field of validity of the regularities we identify. Quite often, we tend to over-generalize regularities inferred from local studies without enough evidence. Again, this does not mean that local studies cannot give access to rather general phenomena and didactic laws. The very powerful concept of didactic contract, for instance, emerged from one of Brousseau's studies known as the Gael's case (Brousseau & Warfield, 1999). This only became a fundamental concept of the TDS, however, because it proved its capacity to make a diversity of students' and teachers' behaviors and interactions beyond this one case understandable and because this understanding and the associated technological discourse in the sense of ATD was able to find its place in the global theory of didactical situations.

Issues of Theoretical Diversity

I would like to come now to another crucial issue when thinking about the relationship between research and action: theoretical diversity. I became especially sensitive to this issue when I entered the ICMI Executive Committee in 1998. As I have explained elsewhere:

> Many times, in recent years, due to my ICMI responsibilities I have been confronted with questions about existing knowledge on particular educational issues that might inform teaching practices, curricular decisions, or teacher education. Faced with such questions, most often I was unable to give a clear answer, and often even unable to orient my interlocutor towards a set of references that would help her (him) develop a coherent and synthetic vision. Of course, things are not so simple in education as in mathematics. We must accept that most of the certainties we acquire are, except for the most general ones, situated both in time and space, and that it is difficult to know their exact domain of validity. The question of how research knowledge may inform practice in particular contexts is a difficult question, still insufficiently addressed. Nevertheless, the theoretical explosion of the field, the diversity of approaches, constructions, discourses, and the lack of connection substantially increases the difficulties of capitalization and dissemination (Artigue, 2016, p. 262).

In the last decade, I have been involved in different projects developed at the European level in order to address this issue in the framework of what is often known today as the "Networking of theories." These projects have also revealed to what extent theoretical diversity deeply permeates our research practices or, in ATD terms, our research praxeologies (Artigue & Bosch, 2014), making connection efforts directly situated at the level of theoretical objects hopeless. We certainly underestimated this point until recently. Personally, in the last decade, I have learnt the price to pay in order to overcome the current state, the necessary effort of decentration, and the uncompromising questioning required to understand the actual use we make of theoretical frameworks beyond their mere ritual invocation. I have also learned the necessity of developing specific devices that can allow us to take our research practices as objects of study without distorting them and the importance of developing metalanguages to support joint work and communication. One example is the metalanguage of key concerns, initially created in the Technology Enhanced Learning in Mathematics (TELMA) European team (Artigue, 2009), then refined in the project ReMath (Lagrange & Kynigos, 2014), which I also used as a guide when, together with Blomhøj, I investigated what the major didactic approaches have to offer to the conceptualization of inquiry-based learning in mathematics education (Blomhøj & Artigue, 2013).

Seen from the outside, such a form of research may be perceived as just theoretical and without possible practical interest. I would like to reaffirm here my conviction that this is not at all the case. Limiting the current fragmentation of the field and inventing forms of discourse that improve the quality of communication and support capitalization of knowledge is an absolute necessity for us if we want to be able to determine exactly what we know and what we do not know, as is legitimately expected from a mature research field, and if we want to create solid grounds for productive relationships between research and action.

Issues of Values

I have already briefly evoked this issue in a final comment regarding the research on differential equations, but this fundamental issue of values certainly needs more than a small comment. Mathematics education, for better or for worse, is a field in which science and values strongly intertwine. Some years ago, I was asked by UNESCO to pilot the realization of a document on the challenges in basic mathematics education (UNESCO, 2011). The group of experts involved agreed that the main challenge to be addressed was that of "quality mathematics education for all." However, coming to an agreement on what was the exact meaning that should be given to this commonly used expression was another story. We had long discussions that reflected differences in perceptions and values. Of course, these also had an impact on the vision we each had of the types of actions to be promoted in order to progress towards this goal.

Even within my own culture, even for theories with close epistemology, such as the theory of didactical situations and the anthropological theory of the didactic, there is no doubt that the forms of didactical engineering research developed are different. TDS relies on a constructivist vision of learning, which is not the case for ATD. The vision of didactic engineering in ATD, which expresses in terms of finalized and non-finalized study and research paths (Chevallard, 2015) with the role given in these to the dialectics between media and milieu and the opening of trajectories within the global structure of a reference epistemological model, is substantially different from the traditional vision of didactical engineering supported by TDS, which is structured around the search for fundamental situations. Actions inspired by these two research works take rather different forms.

In the mathematics education field, I often have the feeling that values are not questioned enough, that communication is often based on fuzzy consensus, and that the fact that the very diverse epistemologies existing in the field have no reason to lead to compatible decisions in terms of action is not really addressed. Establishing adequate relationships between research and action certainly needs systematic efforts to improve the situation, making more explicit the values underlying research and how these have an impact on research results and the vision of action, while also questioning these values.

Moving Forward

Up to now, I have mainly listed and discussed difficulties and issues faced when trying to make research a source of inspiration for action, using the case of didactical engineering in particular to illustrate my reflection. I would like, however, to express my conviction that the evolution of the field of mathematics education research, both theoretically and empirically; the number of existing realizations at different scales and in diverse contexts; the reflexive work carried out on these; and the communities and institutions established have substantially and productively influenced our vision of the relationship between fundamental research and action. I would like also to insist on the fact that we can today rely on conceptual and methodological tools much more powerful than was the case a few decades ago to address these issues and can therefore move forward. In the next part, I will briefly evoke what I see as major advances in that direction, beyond those already mentioned.

Didactical Engineering and Design-Based Research

Staying within the perspective of didactical engineering, the current research work carried out on the transition from research to development of didactical engineering is one promising avenue. As already explained, didactical engineering has developed in France as a research methodology, despite the fact that initially the exact role that would be given to it was not so clear. In the seminal text by Chevallard mentioned above, for instance, Chevallard distinguishes a priori between engineering work for research, for action, or both of them. For a long time, as already explained, the migration of didactical engineering designs or pieces of them from the research sphere to the action sphere was not an object of study. It developed outside any form of theoretical control, and the negative consequences of this state of affairs have been pointed out. This is no longer the case. I give as an example the research developed by Perrin-Glorian and her colleagues around the idea of second generation didactical engineering (Perrin-Glorian, 2011), but other promising projects have been developed, for instance, within the structure of the Lieux d'Éducation Associés (LÉA), joining schools and research laboratories, recently created by the IFÉ. The Arithmétique et Compréhension à l'Ecole Elémentaire (ACE) project[4] piloted by Sensevy is a good example.

Obviously, beyond the sole concept of didactical engineering that emerged in the French didactic tradition, another important evolution is the increasing role given to design-based research in the field and its associated theoretical and empirical work, with the consideration of scaling up as a major issue requiring specific research and methodology. This evolution is evidenced, for instance, by Cobb's research, for which he was awarded the Hans Freudenthal ICMI medal in 2005,[5] or the research and development work carried out by Swann and Burkhardt, who have been jointly awarded the first Emma Castelnuovo ICMI medal in 2016.[6]

Beyond these two evolutions directly linked to design, more global evolutions of the field offer substantial help to move forward the relationships between research and action in the field. I focus here on three of them.

The Increased Importance Taken by Socio-cultural and Anthropological Perspectives

Socio-cultural and anthropological perspectives allow us to better take into consideration the complexity and diversity of institutional, societal, and cultural conditions and constraints to which didactical systems are submitted. They provide conceptual and methodological tools to identify these and their respective strengths, to understand how they interact and shape the dynamics of didactical systems, and to reflect on how they can be moved when it seems a condition necessary to effective action. Beyond that, they help enlarge our vision of design. I have no doubt that this is indeed the case, for instance, with the conception of didactical engineering recently developed in ATD, especially:

- with the concept of non-finalized study and research path already mentioned, which provides an interesting theoretical framework for the conception of teaching strategies based on project and interdisciplinary work and
- with the dialectics between media and milieu, a powerful tool to take into consideration the important changes in access to information and inquiry practices induced by the technological evolution, especially the internet.

---

[4]http://python.espe-bretagne.fr/ace/.

[5]http://www.mathunion.org/icmi/icmi/activities/awards/past-recipients/the-hans-freudenthal-medal-for-2005/.

[6]http://www.mathunion.org/icmi/activities/awards/emma-castelnuovo-award-for-2016/.

The Development of Research on Teachers' Practices

I have lived the shift of attention of research from the student to the teacher. From unquestioned actors in the didactic relationship, teachers, with their beliefs, knowledge, systems of practices, and professional development, have become major figures of interest for research. The body of knowledge that has been built in that area since at least the early 90s is of the highest importance to improve the links between fundamental research and action. It helps understand the strong limitations of the strategies traditionally used to disseminate research results in the profession. Within this area of research, which is very diverse, I personally find constructions and approaches that address teacher work or activity in a rather global and systemic way to be especially useful for the reflection on relationships between research and action. This has been the case, for instance, in the double approach (didactic and ergonomic) of teachers' practices developed by Robert and Rogalski (Robert & Rogalski, 2002), in which teachers' practices are approached through five interconnected dimensions, including personal, social, and institutional determinations, and in the structuring features of classroom practice framework developed by Ruthven to analyze how teachers integrate or fail to integrate new technologies (Ruthven, 2009), to mention just a few examples.

The shift of attention of research from the student to the teacher has helped understand the exact nature of teaching expertise and better acknowledge its specificity. In many contexts, this has had an impact on the vision of the relationships between researchers and teachers, as attested, for instance, by the development of the idea of "community of inquiry" (Jaworsky, 2008), and as a consequence on the vision of relationships between research and action. Teachers can no longer be considered implementers of resources prepared by others, researchers, or those in charge of the educational transposition of research ideas and constructions. Teachers are themselves authors; they should be considered as such and supported in their authorship activity. The recently published ICMI Study on task design makes this clear (Watson & Ohtani, 2015).

The Development of Instrumental Approaches

Having contributed to the emergence and development of instrumental approaches (Artigue, 2002), I have had many opportunities to think about their affordances from a research perspective, but also in terms of the relationship between research and action. The first affordance I have seen is the fact that these approaches have made visible and understandable the essential processes of instrumental geneses that had nearly escaped the attention of those researching or promoting the educational use of technology. Blind points have thus been revealed, and the detrimental effects of such blindness identified. Distinctions, such as the one between the epistemic and pragmatic valence of techniques and schemes, the fact that technology disrupts the balance between these valences, and the fact that restoring appropriate balance requires new types of tasks, that have been established through didactical engineering research, have shown the profound inadequacy of teacher education in that area. They also have set conditions for the elaboration of educational resources. Beyond that, another interesting point is that this instrumental perspective has been progressively incorporated into a diversity of established theoretical frameworks, such as ATD in my initial work with close colleagues, activity theory for others, and the theory of semiotic mediation for Italian colleagues. Each of these incorporations has led to variations of the instrumental approach, despite the shared reference to the seminal work by Rabardel (2002). These different incorporations influence the resulting propositions in terms of design, as has been shown, for instance, in the ReMath project already mentioned (Lagrange & Kynigos, 2014). Another important point for my purpose here is the shift of attention from the student to the teacher, which, once again, has led to the extension of the approach to teachers' instrumental geneses, both personal and professional, then incorporated into a wider notion of genesis of use (Abboud-Blanchard & Vandebrouck, 2012). It is also the extension of this approach to the documentary work of the teacher, a domain of study today very active (Gueudet, Pepin, & Trouche, 2012). There is no doubt that a better knowledge of this essential dimension of teacher work and how it is affected by the technological evolution is of the highest importance for the relationships between research and action.

The last positive evolution I would like to mention is the development of projects of different scales in a diversity of contexts that provide new empirical bases to the reflection on these difficult issues. In recent years, I have been involved in a variety of European projects[7] aiming at the large-scale dissemination of inquiry-based education in mathematics and science following the publication of the report known as Rocard's report (Rocard et al., 2007). I have seen the intense and creative reflection and work that has gone into these projects to develop a more adequate vision of dissemination processes. I have seen the importance of the empirical work carried out. I have also again experienced up to what point attempts at making research at the service of action are themselves the source of questions for fundamental research.

However ...

However, we cannot deny that such accomplishments must come to grips with growing social and political pressures exerted on both research and educational systems by economic and competitive visions and values of education that are often

---

[7]The Fibonacci, Primas, and Mascil European projects (see their respective websites: www.fibonacci-project.eu, www.primas-project.eu and http://www.mascil-project.eu.)

at odds with ours. Tensions and inconsistencies result from this situation that are imposed upon all educational actors. I could make a long list of such inconsistencies. I will just mention some recently experienced in the frame of the European projects just mentioned. As I have explained, such projects aim to organize the dissemination of inquiry-based practices in mathematics and science education. However, in most countries, this goes along with institutional forms of assessment guided by another logic that are contradictory to the form of mathematical and sciences practices that inquiry-based education wants to promote. They put teachers in a double-bind situation.

Eight years ago, in our plenary lecture with Kilpatrick at ICME-11, we denounced the *diktat* of randomized controlled trials as the best if not the only acknowledged source of knowledge in the field. As Kilpatrick said:

> There are far too many research questions for which either randomized controlled trials would be impossible or an appropriate study would require so many controls as to make the interventions, whatever they are, unrealistic …. When narrow criteria are applied, what happens—in the cases I have seen—is that too much is left to untested opinion and individual experience. Not enough use is made of the professional community's judgment and experience. (Artigue & Kilpatrick, 2008, p. 10)

We could say the same today and this pressure did not at all vanish. As an international community, we must denounce these pressures and inconsistencies and try to counter their negative effects on the establishment of productive relationships between research and action. ICMI has here a fundamental role to play.

Conclusion

In this lecture, I have only addressed very partially the difficult issue of the relationship between fundamental research and action. I have used the filter of didactical engineering, that is to say, a "design" filter, as a guide for the reflection, but I perfectly know that action on didactic systems may take a diversity of forms and that this filter, as any filter, is reductive. The primary reason for my choice is not my personal investment in the development of didactical engineering. Rather, despite the fact that I am deeply convinced that it would be an error to reduce mathematics education to a design science, as has been proposed sometimes, I am convinced that design activities, whatever they are named and considered, have a fundamental role to play in the development of this field of scientific knowledge and in the way the knowledge gained can be put at the service of action. I have pointed out and discussed some of the major issues that arise when the relationship between research and action is looked at through this filter, such as reproducibility, generalization, and values, but part of the discussion has certainly more general value. In my opinion, up to now, these three issues have not found satisfactory answers and need to be addressed more seriously by the community. I have also tried to show that advances in design, more global evolutions of the field as a scientific field, and not least the growing number of projects trying to put research at the service of action in a controlled way make us today better equipped to move forward, but I have also stressed the counter-productive effects of politically related abusive pressures and inconsistencies that must be vigorously denounced.

References

Abboud-Blanchard, M., & Vandebrouck, F. (2012). Analyzing teachers' practices in technology environments from an activity theoretical approach. *International Journal for Technology in Mathematics Education, 19*(4), 159–164.

Artigue, M. (1986). Etude de la dynamique d'une situation de classe: Une approche de la reproductibilité. *Recherches en Didactique des Mathématiques, 7*(1), 5–62.

Artigue, M. (1989). Ingénierie didactique. *Recherches en Didactique des Mathématiques, 9*(3), 281–308.

Artigue, M., & Rogalski, M. (1990). Enseigner autrement les équations différentielles en DEUG. In Commission interIREM Université (Ed.), *Enseigner autrement les mathématiques en DEUG A première année* (pp. 113–128). Lyon: LIRDIS.

Artigue, M. (1992). Functions from an algebraic and graphic point of view: Cognitive difficulties and teaching practices. In E. Dubinski & G. Harel (Eds.), *The concept of function—aspects of epistemology and pedagogy* (pp. 109–132). MAA Notes n 25. Mathematical Association of America.

Artigue, M. (2002). Learning mathematics in a CAS environment: the genesis of a reflection about instrumentation and the dialectics between technical and conceptual work. *International Journal of Computers for Mathematics Learning, 7*, 245–274.

Artigue, M., & Kilpatrick, J. (2008). What do we know? and how do we know it? Plenary lecture at ICME-11, Monterrey, Mexico. http://www.mathunion.org/fileadmin/ICMI/files/Digital_Library/ICMEs/Plenary_1_MA_JK_final_01.pdf

Artigue, M. (Ed.) (2009). Connecting approaches to technology enhanced learning in mathematics: The TELMA experience. *International Journal of Computers for Mathematical Learning, 14*(3).

Artigue, M., & Winslow, C. (2010). International comparative studies on mathematics education: A viewpoint from the anthropological theory of didactics. *Recherches en Didactique des Mathématiques, 30*(1), 47–82.

Artigue, M., & Blomhøj, M. (2013). Conceptualizing inquiry-based education in mathematics. *ZDM—The International Journal on Mathematics Education, 45*(6), 797–810.

Artigue, M. (2014). Perspectives on design research: The case of didactical engineering. In A. Bikner-Ahsbahs, C. Knipping, & N. Presmeg (Eds.), *Approaches to qualitative research in mathematics education* (pp. 467–496). New York: Springer.

Artigue, M., & Bosch, M. (2014). Reflection on networking through the praxeological lens. In A. Bikner-Ahsbahs & S. Prediger (Eds.), *Networking of theories as a research practice in mathematics education* (pp. 249–266). New York: Springer.

Artigue, M. (2016). Epilogue. A didactic adventure. In B. R. Hodgson, A. Kuzniak & J. B. Lagrange (Eds.), *The Didactics of Mathematics: Approaches and Issues* (pp. 253–269). New York: Springer.

Arzarello, F., & Bartolini Bussi, M. G. (1998). The paradigm of modelling by iterative conceptualization in mathematics education research. In A. Sierpinska & J. Kilpatrick (Eds.), *Mathematics as a research domain: A search for identity* (pp. 263–276). Dordrecht: Kluwer Academic Publishers.

Brousseau, G. (1981). Problèmes de didactique des décimaux. *Recherches en Didactique des Mathématiques, 2*(1), 37–127.

Brousseau, G. (1997). *Theory of didactical situations in mathematics.* Dordrecht: Kluwer Academic Publishers.
Brousseau, G., & Warfield, V. M. (1999). The case of Gaël. *Journal of Mathematical Behavior, 18*(1), 7–52.
Brousseau, G., Brousseau, N., & Warfield, V. (2014). *Teaching fractions through situations: A fundamental experiment.* New York: Springer.
Chevallard, Y. (1982). *Sur l'ingénierie didactique.* IREM d'Aix-Marseille. http://yves.chevallard.free.fr/spip/spip/article.php3?id_article=195
Chevallard, Y. (2002). Organiser l'étude 3. Écologie & régulation. In J.-L. Dorier et al. (Eds.), *Actes de la 11e École d'Été de Didactique des Mathématiques* (pp. 41–56). Grenoble: La Pensée Sauvage.
Chevallard, Y. (2015). Teaching mathematics in tomorrow's society: A case for an oncoming counter paradigm. In Sung Je Cho (Ed.), *Proceedings of the 12th International Congress on Mathematical Education* (pp. 173–188). New York: Springer.
ERMEL. (2005). *Apprentissages numériques et résolution de problèmes* (Nouvelle ed.). Paris: Hatier.
Gascón, J., & Nicolás, P. (to appear). Economía, ecología y normatividad en la teoría antropológica de lo didáctico. *Proceedings of the 5th Congress of the Anthropological Theory of the Didactic.* Castro-Urdiales, January 26–30, 2016.
Gueudet, G., Pepin, B., & Trouche, L. (2012). *From text to 'Lived resources': Mathematics curriculum material and teacher development.* New York: Springer.
Isoda, M. (2015). The science of lesson study in the problem solving approach. In M. Inprasitha, M. Isoda, P. Wang-Iverson & B. H. Yeap (Eds.), *Lesson Study. Challenges in Mathematics Education* (pp. 81–106). Singapore: World Scientific Publishers.
Jaworski, B. (2008). Building and sustaining inquiry communities in mathematics teaching development: Teachers and didacticians in collaboration. In K. Krainer & T. Wood (Eds.), *International handbook of mathematics teacher education: The mathematics teacher educator as a developing professional* (Vol. 3, pp. 335–361). Rotterdam: Sense Publishers.
Lagrange, J. B., & Kynigos, C. (Eds.) (2014). Special issue: Representing mathematics with digital media: Working across theoretical and contextual boundaries. *Educational studies in Mathematics, 85*(3).
Nebres, B. F. (2008). Centers and peripheries in mathematics education. In M. Menghini, F. Furinghetti, L. Giacardi, & F. Arzarello (Eds.), *The first century of the International Commission on Mathematical Instruction (1908–2008). Reflecting and shaping the world of mathematics education* (pp. 149-163). Istittuto della enciclopedia Italiana. Roma.
Perrin-Glorian, M. J. (2011). L'ingénierie didactique à l'interface de la recherche avec l'enseignement. Développement des ressources et formation des enseignants. In C. Margolinas et al. (Eds.), *En amont et en aval des ingénieries didactiques* (pp. 57–74). Grenoble: La Pensée Sauvage Editions.
Rabardel, P. (2002). *People and technology—a cognitive approach to contemporary instruments.* https://hal.archives-ouvertes.fr/hal-01020705/document
Robert, A., & Rogalski, J. (2002). Le système complexe et cohérent des pratiques des enseignants de mathématiques: Une double approche. *Revue canadienne de l'enseignement des mathématiques, des sciences et des Technologies, 2*(4), 505–528.
Rocard, M. et al. (2007). *Science education now: A renewed pedagogy for the future of Europe.* European Commission.
Ruthven, K. (2009). Towards a naturalistic conceptualisation of technology integration in classroom practice: The example of school mathematics. *Education & Didactique, 3*(1), 131–149.
Swan, M. (2014). Design research in mathematics education. In S. Lerman (Ed.), *Encyclopedia of Mathematics Education* (pp. 148–152). New-York: Springer.
UNESCO. (2011). *The Challenge of Basic Mathematics Education.* Paris: UNESCO.
Watson, A., & Ohtani, M. (Eds.). (2015). *Task design in mathematics education. An ICMI study 22.* New York: Springer.

Van den Heuven-Panhuizen, M., & Drijvers, P. (2014). Realistic mathematics education. In S. Lerman (Ed.), *Encyclopedia of mathematics education* (pp. 521–525). New-York: Springer.
Wittmann, E. (1998). Mathematics education as a design science. In A. Sierpinska & J. Kilpatrick (Eds.), *Mathematics education as a research domain: A search for identity* (pp. 87–103). Dordrecht: Kluwer Academic Publishers.

Open Access Except where otherwise noted, this chapter is licensed under a Creative Commons Attribution 4.0 International License. To view a copy of this license, visit http://creativecommons.org/licenses/by/4.0/.

Elementary Mathematicians from Advanced Standpoints—A Cultural Perspective on Mathematics Education

Alan J. Bishop

Abstract Many challenges face those of us for whom mathematics education research is our life's work. In some countries where significant attempts are continually being made to reform mathematics teaching, it is often a highly politicised field. While rational arguments and relevant data-gathering are valid parts of a democratic research process, awareness of the broad cultural context is paramount. Despite the challenges that adopting a new cultural perspective brings, episodes and analyses from our sociocultural research field do demonstrate much promise for advancing mathematics educational practices. In particular the relatively new ideas of values and valuing show much research promise. In this paper, referencing Felix Klein's fundamental ideas, I will analyse the twin pluralised notions of 'elementary mathematicians' and 'advanced standpoints'. In addition research focussed on a third key notion, 'pedagogical practices' will be discussed. Finally some of the implications of this three dimensional and culturally oriented research will be presented.

Keywords Culture · Elementary mathematicians · Advanced standpoints

Introduction

The main aim of this paper is to broaden the discussion about the future of research in mathematics education. It derives from the parallels between the ideas of Felix Klein and the growth in research approaches in our current era However despite teachers' best efforts over many years mathematics is still rated as one of the most difficult subjects to teach and thus to learn. This is despite good arguments for making mathematics one of the key STEM subjects to teach in the modern world

---

Felix Klein award.

A.J. Bishop (✉)
Faculty of Education, Monash University, Melbourne, VIC, Australia
e-mail: alan.bishop@monash.edu

© The Author(s) 2017
G. Kaiser (ed.), *Proceedings of the 13th International Congress on Mathematical Education*, ICME-13 Monographs, DOI 10.1007/978-3-319-62597-3_11

(Educational Council, 2015). It would surely therefore be one of the most important subjects to research?

However many challenges face those of us for whom mathematics education research is our life's work. Cuts to University funding, general economic pressures, and unnecessary standards-based evaluations are all contributing to a sense of unease and disillusion in many countries. In this paper, as well as referencing Klein's ideas, I will explore my versions of his twin notions of 'elementary mathematics' and 'advanced standpoint', focussing on their humanistic side. Thus in the title of my paper 'elementary mathematicians' and 'standpoints' are the key ideas. In addition a third key notion, 'pedagogical practices' will complete the basic three-dimensional trio of culturally-based constructs which I believe should structure mathematics education research today.

In a sense this paper will be loosely based around my academic career, as the activities and explorations that engaged me reflected my research involvement with many others in our field. However as the old saying goes: "No man is an island" and in accepting the Felix Klein award for 2015 I am conscious of the many collaborators who have helped me structure and develop my ideas.

Our field of research is not like a highly abstract field of theoretical physics, for example where one mathematical mind can achieve much, as Stephen Hawking's has and indeed as Felix Klein's did. Education is naturally much more inclined to the sociocultural fields of people, socio-political groups and multidisciplinary thinking, as is mathematics education.

As the recognition of the team-work of Jill Adler and her colleagues shows, by her being awarded the Hans Freudenthal medal for 2015, quality research in our field today lies with teams and groups of researchers rather than individuals. So if I deserve any reward for my achievements, it is that I have been able/allowed to take advantage of, and have access to, opportunities for research in local, national and international contexts.

I have often been fortunate to have been at the right place at the right time, and working with the right people. So in that spirit and throughout the paper I would like to name especially some key people without whom my research might never had happened.

Thus I initially recognise the contribution of Sir Wilfred Cockcroft in the late 1960s to my early academic career. He was my mathematics Tutor at Southampton University, UK, an intelligent and perceptive academic (Cockcroft, 1982[1]).

He encouraged me to apply for a scholarship to study in the USA. Prof Frank Land who gave me my first research position, was also influential in my time at Hull University (Land, 1962). These leaders, together with another raw research student Donald McIntyre, made it possible for me and others to begin opening up the field of research in mathematics education (Morrison & McIntyre, 1973). They were in a sense my first 'significant others'.

---

[1] In naming key people, I have chosen to include at least one reference to their work that impinged on my own, thus giving some indication of their contribution to my thinking.

Klein and Culture

For me, personally the 1960s were an exploratory time and I had not yet started serious academic writing. I had just returned to the UK from three years at Harvard University, studying with Jerome Bruner and Ed Moise—and having experimented with various new teaching ideas in the schools nearby.

Structurally the systems of teacher education, school mathematics curricula, etc. in the UK, USA, and Europe were growing both politically and in terms of awareness of the need for research. Sputnik also appeared and focussed the minds considerably! It was an exciting international time to be a young researcher. This was where my first ideas of mathematics as a culture were born and where I paid my first respects to my cultural elders.

Moreover, I believe that Felix Klein also had this strong feeling for the idea of mathematics as a culture. He too was clearly concerned to explore mathematics as a form of cultural knowledge, with deep meanings to be understood and valued rather than just as routine knowledge to be accumulated and memorised for examination purposes.

This is how I read his invocation of "elementary mathematics from an advanced standpoint" (Klein, 2004). He did not use the word 'culture'; it had not yet been widely discussed or explored in his context and in his time of the late 1800s and the beginning of the 1900s. But we can see in retrospect from our 'advanced standpoint' of cultural knowledge, that this clearly was what was challenging him—how to choose the right elementary mathematics with which to induct the young students into the language, the world, and the culture of advanced mathematics?

The continuing problem for us, with all the advantages of modern technologies, well-educated teachers and parents etc., is that students are still failing in examinations, still unaware of the wider mathematical world, and still ignorant of the values of mathematical understanding.

This is now our time to take up the challenge from the legacy left by Klein. We can ask: What does the cultural metaphor offer us who are working to improve the mathematical education of young people around the world? In our context, and making the task for researchers like us even harder is that the very people who ought to be the inductees of the elementary mathematicians, namely schoolteachers, are just as unaware of this core idea as their students are.

So with these preliminary thoughts let me build on Klein's ideas, and explore how a cultural perspective on mathematics education can literally change the mathematical world for the better.

Elementary Mathematicians

Initially the field of mathematics education research in those early 1970s when I was really starting out as a researcher concerned the learners, as opposed to "students", their different attributes, the main mathematical challenges for them, their differing skill levels, and their so-called 'abilities'. The research constructs and methodology were principally adapted from the general field of educational psychology, and there was little reference to specifically mathematics learning, nor to theory or data that could relate to the internationally perceived problems of mathematics education.

Klein's analysis and ideas were summed up by his notion of "elementary mathematics" and this reflected the debates that were going on in his university context. In particular he incurred much criticism from within school and university mathematics departments, on the grounds that there was nothing elementary about his mathematical agendas.

Nor was it clear how his ideas could solve the major problems of the so-called 'dropout' curriculum. That is where everyone drops out of the curriculum at some stage in their mathematics education, either through the curriculum being too difficult, or too irrelevant to the rest of their lives, or whatever.

Others attending this ICME conference are more qualified than I to comment on the validity of what I would call the mathematical arguments underlying his choices and sequences of mathematical subjects. However, what is clear to me is that he was focussed on what he saw as the chief missing cultural ideas needing to be included in the school and university mathematics curriculum. For example he discussed the role to be played by functions within algebra, the structural power of group theory, and the whole field of different geometries.

The paradox was that at this very same time the popular slogan was "Mathematics for all" and mathematics education researchers were starting to tackle the difficult questions of how best to teach mathematics for "all"? In fact at the same time university mathematicians were trying to redefine what should be the "mathematics" to be taught to all?

So what was happening to those young learners who I have called "elementary mathematicians"? Basically and conceptually they were in the middle of this debate! Moreover the choice of this label I have made not just to reference Klein's ideas, but also to demonstrate the sociocultural nature of our current educational research.

My argument is that if we want to make more progress in our research field, we need to address the social and cultural positioning of the various players in the main game, and see the fundamental commonality of the notion of "elementary". Whether they are Aboriginal elders, school groups, immigrants, or second language learners of all ages, they all belong in sociocultural communities and in "a place" wherever and of whatever kind. Understanding that idea is crucial for beginning to understand what "elementary" can mean in any context.

The sociocultural nature of this research did help to mend the fences between the mathematics communities and the growing mathematics educator communities. The title of my book called "Mathematical enculturation, a cultural perspective on mathematics education" (Bishop, 1988), grew out of the belief that all students are being enculturated into the mathematics culture. In that sense, for good or for ill, they are learning the mathematical culture by being elementary mathematicians.

Advanced Mathematical Education Standpoints

From the research standpoint we have now, advanced standpoints if you like, gives us opportunities to revisit some old, and ask some new, research questions, just as Felix Klein's analysis did in his context. But for this generation the research focus has not been on the mathematical topics and the curriculum nor on the mathematical topics that Klein and others have identified.

Of course I recognise that the curriculum is a vitally important part of the change process. But for me, and others, the focus of the 1970s and 1980s was firmly on mathematical education as a sociocultural field, with a broad enculturating perspective. Moreover, despite facing the necessary challenges that adopting the new cultural perspective brought, episodes and analyses from our sociocultural research field demonstrated much promise for advancing educational practices.

I now propose to explore the notion of advanced standpoints by documenting some of the major research trends from the 1970s through to the present day. This also allows me to pay my respects to some of the influential colleagues with whom I have collaborated.

I have already noted that in the 1970s there was a growing awareness of the deficiencies of the research models and approaches based on chiefly psychological methodologies. In particular the voices of academics in several developing countries were being heard as they brought to the attention of the mainstream (Western in the main) researchers the view that there were other ways to think of mathematics.

Particularly for me this was a dramatic time as I visited Papua New Guinea to work with a colleague there, Glen Lean (Clarkson, 2008). This was the time when his research work came to be recognised worldwide. He had collected data on over 2000 different counting systems from around Papua New Guinea, a task of huge importance for the region and for the world of mathematics education in general (Lean, 1994).

It was not just a matter of documenting the counting systems, Lean was concerned to note the many ways in which counting was embedded in the local cultures and languages. Far from being a simplified abstract system, counting was revealed by Lean as one of the key mathematical foundational constructs of the societies in Papua New Guinea, and we can claim, everywhere.

This ethnographic database still exists thanks to the very difficult task undertaken by Kay Owens in Dubbo, Australia (Bishop, 1999; Owens, 2001). Allied to the work on counting there was a huge interest in other anthropological data, and later its methods. Ken Clements and Lloyd Dawe (my first Ph.D. student) were active in this developmental work in Australia and the South Pacific (Clements & Ellerton, 1996; Dawe, 1986).

The educational implications of this research work had still to be worked out and colleagues in Mozambique and Brazil took great interest in it. Paulus Gerdes led one of these very effective teams and in particular they were accumulating other aspects of anthropological data that could be related to a more localised and relevant mathematics education instead of the out-dated colonial versions of mathematics instruction. It showed that subjects such as practical geometry and probability and statistics were likely to be much more relevant (Gerdes, 1986).

More than that however it showed that mathematical knowledge was/is not an abstract, universal field—it did/does have strong connections to the particulars of the society, especially when education is being considered. In fact it is important to recognise here that all societies have developed their own mathematics. "Ethnomathematics" was the term developed by Ubiratan D'Ambrosio to encompass both local mathematical ideas and so-called universal Western ideas (D'Ambrosio, 1985). This work was part of a much broader push to focus on other student groups that were not succeeding with the standard mainstream version. This issue relates back to the earlier work based on psychology, and convinced many researchers that if they were to have any influence on their country's mathematics education they would need to address the issue of how to relate "Street mathematics and school mathematics"; the title of an influential book by Nunes, Schliemann, and Carraher (1993).

Research such as this also showed the many positive ways that the local mathematics enabled, and was part of, the "normal" everyday life in those societies. This research of the 1970s blossomed in the 1980s and culminated in the 1988 ICME held in Hungary. In my view the great achievement of that decade was the so-called "Fifth Day Special" in that ICME conference, which brought together for one day more than 200 people and resulted in the UNESCO publication "Mathematics, Education, and Society" (Keitel, Damerow, Bishop, & Gerdes, 1989). Groups worked on papers concerned with political issues, societal relationships and particularly on sociocultural aspects.

Working in planning this Fifth Day Special with colleagues Christine Keitel, Peter Damerow, and Paulus Gerdes was an inspiration to me. Although we did not achieve all that we had hoped for, nevertheless the fact that we could bring together this number of colleagues to work on alternatives to the mainstream, marks the 1980s and the early 1990s as the turning point of change completely.

Political and critical dimensions of mathematics education started to come to the fore as a part reaction to this day's debate. In particular some felt the political had not been foregrounded enough. This gave rise to the Political Dimension of

Mathematics Education (PDME) group that was active for some years through the 1990s. Key figures in this group included Richard Noss and Celia Hoyles (Noss et al., 1990). In some ways this PDME group was a forerunner of the now robust Mathematics Education and Society group.

There was still some way to go in changing the mainstream system, but nevertheless at ICME in Budapest there were many images of progressive ways to develop the systems and to democratise mathematics education more completely. Furthermore, I would like to think that Felix Klein would have found himself thoroughly immersed in the debates and discussions concerning elementary mathematicians from advanced standpoints!

Pedagogical Practices in Relation to Values and Valuing

As the cultural metaphor became widespread, so did the growth of contact between educational researchers. Thus international comparative studies of mathematics education in different countries became a significant form of study. In fact this kind of comparative study was something that ICMI had done at its inception and carried on for many years.

Indeed, it was always of great interest to see how the "neighbours" were dealing with many similar problems that were emerging. Comparisons of textbooks and examination papers were also a source for research ideas. Teacher education institutions also played a strong role at this stage when it was clear that just random collections of textbook practices were not enough as data for serious, even scientific study.

Nevertheless, the roles of journals and books were highly significant in that phase. When Hans Freudenthal invited me in 1983 to take over the editorship of the journal Educational Studies in Mathematics (ESM), I could not believe what it would lead me into, with my own research. Suddenly I found myself deeply involved with international colleagues working throughout the world who were writing excellent research papers, and I had to be the one who finally had to make the decision on whether or not to publish them! What this did for me was to emphasise the social and cultural contexts of mathematics education research and of course I could not have done this work without the help of a skilled and dedicated Editorial Board.

But then I became aware of another significant issue. Many colleagues were now writing excellent research papers but, at that time, there were few opportunities to publish books in our field. I approached Kluwer (the then publisher of ESM before they were taken over by Springer) and we began the Mathematics Education Library, with the first book being by Hans Freudenthal. The publishers were understandably nervous about this venture especially as the title of Freudenthal's book was "Didactical phenomenology of mathematical structures" (Freudenthal, 1983). However, my argument was "Trust me" and they did. We now have more than 50 books in that Mathematics Education Library series.

Sometime later, Kluwer developed another publication initiative, which was the idea of a Handbook of Mathematics Education. The first one was a collaboration between myself as lead editor and Jeremy Kilpatrick, Christine Keitel, Ken Clements and Colette Laborde (Bishop, Clements, Keitel, Kilpatrick, & Laborde, 1996), and later in the second edition Frederick Leung joined the ranks of editors (Bishop, Clements, Keitel, Kilpatrick, & Leung, 2003). Once again it was a huge pleasure to work with these impressive scholars and we realised then the significant role that such publications could play in shaping the social and cultural contexts of our field. Since then the publication scene has matured and grown to reflect the multiplicity of research approaches in our field.

Turning to one of the significant groups that I was a part of during the 1980s, called BACOMET, was a powerful intellectual group, mainly European based, whose research activity spread over several years. The leaders were Bent Christiansen, Geoffrey Howson and Michael Otte and the full title of the group was "Basic Components of Mathematics Education for Teachers" (Bishop, Mellin-Olsen, & van Dormolen, 1991; Christiansen, Howson, & Otte, 1986). My specific interest in that group was with the teaching of values and multicultural aspects of teacher education, and these came in through chasing my ideas of mathematical enculturation. In fact the idea of 'mathematical acculturation' also came to the fore in this period.

Originally this idea seemed to relate most strongly to the elementary mathematicians working in a culture different from the mainstream; for example in Papua New Guinea, Brazil or Mozambique or indeed with significant indigenous groups in South Africa or New Zealand, where the extraordinary work of colleagues such as Bill Barton are noted with thanks (Barton, 2008). Also Tamsin Meaney's work with indigenous Australians is another significant contribution (Meaney, Trinick, & Fairall, 2012).

However, one could argue that for any school learner the teaching and learning process would be one of acculturation. Immigrant students were and are a prime example. This profound idea engaged me in some exciting new research, thanks to working closely with colleagues such as Guida d'Abreu, Norma Presmeg, Nuria Gorgorio and Marta Civil (Abreu, Bishop, & Presmeg, 2002). One consequence of this re-focus on learners' situations developed the idea of the learners being in a culture conflict situation. School and home cultures are often in conflict, particularly for learners who are new to the country concerned.

Over the years I have had some interesting and challenging interactions with a number of other colleagues from the Scandinavian countries beginning with Stieg Mellin-Olsen from Norway (Mellin-Olsen, 1987), and continuing with Ole Skovsmose from Denmark (Skovsmose, 1994). They have continued to open up the political dimensions of the debate as have my two colleagues at Monash University whose research is related to this political discussion, more specifically the politics of gender; namely Gilah Leder and Helen Forgasz (Leder & Forgasz, 1996).

With another group currently I have preferred to focus more on values and valuing—normally ignored, and only partially understood educationally. My original collaborators Philip Clarkson and Wee Tiong Seah have brought their own perspectives to this work on values (Clarkson, Bishop, & Seah, 2010); Clarkson with his various research studies on language and mathematics learning (Clarkson, 1991), and Seah with his wide intercultural perspective (Seah, 2008). More recently other colleagues have joined us in our values research: Annica Andersson from Denmark and Penelope Kalogeropoulos from Monash University, Australia. Andersson's work is based on socioculture and place (Andersson, 2012) while Kalogeropoulos is focussing more on students who are disengaged from school mathematics learning (Kalogeropoulos, 2016). In each case we can see how powerful is the notion of values—what values are controlling the actions of those students who are engaged or disengaged from the learning tasks? Indeed are they pursuing different values and if so what are they? As I noted above many challenges and tensions face those of us who work in the field of mathematics education and how they influence the mathematics discussions of the day. One area of these challenges concerns developing appropriate methodologies for our research. This is an issue that we as a group are still wrestling with, exploring at present whether using role-playing will give us, research colleagues and fellow teachers more insights into values and valuing (Clarkson in press).

Once again it is important to state that Felix Klein did not choose to discuss the idea that mathematics is a culture with its own values. However, it is also clear to me that he would have embraced that idea. He was neither anthropologist nor historian but once again he was ahead of his time, and once again I am sure he would have fitted into the sociocultural genre of our current educational struggles touched on here.

And Finally

The challenge of creating a satisfactory mathematics education for all learners has a long and fundamental history—indeed one can say that the issues and challenges are always with us, only the manner of understanding and dealing with them changes.

Felix Klein had little formal training in mathematics education but he was not short of ideas about the teaching of the subject. He had heard about the teaching methods of Maria Montessori and he recommended using models and small objects as vehicles for developing important geometrical notions and images.

So we can see that mathematics education has embraced diverse theoretical 'standpoints' in the quest to improve the teaching of mathematics. The focus on sociocultural values is one of the latest standpoints, and those of us working in this field believe strongly in its power. However, the best research is not just reflective, though that is important, but should also be projective—offering leadership to the mathematics education community in a parallel approach with medical research, which always has its empirical 'eye' on any new research.

I stated at the start of this paper that my chief aim was to broaden the discussion about the future of research in mathematics education. I have based my ideas upon the notions of "elementary mathematicians" and "advanced mathematical standpoints". I hope I have shown that reworking the ideas of Felix Klein this way could make a significant contribution to your own research. The ideas are that mathematics is a culture with its own norms, values, language and customs.

Finally I feel honoured not just to have been awarded this Klein medal but also to see how much of our current research on culture and values resonates with Klein's ideas. In this paper I have described the nature of some of the groups working in the sociocultural field of mathematics education. I have also noted some of the key colleagues with whom I have worked over my lifetime.

The danger in naming colleagues like this is that some others will inevitably feel excluded! They know who they are and I know who they are, and I apologise to them. I am grateful for their collaboration. It is only space and time that have prevented me from referring specifically to them. Thank you, and good luck to all researchers out there!

Acknowledgements Much of what I have written here tacitly recognizes the support coming from my wife and partner Jennifer. Her own study into language education, and especially the international language Esperanto, relates strongly to what I have written here.

I also wish to recognize the work of my colleague and friend Phil Clarkson with whom I have shared many creative debates both on and off the golf course!

References

Abreu, G., Bishop, A. J., & Presmeg, N. (2002). *Transitions between contexts of mathematical practices*. Dordrecht: Springer.
Andersson, A. (2012). A philosophical perspective on contextualisations in mathematics education. In O. Skovsmose & B. Greer (Eds.), *Opening the cage: Critique and politics of mathematics education* (pp. 309–324). Rotterdam: Sense Publications.
Barton, B. (2008). *The language of mathematics*. Dordrecht: Springer.
Bishop, A. J. (1999). A tribute to the research work of Dr. Glendon Lean Retrieved from https://www.merga.net.au/documents/__glenlean.pdf
Bishop, A. J. (1988). *Mathematical enculturation: A cultural perspective on mathematics education*. Dordrecht: Kluwer.
Bishop, A. J., Clements, M., Keitel, C., Kilpatrick, J., & Laborde, C. (Eds.). (1996). *International handbook of mathematics education*. Dordrecht: Kluwer.
Bishop, A. J., Clements, M., Keitel, C., Kilpatrick, J., & Leung, F. K. (Eds.). (2003). *Second international handbook of mathematics education*. Dordrecht: Springer.

Bishop, A. J., Mellin-Olsen, S., & van Dormolen, J. (1991). *Mathematical knowledge: Its growth through teaching*. Dordrecht: Kluwer.

Christiansen, B., Howson, A. G., & Otte, M. (1986). *Perspectives on mathematics education*. Dordrecht: Reidel.

Clarkson, P. C. (1991). *Bilingualism and mathematics learning*. Geelong: Deakin University Press.

Clarkson, P. C. (2008). In conversation with Alan Bishop. In P. C. Clarkson & N. Presmeg (Eds.), *Critical issues in mathematics education: Major contributions of Alan Bishop* (pp. 13–28, 69–70, 107, 149–150, 189–190, 229–230). Dordrecht: Springer.

Clarkson, P. C. (Ed.) (in press). Mathematical values: Scanning and scoping the field. Dordrecht: Springer.

Clarkson, P. C., Bishop, A., & Seah, W. T. (2010). Mathematics education and student values: The cultivation of mathematical wellbeing. In T. Lovat, R. Toomey, & N. Clement (Eds.), *International research handbook on values education and student wellbeing* (pp. 111–136). Dordrecht: Springer.

Clements, M., & Ellerton, N. (1996). *Mathematics education research: Past, present and future*. Bangkok: UNESCO.

Cockcroft, W. (1982). *Mathematics counts: Report of the committee of inquiry into teaching mathematics in schools*. London: HMSO.

D'Ambrosio, U. (1985). Socio-cultural bases for mathematics education. In M. Carss (Ed.), *Proceedings of the Fifth International Congress on Mathematics Education* (pp. 1–6). Boston MA: Birkhäuser.

Dawe, L. (1986). Teaching and learning mathematics in a multicultural classroom: Guidelines for teachers. *Australian Mathematics Teacher, 42*(1), 8–12.

Educational Council. (2015). *National STEM school education strategy 2016–2026*. Canberra Australia: Educational Council.

Freudenthal, H. (1983). *Didactical phenomenology of mathematical structures*. Dordrecht: Kluwer.

Gerdes, P. (1986). On culture, geometrical thinking and mathematics education. *Educational Studies in Mathematics, 19*(2), 137–162.

Keitel, C., Damerow, P., Bishop, A. J., & Gerdes, P. (1989). *Mathematics, education, and society*. Paris: UNESCO.

Klein, F. (2004). *Elementary mathematics from an advanced standpoint: Arithmetic, algebra analysis* (E. R. Hedrick & C. A. Noble, Trans. from the third German edition). New York: Dover.

Kalogeropoulos, P. (2016). *The role of value alignment in engagement and (dis)engagement in mathematics learning* (Ph D thesis). Monash University.

Land, F. W. (1962). *The language of mathematics*. London: John Murray.

Lean, G. A. (1994). *Counting systems of Papua New Guinea and Oceania* (PhD thesis), Papua New Guinea University of Technology, Lae, Papua New Guinea.

Leder, G., & Forgasz, H. (1996). Research and intervention programs in mathematics education: A gendered issue. In A. J. Bishop, M. Clements, C. Keitel, J. Kilpatrick, & C. Laborde (Eds.), *International handbook of mathematics education* (pp. 945–985). Dordrecht: Kluwer.

Meaney, T., Trinick, T., & Fairall, U. (2012). *Collaborating to meet language challenges in indigenous mathematics classrooms*. Dordrecht: Springer.

Mellin-Olsen, S. (1987). *The politics of mathematics education*. Dordrecht: Kluwer.

Morrison, A., & McIntyre, D. (1973). *Teachers and teaching*. Harmondsworth: Penguin.

Noss, R., Brown, A., Dowling, P., Drake, P., Harris, M., Hoyles, C., et al. (1990). *Political dimensions of mathematics education: Action and critique*. London: University of London.

Nunes, T., Schliemann, A. D., & Carraher, D. W. (1993). *Street mathematics and school mathematics*. Cambridge: Cambridge University Press.

Owens, K. (2001). The work of Glendon Lean on the counting systems of Papua New Guinea and Oceania. *Mathematics Education Research Journal, 13*(1), 47–71.

Seah, W. T. (2008). Valuing values in mathematics education. In P. C. Clarkson & N. Presmeg (Eds.), *Critical issues in mathematics education: Major contributions of Alan Bishop* (pp. 239–252). Dordrecht: Springer.

Skovsmose, O. (1994). *Towards a philosophy of critical mathematics education*. Dordrecht: Springer.

Open Access Except where otherwise noted, this chapter is licensed under a Creative Commons Attribution 4.0 International License. To view a copy of this license, visit http://creativecommons.org/licenses/by/4.0/.

Design and Development for Large-Scale Improvement

Hugh Burkhardt and Malcolm Swan

Abstract This chapter describes the Shell Centre team's "engineering research" approach to the improvement of practice through researched-based design and development of tools for teaching and learning mathematics, for professional development and for supporting large-scale change. The contributions of projects over the past 35 years to the development of design principles and tactics are outlined and illustrated. The roles of tasks of different kinds in learning and assessment are explained, with particular reference to the design of tests, and of formative assessment lessons for concept-development and problem solving. The chapter concludes with a look at the barriers to turning success at classroom level into large-scale change—and how this challenge can be tackled.

Keywords Design · Engineering · Strategy · Formative · Assessment · System

Introduction

The creation of the Emma Castelnuovo Award by ICMI is an important milestone in linking research and practice in mathematics education. The core focus of academic research is on deeper understanding of a field and its phenomena: in mathematics education, exceptional contributions are recognized by the Felix Klein and Hans Freudenthal Awards. But other fields with direct impact on people's lives —medicine and engineering, for example—balance this insight-focused research with research-based design and development of new products and processes that enable practitioners to tackle more effectively the problems of practice. A large part of medical research, for example, is focused on developing new and more effective

---

Emma Castelnuovo Award.

H. Burkhardt (✉) · M. Swan (Deceased)
School of Education, Shell Centre, University of Nottingham, Jubilee Campus, NG8 1BB, Nottingham, UK
e-mail: Hugh.Burkhardt@nottingham.ac.uk

medicines, devices and procedures. Equally, our lives are filled with products of engineering research and development that embody the new fundamental insights research provides. No such balance exists in education, where impact-focused research remains relatively rare. Its importance is recognized by this new award.

This chapter is primarily an account of what we and our colleagues at the Shell Centre have done over the last 35 years to develop and exemplify this "engineering research" approach (Burkhardt, 2006). Towards the end, we will discuss strategic changes in mathematics education research that would encourage a better balance of *insight-focused* and *impact-focused research*, giving the direct serving of practice the priority it deserves.

The Shell Centre Approach

The Shell Centre for Mathematical Education was founded as a professional development centre in 1968 by Nottingham's professors of pure and applied mathematics, Heini Halberstam and George Hall. The vision at that time was that improving teachers' understanding of mathematics and its applications was the key to improving student learning. By the time one of us (HB) was appointed director in 1976, it was becoming recognized that the challenges were much broader than 'knowing more maths', so a radically different 'brief' for the Shell Centre was agreed:

> To work to improve the teaching and learning of mathematics regionally, nationally and internationally.

This ambitious challenge had a chain of implications:

- The focus should be direct impact on practice in classrooms.
- Large-scale impact can only be achieved through reproducible materials.
- Developing these well needs engineering-style research, which other fields have shown can produce both better products and new research insights.
- Good engineering implies a focus on design—strategic, structural, technical—and on systematic development in appropriate contexts.

This led to a search for outstanding designers: the other author (MS) was invited to join the Centre in 1979.

What distinguishes these different aspects of design?

Strategic design (Burkhardt, 2009) is concerned with the "fit" of a design with the system it aims to serve: finding "points of leverage", for example high-stakes testing; devising models of change that work well; guiding policy in a way that satisfies the needs of all the key groups, including policy makers. Poor strategic design is a common source of failure of initiatives to achieve their goals.

Structural design aims to ensure that a tool fits both the 'user' and the 'job' being addressed—just as a knife has a handle and a cutting edge, so materials for teaching problem solving should support the teacher-user in helping the students to develop strategies for solving non-routine problems.

Technical design of the product relies on a combination of input from prior research and design creativity that injects the "surprise and delight" that, along with a sound research basis, epitomizes excellence and gives pleasure to users, both teachers and students.

The other essential element for turning designs into products that are both educationally ambitious and reliably effective in the hands of diverse users is the same as in any impact-focused research-based field:

Systematic development through trials in realistic conditions with the rich and detailed feedback needed to guide improvement. For us this has meant direct observation of trials with reports to the designers based on protocols structured to focus on the key events.

These principles have guided a sequence of linked design and development projects through which the Shell Centre team has developed tools and processes for classroom teaching and learning, formative and summative assessment, and teacher professional development. In the next section we explain the key roles that tasks play. In Section "Developing Design" we describe how specific projects have led us to identify new design principles and tactics. Sections "Developing Conceptual Understanding and Logical Reasoning" and "Developing Strategies for Problem Solving" describe the design of formative assessment lessons to support concept development and problem solving, respectively. In the last decade we have come to see that major obstacles to progress lie at levels "above" the classroom. Section "Tools for Supporting Systemic Change" describes our work on tools and processes to advance systemic improvement.

Descriptions alone cannot adequately communicate design ideas or products; exemplification is essential—but, in a book like this, inevitably brief. The website *emma.mathshell.org* gives examples in full, section by section, along with sketches of all the main Shell Centre Projects.

Building an International Community

The above approach is broadly shared by some other design teams around the world, though it remains rare in the huge body of education research. While there have always been international exchanges of ideas, it seemed to us that the profession would benefit from coming together more formally to share common challenges and opportunities. After discussions over a decade or so, we launched the International Society for Design and Development in Education (ISDDE) at a conference in Oxford in 2005. Since then annual conferences have been held in different parts of the world. The Society currently has about 100 Fellows. Its goals are to:

- Build a design community—this now exists
- Raise standards in design and development—there has been progress, learning together
- Increase influence on policy—this remains a, perhaps *the*, major need and challenge.

Educational Designer was established by ISDDE to share expertise. We decided that a peer-reviewed e-journal format was best because it allows articles to combine relatively brief and readable text with the rich exemplification needed in talking about design, accessed through internal links. Much of the design detail that is perforce squeezed out of this chapter can be found in articles in the journal.

ISDDE is focused on education in mathematics, science, engineering and technology. There are fundamental reasons why design is more important here. In teaching the humanities, teachers master a modest number of lesson *genres* into which they insert texts which they choose from the varied literature of their subjects, producing an infinite variety of lessons. The original literature in subjects such as mathematics and science is too technical for use in school—hence the need for the detailed design of coherent, linked lessons that bring students to grips with various aspects of understanding and doing mathematics—not just *lesson genres*, though these are important.

Tasks in Mathematics Education

Tasks play at least four important roles in mathematics education:

- **Providing 'microworlds' for investigation**: as a stimulus for learning; for developing understanding; for learning strategic methods for tackling complex non-routine problems.
- **Summarizing curriculum goals**. Analytic domain descriptions—"national curricula" or "standards"—are highly ambiguous; complementing them with an exemplar set of tasks covering the target types of performance makes the learning goals much clearer.
- **Assessing students' performance** through tests and coursework for monitoring progress, for selection or for accountability purposes, or through formative assessment in classrooms.
- **Providing targets for performance**. Mathematicians set research targets in terms of tasks: Prove Fermat's last theorem or the 4-colour map problem, solve 'Hilbert problems' or the 'Travelling salesman problem'. Teachers use tasks from 'past exam papers', often over-concentrating their teaching on those task-types.

Tasks and their design is a recurring theme throughout our work.

Task Difficulty

It is important to choose tasks that the students find interesting and challenging—but not impossible! It is known from research that the difficulty of a task depends on various factors, notably its:

- **complexity**—the number of variables, the variety and amount of data, and the number of modes in which information is presented, are some of the aspects of complexity that affect the difficulty a task presents.
- **unfamiliarity**—a non-routine task is more difficult than a similar task one has practised solving; the student has to understand the structure of the situation, work out how to tackle it, and do so while monitoring progress.
- **technical demand**—a task that requires sophisticated mathematics for its solution is more difficult than one that can be solved with more elementary mathematics.
- **student autonomy**—guidance from an expert (usually the teacher), or from the task itself (e.g., by structuring or "scaffolding" it into successive parts) makes a task easier than if it is presented without such guidance.

Assessments of student performance need to take these factors into account. For example, they imply that, in order to design a task for a given level of difficulty, a relatively *complex non-routine* task that students are expected to solve without guidance needs to be *technically* easier than a short exercise that develops or tests a well-practised routine skill. Problem solving tasks need to be *conceptually* easier than those that focus on mathematical concepts.

Rich tasks allow students at different levels to provide different correct responses. For such tasks, difficulty also reflects the level at which the student engages with the task. This is similar to the situation in English or History; the same essay question might be posed to a young student or to a college graduate, expecting quite different "good" responses.

'Expert', 'Apprentice' and 'Novice' Tasks

We have found it useful to distinguish three broad types of task: 'expert', 'apprentice' and 'novice' tasks. Each type has a different balance of sources of difficulty.

'Expert Tasks' are problems in the form they naturally arise, in the real world or within mathematics. Relatively complex and non-routine, if the students are to be able to solve them autonomously they must not be technically demanding. Figure 1 shows two expert tasks, accessible in some middle school classrooms—and also good with older students. The difficulty here comes mainly from the *complexity*, with various factors, not all stated, and *unfamiliarity* so the students have to work out what to do, then do it by constructing a chain of reasoning. 'Table Tiles'

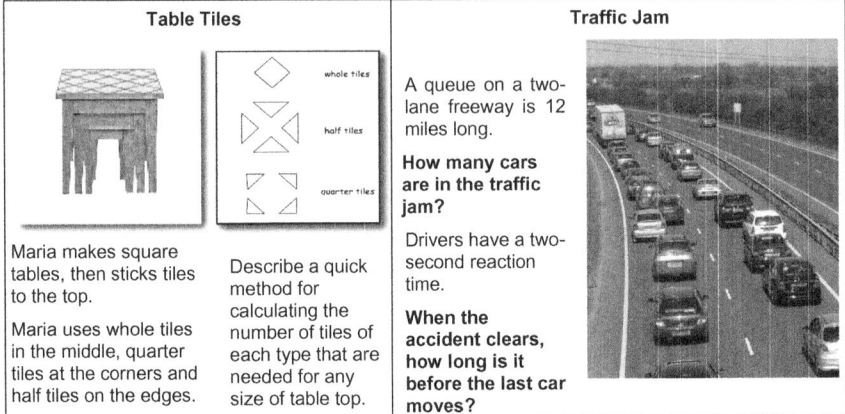

Fig. 1 Two 'expert tasks'

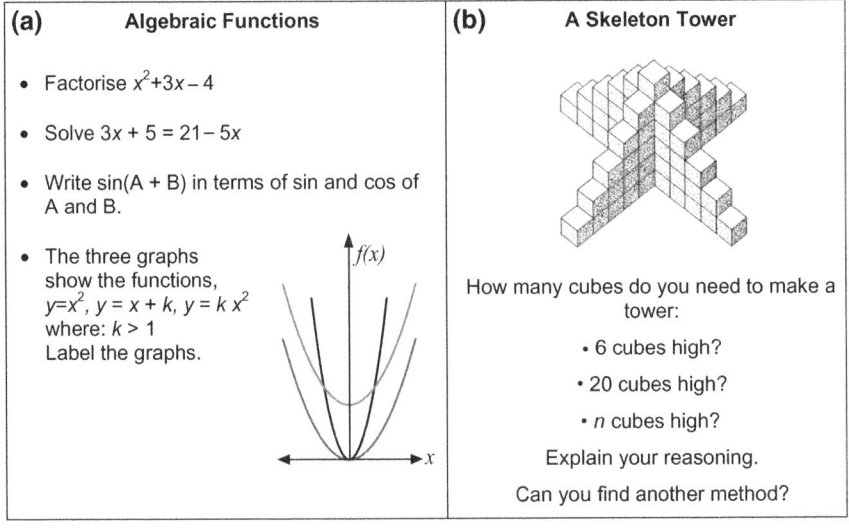

Fig. 2 a Novice tasks. **b** Apprentice task

involves detecting and describing patterns; it includes a "ramp of difficulty": 4 quarter tiles provide the corners of any such table, while the number of half tiles is linear and of whole tiles is quadratic in the table size—reflecting the deep insight that corners are points, sides are lines and the centre is an area. 'Traffic Jam' is about proportional reasoning—arguably the most important modelling tool that students should learn to use in school.

'*Novice Tasks*' are short 'items' with mainly technical demand (as in Fig. 2a). Each is focused on a specific concept or skill, so they can be "up to grade", including content that has recently been taught. Novice tasks are designed to test recall of learned procedures. (Novices are learning the tools of the trade.) *'Apprentice Tasks'* (e.g. Fig. 2b) are expert tasks with scaffolding added to guide the student in a series of steps, reducing complexity, non-routine-ness and student autonomy. (Apprentices learn to solve problems with expert guidance).

The difficulty in 'Skeleton Tower' lies mainly in deciding how to tackle the problem; this is scaffolded by the two specific examples. 6 cubes high is pictured; it can be done by counting which, if recorded, reveals a pattern that can then be extended numerically to 20. Without these steps, this would be an expert task requiring a verbal rule or formula—the last part. Note that expertise involves learning problem solving strategies (Polya, 1945; Schoenfeld, 1985; Swan et al., 1984), including "try some special cases" and "look for patterns and structure"—removing the need for scaffolding like this.

The difficulty of a task has ultimately to be determined by trialing the task with the target group of students. All assessment tasks, whether for use in the classroom or in summative tests, should be developed in this way. To summarize the key point, there is a "few year gap" between the mathematical concepts and skills that a student can use in short imitative novice tasks and those they can use autonomously in solving expert tasks. Students' mathematical expertise is what matters beyond school yet, currently, the curriculum in many countries has only novice tasks—leading to a novice-level mathematics education. To develop factual knowledge, conceptual understanding and strategic competence, a world-class mathematics education needs substantial experience of all three kinds of task—novice, apprentice and expert—in both curriculum and assessment.

Learning Goals and Task Genres

We have learned a lot from 'own language' teaching. This seeks to develop technical fluency (spelling, grammar, syntax), to analyze and create texts in different genres (reports, letters, stories, poems, etc.) and to relate texts to social, historical and cultural contexts. Progress consists in being able to handle more challenging texts in more sophisticated ways. If you change "texts" to "tasks", the top-level goals for mathematics are much the same.

We have recently come to develop a framework that looks in more detail at tasks in terms of the primary purposes of the learning activity they support, the genres of student activity that the task demands, and the type of product that results from the student reasoning involved. It is summarized in Table 1.

In the UK, the US and many other countries the first row, *facts and procedure development*, is dominant in assessment and in many classrooms. (The policy *rhetoric* is often broader). Facts and procedure are easy to assess through tests using novice tasks. Teaching and assessment for the other purposes, however, needs apprentice and expert tasks. *Conceptual understanding* requires chains of reasoning, connections and explanation—as do *problem solving and strategic competence*.

Table 1 An activity-genres framework for design

Purpose of the lesson	Process genres on tasks—the— student:	Student products
Factual knowledge and procedural fluency (e.g. Fig. 2a)	Memorizes and rehearses facts, procedures and notations	Performance
Conceptual understanding and logical reasoning (e.g. Fig. 2b)	Observes and describes phenomena	Description
	Classifies and defines objects	Classification
	Represents, and translates between representations	Representation
	Justifies conjectures, procedures, connections	Explanation
	Identifies and studies structure	Analysis
Problem solving and strategic competence (e.g. Fig. 1)	Formulates models of situations	Model
	Solves non-routine problems	Problem solution
	Interprets and evaluates strategies	Critique

The design of such tasks requires a much broader range of partly-creative design and development skills than official test providers typically possess. Such tasks are therefore lacking in most tests—and therefore in most classrooms. The work described below exemplifies how this need for range and balance can be met. For example, in the Mathematics Assessment Project, described in Sections "Developing Conceptual Understanding and Logical Reasoning" and "Developing Strategies for Problem Solving", we designed formative assessment lessons specifically to address conceptual understanding or strategic problem solving, complementing the procedural curriculum in many schools and showing how higher-level thinking may be taught and assessed. These lessons are now in use by millions of teachers and students across the US; evaluation shows remarkable gains in student learning (Herman et al., 2014).

Developing Design

In this section we shall outline some projects that led us to develop specific design principles and tactics—principles that continue to inform our work. Other projects, outlined in emma.mathshell.org, will be referred to as they arise in what follows.

Testing Strategic Skills (TSS 1980–88) developed a new model of examination-driven gradual change. The stimulus in this collaboration with England's largest examination provider was our pointing out that, of the board's list of 7 "knowledge and abilities to be tested" in mathematics, only 3 were assessed in the actual examinations. The board agreed to a novel strategic design with the following features: introduce one new task type each year, with 2 years notice to schools; provide integrated support in the form of materials; remove from the exam

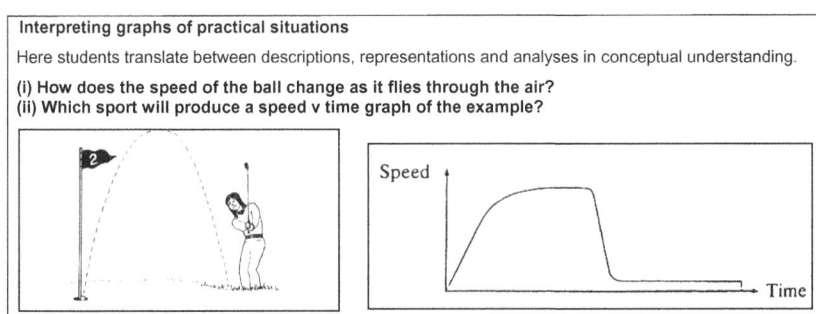

Fig. 3 From *The Language of Functions and Graphs*

syllabus 5% of the mathematical content to compensate. The Shell Centre developed the materials (Swan, Binns, Gillespie, Burkhardt, & The Shell Centre Team, 1987–89; Swan, Pitts, Fraser, Burkhardt, & The Shell Centre Team, 1985, 1984), which comprised 5 exemplar examination tasks,[1] lesson materials for 3 weeks teaching, along with materials for in-school professional development including video and software. The long-term goal was *to move year-by-year towards "tests worth teaching to"*—a target that still remains elusive worldwide. The gradual change model was popular with teachers, students and the board. Two modules were developed, one on problem solving, the other on concept development. *Problems with Patterns and Numbers* (Swan et al., 1984, see Fig. 2b) is concerned with generalization of mathematical situations. *The Language of Functions and Graphs* (Swan et al., 1985, Fig. 3) involves students in translating between representations of practical situations.

A national reorganization of testing ended this promising innovation—as reorganization so often does. 'Replacement unit' models have been used in many places, but the digestible pace of change and coherent well-engineered assessment and support of "TSS" remain rare.

This work led to a collaboration with Alan Schoenfeld's group at Berkeley and others in the US in a series of projects on assessment and large-scale change. This still forms a major strand of our work, some of it described in the sections that follow.

Investigations on Teaching with Microcomputers as an Aid (ITMA 1980–88) This project explored the potential of a single microcomputer with a large monitor in supporting the teaching of non-routine problem solving. Led by Rosemary Fraser and Richard Phillips, the design was largely based on "software microworlds" that stimulate investigation (Fig. 4).

Though inevitably "off-line" for the students, the approach proved powerful in various ways. We shall mention just one: "role shifting". A study of 17 classrooms

---

[1]For expert tasks, it is essential to show the *variety* that can be expected. We used the rubric "The following sample of questions gives an indication of the variety likely to occur in the examination".

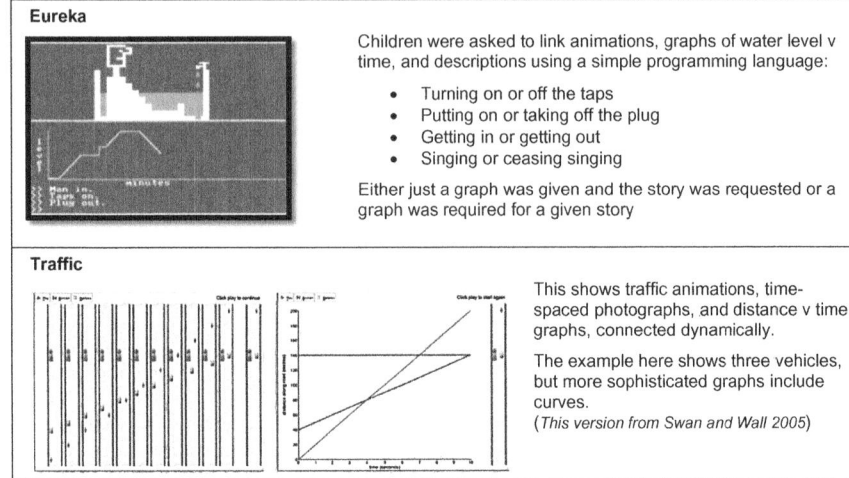

Fig. 4 ITMA microworlds

using ITMA software lessons (Burkhardt et al., 1988) developed a "roles analysis" which showed that the teachers[2] naturally moved from the traditional directive roles (called *manager, explainer, task setter*) into facilitative roles (*counsellor, fellow student, resource*). Students became *explainers* and *task setters*. Designing for role shifting has proved a powerful design tactic in our subsequent work: higher-level discussion and learning happen reliably when students move into teacher roles.

This strand of technology-based work has continued to inform our design more generally. For the UK Government's **World Class Arena** (1999–2005), Daniel Pead led the development of tests of "Problem solving in Mathematics, Science and Technology" that were 'computer + paper' based. This (expensive) format reflected the limitations of AI—still weak after 50 years—in interpreting autonomous student reasoning. This work led to an analysis (see ISDDE, 2012) of the strengths and weaknesses of computers in the five essential aspects of summative assessment: presenting the task (strong); providing a natural working environment for the student (strong for text only subjects, weak for mathematics with its sketches and equations); capturing student responses (fine for text; for mathematics, only for novice tasks—or by scanning written responses); assessing responses (very weak for complex tasks); collecting and reporting scores (very strong).

Diagnostic Teaching (1983–2006) was the guiding principle for a linked sequence of small-scale studies, initially led by Alan Bell. It is based on students revealing their misconceptions through carefully designed "cognitive conflict" situations, then "debugging" them through group and class discussion (Bell, 1993; Swan, 2006). The outcome of this approach to formative assessment is improved

---

[2]With one exception: He loved the lessons but continued to stand by the screen and teach from the front.

Design and Development for Large-Scale Improvement 187

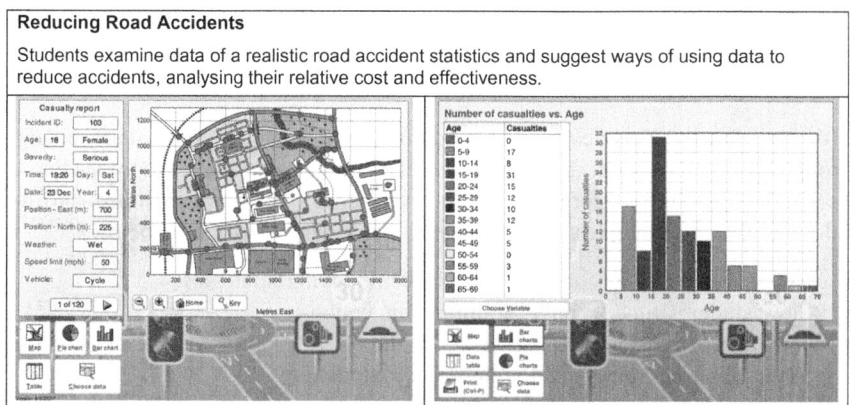

Fig. 5 From Bowland Maths: http://bowlandmaths.org.uk

long-term learning. Since Section "Developing Conceptual Understanding and Logical Reasoning" will describe this ongoing work, we will just point to the research strategy here.

Many studies in education are small-scale investigations of a specific *treatment* —a new approach to teaching a topic, perhaps studied in a few classrooms. If well done, such studies may reveal *trustworthy* insights about that system but without any evidence of *generalizability* (Schoenfeld, 2002). To provide evidence on design *principles*, the same research approach needs to be studied across a range of topics, designers, and teachers—as well as students. That has been the strategic design of the sequence of diagnostic teaching studies; the principles have proved robust.

Bowland Maths (2006–10) is a happier story. The goal was to develop 4-lesson units ("case studies") for 14-year old students that showed the power of mathematics through real (or fantasy) world situations. We developed two of these. *How risky is life?* confronts students with the mismatch between popular ideas of various dangers and the data. In *Reducing road accidents* the students explore a graphical database of detailed accident report data (Fig. 5) in order to advise a town council on what safety measures to take. At a time when the learning goals of the national curriculum are broadening, this enrichment model of change has had some impact. We also developed professional development support modules based on a novel "sandwich model", which we describe in Section "Tools for Supporting Systemic Change".

Developing Conceptual Understanding and Logical Reasoning

In design we use principles from research on learning, notably that students learn through active processing: discussion and reflection in a social classroom leading to the internalization and reorganization of experience. This we developed into the following design principles, which underlie much that we describe below:

- Activate pre-existing concepts and problem solving strategies
- Allow students time to build multiple connections
- Stimulate tension—cognitive conflict—to promote questioning, re-interpretation, reformulation and accommodation
- Devolve problems to students
- Focus on reasoning—not just answers
- Expect students to explain their Interpretations and chains of reasoning
- Include reflective periods of 'stillness', for examining alternative meanings and methods.

We have identified a number of *lesson genres* that contribute to conceptual understanding:

- Interpreting and translating representations—*What is another way of showing this?*
- Classifying, naming and defining objects—*What is the same and what is different?*
- Testing assertions and misconceptions and justifying conjectures—*Always, sometimes or never true?*
- Modifying problems; exploring structure—*What happens if I change this? How will it affect that?*

We will illustrate the first of these genres below with a lesson on percentage change.

Diagnostic Teaching Research and Development

The sequence of diagnostic teaching studies, over several mathematical topics and teaching unit designers, showed that this approach leads to more robust long-term learning than direct instruction. Swan (2006) analysed the average pre-test, post-test and delayed-test scores in those studies. In each case, the intensive discussion and argument among students yielded more substantial long-term learning than standard methods—either exposition or guided discovery led by teachers.

To provide evidence of the generalizability of this result, the same research approach was studied, linked across a range of conceptual topics, and applied by students in non-obvious situations. The principles have proved robust and used in subsequent projects. **Improving the Learning of Mathematics** (Swan & Wall, 2005), for example, was a collaboration with the UK government's Standards Unit which produced curriculum development support materials. This "box" was distributed to all UK schools and colleges, receiving an enthusiastic response from practitioners, researchers, and government inspectors.

Formative Assessment

A large-scale review of research by Black and Wiliam (1998) showed that the use of formative assessment, when well done, leads to remarkable increases in student learning. Wiliam and Thompson (2007) defined it thus:

> Formative assessment is students and teachers using evidence of learning to adapt teaching and learning to meet immediate needs, minute-*to-minute and day-by-day*.

This is, of course, the essence of the Diagnostic Teaching approach. However, making formative assessment central to one's practice is a major departure from the "demonstrate and practice" form of pedagogy that lies at the core of most mathematics teaching, so it is extremely challenging for teachers. Our earlier work led the Bill & Melinda Gates Foundation to invite us to design lesson materials that enable teachers to acquire this expertise.

In the **Mathematics Assessment Project** (2010–14), working with our US partners, we designed 100 "Classroom Challenges"—20 formative assessment lessons for each grade 6 through 10—and refined them on the basis of structured observer reports through two rounds of trialing in US classrooms. Two thirds of the lessons are concept development focused, the others problem solving focused. We describe their design in more detail in Swan and Burkhardt (2014). There have been over 6,000,000 lesson downloads so far from the project website map.mathshell.org alone.

The Design of Concept Development Lessons

These lessons are designed with three complementary objectives:

1. **to reveal to the teacher, and the student, each student's current understanding and misunderstandings** of the central concept—as in all well designed diagnostic assessment.
2. **to move the student's understanding forward by a process of "debugging through discussion,"** in pairs and with the class as a whole—thus integrating diagnosis and treatment. This is crucial: diagnosis alone faces the teacher, again and again, with the considerable design challenge: "What shall I do to help this student?" A common response is to reteach the topic, but faster; it is not

surprising that this rarely helps. The diagnostic teaching approach used in the design of the Classroom Challenges reflects the observation (VanLehn & Ball, 1991) that the key characteristic of successful students is not that they remember procedures precisely but that they can detect and correct their own errors. "Debugging through discussion" develops that (higher level) skill. This more robust, long-term understanding reduces the time needed when re-visiting topics in later years.

3. **to build connections between different conceptual strands**. Mathematics content is best understood as a connected network of concepts and skills—as in other networks, the connections reinforce each other. The linear sequence of lesson-by-lesson teaching naturally develops "strands of learning"—strands that, for most students, have weaknesses, and often breaks, in them. Learning should involve active processing, linking new inputs to the student's existing cognitive structure. Novice tasks alone produce fragmentation; rich tasks help to develop connections.

The design has the following sequence of activities:

- Expose and explore students' existing ideas—*pull back the rug*
- Confront them with their contradictions—*provoke 'cognitive conflict'*
- Resolve conflict through discussion—*allow time for formulation of new concepts*
- Generalize, extend and link learning—*connect to new contexts*.

"*Increasing and decreasing quantities by a percent*" (http://map.mathshell.org/lessons.php?unit=7100&collection=8) is a lesson that shows how this works. It is designed to enable students to detect and correct their own and each other's misconceptions in this often-challenging topic—and to build connections.

During a prior lesson, a sheet of tasks on percent changes is given to students. It includes:

> In a sale, all prices in a shop were decreased by 20%. After the sale they were all increased by 20%. What was the overall effect on the shop prices? Explain how you know.

The vast majority of students (and many adults!) think there is no overall change. *Price—20% + 20% = Price*. You just add % changes.
Real understanding involves knowing that we are combining *multipliers*:
Price × 0.8 × 1.20 = *Price* × 0.96—a 4% reduction.

These are challenging ideas to get across by explanation, or by standard 'demonstrate and practice'—the prevalence of the misconception makes that clear. However, the challenges of understanding percent increases and decreases become much more accessible when students confront them in their own work, as in this lesson.

After the diagnostic pre-assessment, the students are given four cards with carefully chosen numbers (100, 150, 200, 160) for the corners of a poster, and ten arrow cards. Eight of the arrow cards contain expressions like "increase by 50%" or "decrease by 25%"; two are blank. The students' task, in pairs or threes, is to place

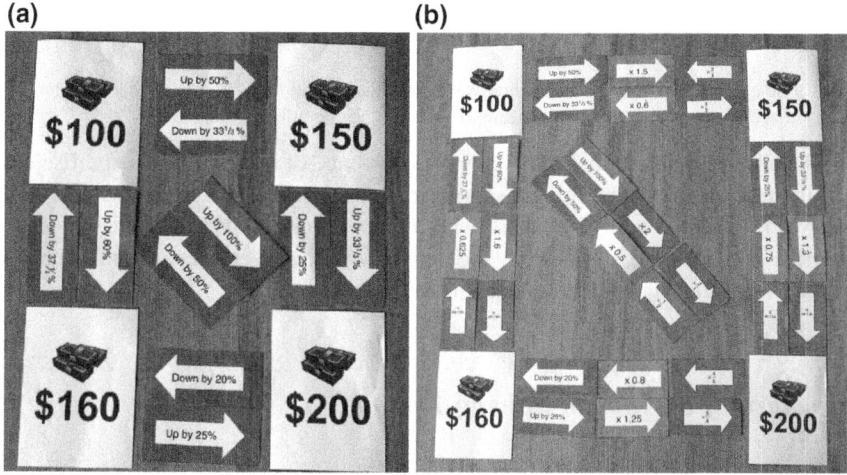

Fig. 6 a The first poster, with "*up* and *down* ...". **b** The second poster, with "multiply by ..."

the arrow cards that correctly indicate the relationships between the 4 numbers. To begin with, standard misconceptions appear: placing the "increase by 50%" arrow between 100 and 150 is straightforward, but many students place the "obvious" reverse, "decrease by 50%" alongside it. Then they discover they need that arrow to connect 200 to 100. This provokes discussion, and provides room for questions from the teacher. Teachers are prompted to ask students to clarify their thinking and share ideas with other students. The result is Fig. 6a.

Then two further sets of arrow cards are distributed, and placed in a similar way: first multiplications by decimals ("×1.5," etc., Calculators are also given out at this point.) and then multiplication by fractions ("×3/2", with "×2/3" for the inverse—a key insight for proportional reasoning). This, shown in Fig. 6b, exemplifies Objective 3 above, linking topics that are initially taught separately. (Linking the two numbers 150 and 160 is kept in reserve, for students who move rapidly through the lesson.) Although aimed at Grade 7, this lesson also provides valuable "stress testing" of understanding in later grades. It shows the importance of building connections, and the way discussion on rich tasks drives this.

This lesson exemplifies a broader design goal: to help students see results from different perspectives. Richard Feynman[3] put it thus: "If you find a result one way, it is worth thinking about. If you can show it in two ways, it may well be true. If you can show it three ways, it probably is." Not proof, but deep understanding.

---

[3]Private communication to HB, then working in physics at Caltech.

Developing Strategies for Problem Solving

> A problem is a task that the individual wants to achieve, and for which he or she does not have access to a straightforward means of solution. (Schoenfeld, 1985)

The ability to tackle such problems is the essence of mathematical expertise. So helping teachers to teach problem solving effectively has been an ongoing strand of our work, starting with *Problems with Patterns and Number*s. Here we shall describe the most recent design: the formative assessment lessons in problem solving from the Mathematics Assessment Project.

Structure of a Problem Solving Lesson

These have a rather different design from the concept-development lessons, though the time-structure is similar. They use, instead of alternative (mis)conceptions, alternative solutions to a single rich problem around which the lesson is built. This is the sequence:

- In a prior lesson the students spend around 20 min tackling an expert task, individually and unaided.
- Before the main lesson the teacher assesses the work, looking for different approaches and, with guidance from the "Common Issues" table, prepares qualitative feedback in the form of questions—sometimes individually but often for the class.
- In the main lesson the students review their own work in the light of the teacher's feedback and write responses.
- Collaborative work in pairs or threes follows, with students working to share ideas and to produce joint solutions.
- Carefully chosen examples of other students' work using different mathematical methods are introduced. The groups are asked to review and critique the various solutions, in their groups, then as a whole class discussion.
- Whole class discussion focuses on the payoff of mathematics at different levels. (The sample student work allows us to show more powerful mathematics than is likely to arise in a typical class.)
- Students improve their solutions to the initial problem, or one much like it.
- Finally, in a period of individual reflection, students write about what they have learned.

We will illustrate this design by sketching two examples. As always, there is no substitute for reviewing the complete lesson guide and, if possible, trying the lesson yourself.

Design and Development for Large-Scale Improvement   193

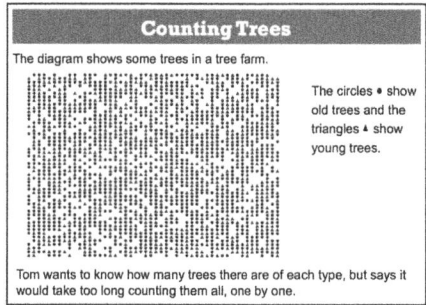 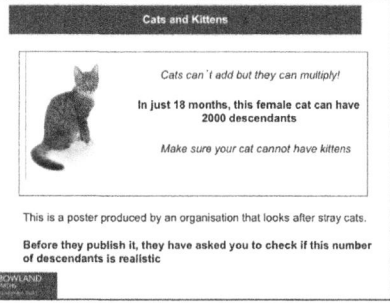

Fig. 7 The tasks from *Counting Trees and Cats and Kittens*

Problem Solving Tasks: Counting Trees and Cats and Kittens

The problem-solving-focused Classroom Challenges make fundamental use of core content but do so in the context of challenging students to use the mathematics in ways that call for being strategic and logical. They emphasize working through and explaining their problem solving processes.

Students are also expected to analyse alternative solutions, each incomplete or incorrect, in order to compare and challenge their approaches. These will stimulate further analysis and development. Consider the "Counting Trees" task in Fig. 7. The fundamental decisions in working on this task are strategic. If you don't count all the trees, then, you need to sample. But how should you do so? How large a sample; where do you take it from; how do you scale up? Do you want to get more sophisticated, and take a few samples, and average results? Of course, the more work one does, the more accurate the results are likely to be—but the less effort one saves.

In this lesson fundamental strategic (and mathematical) considerations emerge as different groups compare their approaches. Furthermore, all of this work is grounded in applications of ratio and proportion: proportionality underlies the notion of sampling and scaling. Hence this problem-solving lesson, like the concept-focused lessons, engages students in linking a range of fundamental concepts.

In the "Cats and Kittens" lesson, an advertisement advising neutering says a female cat can have 2000 descendants in 18 months. It gives the following data: *A female cat can get pregnant at age about 4 months. The pregnancy lasts about 2 months. A typical litter is 4–6 kittens. A cat can have about 3 litters a year, until they are 10 years old.*

The students are asked to work out whether the estimate of 2000 is reasonable? This much more open problem involves overlapping exponential growth but, for middle school students, the essential challenge is finding and using an appropriate representation to organize the calculations. (Student work is shown in emma.-mathshell.org, where there are links to the complete lessons)

The lessons from the Mathematics Assessment Project have proved popular with US teachers—and independent evaluators. On student learning, a report on 9th grade algebra students (Herman et al., 2014) found the average gain in algebra after 8–12 days of instruction using Classroom Challenges was 4.6 months more than the norm for grade 9. How could that be? There are a number of explanations. In content terms, these lessons are *synthetic*: they pull together prior learning and enhance it. But the pedagogy of the lessons, and its impact on the teachers' style, is at least as important.

Tools for Supporting Systemic Change

Many groups around the world now know how to enable typical teachers to teach much better mathematics much more effectively. Nobody knows how to lead school systems to make the changes needed for this to happen on a large scale.

We believe that this is the central challenge of our time. We believe that all the key players (policy makers, the research community, administrators, principals and teachers) play a role in this systemic failure and must be part of the resolution. In this final section, we look at what barriers seem to impede improvement, and how we might help to overcome them (Burkhardt & Schoenfeld, 2003; Burkhardt, 2009, 2015). Despite the limited success so far, we remain hopeful.

With our US partners we have recently taken on a specific challenge: *Can we develop effective system-level tools?* While teaching materials and assessment tools are well-recognized as important, and professional development tools are slowly being accepted, people in leadership roles in school systems (local school districts, states, nations) have not seen the value that tools could provide for them. Working with 10 school system partners in a Mathematics Network of Improvement Communities (NIC), our experience so far shows that we can develop tools to meet specific challenges that the partners specify as important, and that these tools can be helpful. We mention a few in what follows. But it is still early days.

Tools for Professional Development

We start with an area of success. Everyone recognizes that teaching quality is crucial to student learning, that improvement involves qualitatively new challenges for most teachers of mathematics of the kind discussed above, and that ongoing support for teachers in developing their professional expertise is essential. While we have shown that teaching materials can provide powerful support for teachers, they need to be complemented by effective professional development (PD) programs. To implement that on a large scale will require well-engineered tools.

Why do we need materials for professional development? First there is a mismatch of scale: the number of PD leaders with the right expertise in this area is far too few for the number of teachers who need support. Secondly, developing PD that *actually leads to changes in teachers' classroom practice*, and is cost-effective in teacher time, is a challenging design problem. (Evaluations of PD habitually only ask if the participants found it valuable—a very different criterion.)

We have found two key design features for effective professional development. It is activity-based, since active learning by processing issues is as important for teachers as it is for students. It is on-going, since high-quality teaching is the product of decade-long professional learning within a strong theoretical framework like TRU, Teaching for Robust Understanding (Schoenfeld, 2014). This design challenge implies a need for well-engineered materials. Our approach to PD has always been based on teachers learning 'constructively' from carefully designed experiences in their own classrooms. In our "sandwich model" (Swan & Pead, 2008) a group of teachers first meet for a structured discussion of a key issue of pedagogy: 'Handling classroom discussion in a non-directive way', for example. They together prepare a lesson based around teaching materials we design. They each teach the lesson in their own classrooms, or observe each others' lesson, then prepare feedback for the second session together. In this discussion they report back and return to a structured reconsideration of the issues and of the next step in their development.

These modules support the first stages of a route towards our longer-term goal of helping teachers to become part of a professional learning community, using Japanese 'Lesson Study' as a model. In NIC we have developed tools for system leadership on approaches to the design of PD, and of lesson study. The NIC *Classroom observation tool* is designed for use by school principals and others who, despite not having a mathematics background, are required by their systems to observe and evaluate mathematics teachers. Based on the TRU framework, this tool is designed to help 'non-math-ed people' pick out the important things in the classroom, focusing on the nature and quality of what students are asked to do and how they are responding to it. ("A quiet class, working hard" may impress but it is not the core indicator of a good learning environment.)

Strategic Design Opportunities

In looking for ways to overcome systemic obstacles, it's worth looking for 'leverage' points that offer a way to answer the fundamental question facing reform: "*Why should they change?*" Here we list a few responses, starting with those where relatively small changes can have big effects.

Design 'tests worth teaching to'. Though high-stakes examinations are often barriers to progress, they can be and have been powerful levers for improvement (Burkhardt, 2009 gives examples, including TSS). The empirical fact that *What You Test Is What You Get* in most classrooms means that better tests can lead to lessons of higher-quality, as long as teachers are given effective support in meeting the new challenges such tests present. This needs an explicitly specified balance and weighting across the elements in Table 1: factual knowledge and procedural fluency, conceptual understanding and logical reasoning, and problem solving and strategic competence—with comparable weightings in both curriculum and assessment.

Facts and procedures are usually dominant in tests because they are simplest to assess (and to teach). Assessing conceptual understanding is more complex, since it involves chains of autonomous student reasoning. Problem solving also requires extended reasoning, including choices of suitable mathematical tools and their subsequent application. Assessing these needs different design tactics as well as richer tasks: for example, asking students to critique sample responses to a complex task. The design of this broader and deeper kind of assessment depends on task designers with a wider range of skills than test providers have needed for 'novice tests'. The educational disasters that have so often been produced in the process of turning (usually well meaning) intentions into actual high-stakes tests (Burkhardt, 2009) make this a crucial opportunity for progress.

Aim for alignment across curriculum standards, teaching materials, assessment and professional development. This avoids sending mixed signals to teachers. NIC has developed a *Program Coherence Health Check* tool based on comparing the balance of task-types in the various aspects of the system's improvement program —and describing options for improving the alignment.

Plan the pace of change. Politicians try to "fix the teaching mathematics problem" in ways that they wouldn't try in other fields, for example medicine, where gradual improvement is accepted as inevitable. Well-engineered gradual change can work in education, too, while politics-driven "Big Bang" methods typically yield only superficial change. The appropriate strategic design question is "How big a change can teachers carry through effectively, year-by-year, given the support we can make available?" We have observed in Japan and other countries that deep challenges to teacher expertise can, if done well, be exciting for teachers. Developing such long-term professional development practices is vital.

Structural Design Tactics

Moving from the strategic to the structural, the following have proved powerful design tactics.

Use replacement units to support gradual change at a digestible pace. TSS modules (Fig. 3) provide coherent support, integrating assessment, curriculum and professional development materials. Software microworlds, as in ITMA (Fig. 4), help teachers handle inquiry-based learning, with teachers and students naturally shifting roles. "Classroom Challenges" have proved powerful in advancing student learning. Replacement units like these can provide 'protein supplements to a carbohydrate curriculum'.

Use exemplars Descriptions alone tend to be interpreted within the reader's prior experience. We hope the figures in this paper, and in emma.mathshell.org, show the value of task and lesson exemplars in communicating meaning.

Identify target groups be they students, teachers, PD leaders, superintendents, and/or policy makers, and co-develop your tools with them. Who do we need this to work for? Not just the enthusiasts. We found that "second worst teacher in your department" works well with designers as a teacher target group!

Distribute design load "How much guidance shall we give to teachers?" is a key design question. Too little and they won't have enough support; too much and they won't read it. We offer detailed guidance when we are better placed to do so than the teachers we serve. (The 'trials teachers' usually suggest more.)

Design and Development Tactics—and Costs

The following tactics help to make the Shell Centre approach cost-effective.

- **"Fail fast, fail often"**—rapid prototyping with quick feedback allows the design team to learn quickly.
- **Make feedback cost-effective** by getting rich feedback from small samples. We find 3–5 classrooms is large enough to distinguish general from idiosyncratic features, and small enough to allow the rich observational data needed to inform revision.
- **"Design control"** describes our identifying who, after discussions within the team, will take the design decisions in each area of design. The alternative, seeking consensus, is too expensive in time and doesn't always produce good designs.

Research-based design and development is much more expensive than traditional "authoring"—for us, typically US$3000 per task, $30,000 per lesson. But good engineering can ensure that the activities work well and that the materials communicate, enabling typical users to meet ambitious educational goals. Surprisingly, though these sums look large, the cost of using this approach for the whole curriculum would still be negligible within the cost of running a large education system.

The Case for "Big Education"

Other fields accept that big problems in complex systems need big coherent collaborations using agreed common methods and tools, specifically developed for key problems of practice. The CERN Large Hadron Collider and the Human Genome Project are two obvious examples. Most medical research is of this kind. We argue (Burkhardt, 2015) that research in education needs a similar approach if, for example, we want the better evidence on the generalizability of research results that design needs. This is a challenge for a field whose academic value system has encouraged new ideas over reliable research, new results over replication and extension, personal research over team research, disputation over consensus building, academic papers over products and processes—all of which conflict with the goal of well-founded large-scale impact on practice.

…. and finally

The work we have described here is the product of the brilliant individuals we have been fortunate enough to work with over the last 35 years: our colleagues in the Shell Centre team—Alan Bell, Rosemary Fraser, Richard Phillips, Daniel Pead, Rita Crust—and many other researcher-designers in Nottingham and around the world, notably Alan Schoenfeld, Phil Daro, David Foster and the Silicon Valley Math Initiative, Sandra Wilcox and her Michigan State team, Kaye Stacey and other outstanding Australians, and the ITMA team at Marjons. In addition, enormous thanks are due to the teachers in whose classrooms these tools have been trialed and observed. The work has been supported by a variety of willing-to-be guided funding agencies from government, assessment, and the foundations in the UK, the US and the EU.

Finally, we look at the issues that the team has faced over the last 40 years in simply surviving, when so many fine design teams have struggled, often disappearing into other work. The account in this chapter, of coherent strands of research and development over decades, shows that longevity is important; with funding uncertain from project to project, it is not easy to achieve. We have found strategies that can help. First, it is important to diversify the sources of funding—each funding agency has priorities which change over time. For example, in the 1980s following the Cockcroft Report (1982), the UK Government saw the need for R&D to help the system meet the new goals but, with the 1989 introduction of the National Curriculum, government saw implementation as its only concern. However, at this time in the US the NCTM Standards (NCTM, 1989) appeared, which led to a surge of support for R&D over the next 15 years from the National Science Foundation. This pattern, continued across the US, UK and the European Union, along with some luck in the timing, has helped our team survive. This illustrates the second element of strategy—to build long-term relationships across the mathematics education world. We all benefit from the mutual enrichment and support in many ways, including funding. Last but not least, it is strategically important to work on projects that you think are important, looking for the overlap of funding opportunities with challenges that seem to have promise for moving the field forward. To this end, we have been proactive in proposal design—funders rarely understand what they want in any depth and, we have found, are happy to let you convince them to modify their original ideas. These strategies may, with luck, give a dedicated team the time to become good at the deep engineering research that yields products that combine educational ambition with substantial impact on practice—the essence of the Emma Castelnuovo criteria.

References

Bell, A. (1993). Some experiments in diagnostic teaching. *Educational Studies in Mathematics, 24* (1).

Black, P. J., & Wiliam, D. (1998). *Assessment and classroom learning assessment in education, 5*, 7–74.

Burkhardt, H. (2006). From design research to large-scale impact: Engineering research in education. In J. Van den Akker, K. Gravemeijer, S. McKenney, & N. Nieveen (Eds.), *Educational design research*. London: Routledge.

Burkhardt, H. (2009). On strategic design. *Educational Designer, 1*(3). Retrieved from: http://www.educationaldesigner.org/ed/volume1/issue3/article9

Burkhardt, H. (2015). Mathematics education research: a strategic view. In L. English & D. Kirshner (Eds.), *Handbook of international research in mathematics education* (3rd ed.). London: Taylor and Francis.

Burkhardt, H., Fraser, R., Coupland, J., Phillips, R., Pimm, D., & Ridgway, J. (1988). Learning activities & classroom roles with and without the microcomputer. *Journal of Mathematical Behavior, 6,* 305–338.

Burkhardt, H., & Schoenfeld, A. H. (2003). Improving educational research: towards a more useful, more influential and better funded enterprise. *Educational Researcher, 32,* 3–14.

Cockcroft Report. (1982). *Mathematics counts, report to the UK government of the committee of enquiry*. London, UK: HMSO.

Herman, J., Epstein, S., Leon, S., La Torre Matrundola, D., Reber, S., & Choi, K. (2014). In *Implementation and effects of LDC and MDC in Kentucky districts (CRESST Policy Brief No. 13), National Center for Research on Evaluation, Standards, and Student Testing*. Los Angeles: University of California.

ISDDE (2012). Black, P., Burkhardt, H., Daro, P., Jones, I., Lappan, G., Pead, D., et al. (2012). High-stakes examinations to support policy. *Educational Designer, 2*(5). Retrieved from http://www.educationaldesigner.org/ed/volume2/issue5/article16

NCTM. (1989). *Curriculum and evaluation standards for school mathematics*. Reston, VA: National Council of Teachers of Mathematics.

Polya, G. (1945). *How to solve it*. Princeton, NJ: Princeton University Press.

Schoenfeld, A. H. (1985). *Mathematical problem solving*. Orlando, FL: Academic Press.

Schoenfeld, A. H. (2002). Research methods in (mathematics) education. In L. English (Ed.), *Handbook of international research in mathematics education* (pp. 435–488). Mahwah, NJ: Erlbaum.

Schoenfeld, A. H. (2014). What makes for powerful classrooms, and how can we support teachers in creating them? *Educational Researcher, 43*(8), 404–412. doi:10.3102/0013189X1455

Swan, M. (2006). *Collaborative learning in mathematics: A challenge to our beliefs and practices*. London: National Institute for Advanced and Continuing Education (NIACE) for the National Research and Development Centre for Adult Literacy and Numeracy (NRDC).

Swan, M., Binns, B., Gillespie, J., Burkhardt, H., & The Shell Centre Team. (1987–89). In *Numeracy through problem solving: Five modules for teaching and assessment: Design a board game, produce a quiz show, plan a trip, be a paper engineer. Be a shrewd chooser*. Harlow, UK: Longman. Downloadable from http://www.mathshell.com

Swan, M., & Burkhardt, H. (2014). Lesson design for formative assessment. *Educational Designer, 2*(7). Retrieved from: http://www.educationaldesigner.org/ed/volume2/issue7/article24

Swan, M., Pitts, J., Fraser, R., & Burkhardt, H., The Shell Centre Team. (1984). In *Problems with patterns and numbers*. Manchester, U.K: Joint Matriculation Board. Downloadable from http://www.mathshell.com

Swan, M., Pitts, J., Fraser, R., Burkhardt, H, & The Shell Centre Team (1985). In *The Language of Functions and Graphs*. Manchester, U.K.: Joint Matriculation Board. Downloadable from http://www.mathshell.com

Swan, M., & Pead, D. (2008). Bowland maths professional development resources. Online: http://www.bowlandmaths.org.uk/pd. Bowland Trust/Department for Children, Schools and Families.

Swan, M., & Wall, S. (2005). In *Improving the learning of mathematics*. UK: Department for Education and Skills.

VanLehn, K., & Ball, W. (1991). Goal Reconstruction: How Teton blends situated action and planned action. In K. VanLehn (Ed.), *Architectures for Intelligence* (pp. 147–188). Hillsdale, NJ: Erlbaum.

Wiliam, D., & Thompson, M. (2007). Integrating assessment with instruction: What will it take to make it work? In C. A. Dwyer (Ed.), *The future of assessment: shaping teaching and learning* (pp. 53–82). Mahwah, NJ: Lawrence Erlbaum Associates.

Open Access Except where otherwise noted, this chapter is licensed under a Creative Commons Attribution 4.0 International License. To view a copy of this license, visit http://creativecommons.org/licenses/by/4.0/.

Making Sense of Mathematics Achievement in East Asia: Does Culture *Really* Matter?

Frederick K.S. Leung

Abstract East Asian students have persistently performed well in recent international comparative studies of mathematics achievement, and I have been offering explanations from the perspective of the influence of the Confucian Heritage Culture (CHC), which is shared by these high performing East Asian countries. In this paper, two challenges to this cultural explanation will be dealt with: whether these East Asian countries really form a group and whether there is a more direct way to study the influence of culture on mathematics achievement. Three studies on secondary analyses of the TIMSS and PISA datasets are presented to support the assertion that the East Asian countries do form a cultural cluster, and preliminary results of a study that looks into the influence of the English and Chinese languages on students' assessment in mathematics are reported.

Keywords Culture · Confucian heritage culture (CHC) · East Asian countries · Language and mathematics learning · International studies of mathematics achievement

Introduction: The Superior Mathematics Achievement of East Asian Students in International Studies

In the citation for my Hans Freudenthal Medal 2013 by ICMI, my research was recognized as being "in comparative studies of mathematics education and on the influence of culture on mathematics teaching and learning" and one of my

---

Hans Freudenthal award.

F.K.S. Leung (✉)
Faculty of Education, The University of Hong Kong, Pokfulam Road, Hong Kong
e-mail: frederickleung@hku.hk

© The Author(s) 2017
G. Kaiser (ed.), *Proceedings of the 13th International Congress on Mathematical Education*, ICME-13 Monographs, DOI 10.1007/978-3-319-62597-3_13

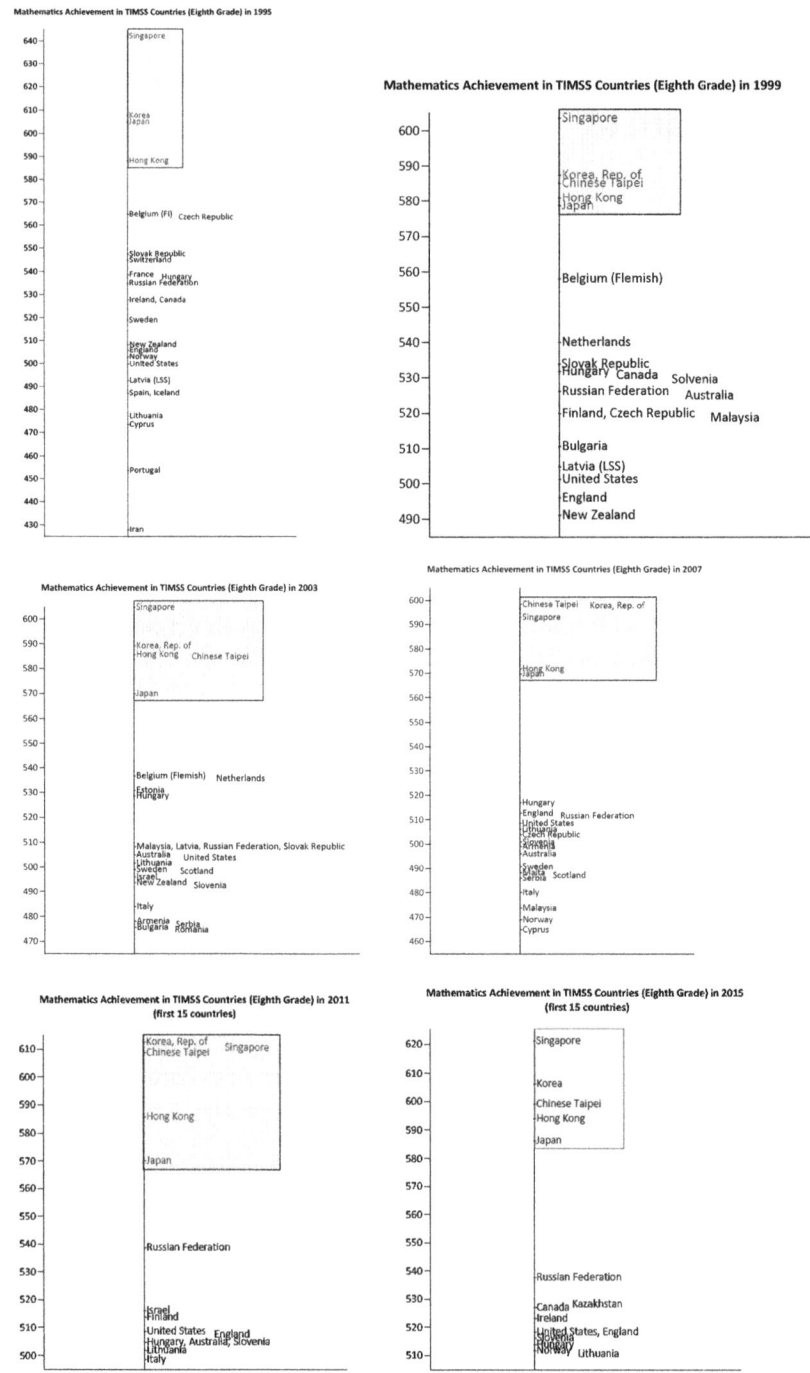

Fig. 1 Performance of East Asian countries in TIMSS

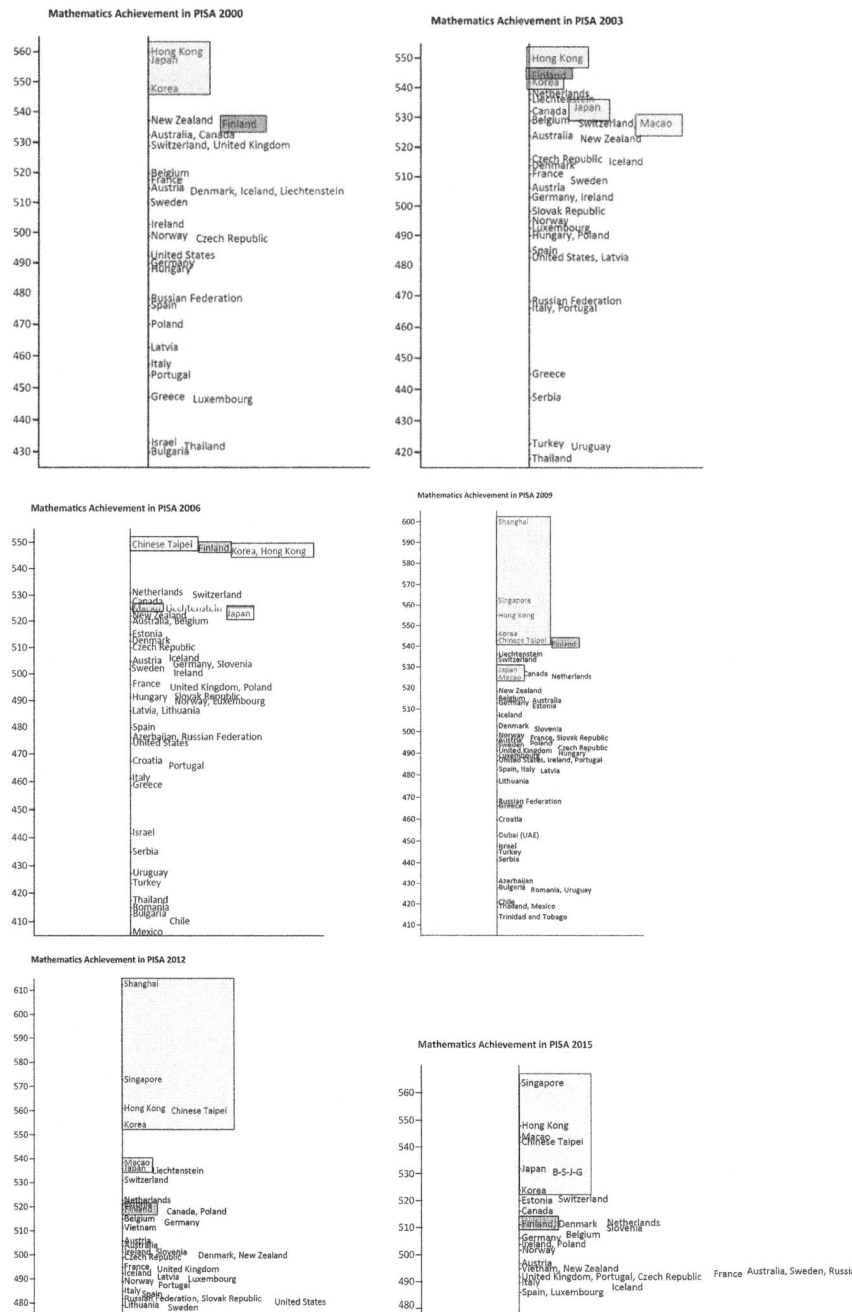

Fig. 2 Performance of East Asian countries in PISA

contributions was "the utilization of the perspective of the Confucian Heritage Culture to explain the superior mathematics achievement of East Asian[1] students in international studies" (ICMI, 2013) such as TIMSS and PISA. As can be seen from Figs. 1 and 2, East Asian countries consistently took up the top positions in the rankings in TIMSS and PISA (Beaton, Mullis, Martin, Gonzalez, Kelly, & Smith, 1996; Mullis et al., 1997, 2000; Mullis, Martin, Gonzalez, & Chrostowski, 2004; Mullis, Martin, & Foy, 2008; Mullis, Martin, Foy, & Arora, 2012; Mullis, Martin, Foy, & Hooper, 2016; OECD, 2001, 2003, 2004, 2007, 2010, 2014, 2016), and this is what I meant by "superior performance" in my previous publications.

A Cultural Explanation of the Superior Mathematics Achievement of East Asian Students

In seeking an explanation for the superior mathematics achievement of East Asian students, I argued that these high-performing East Asian countries shared a common culture, referred to as "Confucian Heritage Culture" (CHC) by Ho (1991; quoted in Biggs, 1996, p. 46). I then examined the common values shared by these CHC countries that are related to education (Leung, 2001).

In putting forth the cultural thesis above, I encountered at least two queries or challenges:

(1) Do these East Asian countries really form a group distinct from other countries? Other than the fact that they rank high in international studies, are there other empirical data that show that they fall into a common group? (For example, Finland has also been performing very well in the recent cycles of PISA. But Finland does not share the East Asian culture.)
(2) Is there a more direct way to study whether culture does affect mathematics achievement (other than simply examining the common values shared by the CHC countries that are related to education)?

Do These East Asian Countries Form a Cluster?

In this section, I present results of three studies based on secondary analyses of the TIMSS and PISA data to show that there is evidence other than ranking in international studies that indicates that the East Asian countries do form a distinct cluster.

---

[1]In this paper, East Asian "countries" refer to the systems of Hong Kong, Japan, Korea, Singapore and Taiwan.

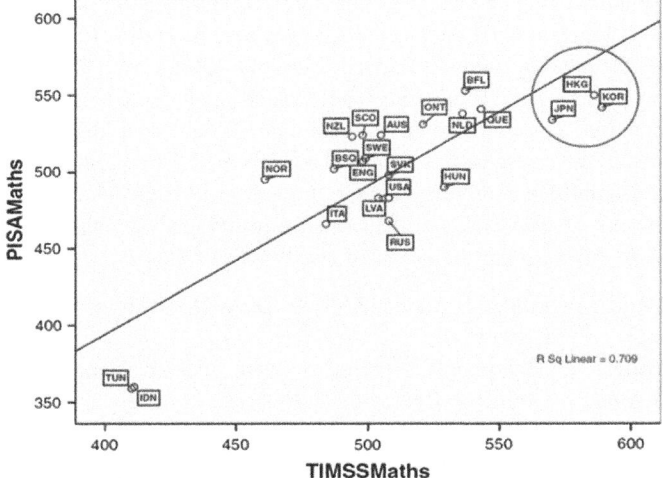

Mathematics country mean scores: PISA 2003 versus TIMSS 2003

Fig. 3 Prediction of PISA results from TIMSS results 2003

The first study by Wu (2009) covers a comparison of PISA and TIMSS 2003 achievement results in mathematics.

Twenty-two countries participated in both TIMSS and PISA 2003,[2] and Wu (2009) compared the performance of these 22 countries between TIMSS and PISA. Wu concluded that "content balance" and "years of schooling" explain the different achievements of countries in TIMSS and PISA 2003. In one of the analyses, Wu predicted these 22 countries' PISA results from their TIMSS results statistically, and the results are shown in Fig. 3.

As can be seen from Fig. 3, Hong Kong, Japan and Korea (the only East Asian countries that participated in the two studies) not only clustered near each other, but also all performed worse in PISA than expected from their TIMSS scores. What common characteristics do these East Asian countries share that cause them not only to perform well in both TIMSS and PISA but also to exhibit the same achievement pattern of performing worse in PISA than in TIMSS?

The second study by Lie (2014) seeks to identify similarities and differences between various groups of countries, referring to their relative weaknesses and strengths.

---

[2]TIMSS was conducted once every four years, and PISA was conducted once every three years. TIMSS tested students' achievement in mathematics and science, while PISA tested mathematics, science and reading literacy. Each cycle of PISA focuses on one of the three areas of study. It so happened that both TIMSS and PISA took place in 2003, and the area of focus for PISA 2003 was mathematics.

In a presentation made at the 55th General Assembly meeting of IEA in Wien in October 2014, Lie examined the patterns of the item residuals[3] of the TIMSS 2003 data to study the relative strengths and weaknesses of countries in their mathematics achievements. Lie used correlations and cluster analysis to examine the item residuals patterns, considering the international difficulty of the items and the overall scores of the items for the country. Lie identified patterns showing that there are groups of countries exhibiting similar strengths and weaknesses.

Eight clusters of countries are identified from this analysis, and I assign (arbitrary) labels to describe these clusters of countries as follows:

English-speaking: England, Scotland, New Zealand, Australia, Canada, USA, Ireland
North Europe: Netherlands, Norway, Sweden, Iceland, Denmark
Central Europe: Germany, Switzerland, Austria
South Europe: Portugal, Spain, Greece
Other: Iran, Thailand, Philippines, South Africa, Columbia, Kuwait
Western Europe: Belgium Fl, Belgium Fr, France
East Asia: Hong Kong, Singapore, Japan, Korea
East Europe: Czech Republic, Slovakia, Hungary, Lithuania, Russia, Latvia, Cyprus, Romania, Bulgaria, Israel.

Of interest to us is that once again the East Asian countries of Hong Kong, Singapore, Japan and Korea cluster together. That is, they do not only perform well in TIMSS 2003, they also share the same patterns of strengths and weaknesses in the performance.

The third study (Guo, 2014) examines the relation between time for studying mathematics and achievement.

In this paper, Guo (2014) examined the relationship between out-of-school study time and mathematics achievement based on the PISA 2012 data. In Guo's analysis, mathematics achievement was plotted against out-of-school study time for each of the PISA 2012 countries. Guo found that a quadratic relation between achievement and study time is a better fit of the data than a linear relation is, and he computed and plotted the quadratic relation for each country. Each of these quadratic graphs then represents a relation between mathematics achievement and the out-of-school study time that is characteristic for that country. Guo then ran a cluster analysis for these characteristic relations and obtained four different clusters of countries (see Fig. 4). Interestingly, Hong Kong, Japan, Korea, Shanghai (China), Singapore and Taiwan again fall into the same cluster, showing that the kind of relationship between mathematics achievement and out-of-school study time is in some way similar among these East Asian countries and cities.

---

[3]"Item residual" is defined as how much better or worse than expected a national p-value is.

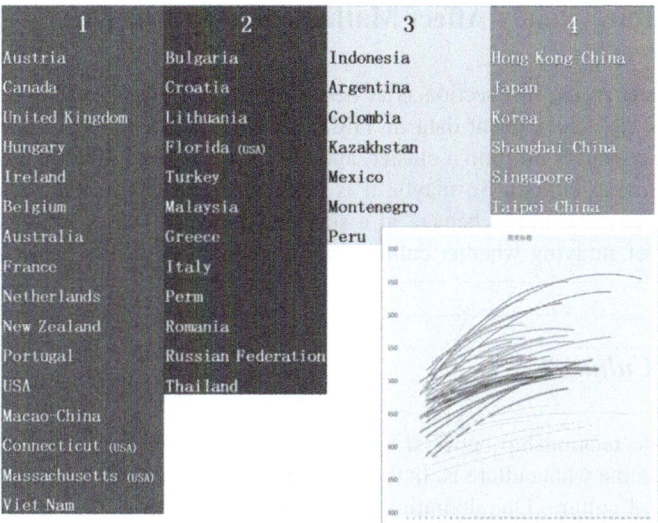

Fig. 4 Clusters of countries by relationship between mathematics achievement and out-of-school study time (*Source* PowerPoint slide from Guo (2014), reproduced with permission)

Discussion

From the three studies discussed, we can conclude that there is further evidence that these East Asian countries do fall into a cluster. These countries or cities are all in East Asia, but there is of course no reason for us to believe that the geographic location in itself will cause these countries to perform in a similar manner. So back to my original question: Could these countries form a cultural cluster?

At a plenary panel during ICME-12 on Mathematics Education in East Asia (see also Leung, 2006), I examined the cultural commonalities of these East Asian countries, namely CHC, and I discussed the following characteristics of CHC, which are deemed to be related to mathematics learning:

(1) Examination culture
(2) Belief in effort
(3) Memorization and practice
(4) Reflection.

To answer the question of whether these East Asian countries really form a cluster, the discussions in this section show that there is further evidence other than the ranking of these countries in international studies that shows that these East Asian countries do form a cluster. Since there is much cultural commonality among these East Asian countries, it is probable that these countries form a *cultural* cluster.

Does Culture Really Affect Mathematics Achievement?

The argument in the last section is at best an inference; that is, from the various analyses of the achievement data in TIMSS and PISA, we can conclude that the East Asian countries do form a cluster, and, on the other hand, these countries also share a common culture. So maybe it is their common culture that has led their students to achieve and to behave in a similar manner. However, is there a more direct way of studying whether culture really affects mathematics achievement?

What Is Culture?

To study the relationship between culture and mathematics achievement, we first need to examine what culture is. In the literature, there are different descriptions and definitions of culture. One definition offered by Smith et al. is that culture is "the fabric of ideas, ideals, beliefs, skills, tools, aesthetic objects, methods of thinking, customs and institutions" (Smith, Stanley, & Shores, 1957, p. 4). Examining these elements of culture defined by Smith et al., I can classify them as belonging to either superstructure or base structure (using a Marxist notion):

Superstructure: ideals, beliefs, ideas, methods of thinking
Base structure: skills, tools, customs, institutions, aesthetic objects.

The Crucial Role of Language

I would further argue that, mediating these two levels of structure, language plays a critical role; i.e., elements at the basic structure shape or determine the elements at the superstructure through the medium of language, since beliefs, ideas, etc., are constructed through the medium of language (see Fig. 5). This is in accordance with a social constructivist view of mathematics learning:

> Social constructivism views mathematics as a social construction ... Human language, rules and agreement play a key role in establishing and justifying the truths of mathematics ... The basis of mathematical knowledge is linguistic knowledge, conventions and rules, and language is a social construction. (Ernest, 1991, p. 42)

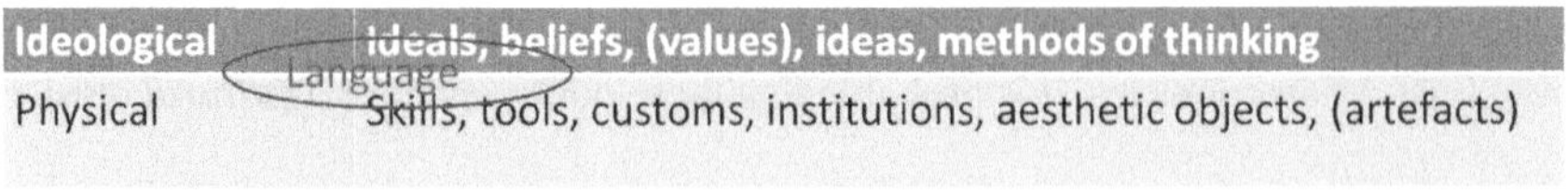

Ideological	ideals, beliefs, (values), ideas, methods of thinking
Physical	Skills, tools, customs, institutions, aesthetic objects, (artefacts)

(Language)

Fig. 5 Elements of culture at the basic structure and superstructure levels mediated by language

All experiences (including the experience of mathematics learning) are mediated by language (Gadamer, 1979), thought and language develop together and conception evolution depends on language experience (Vygotsky, 1986).

We do not know why languages that are so different have evolved from different cultures (the tower of Babel?). We know some basic human experiences are common across cultures; some are shaped by the environment and some are developed within particular historical contexts. These give rise to commonalities and differences in different languages (perhaps many more differences than commonalities). In any case, it is clear that language is the medium and tool for mathematics learning, but this important component of culture has hardly been studied in the past. One possible reason for this lack of attention to the role of language may be the assumption held by many that mathematics is a "universal language" and that it is learned in roughly the same way through different languages.

Language Competence and Mathematics Achievement

In the literature, Aiken (1972) showed that reading vocabulary, reading comprehension, mechanics of language and spelling have higher correlations with arithmetic reasoning than with arithmetic fundamentals at all elementary grade levels. To solve word problems or problem-solving types of questions, the language factor has a more important role than when solving the bare number problems (problems without context). However, do different languages make a difference to mathematics learning? How do we study the relationship between language and mathematics learning and achievement?

There are two approaches to studying the relationship between language and mathematics learning and achievement. The first I would call a theoretical or hypothetical approach. In this approach, we study the question of whether the language structures (e.g., the language used in textbooks and in test items) of different languages differ in complexity and hence have the *potential* to have an impact on students' processing of mathematics and test items. The second approach is an empirical one, where we study the question of whether students in different language groups process mathematics (in their brains) differently: Does the same mathematics content or test item exert the same or different cognitive demands on the students?

In the theoretical approach, we analyse the linguistic structures of mathematics texts or test items in different languages and compare their complexity. In the empirical approach, we compare the "behaviours"[4] of students from different language groups when they are processing the same mathematics contents. In studying

---

[4]By behaviours, we mean whether the brain functions in the same way or in different ways when we process the same mathematics in different languages.

these "internal" behaviours, the functions of the brain are either inferred indirectly from external behaviours (e.g., paper and pencil tests, think-aloud method, clinical observation) or directly observed using neuroscientific equipment.

The Theoretical/Hypothetical Approach

English and the East Asian languages

In the literature, scholars have discussed the features of the East Asian languages[5] (Chinese, Japanese and Korean). East Asian languages are very different from Western languages such as English, and some have argued that there are features of the Chinese language that are favourable to the learning of mathematics. Fuson and Kwon (1991) argued that the irregular systems of number words in the English language affect students' numerical learning. The English language system of number words does not directly name the values of ten and one in two digit numbers as in some East Asian languages such as Korean and Chinese. Some features of English make it even more difficult to see the underlying tens and ones structures and how the first nine numbers are reused to make the decade words.

Leung and Park (2009) explored the interplay between the names of geometric figures in English, Chinese and Korean and students' capability in identifying the geometric figures. Names of some geometric figures in the three languages were analysed, and a test on identifying and defining these geometric figures was administered to Grade 8 students from the three language groups in Hong Kong, Seoul and San Diego. Students were then interviewed on the answers to the test. It was found that the way geometric figures are named and defined in different languages might have had an impact on students' identification of the geometric figures as well as their understanding of the properties of the figures.

Galligan (2001) outlined the differences between the Chinese and English written languages in terms of syntax, orthography, phonology and semantics. She argued that all four linguistic features influence the processing of mathematical text. Galligan's study found some evidence for a Chinese language advantage with respect to number sense and possibly to fractions, logical connectives and relational word problems as well.

At ICME-12, Leung, Park, Shimizu, and Xu (2012) discussed features of the Chinese language, that the Chinese language is logographic (rather than alphabetical and phonetic as in English). Leung et al. observed that the use of "classifiers" in the Chinese language "unscramble the confusion that otherwise surrounds

---

[5]The Japanese and Korean languages are strongly influenced by the Chinese language. Chinese characters (Hanja) were used as the written form of the Korean language before the 15th century, and Kanji (Chinese characters) are still widely used in the Japanese language today.

conservation of numbers" (Brimer & Griffin, 1985, p. 23) and mentioned the advantage of the regular number system in Chinese for learning mathematics. As far as spoken Chinese is concerned, Leung et al. pointed out the monosyllabic nature of the Chinese language, and that the short pronunciation of the numbers zero to ten makes it easy to process mathematics.

As for written Chinese, Leung et al. discussed the logographic nature of the Chinese characters. The orthography of Chinese writing is based on the spatial organization of the components (radicals) of the Chinese characters. Chinese characters possess visual properties such as connectivity, closure, linearity and symmetry, which are faster and easier to be captured by vision (Lai, 2008).

Leung et al. also discussed the different language use in Chinese and English: Ideas are organised differently in the two languages. Using the address of his office as an example, when writing the address in English, one starts with the smallest unit (the room, namely Room 312), then progressively extends it to the building (Runme Shaw Building), the institution (Faculty of Education, the University of Hong Kong), the road where the institution is (Pokfulam Road), the city (Hong Kong) and then finally the largest unit, the country (China). In contrast, in Chinese, one starts with the largest unit, the country (China [中國]), then the city (Hong Kong [香港]), the street (Pokfulam Road [薄扶林道]), the institution (Faculty of Education, the University of Hong Kong [香港大學教育學院]), the building (Runme Shaw Building [邵仁枚樓]), and finally the room (Room 312 [312室]).

How will this different language use affect students' understanding of mathematics? Will a Chinese speaker and an English speaker conceptualize geometric figures differently? For example, when confronted with a complicated geometric figure, will an English student start with the smallest units and view them in the context of the larger units, whereas a Chinese student may start with the larger picture and zoom into the smaller units? These are issues that remain to be studied.

The Empirical Approach

Clinical Studies

Miura, Kim, Chang, and Okamoto (1988) compared the cognitive representations of American, Chinese, Japanese and Korean first graders to determine if there might be variations in those representations resulting from numerical language characteristics that differentiate Asian and non-Asian language groups. Children were asked to construct various numbers using base-10 blocks. Miura et al. found that "Chinese, Japanese and Korean children preferred to use a construction of tens and ones to show numbers—place value appeared to be an integral component of their representations" (Miura et al., 1988, p. 1445), whereas English-speaking children preferred to use a collection of units, suggesting that they represent number as a grouping of counted objects. More Asian children than American children were found to be able to construct each number in two different ways, which suggests greater flexibility of mental number manipulation.

In a study by Lai and Leung (2012) on the visual perceptual abilities of Chinese-speaking and English-speaking children, the Developmental Test of Visual Perception (DTVP) was administered to 41 native Chinese-speaking children (with a mean age of 5 years and 4 months) in Hong Kong and 35 native English-speaking children (with a mean age of 5 years and 2 months) in Melbourne. The results show that Chinese-speaking children significantly outperformed the English-speaking children on general visual perceptual abilities. More interestingly, Chinese-speaking students' performance on visual-motor integration tasks was found to be far better than that of their counterparts, while the two groups of students performed similarly on motor-reduced visual perceptual tasks. Leung and Lai suggested that the written language format of Chinese might have contributed to the enhanced performance of Chinese-speaking children in the visual-motor integration tasks.

Neuroscience Studies

Tang et al. (2006) used fMRI to demonstrate a differential cortical representation of numbers between native Chinese and English speakers. Native English speakers were found to largely employ a language process that relies on the left perisylvian cortices for mental calculation such as a simple addition task. In contrast, native Chinese speakers tended to engage a visuo-premotor association network for the same task. In both groups, the inferior parietal cortex was activated by a task for numerical quantity comparison, but fMRI connectivity analyses revealed a functional distinction between Chinese and English groups among the brain networks involved in the task. These results indicate that the different biological encoding of numbers may be shaped by the visual reading experience during language acquisition and other cultural factors such as mathematics learning strategies and education systems, which cannot be explained completely by the differences in languages per se.

The Influence of the Chinese and English Languages on Students' Processing of Mathematics

Leung reported on an ongoing research project entitled "The influence of the Chinese and English languages on students' processing of mathematics word problems". The project investigates the differences in linguistic structure between the TIMSS mathematics word-problem items in Chinese and in English and the impact of the differences on Chinese- and English-speaking fourth graders in

Taiwan and Australia, respectively, when they process the mathematics items in the two languages.

There is an assumption in international studies of mathematics achievement such as TIMSS and PISA that although these studies test students in different languages, the tests are testing the same mathematics achievement. That is, once the accuracy of the translation of test items is ensured, the mathematics knowledge and the cognitive demands on students measured by items in different languages are equivalent. All ranking of countries in international studies is based on this assumption, but is this assumption valid?

The Research Questions of the Research Project are:

1. What are the differences in linguistic structure between some TIMSS 2011 mathematics word-problem items in Chinese and in English?
2. What are the behavioural differences between Chinese- and English-speaking fourth graders when they process mathematics items in the two languages in terms of their eye movements while working on word problems?
3. How are the linguistic structures of mathematics items in Chinese and in English affecting students' processing of the items?

Design of the Study

There are three stages in the design of the study:

1. Examine the syntactic structures of the TIMSS items

    (a) Item difficulties (performance differences) by different sub-domains under the domain of number
    (b) Quantizing the syntactic structures of the two languages (Chinese and English)
    (c) Linking and comparing the results of a and b.

2. Analyse the linguistic differences
3. Verify 1 and 2 above by eye-tracking studies in Taipei and Adelaide; Primary 4 students from the two different language groups will be selected to join the eye-tracking study.

Social Network Analysis

Social network analysis is employed to study the complexity of the linguistic structures of test items in the two languages. Indices such as centrality, density and prestige score are calculated for analysis, and syntactic dependency network is utilized to determine the elements of the language network. A vertex represents a different linguistic unit such as words or characters, and the edges describe the relations between these units. The indices assess which words play more important roles in helping students to solve mathematics word problems. These indices of the language network are calculated using the social network analysis software UCINET.

All the 65 word problems in the number domain of the TIMSS 2011 Primary 4 test will be analysed. The kinds of linguistic networks (indices) that affect students' processing of the items will be identified. In addition, to explore whether the linguistic structures of the items will affect students' processing of the items, 16 out of the 65 items that students in the two places find easiest and most difficult (according to the average performance of the students in Australia and Taiwan) will be chosen for the eye-movement phase of the study. Figure 6 shows some preliminary results of the social network analysis for an item (in Chinese and in English).

Eye-Tracking Study

Eye tracker monitoring provides a precise record of on-line reading behaviours that are composed of fixations and saccades of the eyes. It is generally assumed that increased time spent on a particular area reflects increased cognitive processing of that information. Through examining records of the eye-tracking, one can assess the relative amount of time spent on a given area of text.

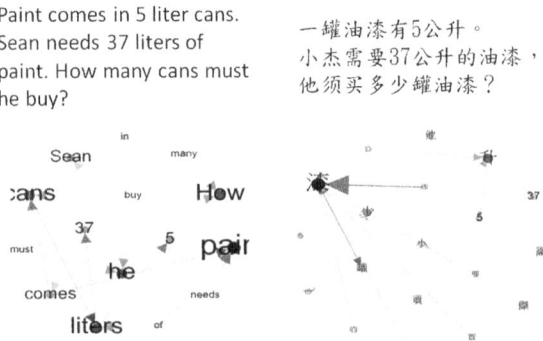

Fig. 6 Preliminary results of the social network analysis

Since the Grade 4 TIMSS items will be analysed in this study, Grade 4 students will be sampled for the eye-tracking study. Purposive sampling of students of different genders and abilities (in terms of both language and mathematics achievement) from schools with different student abilities and social and economic status (SES) background in Taipei and Adelaide will be employed. Some preliminary results of the eye-tracking study record of a student in Taiwan and a student in Australia working on the same item (in Chinese and in English, respectively) are shown in Fig. 7.

Data Analysis

The data set in this study consists of (1) the results of the social-network analysis on the language complexity of the TIMSS mathematics word problems in Chinese and in English, and (2) records of the parts of the word problems in the two languages to which students pay more attention, as measured by the length of time the students

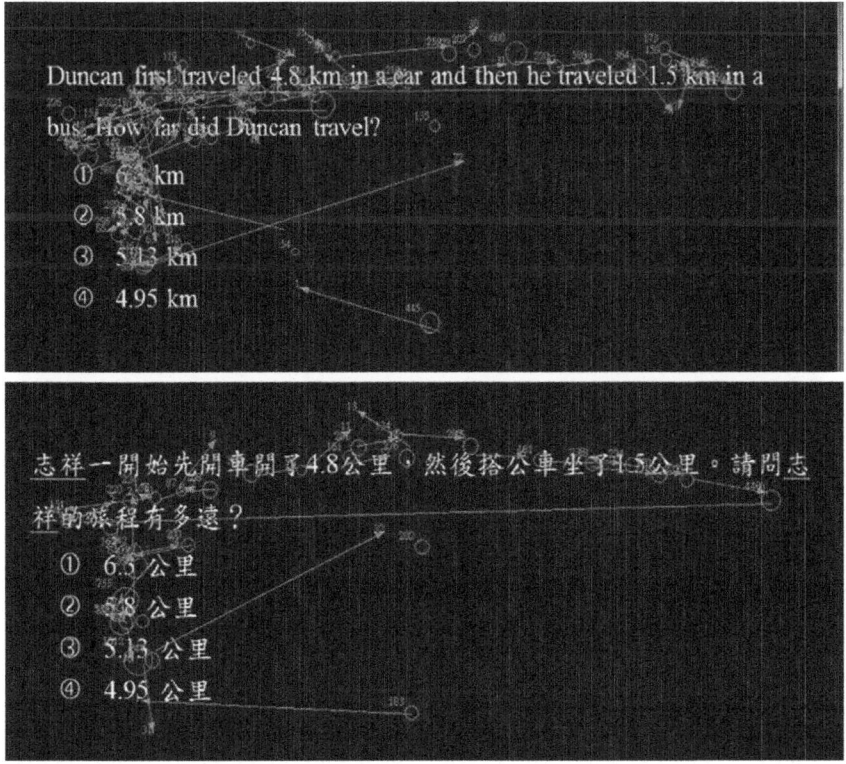

Fig. 7 Preliminary results of the eye-tracking study

fix their eyes on those parts. The two categories of data will be analysed to investigate how the linguistic structures of the TIMSS mathematics items in Chinese and in English are affecting students' processing of the items.

Significance of the Study and Further Research

This study will fill a gap in the knowledge of how language has an impact on mathematics learning and achievement. It may even throw light on why East Asian students outperform their Western counterparts in mathematics achievement in international studies such as TIMSS.

The study may be extended to examining test items in other mathematics domains and including other languages (such as Korean). Other neuroscientific technology (such as near infrared spectroscopic imaging [NIRSI]) may be employed to study brain activity in an even more direct manner.

Conclusion

In this paper, we have argued that as far as mathematics education and achievement are concerned, the East Asian countries do form a cluster distinct from countries in other regions of the world. It has been argued that this cluster of countries shares a common culture, namely the Confucian Heritage Culture (CHC), and there seem to be elements in CHC that explain student achievement in mathematics. In particular, East Asian languages seem to influence mathematics learning and assessment in ways different from Western languages.

To answer the question posed in the title of this paper: "Does culture really matter?" the answer is still: "Probably". However, there is more evidence today that it is probable than there was 20 years ago!

References

Aiken, L. R. (1972). Language factors in learning mathematics. *Review of Educational Research, 42*, 359–385.

Beaton, A. E., Mullis, I. V. S., Martin, M. O., Gonzalez, E. J., Kelly, D. L., & Smith, T. A. (1996). *Mathematics achievement in the middle school years*. Chestnut Hill, MA: International Study Center, Boston College.

Biggs, J. B. (1996). Western misconceptions of the confucian-heritage learning culture. In D. A. Watkins & J. B. Biggs (Eds.), *The Chinese learner* (pp. 45–67). Hong Kong: Comparative Education Research Centre.

Brimer, A., & Griffin, P. (1985). *Mathematics achievement in Hong Kong secondary schools*. Hong Kong: Centre of Asian Studies, The University of Hong Kong.

Ernest, P. (1991). *The philosophy of mathematics education.* London: The Falmer Press.
Fuson, K., & Kwon, Y. (1991). Chinese-based regular and European irregular systems of number words: The disadvantages for English-speaking children. In K. Durkin & B. Shire (Eds.), *Language in mathematical education* (pp. 211–226). Milton Keynes, GB: Open University Press.
Gadamer, H. G. (1979). *Philosophical hermeneutics.* Berkeley: University of California Press.
Galligan, L. (2001). Possible effects of english chinese language differences on the processing of mathematical text: A review. *Mathematics Education Research Journal, 13*(2), 112–132.
Guo, K. (2014). *Relation between time for studying mathematics and achievement.* Paper presented at the First Chinese Mathematics Education Conference. Beijing, April.
Ho, D. Y. F. (1991). *Cognitive socialization in confucian heritage cultures.* Paper presented to Workshop on Continuities and Discontinuities in the Cognitive Socialization of Minority Children. Washington DC: US Department of Health and Human Services, June 29–July 2.
ICMI. (2013). Citation of 2013 Freundenthal Awardee. Retrieved January 23, 2017 from http://www.mathunion.org/icmi/activities/awards/past-recipients/the-hans-freudenthal-medal-for-2013
Lai, M. Y. (2008). In *An exploratory study into Chinese and English children's visual perceptual and their spatial and geometric conceptions in Piagetian tasks* (Unpublished doctoral dissertation). The University of Hong Kong.
Lai, M. Y., & Leung, F. K. S. (2012). Visual perceptual abilities of Chinese-speaking and English-speaking children. *Perceptual and Motor Skills, 114*(2), 433–445.
Leung, F. K. S. (2001). In search of an East Asian identity in mathematics education. *Educational Studies in Mathematics, 47,* 35–52.
Leung, F. K. S. (2006). Mathematics education in East Asia and the West: Does culture matter? In F. K. S. Leung, K. D. Graf, & F. J. Lopez-Real (Eds.) *Mathematics education in different cultural traditions: A comparative study of East Asia and the West, The 13th ICMI Study* (pp. 21–46). New York: Springer.
Leung, F. K. S., & Park, K. M. (2009). The influence of language on the conception of geometric figures. *Proceedings of PME, 33*(1), 418.
Leung, F. K. S., Park, K. M., Shimizu, Y., & Xu, B. (2012). *Mathematics education in East Asia.* Plenary panel, ICME-12, Seoul, 13 July.
Lie, S. (2014). Similarities and differences between groups of countries concerning relative weaknesses and strengths. http://www.iea.nl/fileadmin/user_upload/General_Assembly/55th_GA/GA55_relative_weaknesses_and_strengths_groups.pdf. Accessed January 13, 2017.
Miura, I. T., Kim, C. C., Chang, C. M., & Okamoto, Y. (1988). Effects of language characteristics on children's cognitive representation of numbers: Cross-national comparisons. *Child Development, 59,* 1445–1450.
Mullis, I. V. S., Martin, M. O., Beaton, A. E., Gonzalez, E. J., Kelly, D. L., & Smith, T. A. (1997). *Mathematics achievement in the primary school years.* Chestnut Hill, MA: International Study Center, Boston College.
Mullis, I. V. S., Martin, M. O., & Foy, P. (2008). *TIMSS 2007 international mathematics report.* Chestnut Hill, MA: TIMSS and PIRLS International Study Center, Boston College.
Mullis, I. V. S., Martin, M. O., Foy, P., & Arora, A. (2012). *TIMSS 2011 international results in mathematics.* Chestnut Hill, MA: TIMSS and PIRLS International Study Center, Boston College.
Mullis, I. V. S., Martin, M. O., Foy, P., and Hooper, M. (2016). TIMSS 2015 international results in mathematics. TIMSS and PIRLS International Study Center, Boston College. http://timssandpirls.bc.edu/timss2015/international-results/. Accessed January 13, 2017.
Mullis, I. V. S., Martin, M. O., Gonzalez, E. J., & Chrostowski, S. J. (2004). *TIMSS 2003 international mathematics report.* Chestnut Hill, MA: TIMSS and PIRLS International Study Center, Boston College.
Mullis, I. V. S., Martin, M. O., Gonzalez, E. J., Gregory, K. D., Garden, R. A., O'Connor, K. M., et al. (2000). *TIMSS 1999 international mathematics report.* Chestnut Hill, MA: International Study Center, Boston College.

Organisation for Economic Co-operation and Development. (2001). *Knowledge and skills for life: First results from PISA 2000*. Paris: OECD Publications.

Organisation for Economic Co-operation and Development. (2003). *Literacy skills for the world of tomorrow—Further results from PISA 2000*. Paris: OECD Publications.

Organisation for Economic Co-operation and Development. (2004). *Learning for tomorrow's world—First results from PISA 2003*. Paris: OECD Publications.

Organisation for Economic Co-operation and Development. (2007). In *PISA 2006: Science competencies for tomorrow's World: Volume 1: Analysis*. Paris: OECD Publications.

Organisation for Economic Co-operation and Development. (2010). *PISA 2009 results: What students know and can do*. Paris: OECD Publications.

Organisation for Economic Co-operation and Development. (2014). *PISA 2012 results: What students know and can do (Volume I, Revised edition, February 2014): Student performance in mathematics, reading and science*. Paris: OECD Publications.

Organisation for Economic Co-operation and Development. (2016). *PISA 2015 Results (Volume I): Excellence and equity in education*. Paris: OECD Publications.

Smith, B. O., Stanley, W. O., & Shores, J. H. (1957). *Fundamentals of curriculum development*. New York: Harcourt, Brace and World Inc.

Tang, Y., Zhang, W., Chen, K., Feng, S., Ji, Y., Shen, J., et al. (2006). Arithmetic processing in the brain shaped by cultures. *Proceedings of the National Academy of Science of the United States of America, 103*(28), 10775–10780.

Vygotsky. (1986). *Thought and Language*. Cambridge: MIT Press.

Wu, M. (2009). A comparison of PISA and TIMSS 2003 achievement results in mathematics. *Prospects, 39,* 33–46.

Open Access Except where otherwise noted, this chapter is licensed under a Creative Commons Attribution 4.0 International License. To view a copy of this license, visit http://creativecommons.org/licenses/by/4.0/.

Part III
Reports of the Survey Teams

Digital Technology in Mathematics Education: Research over the Last Decade

Marcelo C. Borba, Petek Askar, Johann Engelbrecht, George Gadanidis, Salvador Llinares and Mario Sánchez Aguilar

Abstract In this survey paper we focus on identifying recent advances in research on digital technology in the field of mathematics education. We have used Internet search engines with keywords related to mathematics education and digital technology and have reviewed some of the main international journals. We identify five sub-areas of research, important trends of development, and illustrate them using case studies: *mobile technologies, massive open online courses (MOOCs), digital libraries and designing learning objects, collaborative learning using digital technology,* and *teacher training using blended learning*. These exemplary case studies may help the reader to understand how recent developments in this area of research have evolved in the last few years. We conclude the report discussing some of the implications that these digital technologies may have for mathematics

---

This paper is a shortened version of the paper by Borba, Askar, Engelbrecht, Gadanidis, Llinares, and Sánchez Aguilar (2016).

M.C. Borba
UNESP, Rio Claro, Brazil
e-mail: mborba@rc.unesp.br

P. Askar
Hacettepe University, Ankara, Turkey
e-mail: petek.askar@gmail.com

J. Engelbrecht (✉)
University of Pretoria, Pretoria, South Africa
e-mail: johann.engelbrecht@up.ac.za

G. Gadanidis
Western University, London, ON, Canada
e-mail: ggadanid@uwo.ca

S. Llinares
Universidad de Alicante, Alicante, España
e-mail: sllinares@ua.es

M.S. Aguilar
Instituto Politécnico Nacional, Mexico City, Mexico
e-mail: mosanchez@ipn.mx

education research and practice as well as making some recommendations for future research in this area.

Keywords Internet · Mobile learning · MOOC · Blended learning · Digital libraries · Learning objects · Collaborative learning

Introduction

> Digital technology has changed the very notion of what being a human means.

Assertions such as the one above can be found in the literature at large. The idea that innovative media tools we use condition our thinking is not new and consequently should not be surprising. McLuhan (1964) determined this link several decades ago with his often-quoted assertion that "the medium is the message". *Mobility, online courses, massive open online courses* (MOOCs) and *touch technology* are terms that are part of our lives as mathematics educators, and many of these terms are part of the daily lives of 45% of the world population that has access to the Internet.

Research struggles to keep up with the pace of change in the world of *digital technology*. We like to determine identifiable phases in the development of using digital technology in mathematics education. The first phase commenced with the introduction of *Logo* as a teaching tool. Academics began to research its use and impact, but before we knew it, "content" software such as *Cabri* or *Geometer's Sketchpad* became available. We had not yet solved all the problems from this first phase of digital technology in mathematics education (Borba, Scucuglia, & Gadanidis, 2014), when the second phase arrived with new notions such as *dragging*, that allowed students to "experiment mathematics". Again, we were still trying to understand the role of computer laboratories in schools, a novelty from both the first and second phases (depending on the country), when the Internet showed up.

We shift our focus from microworlds, to modelling, to computer laboratories, and now to the *relationship revolution* (Schrage, 2001), afforded by online tools, which we refer to as the third phase. This phase, characterized by the Internet, brings us communication changes that dramatically alter the way we relate to one another. This creates new research problems to be addressed (Engelbrecht & Harding, 2005a, b; Borba, Malheiros, & Zulatto, 2010) and prompted us to include *collaborative learning using technology* as one of the current trends of development. In mathematics education, the way the Internet can be used in a *blended learning environment* characterizes the third phase, which introduces online courses and new problems.

Quantitative change in the Internet has generated a change in quality, and expressions such as *Web 2.0* and *broadband Internet* indicate that a new Internet has developed over the last 5–10 years. This new phase, which we are experiencing now and refer to as the fourth phase, brings us *massive open online courses (MOOCs)*—another trend of development, enhanced opportunities for collaborative learning, and the personalization of the Internet through personal devices. This phase also opens new opportunities for storing digital information through the massive increases in storage and computing power, and the emergence of cloud computing; it is in this context that *digital libraries* appear as another trend of development. Along with these developments, a move to *mobile technology* introduces new possibilities in the teaching of mathematics and leads to a further prominent development trend, included in our discussion.

From the discussion above we see that five prominent *trends of development* were identified. In this paper we attempt to contribute to the goal of surveying this changing area in a particular way, in that we focus on these five important sub-areas of current research, reporting developments related to *mobile technologies, massive open online courses (MOOCs), digital libraries and designing learning objects, collaborative learning using digital technology,* and *teacher training using blended learning.*

Methodology

Several publications have attempted to survey this emerging and always-changing area of digital technologies in mathematics education. Eight chapters on digital technology in the Third International Handbook of Mathematics Education (Clements, Bishop, Keitel, Kilpatrick, & Leung, 2013), and the special issue of ZDM on online distance education (Borba & Llinares, 2012) are examples of such publications. These publications attempt to organize the field of research in order to do what Bicudo (2014) calls *meta-analysis*.

To develop the survey we relied on three sources of information: (1) international research journals, including journals in Portuguese and Spanish, (2) Internet search engines with keywords related to mathematics education and digital technology and (3) the knowledge of the authors of this article about the surveyed areas.

Trends of Development

D'Ambrosio and Borba (2010) consider trends of development, such as the use of digital technology in mathematics education, as a response to problems within the region of inquiry in mathematics education. The on-going concerns show that this trend is still growing. But to what problem does this sub-area of investigation respond? It seems that all phases of attempts to introduce digital technology have faced problems related to displacing embedded rules of time and space that we were not aware of when we experienced the "paper-and-pencil" classroom.

We do not claim to present a comprehensive survey of papers, even though we have highlighted a few in each of the chosen categories. We focus on describing five cases of mathematics education in which digital technology is used in different ways.

Use of Mobile Technologies in Mathematics Teaching and Learning

The use of mobile technologies (such as smartphones and tablets) in the teaching and learning of mathematics is gaining a growing interest among educational researchers and practitioners. The characteristics of mobile devices such as portability, availability, access to the Internet, and its wide acceptance among young people and others, have made mobile devices an emerging agent capable of expanding the frontiers of mathematics instruction and learning beyond the walls of the classroom. White and Martin (2014, p. 64) argue that the characteristics of mobile devices (such as capturing and collecting information, communicating and collaborating with others, consuming and critiquing media, constructing and creating personal forms of representation and expression) can be readily mapped onto mathematical, scientific, and engineering practices highlighted in the Common Core Math and Next Generation Standards (NGSS Lead States, 2013).

Research on the possible uses and potentialities of mobile technologies is growing, but in mathematics education, research on this topic is still limited. Nevertheless, we can find research reports (e.g. Crompton & Traxler, 2015; Larkin & Calder, 2015) addressing how this kind of technology could be used in the teaching and learning of mathematics.

Early studies of mobile learning in mathematics date from shortly before 2010 (e.g. Franklin & Peng, 2008), and since then we have witnessed a growth in this type of research, both at international conferences and in specialized journals. Most of the literature reviewed for this survey can be divided into three broad categories: (a) studies on the potential of mobile devices for teaching and learning mathematics; (b) affective studies on the use of mobile devices; and (c) use of mobile devices in mathematics teacher education.

(a) *Studies on the potential of mobile devices for teaching and learning mathematics*

Several studies have focused on exploiting the capabilities of mobile technologies, like portability, mobility, and the capacity to take photos and videos of real phenomena that later can be analyzed and discussed from a mathematical point of view. An example is the work of Wijers, Jonker, and Drijvers (2010), who used a location-based game called *MobileMath* for mobile phones with GPS to allow students to create and explore quadrilaterals and their properties on a real playing field outside the classroom.

(b) *Affective studies on the use of mobile devices*

Some studies have focused on studying the perceptions and emotions that mathematics teachers and students experience when they teach or study mathematics by using mobile devices. For example, Holubz (2015) studied the perceptions of students and teachers about an initiative called "Bring Your Own Device" (BYOD), where the use of the Internet and mobile devices for the study of mathematics is promoted.

(c) *Use of mobile devices in mathematics teacher education*

Finally, we note that a few studies analyse the use of mobile devices in mathematics teacher education. Yerushalmy and Botzer (2011), for instance, discuss theoretical considerations as well as challenges and opportunities underlying the design of inquiry tasks in mobile settings for pre-service and in-service teachers.

As an example that illustrates how mobile devices can be used to promote the learning of mathematical concepts, we consider the work of Crompton (2015). In her study, Crompton proposes a design-based research study in which iPads are used as a means to support elementary students in their learning of the concept of angle.

In this learning context, the students used their mobile devices to identify and photograph angle-like shapes that naturally appeared in their surroundings (for example in a tree stump, in a shoe pattern, or in the corner of a table); the students then analyzed these images using dynamic geometry applications contained in their mobile devices. In this way the students analyzed whether the "natural angles" that they found in their physical environment actually conformed to the mathematical properties of an angle.

The use of mobile devices in the teaching and learning of mathematics is an emerging research area and is expanding and growing quickly. However, we must be cautious: even though mobile devices and their characteristics appear to offer vast opportunities to enrich and transform the practice of mathematics education at all levels, the introduction of these devices in the classroom also pose a number of challenges of a different nature (pedagogical, technical, and management related).

MOOCS in Mathematics Education

Massive Open Online Courses (MOOCs) offer opportunities as well as challenges for distributing knowledge from institutions. Mathematics and mathematics education are not exempt from these new initiatives (Gadanidis, 2013; McCulloch & Rothschild, 2014). In the mathematics education context, MOOCs are "courses" because there are learning objectives, content and resources, facilitators, ways to connect and collaborate, and a beginning and an end to the learning experience. MOOCs typically use a multimedia format and resources are often short videos on specific topics. They are "massive" since there is not a limit to the number of people who can participate. They are "open", since typically no prerequisites exist for taking a MOOC. Furthermore, most MOOCs offer an optional evaluation process. Participants who complete the evaluation process have the option of receiving a certificate of completion, which typically requires a course fee.

MOOCs are built on the assumption of pervasive Internet access. While Internet access is not yet as widespread in developing countries as one might think, the rapid increase in the ease of access to technology suggests that it will be possible in a few years (Borba, Clarkson, & Gadanidis, 2013). This emerging access, coupled with the lack of prerequisites for enrolment, allows MOOCs to reach a large numbers of participants. Since MOOCs allow participants to complete as much or as little of the course as they desire, MOOCs offer self-directed learning opportunities.

In some cases MOOCs facilitate a collaborative professional experience through a virtual social space for discussion, sharing ideas, resources, and opportunities for constructive feedback. Participants in these MOOCs are engaging in the learning process with others. Such collaborative learning (and even assessment) is necessary in large MOOCs, where student- instructor ratio is very high. There is not just a single path in which the network of participants and ideas is developed; engagement can use different modes (e.g. blogs, Twitter, virtual forum) to build a distributed knowledge base.

We present one example of massive professional development initiatives from Costa Rica using an adaptation of the MOOC concept. When new mathematics curricular standards or principles are generated, teachers are considered to be change agents (Llinares, Krainer, & Brown, 2014) and a MOOC may be used to meet the challenge of implementing new curricular standards. The features of the MOOC as a course, and as being open, participatory, distributed, and a life-long networked learning environment, have been adapted for a specific context in Costa Rica, revealing the contextualized nature of MOOCs.

The goal of the adaptation of MOOCs in this initiative in Costa Rica is to support in-service teachers in the gradual implementation of the new curriculum (Ruiz, 2013). The sessions include various thinking tasks through the analysis of high quality mathematical tasks that can be solved in many ways and represented visually, emphasizing conceptual thinking. The modular teaching mini-videos (Unidad Virtual de Aprendizaje—UVA) as a complement of the design of these courses enables presenting the use of technology in mathematics teaching through the solving of real problems.

Digital Library and Designing Learning Objects in Mathematics Education

As stated in the Digital Library Manifesto (Candela et al., 2007), a digital library is potentially a virtual organization, which comprehensively collects, manages, and preserves rich digital content of all forms for its users. Obviously, digital libraries need a digital repository. In the context of education, digital repositories use learning objects to organize their content, which is a different method of organizing learning content than printed materials use.

Learning objects (LO) proposed by IEEE (2002) are elements of a new type of e-learning grounded in the object-oriented approach of computer science. LO can be defined as a digital entity that can be used, reused, and tagged with metadata aimed to support learning.

Accessibility, interoperability, and reusability are the main features of a learning object (Polsani, 2003). Accessibility refers to the tagging of learning objects with metadata. Interoperability refers to the method of sharing learning objects with other technology systems without the need to alter these objects. Reusability refers to the use of learning object in multiple learning environments.

Well-known learning resources in online repositories are MERLOT (Multimedia Educational Resources for Learning and Online Teaching), Wisc-Online, DRI, Khan Academy, and EBA (Digital Repository of Turkey).

The *Multimedia Educational Resource for Learning and Online Teaching* (MERLOT) (https://www.merlot.org/) was founded in 1997. A program of the California State University, it has been widely used internationally. MERLOT is free to use and is sustained through the support of higher education institutions from around the world.

Khan Academy (https://www.khanacademy.org) is a personalized learning resource for all ages; it offers practice exercises, instructional videos, and a personalized learning dashboard enabling learners to study at their own pace in and outside of the classroom. Their math missions guide learners from kindergarten to calculus using state-of-the-art, adaptive technology that identifies strengths and learning gaps.

What will you do in Math Today? (http://researchideas.ca) is an online open repository of resources for mathematics education, created by Gadanidis at Western University, Canada. The portal is supported by various institutions and includes a research-based math text with learning objects categorized as number, pattern and algebra, measurement and geometry, data and probability.

Current learning object studies have been focusing on quality measures, personalization, and mobile learning. Gadanidis, Sedig, and Liang (2004) analyzed the pedagogy and interface design of interactive visualization for mathematical investigation. They concluded that many interactive visualizations do not appear to be well designed, neither from a pedagogical nor from an interface design perspective. Studies have shown that quality assurance of the LORs is a significant factor when predicting the success of repositories (Clements, Pawlowski, & Manouselis, 2015).

Students today often turn to online mathematics learning resources, such as digital libraries and learning objects before consulting a teacher or a textbook. As mathematics educators, we need to develop and organize these resources in such a way that they facilitate access and foster conceptual understanding.

Using Technology in Collaborative Learning

Opinions have been aired that technology enhanced learning (TEL) has not succeeded in revolutionizing education and the learning process (Chatti, Agustiawan, Jarke, & Specht, 2010). One reason that is suggested is that most current initiatives take a technology-push approach in which learning content is pushed to a pre-defined group of learners in a closed environment A fundamental shift toward a more open and student-pull model for learning is needed—a shift toward a more personalized, social, open, dynamic, and knowledge-pull model as opposed to the one-size fits all, centralized, static, top-down, and knowledge-push models of traditional learning.

A *virtual learning environment* (VLE) or *learning management system* (LMS) is a Web-based platform for courses of study, usually within educational institutions. LMSs (or VLEs) could allow participants to be organized into groups; present resources, activities and interactions within a course structure; provide for the different stages of assessment; report on participation; and have some level of integration with other institutional systems.

Personal learning environments (PLEs) can be viewed as the latest step in an alternative approach to e-learning. The concept has been developed in parallel to that of an LMS—the difference being that a LMS is course wide (or institution wide) while a PLE is individual. PLE's may consist of a number of subsystems, such as a desktop application and one or more web-based services. A PLE would integrate formal and informal learning, such as using social networks, and use collaboration possibilities, such as small groups or web services, to connect a range of resources and systems in an individual space.

Related to the concept of a PLE is the idea of a *personal learning network* (PLN). Whereas PLEs are the tools, artifacts, processes and physical connections that allow learners to control and manage their learning, PLNs extend this framework to include an informal learning network of people to connect with for the specific purpose of learning. In a PLN there is an understanding among participants that the reason they are connecting is for the purpose of active learning (Lalonde, 2012).

In the early 21st century, the creation of rich learning *mashups* (mostly web applications that integrate complementary elements from different sources) currently associated with collaborative learning, resulted from advances in digital media. Wild, Kalz, and Palmér (2010) describe mash-ups as "the *frankensteining* of software artefacts and data" (p. 3). They describe the development of a technological framework enabling students to build up their own personal learning environments by composing web-based tools into a single-user experience, getting involved in collaborative activities, sharing their designs with peers, and adapting their designs to reflect their experience of the learning process. Wild et al. (2010) also introduced the term *Mupple* (Mash-up Personal Learning Environment).

Mupples typically consist of distributed web-applications and services that support individual and collaborative learning activities in both formal and informal settings.

PLEs and PLNs have been extensively implemented in teaching computer science students in particular. In mathematics education these approaches have not really been researched sufficiently. Harding and Engelbrecht (2015) investigated PLN clusters that spontaneously formed among students in two fields of study—mathematics and computer science. Students in a cluster use a number of tools to communicate and learn while using social media, mobile phone technology and learning management systems, among other platforms for learning purposes.

Too little has been done with using the concepts of PLEs and PLNs in the teaching of mathematics—a conceptual subject in which we know that collaboration increases the chances of students developing an understanding of the concepts.

Math-for-Teachers as a Blended Course: An Elementary Teacher Education Case from Canada

Blended learning, which combines both online and face-to-face classroom experiences, is becoming common practice in education at all levels (LaFee, 2013; Owen & Dunham, 2015). The online experience can offer students opportunities to revisit and extend ideas and concepts they encountered in the face-to-face classroom. It can also be used as way to "flip" the classroom experience by giving students opportunities to encounter, explore, and reflect on ideas and concepts before they engage with them in the face-to-face classroom. The flipped classroom model also allows instructors more face-to-face time "to dig deeper into the 'why' of the mathematics" (Ford, 2015, p. 370). The online material created by teachers to support a blended model offers some advantages: it can easily be updated to be current and to better match student needs that arise; it can be shared among teachers to provide professional development; it gives parents a window into their children's learning (Ford, 2015; Wilson, 2013).

The mathematics-for-teachers activities are the same mathematics activities we have been developing in K-8 research classrooms for approximately a decade in Canada and in Brazil (Gadanidis & Borba, 2008; Gadanidis, 2012). The online component of our blended program (www.researchideas.ca/wmt) serves a number of purposes: it is a form of research dissemination; it is a collection of math-for-teachers activities; it is a resource freely available to teachers in the field to use in their classrooms; and it is a set of math-for-teachers courses that we offer through the Fields Institute for Research in Mathematical Sciences. The online resource contains a wide variety of content.

Conclusions and Perspective

It seems safe to say that technological change will continue and likely increase in pace. In this context, we can fall into a pattern of chasing the latest innovation rather than charting our own direction, focusing on "what is the latest technology" rather than on "what is worth researching?"

Mobile technology, PLNs, digital learning objects and other artifacts are "stretching" the classroom, transforming the classroom to the extent that it can hardly be recognized as such. A significant part of pre-service mathematics teacher education is done online in many countries (e.g. Brazil, Costa Rica) in that students only meet when writing tests and a few non-mandatory face-to-face meetings (Ruiz, 2013). In this scenario, the regular classroom no longer serves as locus for education. Couches, chairs, tables at students' houses and cafés are the "new classrooms". Flipped classrooms change the notion of what is in and outside of the classroom and also change the roles of students and teachers.

As pointed out in the paper, PLEs and social networks such as *Facebook,* may make it even more difficult to keep the traditional distinction between "inside the classroom" and "outside the classroom" or between "study time" and "leisure time". Different blends are being forged into face-to-face education and online distance education in such a way that it will be interesting to see how much this distinction will be blurred in a few years time.

The trends of development discussed in this paper highlight five important issues in the intersection of e-learning and mathematics education that might serve as contexts for investigating "what might be":

1. Student access to mobile technologies creates a student-mathematics relationship that is not yet widely embraced by mathematics educators, that disrupts the traditional flow of mathematics knowledge from teacher to student, and that is not well understood from a research perspective.
2. The potential of MOOCs to disrupt the institutional and hierarchical nature of traditional education, offering students opportunities to access courses without prerequisites, without fees (unless they require a record of course completion), and the potential of MOOCs to affect access to and the quality of mathematics education is not well understood.
3. The availability of online mathematics learning resources (as the digital libraries and learning objects) means that many students now turn to these resources before they consult a teacher or a textbook, and this raises questions about how the resources are organized to in order to facilitate access and how they are designed pedagogically to foster conceptual understanding.
4. The collaborative and social networking affordances of current technologies raise questions about the design and use of learning management systems as well as personal learning environments and networks.
5. Teacher use of blended learning to extend and supplement classroom learning with online exploration and discussion or to employ a flipped classroom model to make the classroom a place for extension and elaboration rather than direct instruction raises questions about the need to research the various models used.

These five themes are not independent. Firstly, we notice issues related to the nature of *new types of mobile/digital technological* means, favouring the access to knowledge/information of mathematics and mathematics education and modifying the nature of interaction students-knowledge-teacher-context. How do we use the new technological means when the objectives are related to mathematics and mathematics education learning? Secondly, we notice issues related to *how mathematics or knowledge from mathematics education is considered/organized* in this new context (digital libraries, digital repertories, learning objects and inclusive the MOOCS). Finally, we identify issues related to the nature of the *interaction among persons and between persons and mathematical and mathematics education knowledge* when they are learning.

These three cross-cutting aspects define two *dimensions* in the research in this field: (i) when the focus is on how the new mobile/digital technological means define new forms of organizing knowledge and facilitating the access to knowledge (the learning objects, MOOCs, digital library, digital repository and so on), and (ii) when the focus is on how the use of new mobile/digital technological means determines the nature of the interactions between humans, and between humans and knowledge in the learning contexts. So, these dimensions generate epistemological issues (about the nature of mathematical and mathematics education knowledge) and issues about social and individual aspects of learning, as well as issues about the role of interaction in this learning.

These trends that we emphasize and describe belong to the fourth phase in the development of using digital technology in mathematics education (as mentioned in the introduction). This phase is shaped by fast Internet and integrated with various procedures and practices from the other three phases, as well as from the history (going back over 30 years) of attempting to include digital technology in mathematics education. Most students have already decided that cellular phones make up part of their lives inside or outside the classroom. These devices are definitely part of the collective of "students-with-mobile-phones". Other technologies such as paper and pencil, as well as computer software are also accepted in this collective, but for the most part, the current generation at schools and universities do not see the world without mobile technology.

References

Bicudo, M. A. V. (2014). Meta-análise: seu significado para a pesquisa qualitativa [Meta-synthesis: Its meaning in the qualitative research]. *Revemat: Revista Eletrônica de Educação Matemática, 9*(0), 7–20. doi:10.5007/1981-1322.2014v9nespp7

Borba, M. C., Askar, P., Engelbrecht, J., Gadanidis, G., Llinares, S., & Sánchez Aguilar, M. (2016). Blended learning, e-learning and mobile learning in mathematics education. *ZDM Mathematics Education, 48*(5), 589–610. doi:10.1007/s11858-016-0798-4

Borba, M. C., Clarkson, P., & Gadanidis, G. (2013). Learning with the use of the Internet. In M. A. (Ken) Clements, A. J. Bishop, C. Keitel, J. Kilpatrick, & F. K. S. Leung (Eds.), *Third international handbook of mathematics education* (pp. 691–720). New York: Springer. doi:10.1007/978-1-4614-4684-2_22

Borba, M. C., & Llinares, S. (2012). Online mathematics teacher education: overview of an emergent field of research. *ZDM–The International Journal on Mathematics Education, 44*(6), 697–704. doi:10.1007/s11858-012-0457-3

Borba, M. C., Malheiros, A. P. S., & Zulatto, R. B. A. (2010). *Online distance education* (1st ed.). Rotterdam: Sense Publishers.

Borba, M. C., Scucuglia, R. R. S., & Gadanidis, G. (2014). *Fases das tecnologias digitais em educação matemática: sala de aula e internet em movimento [Phases of digital technologies in mathematics education: The classroom and the internet in motion]* (1st ed.). Belo Horizonte: Autêntica.

Candela, L. et al. (2007). *The DELOS digital library reference model: foundation for digital libraries, version 0.96*. Resource document. European Commission within the Sixth Framework Programme. http://delosw.isti.cnr.it/files/pdf/ReferenceModel/DELOS_DLReferenceModel_096.pdf

Chatti, A. C., Agustiawan, M. R., Jarke, M., & Specht, M. (2010). Toward a personal learning environment framework. *International Journal of Virtual and Personal Learning Environments, 1*(4), 66–85. doi:10.4018/jvple.2010100105.

Clements, M. A. K., Bishop A. J., Keitel, C., Kilpatrick, J., & Leung, F. K. S. (Eds.). (2013). *Third international handbook of mathematics education*. New York: Springer. doi:10.1007/978-1-4614-4684-2

Clements, K., Pawlowski, J., & Manouselis, N. (2015). Open educational resources repositories literature review—Towards a comprehensive quality approaches framework. *Computers in Human Behavior, 51*(Part B), 1098–1106. doi:10.1016/j.chb.2015.03.026

Crompton, H. (2015). Understanding angle and angle measure: A design-based research study using context aware ubiquitous learning. *International Journal for Technology in Mathematics Education, 22*(1), 19–30. doi:10.1564/tme_v22.1.02

Crompton, H., & Traxler, J. (Eds.). (2015). *Mobile learning and mathematics. Foundations, design and case studies*. Florence, KY: Routledge.

D'Ambrosio, U., & Borba, M. C. (2010). Dynamics of change of mathematics education in Brazil and a scenario of current research. *ZDM–The International Journal on Mathematics Education, 42*(3), 271–279. doi:10.1007/s11858-010-0261-x

Engelbrecht, J., & Harding, A. (2005a). Teaching undergraduate mathematics on the Internet. Part 1: Technologies and taxonomy. *Educational Studies in Mathematics, 58*(2), 235–252. doi:10.1007/s10649-005-6456-3

Engelbrecht, J., & Harding, A. (2005b). Teaching undergraduate mathematics on the Internet. Part 2: Attributes and possibilities. *Educational Studies in Mathematics, 58*(2), 253–276. doi:10.1007/s10649-005-6457-2

Ford, P. (2015). Flipping a math content course for pre-service elementary school teachers. *Primus, 25*(4), 369–380. doi:10.1080/10511970.2014.981902

Franklin, T., & Peng, L.-W. (2008). Mobile math: math educators and students engage in mobile learning. *Journal of Computing in Higher Education, 20*(2), 69–80. doi:10.1007/s12528-008-9005-0

Gadanidis, G. (2012). Why can't I be a mathematician? *For the Learning of Mathematics, 32*(2), 20–26.

Gadanidis, G. (2013). Designing a Mathematics-for-All MOOC. In T. Bastiaens, & G. Marks (Eds.), *Proceedings of e-learn: World conference on e-learning in corporate, government, healthcare and higher education, 2013* (pp. 704–710). Chesapeake, VA: Association for the Advancement of Computing in Education (AACE). http://www.editlib.org/p/114923. Accessed July 28, 2015.

Gadanidis, G., & Borba, M. (2008). Our lives as performance mathematicians. *For the Learning of Mathematics, 28*(1), 44–51.

Gadanidis, G., Sedig, K., & Liang, H. N. (2004). Designing online mathematical investigation. *Journal of Computers in Mathematics and Science Teaching, 23*(3), 275–298.

Harding, A., & Engelbrecht, J. (2015). Personal learning network clusters: A comparison between mathematics and computer science students. *Journal of Educational Technology and Society, 18*(3), 173–184.

Holubz, B. J. (2015). Mobilizing mathematics. Participants' perspectives on bring your own device. In H. Crompton & J. Traxler (Eds.), *Mobile learning and mathematic. Foundations, design, and case studies* (pp. 213–222). Florence, KY: Routledge.

IEEE Learning Technology Standards Committee. (2002). Draft standard for learning object metadata. Resource document. IEEE. http://129.115.100.158/txlor/docs/IEEE_LOM_1484_12_1_v1_Final_Draft.pdf. Accessed January 29, 2016.

LaFee, S. (2013). Flipped learning. *The Education Digest*, November Issue, 13–18.

Lalonde, C. (2012). How important is Twitter in your personal learning network? *eLearn Magazine*. September 2012. http://elearnmag.acm.org/featured.cfm?aid=2379624. Accessed August 1, 2015.

Larkin, K., & Calder, N. (2015). Mathematics education and mobile technologies. *Mathematics Education Research Journal*. doi:10.1007/s13394-015-0167-6.

Llinares, S., Krainer, K., & Brown, L. (2014). Mathematics teachers and curricula. In S. Lerman (Ed.), *Encyclopedia of mathematics education* (pp. 438–441). New York: Springer. doi:10.1007/978-94-007-4978-8_111

McCulloch, R., & Rothschild, L. P. (2014). MOOCs: An inside view. *Notices of the AMS, 61*(8), 2–8. doi:10.1090/noti1147

McLuhan, M. (1964). *Understanding media: The extensions of man*. New York: McGraw Hill.

NGSS Lead States. (2013). *Next generation science standards: For states, by states*. http://www.nextgenscience.org

Owen, H., & Dunham, N. (2015). Reflections on the use of iterative, agile and collaborative approaches for blended flipped learning development. *Education Sciences, 5*(2), 85–105. doi:10.3390/educsci5020085

Polsani, P. R. (2003). Use and abuse of reusable learning objects. *Journal of Digital Information, 3*(4). http://journals.tdl.org/jodi/article/viewArticle/89/88

Ruiz, A. (2013). La reforma de la educación matemática en Costa Rica. Perspectiva de la praxis [The mathematics education reform in Costa Rica. Perspective of praxis]. *Cuadernos de Investigación y Formación en Educación Matemática*, Year 8, Special Number. http://revistas.ucr.ac.cr/index.php/cifem/article/view/11125

Schrage, M. (2001). The relationship revolution. http://web.archive.org/web/20030602025739/, http://www.ml.com/woml/forum/relation.htm. Accessed July 12, 2008.

White, T., & Martin, L. (2014). Mathematics and mobile learning. *TechTrends, 58*(1), 64–70. doi:10.1007/s11528-013-0722-5

Wijers, M., Jonker, V., & Drijvers, P. (2010). MobileMath: exploring mathematics outside the classroom. *ZDM–The International Journal on Mathematics Education, 42*(7), 789–799. doi:10.1007/s11858-010-0276-3

Wild, F., Kalz, M., & Palmér, M. (Eds.). (2010). *Proceedings of the 3rd Workshop on Mashup Personal Learning Environments*. Barcelona, Spain. http://ceur-ws.org/Vol-638/. Accessed February 14, 2014.

Wilson, S. G. (2013). The flipped class. A method to address the challenges of an undergraduate statistics course. *Teaching of Psychology, 40*(3), 193–199. doi:10.1177/0098628313487461

Yerushalmy, M., & Botzer, G. (2011). Guiding mathematical inquiry in mobile settings. In O. Zaslavsky, & P. Sullivan (Eds.), *Constructing knowledge for teaching secondary mathematics* (pp. 191–207). New York: Springer. doi:10.1007/978-0-387-09812-8_12

Open Access Except where otherwise noted, this chapter is licensed under a Creative Commons Attribution 4.0 International License. To view a copy of this license, visit http://creativecommons.org/licenses/by/4.0/.

Conceptualisation of the Role of Competencies, Knowing and Knowledge in Mathematics Education Research

Mogens Niss, Regina Bruder, Núria Planas, Ross Turner and Jhony Alexander Villa-Ochoa

Abstract This paper surveys the notions, conceptualisations and roles of mathematical competencies and their relatives in research, development and practice from an international perspective. After outlining the questions giving rise to this survey, the paper first takes a brief look at the genesis of competency-oriented ideas as a prelude to identifying and analysing recent trends. The relationships between different notions and terms concerning competencies and their relatives are discussed, and their roles in the 2015 PISA framework are presented. Two kinds of research, on and by means of mathematical competencies, are surveyed. The impact of competency-oriented notions and ideas on curriculum frameworks and documents in a number of countries is being charted, before challenges to the implementation of such notions in actual teaching practice are identified. Finally the paper takes

---

This is a condensed and edited version of the comprehensive report of the Survey Team, Niss, Bruder, Planas, Turner, and Villa-Ochoa (2016).

M. Niss (✉)
Roskilde University, Roskilde, Denmark
e-mail: mn@ruc.dk

R. Bruder
Technische Universität, Darmstadt, Germany
e-mail: r.bruder@math-learning.com

N. Planas
Universitat Autònoma de Barcelona, Bellaterra, Spain
e-mail: nuria.planas@uab.cat

R. Turner
Australian Council of Educational Research, Camberwell, VIC, Australia
e-mail: ross.turner@acer.edu.au

J.A. Villa-Ochoa
Universidas de Antioquia, Medellin, Colombia
e-mail: jhony.villa@udea.co

stock of the international state-of-the-art of competencies and similar notions, with a focus on the need for further research.

Keywords Mathematical competencies · Mathematical proficiency · Mathematical practices · Mathematical literacy · Educational standards · PISA · Fundamental mathematical capabilities

Introduction: What Are the Issues?

Despite the title of this survey, focusing primarily on research, the authors also find it necessary to consider competencies, knowing and knowledge as they pertain to the development and practice of mathematics education. This is so because these notions are crucial to all aspects of mathematics teaching and learning. In fact, anyone involved in mathematics education in whichever capacity has to relate to the fundamental question:

What does it mean to master mathematics?
And to a number of related but not equivalent questions, such as: What does it mean to possess knowledge of mathematics? To know mathematics? To have insight in mathematics? To be able to do mathematics? To possess competence (or proficiency)? To be well versed in mathematical practices?

These questions reflect different facets of the title of the Survey Team. The former three of them focus on mathematical *products* (concepts, definitions, rules, theorems, formulae, methods, and historical facts), which have accumulated in the mind of "the knower". The educational issue corresponding to these questions is: What does it take for a learner to become a knower of mathematics? The latter three questions focus on the *enactment* of mathematics, i.e. what is involved in carrying out characteristic mathematical processes. The corresponding educational issue is: What does it take for a learner to become a doer of mathematics?

"Knowing" and "being able to do" are two different things. Yet, it goes without saying that the relationships and balances between them are both intimate and delicate. Oftentimes, neither the initiating questions of this paper, nor the answers to them, are stated explicitly in official documents and other writings about mathematics education. So, it may seem natural to ask: why are these questions important at all? Well, they are important because the answers to them—whether explicit or implicit—determine at least three crucial components of mathematics education:

- The purposes and goals of mathematics education ('what do we wish to accomplish?')
- The criteria for and degree of success in mathematics teaching and learning ('how can we know whether and how well we have accomplished what we want?')

- The structure and organisation of mathematics teaching ('teachers' and students' respective activities as well as the framework and materials for teaching and learning').

Markedly different answers to the initiating questions posed above give rise to marked differences in the realisation of these components. In fact, the diversity of mathematics education in different parts of the world can, in large part, be explained by the diversity of answers given to the main question: What does it mean to master mathematics?

Answers to the Main Question

Historical Excursion

Let us begin with a brief historical excursion. Classically, the main question was answered by specifying the mathematical content, including facts, that people should know about and the associated procedural skills that they should have. For example, the Danish national upper secondary curriculum in 1935 specified 38 content items and associated procedural skills in great detail, and also specified the structure and content of the written and oral final exams in considerable detail. However, such conceptions of what it means to master mathematics soon came under attack.

Thus the oft-quoted Spens Report (Board of Education, 1938), published in the UK requested that the subject should not ignore the "considerable truths in which Mathematics subserves important activities and adventures of civilised man. [...] We believe that school Mathematics will be put on a sound footing only when teachers agree that it should be taught as art and music and Physical Science should be taught because it is one of the mail lines which the creative spirit of man has followed in its development" (pp. 176–177).

Here us another voice, that of George Pólya who, in the preface to the first (1945) edition of *How to Solve It*, wrote:

> If [the teacher of mathematics] fills his allotted time with drilling his students in routine operations he kills their interest, hampers their intellectual development, and misuses his opportunity. But if he challenges the curiosity of his students by setting them problems proportionate to their knowledge and helps them to solve their problems with stimulating questions, he may give them a taste for, and some means of, independent thinking [Quoted from the 1957 (2nd) edition, p. v].

The first IEA study on mathematical achievement (1964–1967), the precursor of the TIMSS studies, listed five *cognitive behavior levels* as components of mathematical achievement including content knowledge. The last three of these were: "(a) translation of data into symbols or schema or vice versa; (b) comprehension: capacity to analyze problems, to follow reasoning; and (c) inventiveness: reasoning creatively in mathematics" (Husén, 1967).

Seymour Papert, the inventor of the educational computer language LOGO, wrote in 1972 (pp. 249–250):

> Being a mathematician is no more definable as 'knowing' a set of mathematical facts than being a poet is definable as knowing a set of linguistic facts. [...] *being a mathematician, again like being a poet, or a composer, or an engineer, means* doing *rather than knowing or understanding.* [...] In becoming a mathematician does one learn something other and more general than the specific content of particular mathematical topics? Is there such a thing as a Mathematical Way of Thinking? Can this be learnt and taught? [Italics in the original]

These quotations point to other faces of mathematics than systematically organised subject matter, factual content knowledge and procedural skills, namely to *significant mathematical processes*. So, historically we are faced with rather different conceptualisations of what it means to master mathematics, such as:

- focusing primarily on knowledge and understanding of content, e.g. definitions, concepts, theorems, and theoretical structures;
- focusing primarily on skills pertaining to algorithmic procedures and techniques;
- focusing primarily on the overall enactment of mathematics, i.e. working within and by means of mathematics in intra- and extra-mathematical contexts, especially problem solving;
- focusing primarily on general mathematical thinking and on mathematics as part of human culture, like art and science.

None of these different foci can stand alone, and they are not contradictory. When people think they are, unfruitful controversies arise, cf. the maths wars in some countries. Rather, these foci represent different, albeit mutually dependent, emphases. There are, however, context-dependent balances to be struck amongst them.

Recent Trends

Next, we zoom in on some trends which since the 1980's have put emphasis on the *enactment* of mathematics.

The National Council of Teachers of Mathematics (NCTM) in the USA took the lead in breaking significant new paths in this respect. Already NCTM's *An Agenda for Action: Recommendations for School Mathematics of the 1980s* (1980), recommended, among other things, that "problem solving should be in the focus of school mathematics in the 1980s", that "basic skills in mathematics be defined to encompass more than computational facility", and that "the success of mathematics programs and student learning be evaluated by a wider range of measures than conventional testing" (p. 1).

The highly influential NCTM *Curriculum and Evaluation Standards for School Mathematics*, 1989, stated the following goals for all K-12 students: (1) that they learn to value mathematics; (2) that they become confident in their ability to do mathematics; (3) that they become mathematical problem solvers; (4) that they learn to communicate mathematically; and (5) that they learn to reason mathematically. (p. 5). From these goals four overarching *standards* for mathematics at all grade levels were derived: 'Mathematics as problem solving'; 'mathematics as communication'; 'mathematics as reasoning'; and 'mathematical connections'. The 1989 Standards gave rise to the Math Wars in the USA in the 1990s, because opponents held differing views of what it means and takes to come to grips with mathematics. This was one of the factors behind the publication of NCTM's revised *Principles and Standards for School Mathematics* (2000), which preserved the overarching standards for all school levels, but added one more, 'representations', whereas attitudinal and dispositional aspects were omitted.

Similar conceptions were nurtured and implemented in Australia since the 1980s. Thus, the *National Statement on Mathematics for Australian Schools* (1990) gave emphasis both to mathematical products and processes, involving observing, representing and investigating patterns in social and physical phenomena and between mathematical objects, with a focus on mathematical thinking and mathematical modelling. The document *Mathematics—a curriculum profile for Australian Schools* (1994) focused on what it means to work mathematically: 'investigating'; 'conjecturing'; 'using problem solving strategies'; 'applying and verifying'; 'using mathematical language'; and 'working in context'.

The Danish *KOM Project* (Niss & Jensen, 2002; Niss, 2003) developed the notion of mathematical competence and mathematical competencies, defined as follows:

> "Possessing mathematical competence means to have knowledge about, to understand, to exercise, to apply and relate to and judge mathematics and mathematical activity in a multitude of contexts which actually do involve, or potentially might involve, mathematics", whilst mathematical competenc*ies* are the main constituents in mathematical competenc*e*: "A mathematical competency is insight-based readiness to act purposefully in situations that pose *a particular kind* of mathematical challenge." (p. 43). The project identified eight such competencies and depicted them by way of the so-called 'competency flower' (Fig. 1).

Roughly at the same time, but independently of the KOM Project, projects in the USA worked along similar lines. The National Research Council's (NRC) *Adding It Up: Helping Children Learn Mathematics* (2001) and the *RAND Mathematics Study Panel* (2003) adopted the term *mathematical proficiency*, specifying five interwoven strands: 'conceptual understanding'; 'procedural fluency'; 'strategic competence'; 'adaptive reasoning'; and 'productive disposition' (p. 116). To the NRC team this notion captures what is believed "to be necessary *to learn* mathematics successfully" (our italics), whilst to the RAND panel it captures "what it means to *be competent* in mathematics" (our italics). The RAND Panel also introduced the notion of *mathematical practices*:

Fig. 1 The competency flower

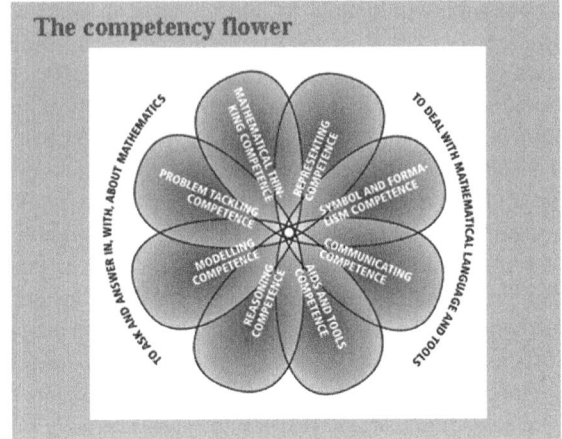

Because expertise in mathematics […] involves more than just possessing certain kinds of knowledge, we recommend […to] focus explicitly on mathematical know-how—what successful mathematicians and mathematics users *do*. We refer to the things that they do as mathematical practices. Being able to justify mathematical claims, use symbolic notation efficiently, and make mathematical generalizations are examples of mathematical practices (p. 29).

This notion was taken further by the *Common Core State Standards Initiative* (2012), in the US, which identified eight mathematical practices: "Make sense of problems and persevere in solving them. Reason abstractly and quantitatively. Construct viable arguments and critique the reasoning of others. Model with mathematics. Use appropriate tools strategically. Attend to precision. Look for and make use of structure. Look for and express regularity in repeated reasoning (pp. 1–2)".

We have now seen that several different people, bodies and agencies have felt the need to insist that mastering mathematics goes beyond possessing mathematical content knowledge and procedural skills. It also involves the *enactment* of mathematics in a broad and comprehensive sense. There are significant similarities between the different conceptualisations of mathematical enactment, but there are characteristic differences as well, not only as far as terminology is concerned but also with regard to the scope of the notions, e.g. concerning the role of attitudinal, dispositional and volitional aspects.

Mathematical Competencies (and Their Relatives)

Obviously, widespread notions such as 'mathematical literacy', 'numeracy', 'quantitative literacy', 'mathematical competence/competencies' and—yes! —'mathematics' are related, yet not identical. How might we characterise relationships among them, for instance by way of a Venn diagram? Of course, there is no unique representation of these relationships. We perceive 'mathematical competencies' as encompassing 'mathematical literacy', which in turn encompasses 'quantitative literacy' (the US term) and 'numeracy' (the UK/Australia term). They are all intersected by the discipline 'mathematics' with its two facets 'mathematical products' and 'mathematical processes'. Thus, 'mathematics' as a discipline is not a subset of 'competencies', 'literacy' or 'numeracy'.

A related issue is what we should value and emphasise in mathematics teaching and learning, considering for example the following list of possible elements: 'thinking mathematically'; 'practical survival skills'; 'number sense'; 'doing mathematics'; 'modelling, mathematising'; 'what mathematicians do'; 'mathematical communication—understanding, expressing'; 'mathematical reasoning'; 'mathematical algorithms and procedures'; 'extracting/defining problems from work/life/world'; 'inferences from data'; 'definitions theorems, proofs'; 'applying mathematics in context'; 'cultural appreciation (nature, history and role of mathematics'.

These elements play rather different parts and are valued rather differently in different contexts. One such context is the notion of mathematical literacy as defined in PISA, here quoted from the 2015 framework:

> Mathematical literacy is an individual's capacity to formulate, employ, and interpret mathematics in a variety of contexts. It includes reasoning mathematically and using mathematical concepts, procedures, facts, and tools to describe, explain and predict phenomena. It assists individuals to recognize the role that mathematics plays in the world and to make well-founded judgments and decisions needed by constructive, engaged and reflective citizens (p. 5).

As appears, the key components in PISA are 'formulate', 'employ' and 'interpret', which are simply code for mathematical modelling. PISA mathematics does not make explicit use of the notion of mathematical competencies in the sense of the KOM Project, but of the derived notion of 'fundamental mathematical capabilities' that takes account of the realities of an international assessment in which elements have to be separable in order to be measurable. Figure 2 offers a diagrammatic summary of the PISA elements.

Aspects of Research Concerning Mathematical Competencies

There are two types of research on mathematical competencies. The first type has competencies as its main *object* of research. The second type employs competencies as an essential *means* of research for some other purpose. Both types comprise theoretical as well as empirical research. Since it is clearly not possible to do justice to the huge body of research in the field, we have confined ourselves to outlining a few selected topics.

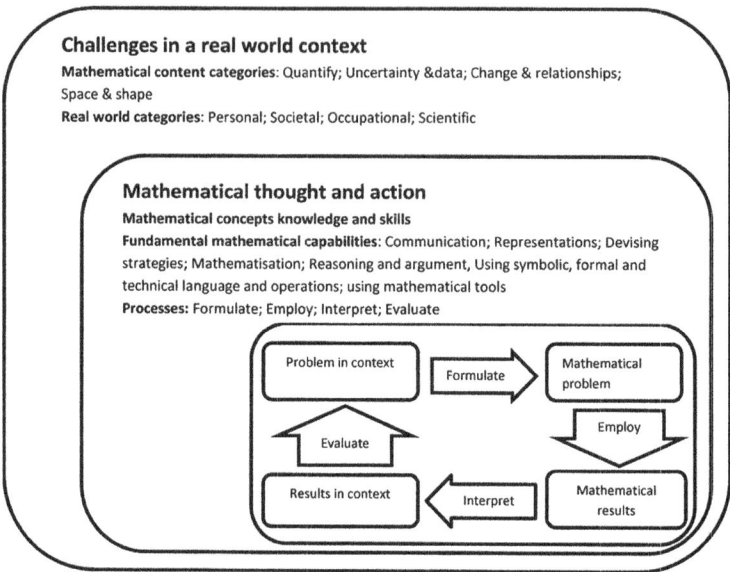

Fig. 2 Main PISA elements 2015

As regards mathematical competencies as an *object* of research, much research —having theoretical perspectives as an integral part—investigates various notions of competence/competency and the specification of their components and facets. Are the notions of a purely cognitive nature or do they include dispositional and affective elements too? To what extent do the notions depend on culture, situations, mathematical levels or domains, and how are they related to universal or particular mathematical practices? There is also research trying to model the structure of the system of competencies.

Other research takes a predominantly empirical perspective on the entire system of competencies, e.g. in order to underpin the existence and significance of the construct, to identify the main dimensions of the competencies as well as stages of their development with individuals or groups, with the aim of fostering and supporting such development. Particular attention has been paid to the professional development of teachers, focusing on their coming to grips with the notion of competencies and supporting them in assessing them.

Still other research focuses on the individual competencies. It has been shown that competencies can be detected and identified empirically in people's actual mathematical activities, albeit with some complications. One such complication is that mathematical competencies are neither developed nor possessed or enacted in isolation. So, in tests it is often difficult to measure particular facets of competencies exactly, for instance because problems often can be solved in rather different ways, invoking rather different sets of competencies. The solution of 1-step tasks cannot really show the difference between deep understanding of mathematics and rote learning of knowledge and procedures. Moreover, competencies are often defined to be overlapping (e.g. problem solving almost inevitably makes use of representations and of work with symbols and formalism). It is important to notice that lots of research has been conducted, also in the past, on individual competencies (e.g. problem solving, modelling, reasoning, representations) without ever referring to the term 'competency'.

A growing body of theoretical and empirical research makes use of mathematical competencies as a *means* of research. For example to underpin new mathematics frameworks or curricula, to capture and understand what happens in actual mathematics teaching, or to create learning environments based on competencies. Competencies are also used to underpin test design, item formats and interpretation of item difficulty. Identifying and analysing misconceptions can also be supported by adopting a competency perspective. The same is true with regard to teachers' beliefs and views of mathematics.

Competencies and National Mathematics Curricula

Competencies and their relatives play a variety of roles in different national curricula, even though several terms other than 'competency' are being used across countries. As it cannot be our ambition to provide geographical coverage of the world's about 200 countries we have selected a sufficiently varied set of countries/regions to provide a fair representation of the spectrum of ideas and issues pertaining to these notions in a curriculum context.

In *Spain*, which is inspired by successive PISA frameworks, the 1989 NCTM Standards, the Common Core State Standards Initiative, and the Danish KOM Project, the role of mathematical competencies appears strong on paper but is weak in terms of actual implementation and practice. Spanish curricula use notions and terms in the intersection of the international sources and ad hoc combinations of them. Thus, Spain employs the term 'sub-competencies', e.g. reasoning mathematically, posing and solving problems, modelling, and communication in, with and about mathematics, for what is elsewhere called competencies. When developing related sub-competencies, the learner is supposed to acquire the ability to understand a chain of mathematical reasoning, (re)formulate a question as a mathematical problem, express oneself mathematically, and to deal with models set up by others.

Since the 1970s *Portugal* has placed problem solving and problem posing—initially viewed and referred to as skills across mathematical content areas—at the heart of mathematics education. Later, the national curricula have evolved to introduce and utilise the notion of mathematical competence and to state the development of mathematical competencies as educational goals for primary and secondary school. Like in Spain, teachers are struggling with competency-based mathematics teaching and with the diversity of terminology around it. Terms such as 'basic content', 'basic skills', 'basic competencies', 'essential competencies', and 'capacities' are used in Portuguese mathematics teacher education programmes. Recent curriculum developments show a tension between what is/should be considered 'content' and what 'capacity', along with debates on whether it is possible to reconcile the two notions, whilst avoiding merely seeing capacities as ways of dealing with specific content knowledge.

In Latin America, *Brazil*, *Colombia*, *Costa Rica*, *Chile*, *Mexico* and the *Dominican Republic*, among others, have experienced curriculum reforms inspired by PISA frameworks and competency-oriented ideas. A common thrust of these reforms has been to focus on learners' recognition of the social role of school mathematics and of real world problem solving in everyday, social and societal life. Despite significant differences amongst countries, they all emphasise the functional use of mathematics. The purpose of education is defined in terms of 'capabilities' (Chile), 'competencies' (Mexico, Colombia and the Dominican Republic) and 'abilities' (Costa Rica). Relating mathematical thinking to specific mathematical topics and processes (problem solving, reasoning, communication and modelling) is commonplace. Chile, the Dominican Republic and Mexico include attitudes in their notions, while Colombia highlights contexts.

In the *Australian* curriculum of 2012, *numeracy* is included as one of the general capabilities: Students become numerate as they develop the knowledge and skills to use mathematics confidently across all learning areas at school and in their lives more broadly. Numeracy involves students in recognizing and understanding the role of mathematics in the world and having the dispositions and capacities to use mathematical knowledge and skills purposefully. The curriculum further contains three content strands: 'number and algebra'; 'measurement and geometry'; and 'statistics and probability', as well as four proficiency strands (close to the ones in "Adding It Up"): 'understanding'; 'fluency'; 'problem solving, including modelling', and 'reasoning'.

In *Korea*, emphasis historically was placed on content, whilst doing mathematics was seen as part of learning that content. In 2011 there was a shift of focus so as to strengthen processes that can be seen as versions of mathematical competencies: Crucial capabilities for members of a complex, specialized, and pluralistic future are believed to be fostered by learning and practicing mathematical processes, including mathematical problem solving, communication and reasoning.

Several South-East Asian countries (*Brunei Darussalam*, *Cambodia*, *Indonesia*, *Lao PDR*, *Malaysia*, *Philippines*, *Singapore*, *Thailand*, *Timor Leste*, *Vietnam*) are introducing new regional assessment programmes and corresponding assessment frameworks. Definitions of mathematics include a clear focus on connecting conceptual and procedural knowledge to its use in daily life whilst emphasising broader goals of the mathematics curriculum, such as 'mathematical thinking and reasoning' and 'problem solving', referencing mathematical competencies in various ways. As particularly regards Singapore, *doing* is perceived as part of *knowing*, as depicted in the well-known regular pentagon having 'mathematical problem solving' as its core, placed in the interior of the pentagon, whereas the five edges —'mathematical concepts', 'processes', 'skills', 'attitudes' and 'metacognition'— can be seen as facets of or prerequisites to problem solving.

The final stations on our excursion to national curricula are the German speaking countries in Europe. As a result of the so-called "PISA shock" in *Germany* in 2000, a multitude of concerted efforts were made to remedy the situation. A first outcome was the *educational standards* for mathematics agreed upon by all German *Länder* (states) in 2003, the core of which consists of six standards—'reasoning'; 'problem solving'; 'modelling'; 'using mathematical representations'; 'dealing with symbolic, formal and technical aspects of mathematics'—that correspond closely to six of the eight Danish KOM Project competencies. These standards/competencies are placed in a three-dimensional structure, having five mathematical content areas ('numbers'; 'measure'; 'space and shape'; 'functional dependencies'; 'data and randomness') and three levels of mastery as in the early PISA frameworks ('reproducing'; 'making connections'; 'generalising and reflecting') as the other two dimensions. *Austria*, too, has adopted a rather complex three-dimensional model of processes ('representing'; 'building models'; 'operating'; 'interpreting'; 'reasoning'), content areas ('numbers and measures'; 'variables'; 'functional dependencies'; 'geometric figures and solids'; 'statistical representation and descriptors'), and finally three levels of mastery ('activation of basic knowledge and skills'; 'creating connections'; 'activation of reflective knowledge'). *Switzerland* in 2007 adopted a national framework, inspired by the NCTM Standards, PISA, and the German educational standards, to harmonise compulsory school education across all cantons. The framework is based on eight fundamental aspects of mathematical action ('knowing, realizing and describing'; 'operating and computing'; 'employing instruments and tools'; 'representing and communicating'; 'mathematising and modelling'; 'reasoning and justifying'; 'interpreting and reflecting on results'; 'investigating and exploring'), combined with mathematical content in a matrix structure for each of the years 4, 8 and 11.

As can be seen, the three German speaking countries have embarked on similar developments, in which competencies/standards are placed in a three-dimensional cluster and also contain dispositional and volitional components. Efforts are being made to empirically measure facets of competencies of individuals and of larger groups of students, while attempting to stepwise reduce the number of activity dimensions so as to reduce overlaps.

This excursion has largely focused on curriculum frameworks and documents. But how strong is the match between goals and wishes expressed in these documents and the practices in mathematics classrooms? This takes us to our next section.

Challenges to Implementation

One major challenge to the implementation in classrooms of competency-oriented ideas, frameworks and curricula is that teachers are not always provided with the professional competencies and didactic-pedagogical resources needed to create classroom cultures in which systematic work to develop students' mathematical competencies is the norm. Throughout the world there seems to be a lack of sufficiently helpful guidelines and support for pre-and in-service teachers.

This is not only an obstacle to the implementation of competencies in everyday classroom practice, but also to research and development committed to influence such practice. We need to better understand how research on mathematical competencies can be transformed into educational action and design of intervention, and vice versa: how can educational action and intervention become objects of research?

Building bridges between research and practice regarding mathematical competencies is notoriously difficult and has been approached in different places. Some approaches (e.g. in Germany) have focused on the design and implementation of professional development programmes for teachers, aiming at developing their theoretical understanding along with their teaching and assessment practice. Catalonia in Spain offers an example of the effective recognition of professional development as critically important for the implementation of competency-based mathematics teaching and learning. The aim is not only to acknowledge but also to work with practicing and future teachers. Thus the task of translating competency research into practice has been addressed in the development project ARC (Application of Resources to the Curriculum). The project was started in order to model, pilot and evaluate "mathematical activities within a competency framework", assuming that this might help teachers meet "the challenge to assist all learners in the development of mathematical competencies by providing validated classroom experiences". Projects involving collaboration of teachers and researchers have been conducted with specific schools, classrooms and activities, so as to make successful teaching public. The projects have brought up the complexity of changing culturally long established practices of mathematics teaching that have become routinised.

Findings from several cycles of these projects suggest further challenges to be taken into account. We have to work more to understand the operational dimensions of mathematical competencies, especially concerning what collaborative action research can contribute, and how. Here, researchers face the need to go beyond anecdotal research-and-development collaboration with particular schools, teachers, classrooms, and students.

Despite the success and relevance of context-bound local initiatives of professional development and research, it is not clear how the improvements obtained under such conditions can be sustained and scaled-up.

Perspectives and Concluding Remarks

We have attempted to present significant, yet necessarily selected, aspects of and challenges to what some call "the competency turn" in mathematics education, research and practice. This has given rise to a number of important observations and conclusions:

- It remains crucial to come to grips with what it means and takes to master mathematics.
- Focusing on the enactment of mathematics in a broad sense is seen as essential in more and more places.
- Conceptualisation of this enactment needs further theoretical clarification and empirical investigations.
- Understanding the relationships and balances between the enactment of mathematics and other components of mathematical insight and knowledge remains a challenge.
- There is a need to clarify the role of attitudinal, volitional and dispositional factors in the conceptualisations and the reality of mathematical competencies.
- Terminological issues continue to cause confusion. To what extent are things called by the same name—e.g. competencies—actually equivalent? And to what extent do things called by different names actually cover different notions? And to the extent they do, what exactly are the relationships between them?
- The lack of a unified conceptual and theoretical framework for competencies, proficiency, processes, practices etc. tends to impede the possibilities of overcoming the challenges identified.
- In summary, a plethora of research and development work will be facing us in the years to come. There is every reason to expect, therefore, that there will be substantial progress to report on in future ICMEs.

References

Australian Education Council. (1990). *A national statement on mathematics for Australian school*. http://files.eric.ed.gov/fulltext7ED428947.pdf. Accessed May 30, 2016.

Australian Education Council/Curriculum Corporation. (1994). A national statement on mathematics for Australian school. http://cataloguw.nla.gov.au/Record/80076. Accessed May 30, 2016.

Board of Education. (1938). *Report of the consultative committee on secondary education with special reference to grammar schools and technical high schools ("The Spens Report")*. London: HM Stationery Office.

Husén, T. (Ed.). (1967). *International study of achievement in mathematics (vols. 1 & 2)*. Stockholm: Almqvist & Wiksell.

National Council of Teachers of Mathematics. (1980). *An agenda for action: Recommendations for school mathematics of the 1980s*. Reston: National Council of Teachers of Mathematics.

National Council of Teachers of Mathematics. (1989). *Curriculum and evaluation standards for school mathematics*. Reston: National Council of Teachers of Mathematics.

National Council of Teachers of Mathematics. (2000). *Principles and standards for school mathematics*. Reston: National Council of Teachers of Mathematics.

National Research Council, & Mathematics Learning Study Committee. (2001). *Adding it up: Helping children learn mathematics*. National Academies Press.

Niss. M. (2003). The Danish KOM project and possible consequences for teacher education. In R. Strässer, G. Brandell, B. Grevholm, & O. Helenius (Eds.) (2003). *Educating for the future. Proceedings of an International Symposium on Mathematics Teacher Education* (pp. 178–192). Gothenburg: Royal Swedish Academy of Science.

Niss, M., Bruder, R., Planas, N., Turner, R., & Villa-Ochoa, J. A. (2016). Survey team on: Conceptualisations of the role of competencies, knowing and knowledge in mathematics education research. *ZDM Mathematics Education, 48*(5), 611–632.

Niss, M., & Jensen, T. H. (Eds.). (2002). *Kompetencer og matematiklæring—Idéer og inspiration til udvikling af matematikundervisning i Danmark. Uddannelsesstyrelsens temahæfteserie nr. 18*. Copenhagen: The Ministry of Education.

Papert, S. (1972). Teaching children to be mathematicians vs. teaching about mathematics. *International Journal of Mathematical Education in Science and Technology, 3*(3), 249–262.

Pólya, G. (1945). *How to solve it?*. Princeton: Princeton University Press.

RAND Mathematics Study Panel. (2003). *Mathematical proficiency for all students. Toward a strategic research and development program in mathematics education*. Santa Monica, CA: RAND.

Open Access Except where otherwise noted, this chapter is licensed under a Creative Commons Attribution 4.0 International License.To view a copy of this license, visit http://creativecommons.org/licenses/by/4.0/.

Assistance of Students with Mathematical Learning Difficulties—How Can Research Support Practice?—A Summary

Petra Scherer, Kim Beswick, Lucie DeBlois, Lulu Healy
and Elisabeth Moser Opitz

Abstract When looking at teaching and learning processes in mathematics education students with mathematical learning difficulties or disabilities are of great interest. To approach the question of how research can support practice, an important step is to clarify the group or groups of students that we are talking about. The following contribution firstly concentrates on the problem of labelling the group of students having mathematical difficulties as there does not exist a single definition. This problem might be put down to the different roots of mathematics education on the one hand and special education on the other hand. Research results with respect to concepts and models for instruction are multifaceted and related to specific content and mathematical topics as well as underlying views of mathematics. Taking into account inclusive education, a closer orientation to mathematical education can be identified and the potential of selected teaching and learning concepts can be illustrated. Beyond this, the role of the teacher and the corresponding teacher education programs are discussed.

Keywords Mathematical learning difficulties · Inclusive education · Special education · Teacher education

---

P. Scherer (✉)
University of Duisburg-Essen, Essen, Germany
e-mail: petra.scherer@uni-due.de

K. Beswick
University of Tasmania, Launceston, Australia
e-mail: kim.beswick@utas.edu.au

L. DeBlois
Université Laval, Quebec, Canada
e-mail: Lucie.DeBlois@fse.ulaval.ca

L. Healy
Anhanguera University of São Paulo, São Paulo, Brazil
e-mail: lulu@baquara.com

E. Moser Opitz
University of Zürich, Zürich, Switzerland
e-mail: elisabeth.moseropitz@uzh.ch

Introduction: Mathematics Learning, Special Education and Inclusion—Setting the Scene

The following paper reports part of the work of the survey team "Assistance of students with mathematical learning difficulties—How can research support practice?" for ICME-13. When starting the work, the important aspects of defining students with mathematical learning difficulties, the role of teachers and teacher education programs as well as effective teaching programs and concepts of what teacher effective means came into the focus. Looking back to the ICME conferences of the last 20 years, we identified contributions in the corresponding topic study groups or discussion groups. It became obvious that we would have to take into account different disciplines; alongside mathematics education, special education, psychology and pedagogy also play important roles. Our aim was

- to describe definitions of mathematical learning difficulties and the problem of labelling,
- to discuss findings related to effective teaching practices and intervention strategies,
- to discuss concepts of assistance in the context of inclusive education, and
- to draw conclusions for teacher education.

Our paper is organized as follows: First, we discuss various definitions and assumptions concerning mathematical learning difficulties or disabilities. In Section "Effective Mathematics Teaching for All Students", we present a synthesis of results of selected meta-analyses and intervention studies, followed by some reflections upon the meaning of inclusive mathematics education and the kinds of learning environments that can support it. We conclude with implications for teacher education and perspectives for future research.

Mathematical Learning Difficulties: Definitions and Usage

In the title of this paper the term "students with mathematical learning difficulties" has been chosen to point to a group of learners perceived as being in particular need of assistance. But who is included in this group? In the first instance, we might interpret the term "students with mathematical learning difficulties" to be synonymous with terms such as "students with mathematical disabilities" or "students with special needs in relation to mathematics", but a closer look at the terms and the contexts in which they are used reveals that they may be associated with different approaches to teaching and learning, and to whether difficulty in learning mathematics is seen essentially as an individual attribute or as a consequence of barriers imposed by society (for this discussion see also Gervasoni & Lindenskov, 2011).

When the question of diagnosis is at the forefront, it is medical models and models which posit achievement as something inherent to the individual that tend to dominate. For example, according to 10th International Classification of Diseases (ICD 10, WHO, 2016), amongst the entries associated with specific developmental disorders of scholastic skills, is the category specific disorder of arithmetical skills (F81.2). This disorder is described as a "specific impairment in arithmetical skills that is not solely explicable on the basis of general mental retardation or of inadequate schooling. The deficit concerns mastery of basic computational skills of addition, subtraction, multiplication, and division rather than of the more abstract mathematical skills involved in algebra, trigonometry, geometry, or calculus" (ICD 10, WHO, 2016).

This definition is used as a basis for the widespread, if heavily criticised, use of the IQ-discrepancy model, where a mathematical learning disorder is diagnosed as a result of a discrepancy between IQ and mathematics performance level. Critics of this model argue that it can lead to over-identification at upper levels and under-identification at lower levels of IQ and that it leaves unspecified the point at which a discrepancy becomes significant (Francis et al., 2005). They also question whether the differences in IQ levels, permit the identification of the particular characteristics of different student groups (Murphy et al., 2007).

In the light of such criticisms, another influential classification system, the fifth Diagnostic and Statistical Manual of Mental Disorders (DSM V) published by the American Psychiatry Association (APA), no longer makes use of the discrepancy model. In previous versions, what was called a mathematics disorder was listed, however, in DSM V, this has been redefined as one of the subtypes of a "specific learning disorder", that is, a neurodevelopmental disorder that impedes the ability to learn or use specific academic skills. The symptoms are described as follows:

> Difficulties mastering number sense, number facts, or calculation (e.g., has poor understanding of numbers, their magnitude, and relationships; counts on fingers to add single-digit numbers instead of recalling the math fact as peers do; gets lost in the midst of arithmetic computation and may switch procedures). Difficulties with mathematical reasoning (e.g., has severe difficulty applying mathematical concepts, facts, or procedures to solve quantitative problems) (DSM V, 2016).

The changes from the DSM IV to the DSM V definitions of learning disorders reflect the lack of consensus as to the precise nature of the so-called specific learning disorders and the problems that arise when learning difficulties in mathematics are treated using exclusively neuropsychological perspectives. Healy and Powell (2013) reviewed some of the critiques and pointed to problems such as the lack of a robust consensus around characteristics of so-called disorders (Mazzocco & Myers, 2003; Gifford, 2005), the use of standard calculation procedures in diagnostic procedures (Ellemor-Collins & Wright, 2007; Gifford, 2005), the assumption that learning can be expected to be homogenous (Dowker, 2004; Ginsburg, 1997) and the failure to recognize environmental and socio-emotional factors (Kaufmann et al., 2013).

In the light of the difficulties associated with definitions which reside in the medical rather than educational community, recent publications of the Eurydice

Network[1] suggest the use of a broader concept of mathematical difficulties, using the term to refer to any group of students with low achievement in mathematics:

> Low achievement is the situation where a child fails to acquire basic skills while they do not have any identified disability and have cognitive skills within the normal range. In those cases, low achievement may be considered as a failure of the education system (European Commission, n. d., p. 4).

This definition stresses how, regardless of the causes, it is important to offer students with mathematical learning difficulties environments that enable them to thrive mathematically. How then might the teaching community intervene in ways that enable students to negotiate the difficulties they experience? To explore this question, the next section focuses on the results of studies into interventions aimed at improving the performance of students with mathematical learning difficulties.

Effective Mathematics Teaching for All Students

In this section we briefly review the results from meta-analyses and consider the findings of particular studies at various levels of schooling, that illustrate the complex conditions surrounding special education teaching before discuss inclusive education.

What Do We Know About Effective Teaching Practices in Mathematics Classrooms?—Intervention Studies

The absence of a generally accepted definition of mathematical learning disabilities implies a cautious approach is adopted when comparing results from different studies, particularly since intervention studies have pursued different objectives arising from different views of teaching and learning mathematics, the choice of the topics for the interventions, and the settings investigated.

According to the meta-analyses of Kroesbergen and Van Luijt (2003) and Gersten et al. (2009), direct instruction, self-instruction or explicit instruction led to practically and statistically important increases in effect size. However, it is not always clear what is meant by "guided instruction" or "explicit instruction". Whilst some authors understand it in the sense of scaffolding, others understand explicit instruction in a narrow way.

Kroesbergen and Van Luijt (2003) observed that the majority of interventions studies have concerned the field of basic arithmetic skills. Some of these studies have examined only the impact of training for procedural competencies like retrieval (Fuchs et al., 2009, 2010). Long-term effects have not been investigated in

---

[1]Network on education systems and policies in Europe http://eacea.ec.europa.eu/education/eurydice/index_en.php.

these studies and no information is available as to whether the students actually improved with fact retrieval, or simply used counting strategies more quickly. Other research (e.g., Andersson, 2010) has underscored the importance of fostering the domains of conceptual knowledge (e.g., place value, base-ten system, relationships within and between arithmetic operations) and procedural knowledge. The research undertaken by Ennemoser and Krajewski (2007), Pedrotty Bryant et al. (2008), Wißmann et al. (2013) and Pfister et al. (2015) showed significant effects for such interventions with primary school students. Intervention studies at secondary level have often focused on direct instruction and "drill and practice" teaching (Maccini et al. 2007). Nevertheless, Woodward and Brown (2006) and Moser Opitz et al. (in press) reported having successfully implemented a middle school program emphasizing conceptual understanding of primary arithmetic and problem solving.

With regard to the settings in which interventions have been implemented, research results are inconsistent. For example, a meta-analysis by Ise et al. (2012) of studies from German-speaking countries showed one-to-one training to have advantages over small group interventions, computer-based programs, and interventions integrated into the classroom. However, Moser Opitz et al. (in press) found a significant effect for an intervention in middle school which was partly integrated in the classroom teaching.

Taken together these results present challenges: First, it is not always clear what is meant by "guided instruction" or "explicit instruction". Second, even if some of the studies focus on conceptual understanding, the interventions reported do not cover the whole range of mathematical domains, but focus on topics that are known to be "stumbling blocks" for many—but not for all—students with learning difficulties in mathematics. Developing interventions for students with learning difficulties in mathematics is, therefore, a "balancing act" between giving guidance and taking into account the learners strategies and concepts; and focussing on well known "stumbling blocks" without forgetting that mathematics means more than arithmetic.

Inclusive Education

The UNESCO International Bureau of Education (2009) defined inclusive education is "an ongoing process aimed at offering quality education for all while respecting diversity and the differing needs and abilities, characteristics and learning expectations of students and communities, eliminating all forms of discrimination" (p. 18).

As with the terms "learning disabilities in mathematics" and "mathematical learning difficulties" there does not exist a common understanding of the term "inclusive education" (Ainscow, 2013) and there are many possible ways of viewing the notion of inclusion. For example, conceptions are likely to be mediated by factors such as the organisation of the school system (differentiated or comprehensive), legal regulations related to the provisions for students eligible for special education, policies related to the progression between grades (exam-based or age-based) and practices related to the organisation of classes (mixed-ability or streaming). Skovsmose (2015) has argued that inclusion represents an example of what he calls a contested concept, a concept that can be given different interpretations that operate in different ways in different discourses. For him, contested concepts represent controversies that can be of a profound political and cultural nature. His view of inclusion is one that rejects the idea of bringing learners into some (politically) presumed "normality". "Instead inclusive education comes to refer to new forms of providing meetings among differences" (Skovsmose, 2015, p. 7). In the remainder of this section, we consider some attempts to construct learning situations that permit such meetings.

Substantial and Rich Learning Environments—Multiple Opportunities

Constructivist and socio-constructivist theories open ways of viewing "knowing" (Ernest, 1994; Von Glasersfeld, 1995) and learning in a social environment (e.g., Wittmann, 2001). For mathematics education, investigative learning and productive practicing are seen as the main elements of these paradigms (e.g., Wittmann, 2001). Productive practicing is to be understood in contrast to bare reproduction of knowledge. It should enable pupils to think, to construct and to extend their knowledge. The teacher has to offer learning situations, that enable the students to make discoveries but this requires that the student possesses powerful tools in the form of (context)-models, schemes, and symbols (Streefland & Treffers, 1990, p. 313f).

With respect to heterogeneous learning groups, several studies have confirmed that investigative learning combined with productive practicing is appropriate for all learners—*especially* for low achievers and children with special needs (e.g., Ahmed, 1987; Moser Opitz, 2000; Scherer, 1999, 2003; Scherer & Moser Opitz, 2010, p. 49 ff.; Trickett & Sulke, 1988; Van den Heuvel-Panhuizen, 1991). According to this view all learners should be confronted with complex learning environments characterised by investigative learning and productive practicing. Such holistic approaches to mathematics teaching and learning require all learners to see relationships between numbers, shapes, and so forth in order to understand mathematical structures (Trickett & Sulke, 1988, p. 112).

Taking into account some of the research reviewed in the Sections "What Do We Know About Effective Teaching Practices in Mathematics Classrooms?— Intervention Studies" and "Inclusive Education", it seems that there still exists scepticism with respect to constructivist or socio-constructivist approaches for students with mathematical learning difficulties. For example, although the results of Kroesbergens's and van Luijt's meta-analysis (2003) suggest that direct instruction could be the most beneficial type of instruction for these students, this conclusion neglects the fact that students with mathematical learning difficulties profit from teaching specific cognitive learning strategies like self-regulated learning (see Mitchell, 2014).

Moreover, to identify children's existing difficulties, it is necessary to give them opportunities to show what they are capable of. More attention should be paid to the creation of substantial and rich mathematical learning environments for inclusive settings, in which different learning trajectories and different forms of interacting with mathematical objects are explicitly recognized (Fernandes & Healy, 2016). The development of such environments is crucially dependent upon differentiation. Learning tasks directed towards levels of difficulty predetermined by the teacher carry the risk that some students are overtaxed or misjudged or fixed at a specific level as viewed by the teacher. Research shows that learning environments that allow *natural differentiation* (ND) can reduce these risks (cf. Wittmann, 2001; Scherer & Krauthausen, 2010). Natural differentiation means that the learning environment provided is substantial and complex and offers multiple ways of learning and multiple strategies for solving a given problem.

Consistent with natural differentiation, learning environments allowing own productions or free productions (cf. Streefland, 1990) offer various opportunities for students' use of their own strategies and provide their own solutions and thus support suitable differentiation. Examples show that especially students with mathematical learning difficulties often make use of the affordances of such environments and show unexpected competencies (e.g., DeBlois, 2014, 2015; Scherer, 1999).

This more open approach brings in specific requirements for the teacher: In general, classroom practice should require more than getting the correct result or being able to perform an algorithm but also explaining and reasoning about solution strategies, and considering solution strategies and associated reasoning. Teachers "need to know how to use pictures or diagrams to represent mathematics concepts and procedures for students, provide students with explanations for common rules and mathematical procedures, and analyze students' solutions and explanations" (Hill et al., 2005, p. 372).

For a more detailed discussion of the complex field of teacher education—pre-service as well as in-service—teachers' beliefs, their mathematical and didactical knowledge and the awareness of interactions in classroom see Scherer et al. (2016; also Beswick, 2008; Peltenburg and Van den Heuvel-Panhuizen, 2012).

Conclusions and Perspectives

In this paper various aspects of the situation of students with mathematical learning difficulties have been discussed. The separation of mathematics education and special education has given rise to specific requirements and problems for research which are further complicated by the different conditions in different countries.

Exploring the different ways in which students with mathematics learning difficulties are identified and described in different areas, suggests that many factors can interact to impede the mathematical development of learners and, rather than dichotomising learners into those who experience mathematical learning difficulties and those who do not, it might be more useful to adopt approaches to mathematics education that recognise and value the diversity of learners' mathematical experiences. This is in contrast to treating differences in learning trajectories as evidence of a deficiency or disorder that necessarily impedes learning or justify segregation. A starting point in constructing a more inclusive mathematics curriculum, therefore, involves envisioning learning scenarios designed to facilitate multiple ways of interacting with mathematical objects, and relationships that respect the diverse experiences (sensory, cognitive, socio-emotional and cultural) and identities of the students with whom we work (Healy et al., 2013). There is need for more evidence-based research in this area.

For teacher education programs, first, it is necessary to distinguish between the needs of teachers and needs of pre-service teachers. For pre-service teachers, there is a need to create situations that help them to distance themselves from their own experiences of learning mathematics as school students (DeBlois & Squalli, 2002). In addition, curriculum, beliefs, personal decompression of mathematical knowledge (Proulx & Bednarz, 2008) and social activities must be discussed in order to manage needs of students with mathematics learning difficulties. The challenge for the teacher is to interpret the events that happen in the classroom in order to make pedagogical and didactical choices (DeBlois, 2006).

In this paper, the focus has mainly been on students, but the challenge of providing a quality mathematics education all goes way beyond the classroom level and involves a rethinking of the institutional structures which mediate both teaching and learning, structures such as curriculum and assessment for example. Experience tells us that it is more efficient to build an accessible building from scratch than to attempt to adapt inaccessible buildings. Can we learn from this as we attempt to build inclusive school mathematics? Perhaps the question is not how we can assist students with mathematical learning difficulties, but how we can learn to build a mathematics education system that no longer disables so many mathematics students.

References

Ahmed, A. (1987). *Better mathematics. A curriculum development study*. London: HMSO.
Ainscow, M. (2013). Developing more equitable education systems: Reflections on a three-year improvement initiative. In V. Farnsworth & Y. Solomon (Eds.), *Reframing educational research* (pp. 77–89). London: Routledge.
Andersson, U. (2010). Skill development in different components of arithmetic and basic cognitive functions: Findings from a 3-year longitudinal study of children with different types of learning difficulties. *Journal of Educational Psychology, 102*(1), 115–134.
Beswick, K. (2008). Influencing teachers' beliefs about teaching mathematics for numeracy to students with mathematics learning difficulties. *Mathematics Teacher Education and Development, 9*, 3–20.

DeBlois, L. (2006). Influence des interprétations des productions des élèves sur les stratégies d'intervention en classe de mathématiques. *Educational Studies in Mathematics, 62*(3), 307–329.
DeBlois, L. (2014). Le rapport aux savoirs pour établir des relations entre troubles de comportements et difficultés d'apprentissage en mathématiques. Dans *Le rapport aux savoirs: Une clé pour analyser les épistémologies enseignantes et les pratiques de la classe.* Coordonné par Marie-Claude Bernard, Annie Savard, Chantale Beaucher. http://lel.crires.ulaval.ca/public/le_rapport_aux_savoirs.pdf. Accessed April 20, 2016.
DeBlois, L. (2015). Classroom interactions: Tensions between interpretations and difficulties learning mathematics. Proceedings of the 37th Annual Meeting Canadian Group in Mathematics Education. University of Alberta, Edmonton (Alberta). 171–186. http://www.cmesg.org/wp-content/uploads/2015/05/CMESG2014.pdf. Accessed April 20, 2016.
DeBlois, L., & Squalli, H. (2002). Implication de l'analyse de productions d'élèves dans la formation des maîtres. *Educational Studies in Mathematics, 50*(2), 212–237.
Dowker, A. D. (2004). *What works for children with mathematical difficulties?* London, UK: Department for Children, Schools and Families.
DSM V. (2016). Diagnostical and statistical manual of mental disorders. (5th ed.). http://dsm.psychiatryonline.org/doi/full/10.1176/appi.books.9780890425596.dsm01. Accessed April 13, 2016.
Ellemor-Collins, D., & Wright, R. J. (2007). Assessing pupil knowledge of the sequential structure of number. *Educational and Child Psychology, 24*(2), 54–63.
Ennemoser, M., & Krajewski, K. (2007). Effekte der Förderung des Teil-Ganzes-Verständnisses bei Erstklässlern mit schwachen Mathematikleistungen. *Vierteljahresschrift für Heilpädagogik und ihre Nachbargebiete, 76*(3), 228–240.
Ernest, P. (1994). Constructivism: Which form provides the most adequate theory of mathematics learning? *Journal für Mathematik-Didaktik, 15*(3/4), 327–342.
European Commission (n. d.). *Addressing low achievement in mathematics and science. Final Report of the Thematic Working Group on Mathematics, Science and Technology (2010–2013)* http://ec.europa.eu/education/policy/strategic-framework/archive/documents/wg-mst-final-report_en.pdf. Accessed June 06, 2016.
Fernandes, S. H. A. A., & Healy, L. (2016). The challenge of constructing an inclusive school mathematics. Paper accepted for presentation in *Topic Group 5: Activities for, and research on, students with special needs*. 13th International Congress on Mathematics Education, Hamburg.
Francis, D., Fletcher, J. M., Stuebing, K. K., Lyon, R. G., Shaywitz, B. A., & Shaywith, S. E. (2005). Psychometric approaches to the identification of LD: IQ and achievement scores are not sufficient. *Journal of Learning Disabilities, 38*(2), 98–108.
Fuchs, L. S., Powell, S. R., Seethaler, P. M., Cirino, P. T., Fletcher, J. M., Fuchs, D., et al. (2010). The effects of strategic counting instruction, with and without deliberate practice, on number combination skill among students with mathematics difficulties. *Learning and Individual Differences, 20*(2), 89–100.
Fuchs, L. S., Powell, S. R., Seethaler, P. M., Cirino, P. T., Fletcher, J. M., Fuchs, D., et al. (2009). Remediating number combination and word problem deficits among students with mathematics difficulties: A randomized control trial. *Journal of Educational Psychology, 101*(3), 561–576.
Gersten, R., Chard, D. J., Jayanthi, M., Baker, S. K., Morphy, O., & Flojo, J. (2009). Mathematics instruction for students with learning disabilities: A meta-analysis of instructional components. *Review of Educational Research, 79*(3), 1202–1242.
Gervasoni, A., & Lindenskov, L. (2011). Students with "special rights" for mathematics education. In B. Atweh, M. Graven, & P. Valero (Eds.), *Mapping equity and quality in mathematics education* (pp. 307–324). New York, NY: Springer.
Gifford, S. (2005). Young children's difficulties in learning mathematics. Review of research in relation to dyscalculia. *Qualifications and Curriculum Authority (QCA/05/1545)*. London, UK: Department for Children, Schools and Families.

Ginsburg, P. H. (1997). Mathematics learning disabilities: A view from developmental psychology. *Journal of Learning Disabilities, 30*(1), 20–33.

Healy, L., Fernandes, S. H. A. A. & Frant, J. B. (2013). Designing tasks for a more inclusive school mathematics. In *Proceedings of ICMI Study 22—Task Design in Mathematics Education* (vol. 1, pp. 63–70.). Oxford.

Healy, L., & Powell, A. B. (2013). Understanding and overcoming "Disadvantage" in learning mathematics. In M. A. Clements, A. Bishop, C. Keitel, J. Kilpatrick, & F. Leung (Eds.), *Third international handbook of mathematics education* (pp. 69–100). Dordrecht, The Netherlands: Springer.

Hill, H. C., Rowan, B., & Ball, D. (2005). Effects of teachers' mathematical knowledge for teaching on student achievement. *American Educational Research Journal, 42*(2), 371–406.

Ise, E., Dolle, K., Pixner, S., & Schulte-Körne, G. (2012). Effektive Förderung rechenschwacher Kinder: Eine Metaanalyse. *Kindheit und Entwicklung, 21*(3), 181–192.

Kaufmann, L., Mazzocco, M. M., Dowker, A., von Aster, M., Göbel, S. M., Grabner, R. H., Henik, A., Jordan, N. C., Karmiloff-Smith, A. D., Kucian, K., Rubinsten, O., Szucs, D., Shalev, R., & Nuerk, H. C. (2013). Dyscalculia from a developmental and differential perspective. *Frontiers in Psychology. 4,* AUG, Article 516.

Kroesbergen, E. H., & Van Luit, J. E. H. (2003). Mathematical interventions for children with special educational needs. *Remedial and Special Education, 24,* 97–114.

Maccini, P., Mulcahy, C. A., & Wilson, M. G. (2007). A follow-up of mathematics interventions for secondary students with learning disabilities. *Learning Disabilities Research & Practice, 22*(1), 58–74.

Mazzocco, M. M., & Myers, G. F. (2003). Complexities in identifying and defining mathematics learning disability in the primary school age years. *Annals of Dyslexia, 53,* 218–253.

Mitchell, D. (2014). *What really works in special and inclusive education. Using evidence-based teaching strategies.* New York: Routledge.

Moser Opitz, E., Freesemann, O., Grob, U., Prediger, S., Matull, I., & Hußmann, S. (accepted). Remediation for students with mathematics difficulties: An intervention study in middle schools. *Journal of Learning Disabilities.* doi:10.1177/0022219416668323.

Moser Opitz, E. (2000). *Zählen – Zahlbegriff – Rechnen Theoretische Grundlagen und eine empirische untersuchung zum mathematischen erstunterricht in sonderklassen.* Bern: Haup.

Murphy, M. M., Mazzocco, M. M., Hanich, L. B., & Early, M. C. (2007). Cognitive characteristics of children with mathematics learning disability (MLD) vary as a function of the cutoff criterion used to define MLD. *Journal of Learning Disabilities, 40*(5), 458–478.

Pedrotty Bryant, D., Bryant, B. R., Gersten, R., Scammacca, N., & Chavez, M. M. (2008). Mathematics intervention for first- and second-grade students with mathematics difficulties: The effects of Tier 2 intervention delivered as booster lessons. *Remedial and Special Education, 29*(1), 20–32.

Peltenburg, M., & Van den Heuvel-Panhuizen, M. (2012). Teacher perception of the mathematical potential of students in special education in the Netherlands. *European Journal for Special Needs Education, 27*(3), 391–407.

Pfister, M., Stöckli, M., Moser Opitz, E., & Pauli, C. (2015). Inklusiven Mathematikunterricht erforschen: Herausforderungen und erste Ergebnisse aus einer Längsschnittstudie. *Unterrichtswissenschaft, 43*(1), 53–66.

Proulx, J., & Bednarz, N. (2008). The mathematical preparation of secondary mathematics schoolteachers: Critiques, difficulties and future directions. Topics Group 29. *ICME 11.* Monterrey (Mexico). http://tsg.icme11.org/document/get/83. Accessed 5 May 2016.

Scherer, P. (1999). *Entdeckendes Lernen im Mathematikunterricht der Schule für Lernbehinderte —Theoretische Grundlegung und evaluierte unterrichtspraktische Erprobung* (2nd ed.). Heidelberg: Edition Schindele.

Scherer, P. (2003). Different students solving the same problems—the same students solving different problems. *Tijdschrift voor Nascholing en Onderzoek van het Reken-Wiskundeonderwijs, 22*(2), 11–20.

Scherer, P., Beswick, K., DeBlois, L., Healy, L., & Moser Opitz, E. (2016). Assistance of students with mathematical learning difficulties: How can research support practice? *ZDM—Mathematics Education, 48*(5), 633–649.

Scherer, P., & Krauthausen, G. (2010). Natural differentiation in mathematics—the NaDiMa project. *Panama-Post, 29*(3), 14–26.

Scherer, P., & Moser Opitz, E. (2010). *Fördern im Mathematikunterricht der Primarstufe*. Heidelberg: Springer.

Skovsmose, O. (2015). Inclusion: A contested concept. In: *Anais do VI Seminário Internacional de Pesquisa em Educação Matemática*—VI SIPEM, Pirenópolis, Goiás, Brasil.

Streefland, L. (1990). Free productions in learning and teaching mathematics. In K. Gravemeijer, M. van den Heuvel, & L. Streefland (Eds.), *Contexts free productions tests and geometry in realistic mathematics education* (pp. 33–52). Utrecht: Freudenthal Institute.

Streefland, L., & Treffers, A. (1990). Produktiver Rechen-Mathematik-Unterricht. *Journal für Mathematik-Didaktik, 11*(4), 297–322.

Trickett, L., & Sulke, F. (1988). Low attainers can do mathematics. In D. Pimm (Ed.), *Mathematics, teachers and children* (pp. 109–117). London: Hodder and Stoughton.

UNESCO International Bureau of Education. (2009). *Inclusive education: The way of the future*. In International Conference on Education, 28th Session, Geneva, UNESCO Paris, November 25–28, 2008. Retrieved from http://www.ibe.unesco.org/fileadmin/user_upload/Policy_Dialogue/48th_ICE/ICE_FINAL_REPORT_eng.pdf

Van den Heuvel-Panhuizen, M. (1991). Ratio in special education. A pilot study on the possibilities of shifting the boundaries. In L. Streefland (Ed.), *Realistic Mathematics Education in Primary School. On the occasion of the opening of the Freudenthal Institute* (pp. 157–181). Utrecht: Freudenthal Institute.

Von Glasersfeld, E. (1995). A constructivist approach to teaching. In L. P. Steffe & J. Gale (Eds.), *Constructivism in education* (pp. 3–15). Hillsdale, NJ: Erlbaum.

WHO—World Health Organisation. (2016). *International Statistical Classification of Diseases and Related Health Problems 10th Revision*. http://apps.who.int/classifications/icd10/browse/2016/en#/F81.9. Accessed April 13, 2016.

Wißmann, J., Heine, A., Handl, P., & Jacobs, A. M. (2013). Förderung von Kindern mit isolierter Rechenschwäche und kombinierter Rechen- und Leseschwäche: Evaluation eines numerischen Förderprogramms für Grundschüler. *Lernen und Lernstörungen, 2*(2), 91–109.

Wittmann, E. C. (2001). Developing mathematics education in a systemic process. *Educational Studies in Mathematics, 48*(1), 1–20.

Woodward, J., & Brown, C. (2006). Meeting the curricular needs of academically low-achieving students in middle grade mathematics. *The Journal of Special Education, 40*(3), 151–159.

Open Access Except where otherwise noted, this chapter is licensed under a Creative Commons Attribution 4.0 International License. To view a copy of this license, visit http://creativecommons.org/licenses/by/4.0/.

Mathematics Teachers Working and Learning Through Collaboration

Barbara Jaworski, Olive Chapman, Alison Clark-Wilson,
Annalisa Cusi, Cristina Esteley, Merrilyn Goos, Masami Isoda,
Marie Joubert and Ornella Robutti

Abstract The authors of this paper were tasked by ICME-13 organisers with conducting a survey on the topic "Mathematics Teachers Working and Learning through Collaboration". Four research questions guided the survey, concerned with: the *nature* of collaborative working; the *people* who engage collaboratively; the *methodological* and *theoretical perspectives* used; what *learning* could be observed and how it related to collaboration? The resulting survey drew from a wide range of sources, identifying papers relevant to the topic—316 papers were identified, analysed against a set of criteria and organised into three major themes, each relating to one or more of our research questions: *Different contexts and features of mathematics teachers working in collaboration*; *Theories and methodologies framing the studies; Outcomes of collaborations*. In addition to the papers revealed by the survey, the team sought contributions from projects around the world which are not represented in the published literature. Members from these projects offered 'narratives' from the work of teachers in the projects. This paper reports on the nature of the projects revealed by the survey and the narratives, their theoretical and methodological focuses, and the range of findings they expressed. While we offer a

---

B. Jaworski (✉)
Loughborough University, Loughborough, England, UK
e-mail: b.jaworski@lboro.ac.uk

O. Chapman
University of Calgary, Calgary, Canada
e-mail: chapman@ucalgary.ca

A. Clark-Wilson
UCL Institute of Education, University College London, London, UK
e-mail: a.clark-wilson@ioe.ac.uk

A. Cusi · O. Robutti
University of Turin, Turin, Italy
e-mail: annalo@tin.it

O. Robutti
e-mail: ornella.robutti@unito.it

© The Author(s) 2017
G. Kaiser (ed.), *Proceedings of the 13th International Congress on Mathematical Education*, ICME-13 Monographs, DOI 10.1007/978-3-319-62597-3_17

significant range of factors and findings, resulting from a very considerable work, we are aware of limitations in our study: we missed relevant papers in journals outside our range; papers reviewed were usually not authored by teachers so the teachers' voice was often missing; narratives came from projects with which we were familiar, so we missed others. The survey team is in the process of initiating an ICMI study which can take this work into these missing areas. This paper follows closely the presentation made by the survey team at the ICME-13 congress. In presenting findings from the survey, we have tried to provide examples from and make reference to the survey papers. Because the set of references would be too large to fit within our word limit, we have had to reduce the number of references made. However, readers can find a full set of references in a more detailed paper, Robutti et al. in (ZDM Mathematics Education, 48(5), 651–690, 2016).

ICME-13 Theme—Mathematics Teachers Working and Learning Through Collaboration

Introduction

This paper, produced for the Proceedings of ICME-13, follows as closely as possible our presentation at the congress. It is limited by a necessary imposed length restriction. This means that we have not been able to include all of the citations from our surveyed sources that exemplified our findings. For these we refer the reader to our full paper, Robutti et al. (2016).

Our theme zooms in on the wider professional development scene to focus on *the learning* that occurs when *teachers of mathematics work together collaboratively*, for the mathematical learning of students, which motivates their teaching. We were tasked by ICME to identify and characterise important new knowledge, recent developments, new perspectives, and emergent issues with respect to our theme.

C. Esteley
National University of Cordoba, Córdoba, Argentina
e-mail: esteley@famaf.unc.edu.ar

M. Goos
The University of Queensland, Brisbane, Australia
e-mail: m.goos@uq.edu.au

M. Isoda
University of Tsukuba, Tsukuba, Japan
e-mail: isoda@criced.tsukuba.ac.jp

M. Joubert
African Institute of Mathematical Sciences, Cape Town, South Africa
e-mail: marie@aimssec.ac.za

The Latin word "collaborāre" means "To work in conjunction with another or others, to co-operate" (OED). Collaboration involves mathematics teachers engaging in joint activity, common purpose, critical dialogue and inquiry, mutual support; addressing issues that challenge teachers professionally and reflecting on their role in school and in society. We address *what* co-learning occurs and *how* it occurs.

Teachers *working* includes all the dimensions of teaching in and beyond face-to-face activity with students in the classroom:

- the didactics and pedagogy of creating the classroom environment;
- the evaluation of students' mathematical learning;
- the professional development activity through which teachers learn to teach;
- the institutional demands of school and educational system;
- the societal demands of parents, employers and politicians.

Many of the papers that address collaborations involving teachers also speak about *communities* of teachers. Community can be an informal term or be defined theoretically as, for example, *Community of Practice*, or *Community of Inquiry*. Communities of teachers working together can include various 'others'; teacher educators, researchers, didacticians, school leaders, parents and so on.

Research questions guiding the survey are:

1. What is the *nature* of collaborative working (to include the different roles that teachers can play) and how does this relate to situation, culture and context?
2. Who are the *people* who engage collaboratively to promote the effective learning and teaching of mathematics, what are their roles, and how do they relate to each other within the different communities?
3. What *methodological* and *theoretical perspectives* are used to guide and inform collaborative working and learning?
4. What *learning* can be observed and how does it relate to collaboration?

Methodology Adopted for This Survey

The team searched the mathematics education literature for *journals, books, handbooks*, and *proceedings* relevant to the topic, from 2005 to 2015. They first located *articles/chapters/reports* in which *collaboration* is an explicit part of the research design and its influence on mathematics teachers' learning/working practices. Their strategies to select the papers involved manual and automated processes applied to titles, indexes, abstracts and full-texts. A second selection focussed on collaboration, its processes and products. 316 sources were identified that concerned research from across the world including 23% from North America,

9% from South America, 31% from Europe, 4% from Africa, and 24% from Australasia. Eighteen papers reflected cross-continent collaborations, for example between the Netherlands and Indonesia, between Taiwan and Portugal, or between Germany and Chile. The data were entered into a spreadsheet in which we captured both demographic information and a range of factors relating to projects, teachers, researchers, forms of collaboration, impact on students, nature of learning, and so on. Overall we had 8500 cells of data. For more detail see (Robutti et al., 2016).

Analysis of the spreadsheet data led to identification of *fundamental themes* that could frame the topic of collaboration:

- Theme 1—*Different contexts and features of mathematics teachers working in collaboration* [Addressing RQs 1 and 2]
- Theme 2—*Theories and methodologies framing the studies* [Addressing RQ 3]
- Theme 3—*Outcomes of collaborations* [Addressing RQs 1, 2, 3, 4].

The comparison and contrast between the different sources led to the identification of the following dimensions and related sub-dimensions connected to the themes.

Theme 1	1. Initiation, foci and aims 2. How the collaboration was conceived and organised 3. Scale of collaborations 4. Composition and roles
Theme 2	1. Theories that frame the studies 2. Methodologies of work with teachers
Theme 3	1. Reflections on collaborating 2. Impacts on teachers' knowledge, thinking, and practice

Theme 1: Different Contexts and Features

Initiation, Foci and Aims

We found a diversity of forms of initiation, foci and aims, including:

- Initiatives mandated by ministries and national/regional institutions;
- Collaborations supported by ministries and national/regional institutions;
- Research collaborations initiated by researchers;
- Professional development initiated by researchers/didacticians;
- Within-school collaborations without involvement of 'others'.

The foci for projects fall into two broad categories. The first refers to innovation about mathematical content, the development of new curricula, different pedagogical approaches, and the integration of new tools and resources—with aims to promote the development of teachers':

(a) awareness of students' different learning trajectories;
(b) necessary competencies to foster students' learning;
(c) understanding how teaching and learning resources could support/inhibit learning.

As an example of the first kind of aim (a), Fried and Amit (2005) report from a project, in Israel, which involved 31 schools and 82 teachers, who were "encouraged to discuss teaching approaches required by the students at each grade level and the relationships between the different stages of the development" (p. 419).

The paper from Chen and Chang (2012) reports an example of the second kind of aim (b). They describe the development of a small professional learning community in China, focused on improving teachers' discourse-based assessment practice from convergent formative assessment to more divergent formative assessment.

The project presented by Lin, Chen, Hsu, and yang (2013) exemplifies the third kind of aim (c). They report how a group of teachers, in Taiwan, was involved in the design of instruction materials to make them aware of the ways in which these materials could support/inhibit learning.

The second category of focus refers to the different practices designed to foster teachers' professional learning. The corresponding aims are to evaluate the implementation of specific processes and tools as professional development programmes for mathematics teachers. Krammer et al. (2006), for example, report on a project, developed in Germany and Switzerland, which aimed to examine the conditions and effectiveness of web-based professional development with classroom videos to support mutual exchange, shared reflection and reciprocal analyses of instruction.

How the Collaboration Was Conceived and Organised

We analysed this dimension focusing on two main questions:

- How is collaborative work within the community conceived?
- How is collaborative work within the community activated?

As regards the first question, the team found two ways in which communities developed:

(i) as a declared objective of the collaborations
(ii) as a methodological approach for teacher education.

In the project reported by Potari (2013), developed in Greece, the creation of a community is an object of collaboration (i). The objectives of the mathematics educators were "not to transmit knowledge from the research to the teachers but to form a community of inquiry where the teachers use research as a tool for their inquiry" (p. 509). Collaboration is the central methodological approach of the study by Martins and Santos (2012), who report from a project in Portugal. The described collaborative work was aimed at evolving teachers' abilities to reflect over time, through the stimuli from other people involved in these reflections (mentor, tutor, supervisor, critical friend).

As regards the second question, the creation of collaborative contexts for the teachers within a community occurred though a number of approaches from which common characteristics were

- Cycles of activities, such as study, design, implementation, analysis, re-design, re-implementation;
- Fundamental roles played by expert figures such as other teachers, teacher educators, mathematicians, and researchers;
- Teachers' engagement: in terms of challenge, solidarity, accountability, trust, respect;
- Activation of processes of reflection, sometimes with reference to theoretical lenses.

One project which stood out in demonstrating all of these characteristics was Brodie and Shalem (2011).

The Scale of Collaborations

Scale varied considerably across the papers reviewed as shown in Fig. 1.

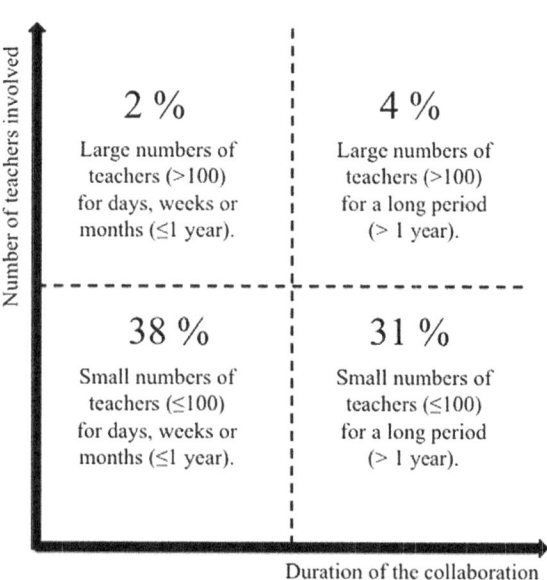

Fig. 1 Number of teachers versus time

It can be seen that large-scale projects in both teacher numbers and length of project were far fewer than those at much smaller scale.

Composition of Collaborative Groups and Participants' Roles

Given the sources of the reviewed papers, participants almost always included mathematics education researchers working with pre- or in-service teachers (who may also be researchers); sometimes, also included were school principals, education assistants and officers; curriculum leaders masters and doctoral students; community leaders; members of specific cultural communities.

Participants' roles were not always clearly described—where they were, key features included teachers designing and piloting curriculum revisions in collaboration with researchers; aiming for 'equal status' between participants; the principle of 'co-learners in partnership' for all participants, and, for teachers, a shift from 'teacher-participants' to 'teacher-researchers'; teachers as mentors to other teachers; dynamic evolution of participants' roles over time. For example, Hospesovà, Machàckovà, and Tichà (2006), in the Czech Republic, aimed for equal status between different participants; Goodchild (2008), in Norway, aimed to develop the principle of 'co-learners in partnership' for all participants.

Theme 2: Theories and Methodologies Framing the Studies

Theories

Many papers did not declare explicitly the theoretical perspectives behind a project. Of those that did, four perspectives were evident: *Community* (of *Practice* or of *Inquiry*) (69%), *Activity Theory* (20%), *Metadidactical Transposition* (6%) and *Valsiner's Zone Theory* (5%). These fell into two pairs with characteristics in common.

The first pair of frameworks is concerned with **knowledge growth** promoted by mediation of tools, signs, artefacts and/or tensions and contradictions between different elements of mediation.

1. **Communities**, involving

    - Communities of *Practice* (e.g., Wenger, 1998): with constructs of *engagement, imagination, alignment*. For example, Goos and Bennison (2008) showed how a community of practice, focused on "becoming a teacher of secondary school mathematics" emerged during a pre-service teacher education program and was sustained after students graduated (p. 43).

- Communities of *Inquiry* (e.g., Jaworski, 2006): expanding on Wenger with a construct of *critical alignment*. For example, Jaworski's (2008) research on creating a community of inquiry with teachers led to critical alignment in practice, treating issues and tensions as central to teachers' lives and work.

2. **Activity Theory** (e.g., Engeström, 1999) with a construct of *expansive learning*. For example, Sakonidis and Potari (2014) used Activity Theory to analyse their joint activity as teacher educators collaborating with teachers, and thus became aware of not only the transformative nature of teachers' professional learning but also aspects of their own practice that allowed further development to occur.

The second pair of frameworks is concerned with evolution of **teacher identities** promoted/constrained by professional development programs and/or institutional contexts.

3. **Valsiner's Zone Theory** (from Vygotsky e.g., Valsiner, 1997):The Teacher's Zone of Proximal Development is expanded by the Zone of Free Movement and Zone of Promoted Action. For example, Goos's (2013) analysis of two case studies of teacher learning and development identified the significance of productive tensions within a teacher's zone system as a potential trigger for change.
4. **Metadidactical Transposition** (Aldon et al., 2013) from the Anthropological Theory of Didactics (Chevallard, 1985) that traces the evolution of teaching practices as teachers and researchers collaborate.

Methodologies

We found here two sorts of methodologies: the *research methodologies* which framed the research in the studies (these were mainly qualitative e.g., classroom observation, interviews); *developmental methodologies* which framed the nature of development in the study (for example, Collaboration, Professional Development, Change and Communities of inquiry). Within the latter, the following were most evident: Lesson Study; Action/Design/Developmental Research; Narrative Analysis; and Other (including Professional Learning Communities, Video Clubs, Online communities). The most commonly used of these were:

- *Lesson Study* (LS)—very few papers reported on LS *in Japan* (more on this below). However, 20% of surveyed papers reported on lesson study from USA, Australia, UK, countries in Asia, Africa, Europe, South America. Often LS involved a cycle of plan, teach/observe, post-lesson discussion and reflection, and promoted the role of "knowledgeable others". Takahashi (2014) showed how knowledgeable others are crucial for bringing new knowledge from research and the curriculum to the lesson study team and for making the connection between research and practice.

- The challenge of embedding Japanese Lesson Study in other cultural contexts was evident: for example, Kusangi (2014) argues that lesson study in Indonesia has been bureaucratized because of the centralized regulation of teaching and the emphasis placed on preparing students for national examinations.
- *Action research, Design Research, Learning Study* and *Developmental Research* all involved iterative cyclic processes, which varied according to the nature and context of the project.
- *Narrative analysis*: a tool for both research and development in which teacher narratives often revealed evidence of learning. For example, Ponte, Segurado, and Oliveira (2003) link the uses of narrative, as a tool, to understanding mathematics teachers' learning through collaboration.

Theme 3: Research Findings and Knowledge Generated

Here we focus on data from survey articles and from project narratives.

Data from Survey Articles

Almost all the articles revealed by the survey reported on research projects. This theme draws together the findings from these research projects, also drawing on the reports of about 50 'teacher enquiries' funded by the National Centre for Excellence in Teaching Mathematics (NCETM) in the UK. The NCETM enquiries were initiated and coordinated by teachers, and when they finished, a report was produced by the teachers.

The reported findings fall into two main areas: *Reflections on Collaborating* and *Impacts on teachers' knowledge, thinking and practice*

In *Reflections on Collaborating*:

- *Supporting Factors* fell into three main areas: *roles of participants; shared goals; and a feeling of safety.* In terms of the roles of participants, *diversity, collaborating with other teachers*, and *shifting roles over time* all seemed to be important. For example, in one of the NCETM reports, Wynne-Jones (2013) explained the value of collaborating with other teachers: "Working alongside a range of colleagues gave a much wider perspective. One person's ideas alone would not have had the depth generated by teachers working in groups." In another, Dowling (2013), a teacher wrote about feeling safe: "You need to risk it and don't get scared".
- *Inhibiting Factors* related to *Ownership, Time* and the *Value of Collaboration* (per se). For example, within problems of *ownership*, Besamusca and Drijvers (2013) focused on issues to do with *responsibility*, Hospesovà et al. (2006) focused on teachers feeling *uncomfortable*. To do with problems of *time*, Berg

(2011) focused on shortage of time in day to day activities, Cavanagh and McMaster (2015) focused on the time needed to establish a functional community. Jaworski (2008) focused on the day to day constraints of school life and on the value of collaboration, Campbell (2009) questioned the value of collaboration per se.

In *Impacts on teachers' knowledge, thinking and practice.*

- Changes in *knowledge, thinking and beliefs* referred first to learning to work within a community. Some authors wrote about individuals learning how to participate in the community, others reported learning to work together, describing, for example, how the nature of their collaborative work changed over time (Menezes, 2011). An important aspect of impacts on knowledge and beliefs seemed to be *reflection*. A number of papers reported that the participants had learned to reflect better, with Pires and Martins (2009) claiming that "[t]he program has … allowed the development of their ability to reflect (oral and written) on practices" (p. 47).

Impacts on *practice* mainly referred to better teaching, with a number of papers claiming that participating teachers had improved their teaching in general. It seems that many teachers developed confidence in the classroom, and with this, a willingness to try new approaches (Warren, 2008). In more specific terms, a number of projects reported on *better questioning* in the classroom and a number of others on *better sequencing* of lessons. A large number of the NCETM enquiries reported on specific changes in teaching practice, such as the use of concrete materials and equipment. Riley (2013), for example, reported on the teachers taking part in her project, saying "practical equipment is now being used far more effectively to support the learning and embedding of children's knowledge, skills and understanding of number systems". As a further example, Alldis and Whitney (2013) in their NCETM report quoted a teacher explaining how her teaching had changed: "I spend more time allowing pupils to explore their mathematics observing their discussions with peers and ensuring when pupils come to a solution that they must be able to 'convince' another member of the group and not simply accept an answer." Finally, there appeared to have been an impact on the way many teachers were now *considering their students' thinking*. Posthuma (2012), for example, reported that teachers in her project had learned the importance of attending to student thinking.

Data Directly from the Projects

As our survey progressed and revealed more and more papers, we were increasingly more aware that very much collaborative activity is not (or not yet) reported in published journals. We therefore decided to invite participants from known projects around the world to send us data in the form of narratives. Findings in this section are taken from the narratives we received. Crucially, these narratives highlight the authentic voice of the teachers. Examples include:

The East Africa Mathematics Education and Research Network, with office bearers from Kenya, Tanzania, Rwanda and Uganda, involved collaboration with local partners to support and enable teachers to meet and work together towards improvement of mathematics teaching. Teachers learned collaboratively on issues and topics of significance to them. Asked to identify what was the most significant aspect for them in the process of learning with and from teachers, participants noted the involvement of participants in tackling tricky mathematical topics; freedom in asking and answering questions without being ridiculed; and knowing how to prepare teaching aids for the topic, such as a protractor.

(Contributed by Anjum Halai)

The Focus on Primary Mathematics, a three-year initiative that included teachers from two government schools in Cape Town, which aimed to bring together the two schools to work collaboratively to improve teaching and learning in mathematics in Grades 1–3. Feedback from teachers on *what worked well* included:

- The mathematical learning of the children in their school was better. The children were more confident, enjoyed mathematics more and were achieving better results.
- The initiative had encouraged teachers to discuss, share and collaboratively plan mathematics teaching and through this they had begun to do the same in other subjects.
- They had become much more comfortable with sharing their own lack of mathematical knowledge and understanding and had learnt how to ask questions.
- Their own willingness to open up and be honest about their shortcomings in terms of mathematical understanding had encouraged their colleagues to do the same.

Teachers gave up time to participate in formal training programmes run during school holidays and regularly participated in professional development sessions (Contributed by Marie Joubert).

Two examples from Latin America, produced by teachers or researchers from Brazil and Argentina. The first was written by, Yuriko, Ana, Fabio and Roberta, four members of a collaborative group from Brazil. The Brazilian group was initiated in 2014 and was motivated by an innovative governmental programme spearheaded by a Regional Secretary of Education. Quotes from the participants include:

- *Yuriko (Researcher)* "The group started in 2014 with 12 teachers from 5 public schools ... The innovative feature of the meetings is the collective planning of lessons ... Then, the participants visit each other's classrooms to observe, to register in audio/video media [the class] ...".
- *Ana (Pedagogical coordinator)*: "The theme of the study chosen by the group was Fractions ... During the preparation of lessons, the participants realized the need for a didactical sequence with the manipulation of concrete material...".

- *Fabio (Teacher):* "I learned many conceptual interpretations of *Fraction*, the potential of manipulation of concrete material in the learning of my students. Seeing the video of my lessons was important. I learned how to ask questions."
- *Roberta (Teacher)*: "I think that there was not enough orientation about how to use the Learning Situations in SEE Material and to explore them to improve the classroom practices. Doing so is possible only with the learning community with the understanding brought by group discussions with colleagues and different visions of researchers …".

The second was written by María, a teacher who was part of a collaborative group in Argentina initiated in 2003 by three teachers and a researcher with the aim of producing mathematical modelling activities for the teacher's classroom.

- *Teacher María*: "… a collaborative group which brings together mathematics educators and teachers at the secondary level. Strange situation this meeting of people with different backgrounds respective to work and training that, as an experienced teacher, appeared as a novelty to me".

Finally, we present a few details about Lesson Study as it has developed in Japan and which has been emulated in different ways around the world as indicated above.

Japanese Lesson Study

Lesson study developed as a reproducible science of teaching with theory and practice. In Japan, National Level LS for Leading Curriculum Reform and Innovation of Teaching has developed since 1872. Origins of Lesson Study go back to the rapid, top down establishment of a modern education system in Japan. A search for innovative teaching methods led to the ideas of Pestalozzi, a Swiss educator, to foster a dialogical style of classroom communication. A national system of teacher-led professional development helped teachers to master this approach. LS is maturely established throughout Japan, and therefore central to teachers' practice and developmental activity. A limited number of sources, such as Isoda, Stephens, Ohara, and Miyakawa (2007) and Inprasitha, Isoda, Wang-Iverson, and Yeap (2015), report on Japanese LS.

Concluding Discussion

A Rich and Diverse Picture, but Lacking Teachers' Voice

The sources revealed by the survey provided a rich and diverse picture of the collaborative work involving mathematics teachers that is taking place around the world, albeit predominantly reported from the perspective of researchers. The inclusion of writings from non-academic sources, our narratives and reports, provides evidence for widespread collaborative activity. However, overall we found a lack of the teachers' voice providing insights to teacher learning. We include here one quotation in which a teacher's voice was vividly expressed in relation to the voices of didacticians [mathematics education researchers] working with the teachers:

> At the beginning [of the project] I struggled, had a bit of a problem with this because then I thought very much about you [didacticians] coming and telling us how we should run mathematics teaching. That was what I thought, you are the great teachers ... but now I see that my view has gradually changed because I see that you are participants in this as much as we are, even though you are the ones organising this ... . And I think it's much better now, I feel much more comfortable, because now I feel that we are more equal than we were in the beginning, from my point of view (Bjuland & Jaworski, 2009, pp. 34–35).

Strengths and Weaknesses of Survey Methodology

Despite extensive work on this survey, locating and analysing relevant papers and narratives, we are aware that there are still limitations to our work.

- Although our search methodology revealed 316+ papers, we know we missed relevant papers in journals outside our range.
- Papers reviewed were usually not authored by teachers so the teachers' voice was largely missing.
- Our narratives came from projects with which we were familiar and so we missed many more opportunities to include the teacher voice than we accessed.

Questions for Further Research

As we analysed the papers revealed in our search and studied the narratives we received it became clear that the search and its analysis left many questions unaddressed, such as:

- What brings teachers into effective collaborative learning situations?
- How can the teachers' voice be heard in ways that matter?
- What do mathematics teacher educator researchers need to learn?
- How can research support teacher development, as well as recording it?

In recognition of limitations in the survey and the many questions still to address, members of the survey team intend to propose an ICMI study on the topic of the survey: *Mathematics teachers working and learning through collaboration*. If you are interested in participating in such a study, please watch out for an announcement.

References

Aldon, G., Arzarello, F., Cusi, A., Garuti, R., Martignone, F., Robutti, O., et al. (2013). The Meta-Didactical Transposition: a model for analysing teachers education programmes. In L. A. M. & A. Heinze (Eds.), *Proceedings of PME 37* (Vol. 1, pp. 97–124). Kiel, Germany: PME.

Alldis, J., & Whitney, H. (2013). Developing effective continuity in learning from KS1 to 4 in representing mathematically through collaborative research and development (NCETM CTP 4013). https://www.ncetm.org.uk/files/18384131/CTP4013+Final+Report.pdf

Berg, C. V. (2011). In-service teachers' professional development: Which systemic aspects are involved? *Research in Mathematics Education, 13*(2), 223–224.

Besamusca, A., & Drijvers, P. (2013). The impact of participation in a community of practice on teachers' professional development concerning the use of ICT in the classroom. In A. Lindmeier, & A. Heinze (Eds.), *37th Conference of the International Group for the Psychology of Mathematics Education, Kiel, Germany, 2013* (Vol. 2, pp. 81–88).

Bjuland, R., & Jaworski, B. (2009). Teachers' perspectives on collaboration with didacticians to create an inquiry community. *Research in Mathematics Education, 11*(1), 21–38. doi:10.1080/14794800902732209

Brodie, K., & Shalem, Y. (2011). Accountability conversations: Mathematics teachers' learning through challenge and solidarity. *Journal of Mathematics Teacher Education, 14*(6), 419–439. doi:10.1007/s10857-011-9178-8

Campbell, M. P. (2009). Mathematics teachers and professional learning communities: understanding professional development in collaborative settings. In S. L. Swars, D. W. Stinson, & S. Lemons-Smith (Eds.), *31st annual meeting of the North American Chapter of the International Group for the Psychology of Mathematics Education, Atlanta, Georgia*, (pp. 956–964).

Cavanagh, M., & McMaster, H. (2015). A professional experience learning community for secondary mathematics: Developing pre-service teachers' reflective practice. *Mathematics Education Research Journal, 27*(4), 471–490. doi:10.1007/s13394-015-0145-z

Chen, C.-H., & Chang, C.-Y. (2012). An exploration of mathematics teachers' discourse in a teacher professional learning community. In T.-Y. Tso (Ed.), *Proceedings of the 36th Conference of the International Group for the Psychology of Mathematics Education, Taipei, Taiwan*, (Vol. 2, pp. 123–130).

Chevallard, Y. (1985). *La transposition didactique*. Grenoble: La Pensée Sauvage.

Dowling, D. (2013). Hungary for calculation: developing approaches to calculation in the new curriculum using Hungarian methodology as our inspiration (NCETM CTP4213). https://www.ncetm.org.uk/files/20365123/CTP4213+Final+Report.pdf

Engeström, Y. (1999). Activity theory and individual and social transformation. In Y. Engeström, R. Miettinen, & R.-L. Punamäki (Eds.), *Perspectives on activity theory* (pp. 19–38). Cambridge: Cambridge University Press.

Fried, M. N., & Amit, M. (2005). A Spiral Task as a Model for In-service Teacher Education. *Journal of Mathematics Teacher Education, 8*(5), 419–436. doi:10.1007/s10857-005-3850-9

Goodchild, S. (2008). A quest for 'good' research. In B. Jaworski & T. Wood (Eds.), *The international handbook of mathematics teacher education (Vol 4) the mathematics teacher educator as a developing professional* (pp. 201–220). Netherlands: Sense Publishers.

Goos, M. (2013). Sociocultural perspectives in research on and with mathematics teachers: a zone theory approach. *ZDM Mathematics Education, 45*, 521–533.

Goos, M. E., & Bennison, A. (2008). Developing a communal identity as beginning teachers of mathematics: Emergence of an online community of practice. *Journal of Mathematics Teacher Education, 11*(1), 41–60. doi:10.1007/s10857-007-9061-9

Hospesovà, A., Machàckovà, J., & Tichà, M. (2006). Joint reflection as a way to cooperation between researchers and teachers. In *30th Conference of the International Group for the Psychology of Mathematics Education*, (Vol. 1, pp. 99–103).

Inprasitha, M., Isoda, M., Wang-Iverson, P., & Yeap, B. (Eds.). (2015). *Lesson study: Challenges in mathematics education (Vol. 3, series on mathematics education)*. Singapore: World Scientific.

Isoda, M., Stephens, M., Ohara, Y., & Miyakawa, T. (Eds.). (2007). *Japanese lesson study in mathematics: Its impact, diversity and potential for educational improvement*. Singapore: World Scientific.

Jaworski, B. (2006). Theory and practice in mathematics teaching development: Critical inquiry as a mode of learning in teaching. *Journal of Mathematics Teacher Education, 9*(2), 187–211.

Jaworski, B. (2008). Building and sustaining inquiry communities in mathematics teaching development: Teachers and didacticians in collaboration. In K. Krainer & T. Wood (Eds.), *The international handbook of mathematics teacher education* (Vol. 3, pp. 309–330). Rotterdam: Sense Publishers.

Krammer, K., Ratzka, N., Klieme, E., Lipowsky, F., Pauli, C., & Reusser, K. (2006). Learning with classroom videos: Conception and first results of an online teacher-training program. *ZDM, 38*(5), 422–432.

Kusanagi, K. N. (2014). The Bureaucratising of Lesson Study: A Javanese Case. *Mathematics Teacher Education and Development, 16,* 84–103.

Lin, F.-L., Chen, J.-C., Hsu, H.-Y., & Yang, K.-L. (2013). Elaborating stages of teacher growth in design-based professional development. *Proceedings of the 37th Conference of the International Group for the Psychology of Mathematics Education*, (Vol. 3).

Martins, C., & Santos, L. (2012). Development of reflection ability in continuous training in mathematics (PFCM). In T.-Y. Tso (Ed.), *Proceedings of the 36th Conference of the International Group for the Psychology of Mathematics Education, 2012* (Vol. 3, pp. 193 200).

Menezes, L. (2011) Collaborative research as a strategy of professional development of teachers. In B. Ubuz (Ed.), *35th Conference of International Group for the Psychology of Mathematics Education, Ankara, Turkey, 2011* (Vol. 3, pp. 252–232). PME.

Pires, M. V., & Martins, C. (2009). *Olhares sobre um plano de formação contínua em Matemática*. Paper presented at the VI CIBEM or Ibero-American Congress on Mathematics Education, Puerto Monte, Chile,

Ponte, J. P., Segurado, I., & Oliveira, H. (2003). A collaborative project using narratives: What happens when pupils work on mathematical investigations? In A. Peter-Koop, V. Santos-Wagner, C. Breen, & A. Begg (Eds.), *Collaboration in teacher education: Examples from the context of mathematics education* (pp. 85–97). Dordrecht: Kluwer.

Posthuma, B. (2012). *Mathematics teachers' reflective practice within the context of adapted lesson study., 33,* 3. doi:10.4102/pythagoras.v33i3.140

Potari, D. (2013). The relationship of theory and practice in mathematics teacher professional development: An activity theory perspective. *ZDM Mathematics Education, 45*(4), 507–519. doi:10.1007/s11858-013-0498-2

Riley, S. (2013). Extending Teacher Pedagogy of Early Mathematics Development to ensure Arithmetic and Mathematical Proficiency across Key Stage 1 (NCETM CTP 2513). https://www.ncetm.org.uk/files/17095392/CTP2513+Final+Report.pdf

Robutti, O., Cusi, A., Clark-Wilson, A., Jaworski, B., Chapman, O., Esteley, C., et al. (2016). ICME international survey on teachers working and learning through collaboration: June 2016. *ZDM, 48*(5), 651–690. doi:10.1007/s11858-016-0797-5

Sakonidis, C., & Potari, D. (2014). Mathematics teacher educators'/researchers' collaboration with teachers as a context for professional learning. *ZDM Mathematics Education, 46*(2), 293–304. doi:10.1007/s11858-014-0569-z

Takahashi, A. (2014). The role of the knowledgeable other in lesson study: Examining the final comments of experienced lesson study practitioners. *Mathematics Teacher Education and Development, 16*(1), 4–21.

Valsiner, J. (1997). *Culture and the development of children's action: A theory of human development* (2nd ed.). New York: John Wiley & Sons.

Warren, E. (2008). Early childhood teachers' professional learning in early algebraic thinking: A model that supports new knowledge and pedagogy. *Mathematics Teacher Education and Development, 10*, (30–45).

Wenger, E. (1998). *Communities of practice. Learning, meaning and identity.* Cambridge: Cambridge University Press.

Wynne-Jones, G. (2013). Collaborative development of a calculation policy as an opportunity for teacher CPD (NCETM CTP 2713). https://www.ncetm.org.uk/files/17715281/CTP2713+Coppice+Primary+School+Final+report.pdf

Open Access Except where otherwise noted, this chapter is licensed under a Creative Commons Attribution 4.0 International License. To view a copy of this license, visit http://creativecommons.org/licenses/by/4.0/.

Geometry Education, Including the Use of New Technologies: A Survey of Recent Research

Nathalie Sinclair, Maria G. Bartolini Bussi, Michael de Villiers, Keith Jones, Ulrich Kortenkamp, Allen Leung and Kay Owens

Abstract This is a summary report of the ICME-13 survey on the theme of recent research in geometry education. Based on an analysis of the research literature published since 2008, the survey focuses on seven major research threads. These are the use of theories in geometry education research, the nature of visuospatial reasoning, the role of diagrams and gestures, the role of digital technologies, the teaching and learning of definitions, the teaching and learning of the proving process, and moves beyond traditional Euclidean approaches. Within each theme, there is commentary on promising future directions for research.

N. Sinclair
Simon Fraser University, Burnaby, Canada
e-mail: nathsinc@sfu.ca

M.G. Bartolini Bussi
Università di Modena e Reggio Emilia, Reggio Emilia, Italy
e-mail: mariagiuseppina.bartolini@unimore.it

M. de Villiers
University of Stellenbosch, Stellenbosch, South Africa
e-mail: profmd1@mweb.co.za

K. Jones (✉)
University of Southampton, Southampton, UK
e-mail: d.k.jones@soton.ac.uk

U. Kortenkamp
University of Potsdam, Potsdam, Germany
e-mail: ulrich.kortenkamp@uni-potsdam.de

A. Leung
Hong Kong Baptist University, Hong Kong, Hong Kong SAR
e-mail: aylleung@hkbu.edu.hk

K. Owens
Charles Sturt University, Bathurst, Australia
e-mail: kowens@csu.edu.au

Introduction

Recent research in geometry education was identified by the IPC of ICME-13 as being especially important and warranting an in-depth review pinpointing new knowledge, new perspectives, significant realisations and emerging challenges in a comprehensive and synthetic way, paying specific attention to the evolution since the previous ICME in 2012. In conducting the review (for a fuller report, see Sinclair et al., 2016), our first task was to identify major threads in the recent research literature relating to geometry education. We undertook this phase of work by generating a list of key threads in recent research based on our collective knowledge and experience. Each team member proposed a list of possible threads for consideration. Some suggestions were combined to produce a broader thread and for some there was not sufficient research literature on which to draw. This iterative process led to the identification of seven major research threads: the use of theories in geometry education research, the nature of visuospatial reasoning, the role of diagrams and gestures, the role of digital technologies, the teaching and learning of definitions, the teaching and learning of the proving process, and moves beyond traditional Euclidean approaches.

With the major research threads identified, pertinent research was identified in peer-reviewed journal articles, international peer-reviewed conference proceedings (such as PME and CERME) and leading research handbooks. For each thread, we produced a comprehensive annotated bibliography and this led to a further refinement in the detail of the research threads. While paying specific attention to research undertaken since 2012, we refer back to seminal publications including ones prior to 2008. In what follows, we summarize the key findings for each thread of the review survey, beginning with the use of theories in geometry education research. We conclude with a final section in which we highlight some overall issues and opportunities and discuss future directions for research.

The Use of Theories in Geometry Education Research

The development and refinement of theories of teaching and learning is one of the key aims of research in education in general, and for research in mathematics education in particular. In geometry education research, this focus on theory includes the developing and refining of theories that are specifically about the teaching and learning of geometry, as well as the application and development of more general theories to the specifics of geometry education.

Theories that are specifically about geometry education and that continue to be evident in research include the van Hiele model (1986), the theory of figural concepts (Fischbein, 1993; Mariotti & Fischbein, 1997), and the theory of figural apprehension (Duval, 1998). Researchers are continuing to develop, refine and apply these theories. More recently, the theory of geometric work has been developed (c.f., Kuzniak, 2014). This aims at networking the theoretical ideas of figural, instrumental, and discursive geneses of geometric understanding by characterizing different geometrical paradigms and accounting for interaction between the epistemological and cognitive levels.

Theories being applied to geometry education include prototype phenomenon (Gal & Linchevski, 2010), semiotic mediation (Bartolini Bussi & Baccaglini-Frank, 2015; Bartolini Bussi & Mariotti, 2008), variation (Gu, Huang, & Marton 2004; Huang & Li, 2016), the cK¢ (conception, knowing, concept) model (Balacheff, 2013), discursive, embodied, ecocultural and material perspectives (Ng & Sinclair, 2015a, b; Owens, 2014, 2015) and instrumental genesis (evident in research on the use of digital technologies) (Hegedus & Moreno-Armella, 2010). These theoretical approaches illustrate the wide scope of geometry education research.

Overall, during the past decade, there has been increased focus on embodied and discursive theories in research on the teaching and learning of geometry, with a concomitant research emphasis on theories relating to visuospatial reasoning, the role of gestures and diagrams, and the use of digital technologies.

The Nature of Visuospatial Reasoning

Our review survey identified how attention is increasingly focusing on forms of visuospatial reasoning (Healy & Powell, 2013; Owens, 2015; Rivera, 2011), variously referred to as visualization, visualising, spatial thinking, spatial reasoning, visuospatial thinking and visual reasoning. Here we use the term visuospatial reasoning to emphasize the spatial, visualizing (imagistic and as representations that others can see), and reasoning aspects of the visuospatial. While visuospatial reasoning is arguably relevant in all areas of mathematics, it has particular significance in the teaching and learning of geometry.

Overall, we can summarize our review survey on this thread by highlighting three developments. First, visualising (mentally and physically) is well-recognized as important in mathematics education but may not always be given sufficient emphasis in curriculum and teaching, perhaps because it is not straightforward to assess. Second, reasoning involves thinking about, and making decisions based on, visuospatial perception and understanding, both of which are influenced by prior knowledge and context of learning. Third, there is evidence that visuospatial reasoning in geometry can be improved through experience from perception to higher levels of reasoning (relevant here is the notion of cognitive malleability).

There are a number of educational implications of the recent research. First, the idea of locating: this is where younger children use geometric features and landmarks to find their way around larger spaces through spatial as well as visual perception for decision-making. Here, cultural studies support cognitive science studies. Second, the idea of transformation: here experience of mental rotation has been shown to improve algebraic manipulation, illustrating the value of spatial, as compared to object, visualizing. Overall, the value of a spatially-enriched education is being recognized more generally, including its value in counteracting the impact of gender, culture, experience, and capability differences.

Research increasingly emphasizes the need for good visuospatial working memory in geometry and more widely (Giofrè, Mammarella, Ronconi, & Cornoldi, 2013). Classroom activities such as origami, pop-up engineering, quality block play, various practical activities, and specific forms of technology have all been shown to enhance visuospatial reasoning. Across the research domains of education and the cognitive sciences, there is converging agreement on the importance and malleability of visuospatial reasoning.

On top of this, while drawing provided evidence of learning (as well as being a mediating tool in learning), there is undoubted complexity of reasoning about visuals and different impacts of different representations. What is more, different activities create different imaginal, formational and transformational visuospatial reasoning. Alongside research on how the processes of visual perception and perception-based knowledge influence learning is evidence that the impact of Western-style education may limit visuospatial reasoning of indigenous, colonised groups while the use of technologies such as dynamic geometry environments (DGEs) can assist learning in developing communities as well as developed (Owens, 2015).

The Role of Diagrams and Gestures in Geometry Education

Research highlighting the role of diagrams and gestures has largely emerged out of recent emphases on the semiotic and embodied nature of geometry thinking and learning (e.g. Bartolini Bussi & Baccaglini-Frank, 2015; Bartolini Bussi & Mariotti, 2008; Ng & Sinclair, 2015a, b). Here we consider historical-cultural perspectives that underscore the role of semiotic processes and artefacts in geometry teaching and learning. In doing so, we highlight embodiment perspectives that stress the important roles of gestures and diagrams in geometry teaching and learning.

In addition to research with school-aged pupils, researchers have begun to study the role of gestures and diagrams in the work of professional mathematicians (Barany & MacKenzie, 2014; Hare & Sinclair, 2015). This work corroborates some of the claims of Châtelet (2000) that diagrams are more than representations of existing knowledge while also providing more detailed and real-time evidence of the meanings that gesturing and diagramming help to create, even in highly advanced mathematics.

This work, when combined with the studies noted above on the semiotic and embodied nature of geometry thinking and learning, provides a clear indication of the importance of encouraging learners to engage in more gesturing and diagramming. Existing research (referenced above) suggests that the more teachers gesture, the more do their students. Future research could provide insight into the types of gestures that might be helpful, as well as the modalities in which students are invited to gesture. Gestures, imitation, and explanation are also important in indigenous communities making use of their strong capabilities in visuospatial reasoning (c.f., Owens, 2015).

The Role of Digital Technologies in Geometry Education

New technological developments over the past decade have led to new challenges in the use of technology in the teaching and learning of geometry. This is despite the role of technologies in teaching and learning not being understood completely, nor being explored in sufficient detail since the introduction of DGEs (such as Cabri-Géomètre and Geometer's Sketchpad) in the early 1990s. As such, this demonstrates the importance of three areas of research: (1) the introduction and design of new technology, both hardware and software, (2) theory and methodology for a better understanding of the role of existing and emerging technology, and (3) empirical studies on the use of technology in teaching and learning.

Technology in geometry education has become relatively mainstream, yet there is still not enough research into its specific effects. This is due, in part, to the way that some technologies, such as DGEs, change geometric representations and discourse quite significantly, as compared with paper-and-pencil approaches. As a result, articulation in the classroom, with textbooks, physical manipulatives, and especially assessment (Venturini & Sinclair, 2016) can be quite challenging.

Trends for digital geometry tools include (1) how geometry on the web is encountering various technological difficulties associated with the need to replace Java; (2) the issue of interface design and how users interact with onscreen geometry (Jackiw & Sinclair, 2009; Kortenkamp & Dohrmann, 2010; Laborde & Laborde, 2014; Mackrell 2011; Schimpf & Spannagel, 2011; (3) the rise of mobile devices and touch technology; and (4) new modes of interaction such as multi-touch and multi-collaboration.

In terms of specific digital tools and concepts, there are many issues—ranging from detailed matters such as the use of onscreen 'sliders' in geometry to wider socio-cultural aspects such as how technology has social impact. Across geometry education, the issues range from the role of digital technologies in developing learners' spatial capabilities and capability with 3D geometry (the latter being available digitally but currently restricted to 2D projection) to the issue of task design and how tasks change with availability of technology (Leung & Baccaglini-Frank, 2016). On top of this, there are issues of assessment, feedback and learning analytics, where new approaches are emerging. What is more, teacher education and professional development continue as challenging and important tasks for the mathematics education community in general, and for geometry educators in particular.

The role of technology is just beginning to be understood. At the same time, technology continues to evolve and rapidly change the everyday world and the classroom. Students and teachers are increasingly using digital tools throughout the day (and beyond school, Carreira, Jones, Amado, Jacinto, & Nobre, 2016). It is increasingly necessary to understand better how new and emerging digital tools can be used effectively.

The Teaching and Learning of Geometric Definitions

The importance of definitions in geometry education (c.f. de Villiers, Govender, & Patterson, 2009; Mariotti & Fischbein, 1997; Smith, 2010) is reflected in the research literature, with a number of studies on this theme appearing over the past decade. Such research has focused both on understanding the process of defining and on the need for definitions. Overall, the majority of studies have concentrated on descriptive (a posteriori) defining; for example, defining circles, triangles, quadrilaterals, and polyhedra after exploring their properties with the use of DGEs, paper-folding, and/or pencil-and-paper construction (c.f. Choi and Oh 2008; Fujita & Jones, 2007; Salinas, Lynch-Davis, Mawhinney, & Crocker, 2014; Usiskin, Griffin, Witonsky, & Willmore, 2008; Zandieh & Rasmussen, 2010).

It appears that the fundamental issue of understanding the need for axioms and for accepting some statements as definitions to avoid circularity has been largely under-researched in the mathematics education community (though see Fujita, Jones, & Miyazaki, 2011; Miyazaki, Fujita, & Jones, 2017). Another under-researched area seems to be exploring the existence of a mathematical choice between defining (and classifying) the quadrilaterals hierarchically or in partitions (compare de Villiers 1994; Usiskin et al. 2008). A specific research question in this regard might be the extent to which students and teachers understand (or how to develop such understanding) that choosing a hierarchical definition over a partition leads to a more economical (shorter) definitions more concise formulation of some theorems, simplified deductive structure (by decreasing the number of proofs required), assists in problem solving, etc.

The potential of DGEs, and some use of analogy also, in developing understanding for definitions have been explored in several studies with triangles and quadrilaterals (c.f. Kaur, 2015). Such approaches appear to have assisted students in developing more robust, dynamic concept images than the traditional prototypical, static images that tend to prevent inclusive definitions. Nevertheless, everyday language and prototypical conceptions remain an issue especially in regard to class-inclusion as well as students' understanding the constraints of a DGE figure. A paucity of research on the use of symmetry concepts in choice of definitions, and on engaging students in the process of the constructive (a priori) defining of new concepts, means that these are ripe areas for future research.

The Teaching and Learning of the Proving Process in Geometry

Much research over the past decade has focused on studying the teaching and learning of the proving process, particularly in light of the increasing use of educational technology. Researchers have turned their attention to the following issues, many of them of perennial interest: what is and what constitutes a mathematical proof; how to interpret proof as an explanation that convinces others, and what makes something convincing; what kind of pedagogy and pedagogical tools are conducive to the construction of proof; and so on.

In terms of what is and what constitutes a mathematical proof, recent studies suggest that (geometrical) proof is bounded socio-culturally and is intimately related to the perceptual world. In terms of alternative frameworks for what is a 'geometrical proof', research suggests this is closely tied with the corresponding conjecture or hypothesis; in particular, with how the conjecture or hypothesis came about. In that, empirical-based argument may play a role in the formation of geometrical proof with respect to convincing or explaining.

The use of DGEs has been playing a vital epistemic role in studies that have probed the process of generating geometrical conjectures (c.f. Leung, 2008; Leung, Baccaglini-Frank, & Mariotti, 2013). For example, through the lens of the theory of semiotic mediation (TSM), the conjecture production process is a semiotic process that involved a transformation from personal signs to mathematical signs. Here, feedback and mediation from the technological tool serve as means for boundary crossing between the empirical and theoretical contexts in the proving process. In particular, the DGE drag-mode instigates the complex interplays between inductive, abductive and deductive reasoning in the transition between empirical and theoretical proof perspectives. Studying and categorizing DGE dragging modalities/strategies have been a core focus attempting to conceptualize proof and explanation when using a DGE. Studies have explored the role of DGE as an epistemic tool, in particular dragging, to open up a quasi-empirical dimension to the nature of proof, even indirect proof.

Pedagogies such as tool-based task design, inquiry-based learning, mathematical discussion, problem modification, geometrical construction, and a focus on gesturing, have been introduced to improve the conjecture formation processes. This has included the following: use of the shift-problem approach (empirical proof schemes, external conviction proof schemes, and deductive proof schemes); modification of textbook problems into DGE investigations; use of flow-chart proving (Miyazaki, Fujita, & Jones, 2015); use of the lens of cognitive unity to address the tension between carrying out a geometrical construction and constructing the related proof; and the interplay among gestures, discourse and diagram in geometric reasoning.

Moves Beyond Traditional Euclidean Approaches to Geometry

In the teaching and learning of 3D geometry, research indicates that students exhibit similar prototypical predispositions to 3D objects as they do with 2D objects (c.f. Sarfaty & Patkin, 2013). Physically building, constructing and drawing 3D objects such as polyhedra, and/or exploring them dynamically with 3D DGEs appear to develop better concept images and understanding of their properties. There are few studies on engaging students in extending interesting 2D results to 3D; for example, triangle concurrencies, Pythagoras, Varignon's or Viviani's theorem. The use of analogy when moving from 2D to 3D and higher dimensions could be more extensively explored using analogous concepts for triangle, square, circles, perpendicular bisector, angle bisector, etc.

Experimental studies on spherical and hyperbolical surfaces have used specific manipulatives such as spheres or DGEs to explore and prove results and properties of non-Euclidean objects, and in most cases, contrasting/comparing them with equivalents from Euclidean geometry. Guven and Baki (2010) theorised van Hiele levels of understanding for spherical geometry similar to 2D, which appeared to be reasonably confirmed by a Guttman scalogram analysis, though future studies would be useful. There have been studies focused on a Turtle geometry model of the hyperbolic surface (Arzarello, Bartolini Bussi, Leung, Mariotti, & Stevenson, 2012) and on topological surfaces (the Mobius strip, the torus and the Klein bottle) using a DGE (Hawkins & Sinclair, 2008). In contrast, there has been little or no research on the teaching and learning of fractals over the past 10 years.

Concluding Comments and Future Directions

The seven themes that we identified in our survey review (aided by survey team member Alexey Zaslavsky, Russia, and ICME-13 IPC liaison Behiye Ubuz, Turkey) reflect both traditional research interests in the teaching and learning of geometry as well as new areas of growth. During the past decade, there has been increased focus on embodied and discursive theories in research on the teaching and learning of geometry, with a concomitant research emphasis on visuospatial reasoning, on the use of gestures and diagrams and on digital tools. The effectiveness of certain digital tools, such as DGES, as well as their increased availability, has also affected researched on topics that span the k-16 geometry curriculum (from early experiences with dynamic triangles to later explorations in spherical geometry) as well as major areas of research such as the proving process and the use and role of definitions. There has also been a broadening of the traditional scope of geometry, both in terms of cultural perspectives and also in terms of concepts and activities that do not follow the typical Euclidean development—including the Euclidean approach to definitions.

We expect to see continued growth in these areas, and we also hope to see increased research interest in the teaching and learning of geometry since it is a topic whose significance may have become under-recognized through an increased policy emphasis on number and algebra. A valuable focus of future research might be to investigate how geometric ways of thinking, including visuospatial reasoning and diagramming, may serve not only to improve geometric understanding, but also mathematical understanding more generally, and may even broaden the range of learners who might become interested in, and excel at, mathematics.

References

Arzarello, F., Bartolini Bussi, M. G., Leung, A., Mariotti, M. A., & Stevenson, I. (2012). Experimental approach to theoretical thinking. In G. Hanna & M. de Villers (Eds.), *Proof and proving in mathematics education* (pp. 97–137). New York: Springer.

Balacheff, N. (2013). cK¢, a model to reason on learners' conceptions. In M. V. Martinez & A. C. Superfine (Eds.), *Proceedings of PME-NA 35* (pp. 2–15). IL, United States: Chicago.

Barany, M., & MacKenzie, D. (2014). Chalk: Materials and concepts in mathematics research. In C. Coopmans, J. Vertesi, M. Lynch, & S. Woolgar (Eds.), *Representation in scientific practice revisited* (pp. 107–130). Cambridge, MA: MIT Press.

Bartolini Bussi, M. G., & Baccaglini-Frank, A. (2015). Geometry in early years: Sowing the seeds towards a mathematical definition of squares and rectangles. *ZDM: International Journal on Mathematics Education, 47*(3), 391–405.

Bartolini Bussi, M. G., & Mariotti, M. A. (2008). Semiotic mediation in the mathematics classroom: Artefacts and signs after a Vygotskian perspective. In L. English et al. (Eds.), *Handbook of international research in mathematics education* (pp. 720–749). Mahwah, NJ: Erlbaum.

Carreira, S., Jones, K., Amado, N., Jacinto, H., & Nobre, S. (2016). *Youngsters solving mathematical problems with technology*. New York: Springer.

Châtelet, G. (2000). *Figuring space: Philosophy, mathematics, and physics*. Dordrecht: Kluwer.

Choi, K., & Oh, S. K. (2008). Teachers' conceptual errors related to the definitions in the area of geometry of elementary school mathematics. *Journal of the Korean Society of Mathematical Education. Series A. The Mathematical Education, 47*(2), 197–219.

de Villiers, M. (1994). The role and function of a hierarchical classification of quadrilaterals. *For the Learning of Mathematics, 14*(1), 11–18.

de Villiers, M., Govender, R., & Patterson, N. (2009). Defining in geometry. In T. Craine & R. Rubinstein (Eds.), *Understanding geometry for a changing world* (pp. 189–203). Reston: NCTM.

Duval, R. (1998). Geometry from a cognitive point of view. In C. Mammana & V. Villani (Eds.), *Perspectives on the teaching of geometry for the 21st Century: An ICMI study* (pp. 37–52). Dordrecht: Kluwer.

Fischbein, E. (1993). The theory of figural concepts. *Educational studies in mathematics, 24*(2), 139–162.

Fujita, T., & Jones, K. (2007). Learners' understanding of the definitions and hierarchical classification of quadrilaterals: Towards a theoretical framing. *Research in Mathematics Education, 9*, 3–20.

Fujita, T., Jones, K., & Miyazaki, M. (2011). Supporting students to overcome circular arguments in secondary school mathematics. *Proceedings of PME35* (Vol. 2, pp. 353–60).

Gal, H., & Linchevski, L. (2010). To see or not to see: Analyzing difficulties in geometry from the perspective of visual perception. *Educational Studies in Mathematics, 74*(2), 163–183.

Giofrè, D., Mammarella, I. C., Ronconi, L., & Cornoldi, C. (2013). Visuospatial working memory in intuitive geometry, and in academic achievement in geometry. *Learning and Individual Differences, 23,* 114–122.

Gu, L., Huang, R., & Marton, F. (2004). Teaching with variation. In L. Fan et al. (Eds.), *How Chinese learn mathematic* (pp. 309–345). Singapore: World Scientific.

Guven, B., & Baki, A. (2010). Characterizing student mathematics teachers' levels of understanding in spherical geometry. *International Journal of Mathematical Education in Science and Technology, 41*(8), 991–1013.

Hare, A., & Sinclair, N. (2015). Pointing in an undergraduate abstract algebra lecture: Interface between speaking and writing. *Proceedings of PME39* (vol. 3, pp. 33–40).

Hawkins, A., & Sinclair, N. (2008). Explorations with Sketchpad in topogeometry. *International Journal of Computers for Mathematical Learning, 13*(1), 71–82.

Healy, L., & Powell, A. (2013). Understanding and overcoming 'disadvantage' in learning mathematics. In M. Clements et al. (Eds.), *Third international handbook of mathematics education* (pp. 69–100). New York: Springer.

Hegedus, S. J., & Moreno-Armella, L. (2010). Accommodating the instrumental genesis framework within dynamic technological environments. *For the Learning of Mathematics, 30*(1), 26–31.

Huang, R., & Li, Y. (Eds.). (2016). *Teaching and learning mathematics through variation*. Rotterdam: Sense.

Jackiw, N., & Sinclair, N. (2009). Sounds and pictures: Dynamism and dualism in dynamic geometry. *ZDM: International Journal on Mathematics Education, 41*(4), 413–426.

Kaur, H. (2015). Two aspects of young children's thinking about different types of dynamic triangles. *ZDM: International Journal on Mathematics Education, 47*(3), 407–420.

Kortenkamp, U., & Dohrmann, C. (2010). User interface design for dynamic geometry software. *Acta Didactica Napocensia, 3*(2), 59–66.

Kuzniak, A. (2014). Understanding the nature of the geometric work through its development and its transformation. In S. Rezat, M. Hattermann, & A. Peter-Koop (Eds.), *Transformation: A fundamental idea of mathematics education* (pp. 311–325). New York: Springer.

Laborde, C., & Laborde, J.-M. (2014). Dynamic and tangible representations in mathematics education. In S. Rezat, M. Hattermann, & A. Peter-Koop (Eds.), *Transformation: A fundamental idea of mathematics education* (pp. 187–202). New York: Springer.

Leung, A. (2008). Dragging in a dynamic geometry environment through the lens of variation. *International Journal of Computers for Mathematical Learning, 13*(2), 135–157.

Leung, A., & Baccaglini-Frank, A. (Eds.). (2016). *Digital technologies in designing mathematics education tasks*. New York: Springer.

Leung, A., Baccaglini-Frank, A., & Mariotti, M. A. (2013). Discernment in dynamic geometry environments. *Educational Studies in Mathematics, 84*(3), 439–460.

Mackrell, K. (2011). Design decisions in interactive geometry software. *ZDM: International Journal on Mathematics Education, 43*(3), 373–387.

Mariotti, M. A., & Fischbein, E. (1997). Defining in classroom activities. *Educational Studies in Mathematics, 34,* 219–248.

Miyazaki, M., Fujita, T., & Jones, K. (2015). Flow-chart proofs with open problems as scaffolds for learning about geometrical proof. *ZDM: International Journal on Mathematics Education, 47*(7), 1211–1224.

Miyazaki, M., Fujita, T., & Jones, K. (2017). Students' understanding of the structure of deductive proof. *Educational Studies in Mathematics, 94*(2), 223–239.

Ng, O., & Sinclair, N. (2015a). Young children reasoning about symmetry in a dynamic geometry environment. *ZDM: International Journal on Mathematics Education, 47*(3), 421–434.

Ng, O., & Sinclair, N. (2015b). 'Area without numbers': Using Touchscreen dynamic geometry to reason about shape. *Canadian Journal of Science, Mathematics and Technology Education, 15*(1), 84–101.

Owens, K. (2014). Diversifying our perspectives on mathematics about space and geometry. *International Journal of Science and Mathematics Education, 12*(4), 941–974.

Owens, K. (2015). *Visuospatial reasoning: An ecocultural perspective for space, geometry and measurement education.* New York: Springer.

Rivera, F. (2011). *Towards a visually-oriented school mathematics classrooms: Research, theory, practice, and issues.* New York: Springer.

Schimpf, F., & Spannagel, C. (2011). Reducing the graphical user interface of a dynamic geometry system. *ZDM: International Journal on Mathematics Education, 43*(3), 389–397.

Salinas, T. M., Lynch-Davis, K., Mawhinney, K. J., & Crocker, D. A. (2014). Exploring quadrilaterals to reveal teachers' use of definitions: Results and implications. *Australian Senior Mathematics Journal, 28*(2), 50–59.

Sarfaty, Y., & Patkin, D. (2013). The ability of second graders to identify solids in different positions and to justify their answer. *Pythagoras, 34*(1), 1–10.

Sinclair, N., Bartolini Bussi, M., de Villiers, M., Jones, K., Kortenkamp, U., Leung, A., et al. (2016). Recent research on geometry education: An ICME-13 Survey Team report. *ZDM: International Journal on Mathematics Education, 48*(5), 691–719.

Smith, J. T. (2010). Definitions and non-definability in geometry. *American Mathematical Monthly, 117*(6), 475–489.

Usiskin, Z., Griffin, J., Witonsky, D., & Willmore, E. (2008). *The classification of quadrilaterals: A study of definition.* Charlotte, NC: InfoAge.

van Hiele, P. M. (1986). *Structure and insight: A theory of mathematics education.* New York: Academic Press.

Venturini, M., & Sinclair, N. (2016). Designing assessment tasks in a dynamic geometry environment. In A. Leung & A. Baccaglini-Frank (Eds.), *Digital technologies in designing mathematics education tasks* (pp. 77–98). New York: Springer.

Zandieh, M., & Rasmussen, C. (2010). Defining as a mathematical activity. *Journal of Mathematical Behavior, 29,* 55–75.

Open Access Except where otherwise noted, this chapter is licensed under a Creative Commons Attribution 4.0 International License. To view a copy of this license, visit http://creativecommons.org/licenses/by/4.0/.

Part IV
Reports from the Thematic Afternoon

European Didactic Traditions in Mathematics: Aspects and Examples from Four Selected Cases

Werner Blum, Michèle Artigue, Maria Alessandra Mariotti,
Rudolf Sträßer and Marja Van den Heuvel-Panhuizen

Abstract In this paper, we report on the presentations and activities from the strand on "European Didactic Traditions" during the Thematic Afternoon at ICME-13. The focal point of the first hour of this afternoon were four key features that were identified as common in all European traditions and the second and third hours were devoted to the presentation of concrete examples from four specific traditions, organised in four parallel sessions.

Introduction

Across Europe, there have been a variety of traditions in mathematics education, both in the practice of learning and teaching at school and in research and development, that have resulted from different cultural, historical, and political backgrounds. Nevertheless, several of these traditions share some common features

---

W. Blum (✉)
University of Kassel, Kassel, Germany
e-mail: blum@mathematik.uni-kassel.de

M. Artigue
University Paris Diderot, Paris, France
e-mail: michele.artigue@univ-paris-diderot.fr

M.A. Mariotti
University of Siena, Pisa, Italy
e-mail: mariotti21@unisi.it

R. Sträßer
JLU Giessen, Giessen, Germany
e-mail: rudolf.straesser@uni-giessen.de

M. Van den Heuvel-Panhuizen
Utrecht University, Utrecht, Netherlands
e-mail: m.vandenheuvel-panhuizen@uu.nl

beyond historic and cultural differences, one of these being the use of the word *didactic* to denote the science of teaching and learning (*didactiek* in Dutch, *didactique* in French, *didáctica* in Spanish, *didattica* in Italian, *didaktika* in Czech, *dydaktyka* in Polish or *didaktik* in Swedish, Danish, and German) rather than *education*, which is common in Anglo-Saxon traditions. These didactic traditions can be traced back as far as to Comenius' *Didactica Magna* in the 17th century. They share in particular the following common features: a strong connection with mathematics and mathematicians, the key role of theory, the key role of design activities for learning and teaching environments, and a firm basis in empirical research. Other common features (such as an important role of proofs and proving or of the interplay between mathematics and the real world) can be considered part of those four features.

In the following (in Section "The Four Key Features"), we will elaborate a bit more on those four features. This was the main part of the first hour of the Thematic Afternoon at ICME-13. The features will be made more concrete by referring to four selected cases of European traditions in the didactics of mathematics: the Netherlands, France, Italy, and Germany. We will report (in Section "The Four Cases") briefly on the activities that have taken place in the second and third hour of the Thematic Afternoon in four parallel sessions devoted to these four cases. In these sessions, some distinct and specific characteristics beyond the communalities captured by the four features became clearer for each of the four countries. More details of these country-specific activities can be found on the website of ICME-13.

The Four Key Features

The Role of Mathematics and Mathematicians

Here we will highlight the role in the didactics of mathematics that some outstanding mathematicians have played in those four countries by their involvement in educational issues such as designing curricula for school or for teacher education and writing textbooks, and by their fostering of the development of mathematics education as a research field. In this respect, a prominent exemplar is Felix Klein (see Tobies, 1981), who also had a great influence on other mathematicians who had the opportunity of getting to know his work during their visits to Germany as researchers.

An important occasion for international comparison of different experiences in the didactics of mathematics was the Fourth International Congress of Mathematicians, which took place in Rome from 6 to 11 April 1908. During this congress, the International Commission on the Teaching of Mathematics (Commission Internationale de l'Enseignement Mathématique, Internationale Mathematische Unterrichtskommission, Commissione Internazionale dell'Insegnamento Matematico) was founded (details of the history of this institution can be retrieved at http://www.icmihistory.unito.it/timeline.php).

After a dramatic interruption due to the Second World War, mathematicians were again involved in instances of reforming. In many countries, the ideas and principles of the so-called New Math were shared. We can recognise a common interest in reforming curricula, which may be related to the impact of a new generation of mathematicians on the reorganisation of mathematics initiated by the Bourbaki Group. Thus, although the concrete results of the New Math movement were very different in various countries, a common feature was that substantial innovation entered into school practice through the active involvement of eminent figures such as Gustave Choquet and Jean Dieudonné in France, Emma Castelnuovo in Italy, and Hans Freudenthal in the Netherlands. Castelnuovo is an interesting exemplar of actions coming from inside the school, showing how the particular structure of the Italian school system could allow innovation coming from teachers.

In the context of this reform, new perspectives developed that moved the focus of reflection from issues concerning mathematical content and its organisation in an appropriate curriculum to issues concerning the description and explanation of the learning and teaching of mathematics, giving birth to a new scientific discipline, the didactics of mathematics, that rapidly developed through active international interaction. In some cases, for instance in France and Italy, it is possible to recognise again the strong influence of the mathematicians community, since the first generation of researchers in the didactics of mathematics consisted for the most part of academics affiliated with mathematics departments. This observation does not ignore the existence of a recurrent tension between mathematicians and researchers in mathematics education.

In summary, some common features that can be considered the core of the European tradition of didactics of mathematics can be directly related also to the uninterrupted and fruitful commitment of mathematicians to educational issues and in their intent to improve the teaching and learning of mathematics. An example is the strong role that proofs and proving have in all European traditions.

The Role of Theory

The word *theory* in mathematics education denotes a diversity of objects, from very local constructs to structured systems of concepts; some are "home-grown" while others are "borrowed" with some adaptation from other fields, and some have developed over decades while others have emerged only recently. This diversity can also be observed in the four European traditions under consideration.

The French tradition is certainly the most theoretical of these. It has three main pillars: Vergnaud's theory of conceptual fields (see Vergnaud, 1991), Brousseau's theory of didactical situations (TDS) (see Brousseau, 1997), and the anthropological theory of the didactic (ATD) that emerged from Chevallard's theory of didactic transposition (see Chevallard & Sensevy, 2014). These developed over decades with the conviction that mathematics education should be a scientific field of research with fundamental and applied dimensions supported by genuine theoretical constructions and appropriate methodologies giving an essential role to the observation and analysis of didactic systems and to didactical engineering. These theories were first conceived as tools for the understanding of mathematics teaching and learning practices and processes, taking into consideration the diversity of conditions and constraints that shape them, and for identifying associated phenomena, such as the "didactic contract." The three theories are also characterised by a strong epistemological sensitivity. Over the years, this theoretical landscape has been continuously enriched by new constructions and approaches, but efforts have always been made to maintain its global coherence.

The Dutch tradition is less diversified as it has developed around a single approach known today as Realistic Mathematics Education (see Van den Heuvel-Panhuizen & Drijvers, 2014). It also emerged in the seventies with Freudenthal's intention to give mathematics education a scientific basis. Similar to the French case, this construction was supported by a deep epistemological reflection: Freudenthal's didactical phenomenology of mathematical structures (see Freudenthal, 1983). In this tradition, theoretical development and design are highly interdependent. This is visible in the RME structure, which is made of six principles clearly connected to design: the activity, reality, level, intertwinement, interactivity, and guidance principles. Through design research in line with these principles, many local instruction theories focusing on specific mathematical topics have been produced. RME is still in conceptual development, benefiting from interactions with other approaches such as socio-constructivism, instrumentation theory, and embodied cognition theory.

In the Italian tradition, conversely, it is not possible to identify theories that would have similarly emerged and developed, despite a long-term tradition of action research, collaboratively carried out by mathematicians interested in education and by teachers. Progressively, however, a specific research trend has emerged from this action research and consolidated within a paradigm of research for innovation, leading to the development of specific theoretical frames and constructs (see Arzarello & Bartolini Bussi, 1998). Boero's construct of field of experience, Bartolini Bussi and Mariotti's theory of semiotic mediation, and Arzarello's constructs of semiotic bundle and action, production, and communication (APC) space represent this trend well.

In Germany, scholars since the early seventies have aimed to create the field of mathematics education as a scientific discipline, as shown by articles published in ZDM in 1974 and also the efforts made by Hans-Georg Steiner to establish an international debate on the theory of mathematics education and the underlying philosophies and epistemologies of mathematics within an international TME group he founded in 1984. However, it would be difficult to identify a specific German way of approaching theoretical issues in mathematics education even though, when seen from the outside, the interactionist approach initiated by Heinrich Bauersfeld seems to have been influential at an international level. Research in Germany currently uses a large variety of "local" theories and of corresponding research methods and of corresponding research methods; more information on these theories and methods can be found in Chapter "German-Speaking Traditions in Mathematics Education Research" in this volume.

Thus, the theoretical landscape offered by these four traditions is diverse and heterogeneous. Considering that such diversity is inherent to this field of research, the European community of research in mathematics education has developed specific efforts to build connections, an enterprise today known as "networking between theories." Not surprisingly, researchers from these four traditions are particularly active in that area.

The Role of Design Activities for Teaching and Learning Environments

Design activities in mathematics education can involve the design of tasks, lessons, teaching sequences, textbooks, curricula, assessments, and ICT-based material or programs for teacher education and can be done by teachers, educators, textbook authors, curriculum and assessment developers, ICT designers, or researchers. Such activities can be ad hoc or research based. Without design, no education is possible. It is through designed instructional materials and processes, in which the intended *what* and *how* of teaching is operationalised, that learning environments for students can be created. As such, educational design forms a meeting point of theory and practice through which they influence each other reciprocally. All four European didactic traditions reflect this.

In France, the design of mathematical tasks, situations, or sequences of situations is essential to didactic research and is controlled by the theoretical frameworks underlying this research (see Section "The Role of Theory"). This is clearly reflected in the methodology of didactical engineering within the theory of didactical situations that emerged in the early eighties. Designs are grounded in epistemological analyses, and situations are sought that capture the epistemological essence of the mathematics to be learned. In the last decade, the anthropological theory of the didactic has developed its own design perspective that gives particular importance to identifying issues that question the world and have strong mathematical potential. Design as a development activity mostly takes place within the IREMs. Dissemination happens through the publications of these institutes, professional journals, curricular resources, and some textbooks. Up to now, only a few research projects were aimed at upscaling.

Within the German didactic tradition, two periods can be distinguished. Before the seventies and eighties, design activities were mostly meant for developing learning and teaching environments for direct use in mathematics education. These design activities belonged to the long German tradition of *Stoffdidaktik*, which focused strongly on mathematical content and course development, with less attention on course evaluation. In the seventies, an empirical turn occurred, resulting in design activities done to study the effect of specified didactical variables through classroom experiments. Course development became less prominent, but this was—in one strand of German didactics of mathematics—counterbalanced by defining didactics of mathematics as a "design science" with a strong focus on mathematics. Currently, both approaches to design activities can be found in Germany and have evolved into a topic-specific didactical design research connecting design and empirical research.

In Italy, the role of design has also changed over time. Characteristic for the period from the mid sixties to the mid-eighties were a deep epistemological concern and a strong pragmatic interest in improving classroom mathematics teaching. The theoretical reflection on didactical suggestions and their effectiveness was not so strong. The focus was on the content and its well-crafted presentation in practice, based on conceptual analyses. The period from the mid eighties to the present can be characterised by long and complex processes targeting the development of theoretical constructs based on teaching experiments, with the design of teaching and learning environments simultaneously both as an objective and as a means of the experimentation.

In the Netherlands, a strong tradition in design can be found. Making things work, looking for pragmatic solutions, creativity, and innovation are typical features of the Dutch culture. This emphasis on design can also be found in mathematics education. At the end of the sixties, the reform of mathematics education started with designing an alternative for the mechanistic mathematics education that then prevailed. Initial design activities were practice-oriented. The theory development that resulted in Realistic Mathematics Education (see Section "The Role of Theory") grew from this practical work and later guided further design activities. Design implementation, including contexts, didactical models, longitudinal teaching-learning trajectories, textbook series, examination programs, mathematics events, and digital tool and environments, has been realised through a strong infrastructure of conferences, journals, and networks.

The Role of Empirical Research

As discussed in Section "The Role of Design Activities for Teaching and Learning Environments", designing learning environments for mathematics has been an important activity in all four countries. This created the need to legitimise such environments. One way to do this has been to show the effectiveness of these environments by means of empirical research (whatever "effectiveness" may mean). Thus, with various institutional settings and with varying visibility, empirical research has an important role in the didactics of mathematics for all four cases. Because of the complexity of the field, direct cause-effect research (mimicking classical natural science research) was soon found difficult, if not impossible. Nevertheless, partly as a fall-out from the need to design learning environments, empirical research in European didactics of mathematics developed a variety of questions, aims, topics, and research methods such as statistical analysis with the help of tests and questionnaires, content analysis of curricula and textbooks, and classroom analysis with the help of videos and observation sheets that was sometimes followed by transcript analysis (often with concepts from linguistics). More recently, triangulation and mixed methods complemented the range of research methods used in empirical research in the four countries.

A major division in the plethora of empirical research in the four cases is the difference between large-scale research and small and medium-sized case studies. The COACTIV study in Germany is a prototype of large-scale research. It was designed to investigate teacher competence as a key determinant of instructional quality in mathematics (for more details on COACTIV, see Kunter et al., 2013). A contrasting example is Mithalal's case study on 3D-geometry. Using Duval's *déconstruction dimensionnelle* and the theory of didactical situations as the theoretical framework (see Section "The Role of Theory"), the study took a qualitative approach to analyze the students dealing with the reconstruction of a drawing showing a 3D-configuration (for details see https://hal.archives-ouvertes.fr/tel-00590941).

Large-scale research can be further distinguished from medium- or small-scale research along the following lines: Large-scale studies tend to make differences within a representative sample an argument, while small- or medium-scale studies tend to make specialities of the "case" an argument. In addition to this, empirical research can be distinguished along methodological lines: Quantitative studies tend to use (sophisticated) statistical techniques to arrive at general "laws," while qualitative studies tend to use techniques from content analysis to better understand the phenomena.

If we look into purposes of empirical research, we find commonalities and differences in the four European cases. Prescriptive studies, which tend to show how things *should* be, are found in all four countries, as are descriptive studies, which tend to give the best possible description and understanding of the domain under study while not being primarily interested in changing the domain. We find experimental studies on theories on the didactics of mathematics, which are undertaken to develop or elaborate a theory and put it to a test, in Italy, France, and the Netherlands (less frequently in Germany), while illustrations of an existing theory (as a sort of "existence proof") can be found in all four countries.

Another distinction is action research as opposed to fundamental research. Action research is deeply involved with the phenomena and persons under study and has the main aim of improving the actual teaching and learning; this is widespread in Italy and the Netherlands. In contrast to this, fundamental research tends to prioritise understanding of the phenomena under study and has the major aim of improving theoretical concepts; this type of research can be found in all four European countries. An additional purpose of empirical research can be specific political interests (in contrast to the development of science or in addition to an interest in scientific progress and curriculum development); this type of research can be found especially in Germany.

The Four Cases

The Case of France

The two hours devoted to the French tradition made it possible to enter more deeply into its history, to present some of its achievements, and to reflect on its interactions with other educational cultures. The historical introduction presented by Michèle Artigue and Luc Trouche situated the context of emergence of this tradition and came back to its three pillars (see Section "The Role of Theory") using excerpts from the interviews with Brousseau, Chevallard, and Vergnaud that had been realised for this occasion. It also showed its current dynamism and the productive connections this tradition has established with connected fields such as cognitive ergonomy, leading to original constructions such as the double approach (ergonomic and didactic) of teachers' practices or the instrumental approach. Two case studies prepared by Aurélie Chesnais and Viviane Durand-Guerrier on axial symmetry and by Marianna Bosch and Hamid Chaachoua on algebra were then used to show the progressive development and capitalisation of research within this tradition and how its vision of the field, epistemological sensitivity, and theoretical constructions have led to original perspectives and results. While describing the evolution of research problematics and approaches on these themes over decades, the presenters made clear the attention paid by the French research community to the progressive structuration and capitalisation of didactic knowledge.

The second hour, devoted to influences and interactions, started with an animation prepared by Patrick Gibel showing how the supervision or co-supervision of foreign PhD students has contributed to the dissemination of the French tradition since the seventies. The session was then led by four researchers: Christine Knipping (Germany), Michela Maschietto (Italy), Faïza Chellougui (Tunisia), and Avenilde Romo Vazquez (Mexico), who all have prepared their PhD in co-supervision with a French researcher. Christine Knipping, who acted as a critical friend, looked at the French tradition through the lenses of validation and proof. Her main points were cohesion, interchanges (both within the French community and beyond), and dissemination, examining the specific role played by PhD students. Michela Maschietto described her personal journey from fellowship and doctorate in Paris back to Italy and the starting of new and very productive collaborations, combining French and Italian approaches towards material and digital tools. Faïza Chellougui showed the importance of collaboration with French didacticians in the development of didactic research in her country and more globally in francophone Africa, emphasising the specific role played by the EMF structure. Avenilde Romo Vazquez reviewed the long-term history of interaction between France and Latin America, and more specifically France and Mexico, in mathematics education. The four researchers made clear the influence of the French tradition and the collaborative work with French researchers on their personal development, but they also showed how, in return, the French tradition benefits from these international connections, which open it to new questions and constructions. The three interviews mentioned above and a document analyzing the history of didactic interactions with eight countries from francophone Africa, Latin America, and Asia, prepared for this thematic afternoon, are accessible on the website of the CFEM, the French sub-commission of ICMI (http://www.cfem.asso.fr/cfem/ICME-13-didactique-francaise).

The Case of Italy

Starting from a short historic overview, a variety of voices illustrated specific aspects of the Italian trend in *didattica della matematica*, from both inside and outside the community of Italian didacticians. The historic overview highlighted the continuous and increasing interest and involvement of the community of mathematicians in educational issues, in particular the role played by special figures in the emergence and the development of mathematics education as a scientific and autonomous discipline: from Federigo Enriquez to Emma Castelnuovo and from Bruno de Finetti to Giovanni Prodi.

As far as the inner voices are concerned, the contributions of Paolo Boero and of Mariolina Bartolini Bussi highlighted crucial features that shaped Italian didactics and, more specifically, the emergence of studies on mathematics learning and teaching. Some of these features have been related to local conditions, for instance, the high degree of freedom left to the teacher in the design and the realisation of didactic interventions in the Italian school system. Such a freedom has allowed active innovations realised by individuals or by groups of teachers, but has also provided researchers with an environment where basic research can involve long-term teaching experiments and a stable collaboration with school teachers. The two speakers gave examples where collaborations between mathematicians and school teachers responding to innovation issues have led to stable research teams from which the Italian research community has stemmed. These teams have specifically focused on whole-class interaction (beyond the more popular studies on individual problem solving and small-group cooperative learning), the teacher's role as a guide (beyond the more popular focus on learners' processes), long-term processes (beyond the more popular studies on short-term processes), and manipulation of concrete artifacts (e.g., abacuses, curve drawing devices, and tools for perspective drawing) without overlooking the theoretical aspects of mathematical processes.

The specificity of the Italian case was also highlighted in comparison with the reality of other countries. The fruitfulness of this comparison was presented by Nadia Douek and Bettina Pedemonte, who developed their PhD dissertations under the co-direction of both an Italian and a French supervisor. They reported and commented on specific collaboration experiences between French and Italian research communities and thus presented a living experience of researchers integrating different perspectives and different methodologies in a challenging but also rewarding way. Special attention was devoted to a collaborative initiative that involves French and Italian researchers: the case of the SFIDA (Séminaire Franco Italien de Didactique de l'Algèbre), which has displayed a rich variety of epistemological and didactic perspectives in its long life.

A final contribution, coming from East Asia, put the Italian tradition under the lens of a completely new eye. The presentation of Xuhua Sun, a colleague from Macau who recently came in contact with the Italian tradition in collaborating with Mariolina Bartolini Bussi on the organisation of the 23rd ICMI study, invited the audience to reflect on institutional and historical aspects of the Italian tradition. The focus on institutional aspects of Italian schooling included some reflections on the Italian cultural background in contrast with the Chinese one: class time, special education setting, teaching and learning freedom, etc. The historic perspective focused on aspects of mathematics and mathematics education of the Italian tradition that had an impact on the Chinese system, including some reflections on the comparison between Italian and Chinese attitudes towards proof.

The Case of the Netherlands

In accordance with Freudenthal's idea of giving reality a central role in mathematics education, the presentation on the Dutch didactic tradition started with a short movie about mathematics in the Netherlands, showing both the richness reality offers to developing mathematical concepts and tools through the process of mathematisation and the many opportunities for applying mathematics to solve real-world problems. In a second movie addressing mathematics education in the Netherlands, some snapshots from past and current classroom situations were presented. In a video interview made by Marc van Zanten, Adri Treffers then reflected on the important sources of inspiration for his ideas on mathematics education, underlining some crucial characteristics of Realistic Mathematics Education (see Sections "The Role of Theory" and "The Role of Design Activities for Teaching and Learning Environments"), in which his focus was on intra-mathematical contexts. First, he mentioned the relevance of own productions, which let students explore relations between numbers and properties of operations along with practicing their knowledge of number facts and basic skills. Second, he emphasised the significance of presenting students problems that they have not studied previously and giving them room to start with informal context-based solutions. Third, he emphasised the need to challenge students with mathematical puzzles in order to trigger students' mathematical thinking. Next, in a video interview made by Michiel Doorman, Jan de Lange told us how surprised he was when he became a teacher and discovered that students did not recognise mathematics in the world around them. He chose an extra-mathematical context and used his hobby, airplanes, to work with his students on glide ratios, vectors, and sine and cosine, all in the same context, and found that this approach even worked with low achievers. His view is that very young children should also be mathematically challenged by asking them good questions. Even simple toys can be a rich context for learning mathematical concepts. Education should make use of children's curiosity.

After that, Paul Drijvers presented the non-routine "driving to Hamburg" problem about making a graph. This problem requires modelling and allows solutions at different levels. Asking the audience to solve this problem by themselves let them experience what RME can mean in practice. Next, Marja van den Heuvel-Panhuizen synthesised some key aspects of RME that were touched on in the first part of the Dutch presentation and followed this by handing out two booklets in which 30 authors from the Netherlands (see http://dspace.library.uu.nl/handle/1874/340527) and 45 authors from 16 other countries (see http://dspace.library.uu.nl/handle/1874/340526) reflected on their experiences with RME.

In the second part of the Dutch presentation, four of these authors from outside of the Netherlands signified what RME brought about in their country. David Webb did this for the USA, Zulkardi and Ratu Ilma Indra Putri for Indonesia, and Sue Hough for England and the Cayman Islands. Finally, Dirk De Bock from Belgium and Cyril Julie from South Africa acted as critical friends, mentioning RME's challenges and indicating opportunities for further development.

The Case of Germany

The two hours devoted to the German tradition were used to present a narrative on the development of the didactics of mathematics in German-speaking countries during the last decades. This sketch was enriched by snippets from longer interviews with Lisa Hefendehl-Hebeker, Hans-Georg Weigand, and Erich C. Wittmann and was followed by comments made by colleagues from Norway and Sweden, Poland, and the Czech Republic (see below).

From the sixties onwards, personal reports from mathematics classrooms and document analysis for curriculum development, subject matter didactics (*Stoffdidaktik*) and statistical, mainly comparative studies (often done by university psychologists) were the starting point for a fresh development in didactics of mathematics. During the sixties and seventies, research into mathematics education was institutionalised by the creation of full university professorships in didactics of mathematics, a scientific society (Gesellschaft für Didaktik der Mathematik), a research journal (*Journal für Mathematik-Didaktik*) and a research institute (Institut für Didaktik der Mathematik at Bielefeld University). The seventies and eighties were marked by an empirical turn to everyday classrooms with more detailed empirical research, especially with qualitative, sometimes linguistic analysis of classroom processes initiated by the Bauersfeld group. Since the eighties, the rather homogeneous field diversified into a plethora of research on a variety of aspects of the teaching and learning of mathematics, including "empirical research, subject matter didactics, applications in mathematics teaching, historical and philosophical investigations, methodological aspects of mathematics education, principles of mathematics education, the epistemological dimension of mathematics education and proving" (Burscheid, Struve, & Walther, 1992, 297–302).

Not including the specific and different development in the German Democratic Republic (GDR, the eastern part of Germany), the beginning of the 21st century and the present situation can be described by three major strands in the didactics of mathematics in the German-speaking countries. The first is *Stoffdidaktik*, which has widened its approach by taking into account individual psychological aspects of teaching mathematics and concentrates on the design of learning environments. The second is case studies, especially classroom studies, which use mostly qualitative methods to reconstruct diverse aspects of everyday teaching and learning with the help of a variety of research methods. As an illustration, Kerstin Tiedemann presented her study, "Helping primary students to learn math—language and interaction." The third is quantitative large-scale evaluation studies (such as TIMSS and PISA) and qualitative large-scale development studies (such as the SINUS and SINUS-transfer study), which are receiving increasing attention and are partly influenced by political concerns and demands. As an example, Stefan Krauss presented a glimpse of the COACTIV study, concentrating on the impact of professional knowledge on student achievement.

During the third hour, three critical friends presented views from outside the German-speaking countries including cooperative activities with German colleagues. Barbro Grevholm (Norway and Sweden) spoke on "Doing empirical research differently: Nordic countries and Germany," Edyta Nowinska (Poland) presented "Perspectives on collaborative empirical research in Germany and in Poland," and Nada Vondrova (Czech Republic) commented on "*Didaktik der Mathematik* and *didaktika matematiky*."

In addition to this presentation in the frame of European traditions, the Thematic Afternoon also had an activity entirely devoted to German-speaking traditions in mathematics education research made up of eight sub-sections; for details see the ICME-13 website, Chapter "German-Speaking Traditions in Mathematics Education Research" in this volume and the monograph Jahnke, Hefendehl-Hebeker & Leuders (2018).

References

Arzarello, F., & Bartolini Bussi, M. G. (1998). Italian trends of research in mathematics education: A national case study in the international perspective. In J. Kilpatrick & A. Sierpinska (Eds), *Mathematics education as a research domain: A search for identity* (pp. 243–262). Dordrecht: Kluwer.

Brousseau, G. (1997). *Theory of didactical situations in mathematics*. Dordrecht: Kluwer.

Burscheid, H. -J., Struve, H., & Walther, G (1992). A survey of research. *Zentralblatt für Didaktik der Mathematik, 24*(7), 296–302.

Chevallard, Y., & Sensevy, G. (2014). Anthropological approaches in mathematics education, French perspectives. In S. Lerman (Ed.), *Encyclopedia of Mathematics Education* (pp. 38–43). New-York: Springer.

Freudenthal, H. (1983). *Didactical phenomenology of mathematical structures*. Dordrecht: Kluwer.

Jahnke, H.-N., Hefendehl-Hebeker, L., & Leuders, T. (Eds, 2018). Traditions in German-speaking Mathematics Education Research. New York: Springer.

Kunter, M., Baumert, J., Blum, W., Klusmann, U., Krauss, S., & Neubrand, M. (Eds, 2013). *Cognitive activation in the mathematics classroom and professional competence of teachers—Results from the COACTIV project*. New York: Springer.

Tobies, R. (1981). *Felix Klein*. Leipzig: Teubner.

Van den Heuvel-Panhuizen, M., & Drijvers, P. (2014). Realistic mathematics education. In S. Lerman (Ed.), *Encyclopedia of mathematics education* (pp. 521–525). Heidelberg/New York/London: Springer.

Vergnaud, G. (1991). La théorie des champs conceptuels [The theory of conceptual fields]. *Recherches en Didactique des Mathématiques, 10*(2–3), 133–170.

Open Access Except where otherwise noted, this chapter is licensed under a Creative Commons Attribution 4.0 International License. To view a copy of this license, visit http://creativecommons.org/licenses/by/4.0/.

German-Speaking Traditions in Mathematics Education Research

Hans Niels Jahnke, Rolf Biehler, Angelika Bikner-Ahsbahs,
Uwe Gellert, Gilbert Greefrath, Lisa Hefendehl-Hebeker,
Götz Krummheuer, Timo Leuders, Marcus Nührenbörger,
Andreas Obersteiner, Kristina Reiss, Bettina Rösken-Winter,
Andreas Schulz, Andreas Vohns, Rudolf vom Hofe
and Katrin Vorhölter

Abstract This paper describes and analyzes developments that have taken place in German mathematics education research during the last 40 years. Which developments and ideas were characteristic for the discussion and how was Germany influenced by and how did it interact with the international community? The themes range from subject matter didactics to large-scale studies.

Keywords Subject matter didactics · Design science · Modelling · *Allgemeinbildung* · Theory · Classroom studies · Educational research · Large-scale studies

---

H.N. Jahnke (✉) · L. Hefendehl-Hebeker
Universität Duisburg-Essen, Essen, Germany
e-mail: njahnke@uni-due.de

L. Hefendehl-Hebeker
e-mail: lisa.hefendehl@uni-due.de

R. vom Hofe
Universität Bielefeld, Bielefeld, Germany
e-mail: vomhofe@math.uni-bielefeld.de

M. Nührenbörger
Technische Universität Dortmund, Dortmund, Germany
e-mail: marcus.nuehrenboerger@tu-dortmund.de

B. Rösken-Winter
Humboldt-Universität zu Berlin, Berlin, Germany
e-mail: bettina.roesken-winter@hu-berlin.de

G. Greefrath
Universität Münster, Münster, Germany
e-mail: greefrath@uni-muenster.de

Introduction

First, what do we mean by *traditions* in this paper? As most readers know, in 1976, ICME-3 took place in Germany in the city of Karlsruhe, and ICME returned to Germany exactly 40 years later. Thus, it is quite natural to ask which developments have taken place in German mathematics education research during these 40 years, which developments and ideas were characteristic, which people proved to be influential, and how was Germany influenced by and how did it interact with the international community. Thus, the present paper will be confined to this period. However, since there was a great period of educational thinking in the 19th century, it will also digress a bit into the era of W. v. Humboldt from around 1800.

"German speaking" encompasses more than just Germany. Austria and Switzerland belong to the family, and the former German Democratic Republic (GDR) has its own traditions that are still influential. In preparing this event, the authors discussed these problems seriously. We felt that we should limit ourselves and confine the paper to Germany, with small references to Austria and the former GDR.

The paper splits German mathematics education research into eight sub-themes ranging from subject-matter didactics to large-scale studies without any claims that these sub-themes exhaust the whole field.

R. Biehler
Universität Paderborn, Paderborn, Germany
e-mail: biehler@math.uni-paderborn.de

K. Vorhölter
Universität Hamburg, Hamburg, Germany
e-mail: katrin.vorhoelter@uni-hamburg.de

A. Bikner-Ahsbahs
Universität Bremen, Bremen, Germany
e-mail: bikner@math.uni-bremen.de

A. Vohns
Alpen-Adria-Universität Klagenfurt, Klagenfurt, Germany
e-mail: andreas.vohns@aau.at

U. Gellert
Freie Universität Berlin, Berlin, Germany
e-mail: uwe.gellert@fu-berlin.de

G. Krummheuer
Goethe Universität Frankfurt am Main, Frankfurt, Germany
e-mail: krummheuer@math.uni-frankfurt.de

T. Leuders · A. Obersteiner · A. Schulz
Pädagogische Hochschule Freiburg, Freiburg, Germany
e-mail: leuders@ph-freiburg.de

Subject-Matter Didactics (German: *Stoffdidaktik*)

In the development of the didactics of mathematics as a professional field in Germany, subject-related approaches played an important role. Felix Klein created a model that has been referred to for a long time. A general goal was to develop approaches for representing mathematical concepts and knowledge in a way that corresponded to the cognitive abilities and personal experiences of the students while simultaneously simplifying the material without disturbing the mathematical substance. A fundamental claim was that such simplifications should be "intellectually honest" and "upwardly compatible" (Kirsch, 1977). Concepts and explanations should be taught to students with sufficient mathematical rigor in a manner that connects with and expands their knowledge of the subject. For this reason, subject-matter didactics placed value on constructing viable and robust mental representations (*Grundvorstellungen*) to capture mathematical concepts and procedures as they are represented in the mental realm. In the 80s, views of the nature of learning as well as objects and methods of research in mathematics education changed and the perspective was widened and opened towards new directions and gave more attention to the learners' perspective. This shift of view issued new challenges to subject-related considerations that have been enhanced by the recent discussion about professional mathematical knowledge for teaching.

The session started with an overview lecture on the main issues of subject-matter didactics given by Lisa Hefendehl-Hebeker and Rudolf vom Hofe entitled, "Subject-matter didactics: Overview of origin, main issues, theory, methods, and fields of application." Subsequent presentations concentrated on two paramount concepts of subject-matter didactics that can serve as guiding orientations in a local and global sense to present mathematical knowledge corresponding to the overarching goals.

The concept of *Grundvorstellungen*, which can be roughly translated as "basic mental models," describes relationships between mathematical content and the phenomenon of individual concept formation. For example, the actions of distributing and measuring provide basic mental models for the operation of division within the domain of natural numbers (partitive and quotitive basic model). Sebastian Wartha and Axel Schulz unfolded this concept in the context of natural numbers and fractions: "Numbers, fractions, operations and representations: *Grundvorstellungen* in primary school."

A. Obersteiner
e-mail: andreas.obersteiner@ph-freiburg.de

A. Schulz
e-mail: andreas.schulz@ph-freiburg.de

K. Reiss
TUM School of Education, Munich, Germany
e-mail: kristina.reiss@tum.de

The tension between *clarity* and *rigour* in calculus has been a main theme in the German tradition of subject-matter didactics and still is an actual problem field, especially in upper secondary school teaching. Blum and Kirsch (1991) suggested more intuitive approaches (at least for basic courses) with the original naïve ideas of function and limit and sequential steps of exactitude, which could be achieved according to the capacity of the learners. In reference to this discussion, Andreas Büchter and Hans Humenberger gave a presentation entitled, "Clarity and rigour in calculus courses."

Design Science

Within the German-speaking tradition, considering mathematics education as a design science primarily draws on the work of Wittmann. He underlined the role of substantial learning environments while elaborating on how mathematics education can be established as a scientific field in its own right. From their very nature, substantial learning environments contain substantial mathematical content, even beyond school level, and also offer rich mathematical activities for (pre-service) teachers on a higher level. Exploring the epistemological structure reflected in such learning environments or reflecting didactical principles while testing the learning environments in practice adds to a deeper understanding of both the mathematics involved and students' learning processes.

The main objective of design science has been developing feasible designs for conceptual and practical innovations, involving the teachers (and educators as well) actively in any design process, for example, designing teaching concepts and learning units, tasks, examples and materials for different lessons, curricula, assessments, and programs for teacher education. In this sense, the development of substantial learning environments can be seen from a twofold perspective: First, designing such learning environments should be based on substantial mathematics, meaning that students can be immersed in mathematical processes such as mathematizing, exploring, reasoning, and communicating. Second, investigating substantial learning environments should be the essential starting point of mathematics education research. In collective teaching experiments, the research focus lies on the induced learning processes and children's thinking as well as on the mathematical communication in the classroom. By working together with teachers in schools, the researchers reflect the effects of the designed substantial learning environments. However, researchers are not the only ones who analyze empirical data: Teachers also collect and reflect on their own empirical data and use it to improve their teaching. Bringing these two intentions together allows bridging of theory and practice in mathematical research.

From a broader perspective, the design science approach has played a distinctive role within prominent European traditions concerned with designing and evaluating learning material and processes (such as Realistic Mathematics Education in the Netherlands or the theory of didactical situations in France, for example). On the one hand, different conceptualizations for designing learning environments for students (or teachers) have developed in light of the didactical traditions of each country. On the other hand, these conceptualizations reflect the different theories involved that connect design and research, and balance theory and practice effectively. Nowadays, the variety of approaches used by researchers and teachers to work together collaboratively to promote mathematical learning and to develop substantial learning environments indicate the progress of design science and give insights into different ways of connecting design and empirical research. The following presentations were given: Marcus Nührenbörger and Bettina Rösken-Winter, "Mathematics education as a 'design science': Where did we start?" Susanne Prediger and Paul Cobb (USA); "Trends and developments: German trends in design science and design research at the system level"; Michael Link (Switzerland), Ralph Schwarzkopf, Anna S. Steinweg, and Chun Ip Fung (China), "Designing and researching substantial learning environments: Four examples of design experiments"; Erich Ch. Wittmann, "Design science revisited: Where are we now?"

Modelling

German work on modelling in mathematical education started in the 80s. In his talk about "Mathematical modelling in German-speaking countries: Introduction and overview," Gilbert Greefrath outlined the German discussion of mathematical modelling by presenting definitions, pedagogical aims, typical modelling cycles, and key examples of the German debate on mathematical modelling. In addition, he gave an overview of central pragmatic and specific approaches and addressed current development in research, educational standards, modelling competencies, comparative studies, and final exams. He also discussed the role of technology in mathematical modelling (see Greefrath & Vorhölter, 2016).

Afterwards, four important aspects of the German modelling discussion of the last decades were deepened, subdivided into two parts: Cognitive and empirical approaches and promoting modelling competencies.

In the first part, Rita Borromeo Ferri took a cognitive approach in her presentation on "Classification of modelling cycles: An insight into cognitive processes." In this presentation, she gave a classification of modelling cycles that focused on how these give a better insight into the cognitive processes of learners when solving modelling problems (Borromeo Ferri, 2006). In a short overview, she showed how this knowledge has been used for empirical and theoretical research in the German modelling debate. Focusing on "Quantitative research on modelling: Examples from German-speaking countries," Dominik Leiss and Stanislaw Schukajlow-Wasjutinski gave an overview of some empirical approaches in the German modelling discussion. They presented current research projects on mathematical modelling that are meeting the challenge of going beyond case studies to increase the external validity of their results. In the presented studies, a wide spectrum of quantitative research methods ranging from correlative analyses to mediation analyses were used. They reported on findings from studies conducted in German-speaking countries on the role of quantitative research methods while searching for the "best" learning environment for teaching modelling in a regular classroom (see Schukajlow et al., 2012).

In the second part, the promotion of modelling competencies in German schools was addressed in two different ways. Katja Maaß focused in her talk on "Mathematical modelling in professional development: Traditions in Germany on professional development courses," addressing topics such as differentiation and assessment when modelling. Based on expert interviews and a desktop analysis, she outlined the most important milestones, thereby showcasing important steps which might be useful for other countries as well (see Maaß & Mischo, 2011). In addition, Katrin Vorhölter gave an overview on "Implementing mathematical modelling in schools" by presenting several projects of the last two decades aiming at the implementation of modelling in Germany. Two kinds of implementation were distinguished: One the one hand, teaching units for promoting modelling competencies during mathematics lessons were presented whose implementations are often accompanied by research. On the other hand, so-called modelling days and weeks, highly requested by teachers but not often systematically researched, were introduced (see Greefrath & Vorhölter, 2016).

Finally, Gloria Stillman gave an "International perspective on the German modelling debate." She pointed out that the German debate has strong historical roots but also shows a healthy vibrancy where new people are continually coming into the field and the field is expanding and broadening in views and its research base.

Allgemeinbildung and Mathematical Literacy

In Germany, the idea that mathematics should be a constitutive component for the cultivation of human beings and, thus, an indispensable part of *Allgemeinbildung* dates back to Wilhelm von Humboldt in the beginning of the 19th century. This constituted a tradition of pedagogical thinking that is still influential in modern times.

In his talk, "Mathematics and *Allgemeinbildung* in the time of W. v. Humboldt," Hans Niels Jahnke showed that during Humboldt's time, the German view on mathematics put an emphasis on its cultural meaning. Humboldt's opinions on mathematics were dominated by a pronounced anti-utilitarianism, a preference for pure mathematics, an affinity between mathematics and aesthetics, and a high esteem of rigorous thinking. The education of the individual should not be regulated by demands from outside and future professional life. Rather, the aims of education should be defined in terms of an individual's needs for self-development. The strong emphasis of the reformers around Humboldt on theoretical thinking and pure science was in their eyes not a denial of the demands of practical life, but the best way to meet these demands. According to them, theoretical thinking is a necessary condition for change. To educate young people in theoretical thinking is the best way to make them 'apt for the future' The second part of the talk showed how the basic ideas of this approach can be identified in modern papers on *Allgemeinbildung* and mathematics by H. Winter.

In his reaction entitled "*Bildung, Paideia,* and some undergraduate programs manifesting them," Michael F. Fried discussed how similar notions are enshrined in the idea of paideia and the classical concept of the liberal arts. He showed that such ideas also work in modern times by hinting at the examples of prominent colleges in North America.

In his talk on "*Allgemeinbildung,* mathematical literacy, and competence orientation," Rolf Biehler gave a sketch of the discussion on *Allgemeinbildung* and mathematical literacy in Germany from the late 60s to today. In terms of mathematics, *Allgemeinbildung* was related to those components of mathematics that are considered to be relevant to the general public. In the 70s, educational goals for *Allgemeinbildung* were condensed in different visions of, for example, a 'scientifically educated human being,' a 'reflected citizen,' an 'emancipated individual being able to critique society,' and a person 'well educated for the needs of the economic system.' Among others, these ideas led to the first approaches of critical mathematics education (Christine Keitel and colleagues, see Damerow et al., 1974). In 1995, Hans-Werner Heymann, "Why teach mathematics," related the discussion of *Allgemeinbildung* in the educational sciences to mathematics education, developing a system of justifications about why mathematics should be taught (Heymann, 2003; see also Biehler, Heymann, & Winkelmann, 1995). Contrary to Heymann's intentions, the public reception of this book focused narrowly on one aspect, namely, that seven years of mathematics would be enough if mathematics education were only devoted to immediate everyday applications. Due to bad results in TIMSS and PISA starting in the late 90s, a new discussion on educational goals in mathematics arose. PISA's conception of mathematical literacy was extended by ideas from the German debate (Humboldt, Freudenthal, Winter, and Heymann) and a new notion of *Mathematische Grundbildung* emerged (Neubrand, 2003) that very much influenced the new national standards in mathematics in Germany (2003, 2012). Last but not least, the challenge of mathematics education given the heterogeneity of students was thematized with some advanced and basic examples stemming from statistical literacy education (http://www.procivicstat.org).

In his reaction, Mogens Niss from Roskilde University, Denmark, related German development to the international development on competence orientation (featuring the KOM project), including the various conceptualizations in the various PISA frameworks (Niss & Højgaard, 2011).

Theory Traditions in German-Speaking Countries

In the 70s and 80s, teacher education was established at universities, and scientific media and a scientific society in mathematics education were founded in the German-speaking countries. This raised the issue of how to develop mathematics education a scientific discipline. At about the same time, the question of how far the didactics of mathematics already had developed as a scientific discipline was intensively discussed. Referring to Kuhn and Masterman, Burscheid (1983) used a four-stage model to identify the developmental stage of the scientific discipline. Critical reactions from Steiner (1983) and Fischer (1983) required more focus on the needs of mathematics education itself. Since the development of any scientific discipline is deeply intertwined with its theoretical work, there was a need to clarify what kinds of theories were adequate for the discipline. This was done by Jahnke (1978) and Bigalke (1984), who proposed the Sneed and Stegmüller's concept a suitable theory concept for the field:

> A theory in mathematics education is a structured entity shaped by propositions, values, and norms about learning mathematics. It consists of a kernel that encompasses the unimpeachable foundations and norms of the theory and an empirical component that contains all possible expansions of the kernel and all intended applications that arise from the kernel and its expansions. (Bigalke, 1984, p. 152, translated, ABB)

In 1984, Hans-Georg Steiner inaugurated a series of five international conferences on Theories of Mathematics Education (TME), pursuing a scientific program that aimed at founding and developing the didactics of mathematics as a scientific discipline on the international level. His program addressed three partly overlapping areas:

> (1) Identification and elaboration of basic problems in the orientation, foundation, methodology, and organization of mathematics education as a discipline; (2) the development of a comprehensive approach to mathematics education in its totality when viewed as an interactive system comprising research, development, and practice; and (3) self-referent research and meta-research related to mathematics education that provides information about the state of the art—the situation, problems, and needs of the discipline —while respecting national and regional differences. (Steiner, 1987, p. 46)

The spirit of TME has been renewed today by the more bottom-up meta-theoretical approach of the networking of theories exploring how research with multiple theories can be conducted (specifically when they have emerged within specific educational systems), where the limits are, and how far new insights can be gained. Addressing networking strategies, this approach takes up the principle of complementarity, which Steiner (1987) worked out in the TME program (ibid., p. 48), being open for the theoretical diversity of the field.

In the 1990s, the research field in German-speaking countries began to investigate various methodologies based on a growing diversity in theory use. As examples, two theory traditions were presented in the session at ICME-13. Building on views of Peirce and Wittgenstein, Dörfler (2016) outlined a semiotic perspective on mathematics as an activity of diagrammatic reasoning and related to it as sign games and their techniques deeply involving rules for acting. Regina Bruder & Schmitt (2016) complemented this more home-grown theoretical view with the theory of learning activity that was developed based on activity theory by Hans Joachim Lompscher to inform the practice of teaching and learning in a school in the GDR. Bruder took up Lompscher's work and adapted it to the needs of teaching and learning mathematics. By applying the two theoretical views, mathematics as diagrammatic reasoning and learning activitiy, to the same data set in her presentation, Bikner-Ahsbahs (2016) readdressed Steiner's concern about complementarity by analyzing the data on the basis of the networking of theories; she wrote:

> Both approaches may enrich each other to inform practice (see TME program): coming from the learning activity we may zoom into (see Jungwirth 2009 cited by Prediger et al. 2009, p. 1532) diagram use, and coming from diagram use we may zoom out (ibid., p. 1532) to embed the diagram use into the whole course of the learning activity. (Bikner-Ahsbahs, 2016, p. 41)

Sociological Perspectives on Classroom Interaction

The specific aspects of sociological perspectives on classroom interaction, as a focus within the German-speaking traditions in mathematics education research, rest on a fundamental sociological orientation in mathematics lessons. This orientation has its origin in the works of Heinrich Bauersfeld and his colleagues at the Institut für Didaktik der Mathematik (IDM) at Bielefeld (Bauersfeld, 1980; Krummheuer & Voigt, 1991). These early studies unfolded the power of sociological description by reconstructing social processes regarding the negotiation of meaning and the social constitution of shared knowledge through collective argumentation in the daily practice of mathematic lessons. The "social" in these interactionist studies of mathematics classroom micro-culture was firmly located in the interpersonal space of those who interact. This space was considered a contingent sphere in which mathematical meaning emerges as the product of processes of negotiation. With respect to the sociological reference theories (primarily) of symbolic interactionism and ethnomethodology, a microsociology of mathematics lessons was created and elaborated. This theoretical approach to the mathematics classroom was based on three assumptions: (1) The mathematics that students learn and the conditions of the learning process are partly open to a process of negotiation of meaning in which the learners and the teacher(s) interactively exchange their definitions of the learning situation. (2) A process of collective argumentation concerning the mathematical content (concepts, terms, procedures, algorithms, etc.) is a constitutive social condition of the possibility of learning of this content. Participation in this process, albeit in different forms, is necessary for success in school mathematics. (3) Increased autonomous participation in such collective argumentation is the indication of successful learning in the mathematics classroom. The results of the empirically based development of a theory of learning in mathematics classrooms show that interaction in mathematics classes occurs in patterns of interaction in which the mathematical content is relevant. The patterns that support learning of mathematics are formats of collective argumentation. By increasing their autonomous participation in formats of argumentation, the learners take part in a process of development towards a full participation in school mathematics practice (Krummheuer, 2007).

The focus on a sociological theory of learning mathematics has been taken up and complemented by other sociological perspectives that have aimed at reconstructing the conditions and the structure surrounding the construction of performance and success in mathematics lessons. From these perspectives, schools and classrooms are not only considered places in which the learning of mathematics occurs, but also as institutional loci in which further societal functions of schooling, such as cultural reproduction and allocation, need to be pursued in parallel. At stake in mathematical activities in the classroom is not only the development of students' knowledge and skills, but also the creation of hierarchies of achievement in mathematics, of differential access to valued forms of mathematics, and of familiarization with work ethic (Gellert, 2008). From such a point of view, issues such as the distribution of knowledge, access, and students' resources are crucial ingredients to the forms the interaction in the mathematics classrooms may take.

During the session on sociological perspectives on classroom interaction, Götz Krummheuer's introductory talk, "Interpretative classroom research: Origins, insights, developments," summarized the development of a theory of learning mathematics. Two reactions to the presentation were prepared by Núria Planas (Spain) and Michelle Stephan (USA). The sociological zoom was then expanded by Uwe Gellert's presentation of "Classroom research as part of the social-political agenda" and of studies of German scholars concerning this matter. A prepared reaction by Eva Jablonka (UK) finalized the program.

Educational Research on Learning and Teaching of Mathematics

Educational research aims at generating knowledge on teaching and learning mathematics. To achieve this goal in the complex research domain, many empirical studies triangulate data, methods, investigators, and theory. This is reflected in the following strategies:

(1) a narrow focus on distinct phenomena concerning learning and teaching mathematics,
(2) an interdisciplinary perspective that integrates different background theories, and
(3) a mixed-method approach that combines different methodological practices.

As early as the 80s, mathematics education was already actively using and developing such research strategies, e.g., Ursula Viet's investigation of the cognitive development of fifth- and sixth-grade students in arithmetic and geometry. The benefits and limitations of such approaches for mathematics education were discussed by considering recent research projects from the last two decades.

In the first part of the presentation, Timo Leuders focused on two interdisciplinary research projects: Between 2000 and 2006, the priority program Educational Quality of Schools initiated more than 30 interdisciplinary cooperations to analyze domain-specific and cross-curricular learning. One of these projects was a multi-step research project by Regina Bruder (mathematics education) and Bernhard Schmitz (educational psychology) that investigated the problem-solving and self-regulatory behavior of students, also connecting the two perspectives theoretically. In another project within the program, Alexander Renkl (educational psychology) and Kristina Reiss (mathematics education) investigated how students' proof competence can be fostered by learning with worked-out examples and self-explanation prompts. For both projects in the presentation, the researchers reported in video interviews on the experiences, advantages, and challenges of their interdisciplinary approach.

The second part of the presentation about flexible mixed-methods approaches by Andreas Schulz started with a complementary perspective on qualitative and quantitative research. He showed that both make use of inductive as well as deductive reasoning and that both can complement and compensate for their strengths and weaknesses within a mixed-methods design. This was illustrated by a video and audio presentation and discussion of two such approaches: Kathleen Philipp made use of a sequential mixed-methods design. She analyzed students' strategies during solving several mathematical problems and developed a competence model about experimental thinking in mathematics. This laid the groundwork for an intervention study that confirmed that experimental competences in mathematics can be fostered effectively. Susanne Prediger and Lena Wessel implemented an integrated/parallel mixed-methods design. They fostered students' understanding of fractions and scaffolded the learning processes by fostering students' abilities to talk about fractions and their meaning. The effectivity of the randomized control study was evaluated by both statistical analyses and qualitative analyses of the teaching-learning processes.

In the international commentary that followed, Kaye Stacey confirmed and illustrated the need for a flexible combination of quantitative and qualitative research to generate both meaningful and reliable evidence for the understanding of learning and teaching in the field of mathematics education. This lead to an engaged discussion with the international audience about the potential and challenges of interdisciplinary research and mixed-methods approaches in mathematics education.

Large-Scale Studies

Large-scale studies assess mathematical competence using large samples. They often compare mathematical competence between groups of individuals within or between countries. The development of sophisticated statistical methods in recent years has encouraged collaborations between researchers from mathematics education on the one hand and from statistics or psychology on the other. This development has also allowed the empirical verification of theoretical models of mathematical competence and competence development.

In Germany, international large-scale studies did not receive much attention before 1995, when Germany took part in the Third International Mathematics and Science Studies (TIMSS) for the first time. The results showed that German lower and upper secondary school students' mathematical performance did not meet the expectations of teachers, educators, and the public. German students performed below the international average and showed acceptable results only for routine problems (Baumert, Bos, & Lehmann, 2000). The results of PISA 2000 (Baumert et al., 2001) were again disappointing and became known as the "PISA shock." The consequences of these studies were intensive debates among educators and stakeholders and the launch of educational programs to improve mathematics instruction at school. Another consequence was the agreement to use large-scale assessments on a regular basis to monitor the outcome of school education.

Assessing students' mathematical competences requires models of what mathematical competence actually is. Initial models were predominantly based on theoretical and normative considerations, but rarely on empirical evidence. In a recursive process, Reiss and colleagues (e.g., Reiss, Heinze, Kessler, Rudolph-Albert, & Renkl, 2007; Reiss, Roppelt, Haag, Pant, & Köller, 2012) developed a model for primary mathematics education that took into account theoretical and normative perspectives and was continuously refined based on empirical evidence. The model suggests five levels of mathematical competence reaching from technical background knowledge and routine procedures to complex mathematical modelling.

To monitor the outcome of educational quality on a regular basis, new institutions have been founded in Germany, such as the Institute for Educational Quality Improvement (IQB, Berlin) and the Center for International Student Assessment (ZIB, Munich). However, the idea of system monitoring is not specific to Germany. Other countries founded similar institutions and developed similar models of mathematical competence to assess students' competences on a regular basis. In Austria, for example, the Federal Institute for Educational Research, Innovation, and Development of the Austrian School System (BIFIE) is responsible for assessments. These assessments are based on a model of mathematical competence that describes mathematical competence in three dimensions (process domain, content domain, and level of complexity). This model is not only used for assessment purposes but is also the basis for developing curricula for the mathematics classroom.

Large-scale studies allow monitoring of the outcome of mathematics education on the system level. The broad empirical data these studies collect have been used to empirically validate theoretical models of mathematical competence and have contributed to a more realistic view of what students are capable of learning at school.

Final Remark

Looking back at the eight themes above, the reader will realize the profound changes that have taken place in German-speaking mathematics education research during the last 40 years. The development comes near to a sort of revolution—not very typical for Germany. The only themes that could have appeared in the program of the Karlsruhe Congress in 1976 are subject-matter didactics and, with qualifications, design science and *Allgemeinbildung*. All other topics, especially modelling, theory traditions, classroom studies, and empirical research represent for Germany completely new fields of activity. Today, they define the stage on which German mathematics educators have to act. Nevertheless, the more traditional fields that are nearer to mathematics, subject matter analysis and elementarization, are still alive and will continue to be areas of intense work so that the common ground of mathematics and education will not be lost.

References

Bauersfeld, H. (1980). Hidden dimensions in the so-called reality of mathematics classroom. *Educational Studies in Mathematics, 11*(1), 23–41.

Baumert, J., Bos, W., & Lehmann, R. (Eds.). (2000). *TIMSS/III—Dritte Internationale Mathematik und Naturwissenschaftsstudie—Mathematische und naturwissenschaftliche Bildung am Ende der Schullaufbahn.* Opladen: Leske + Budrich.

Baumert, J., Klieme, E., Neubrand, M., Prenzel, M., Schiefele, U., Schneider, W., et al. (Eds.). (2001). *PISA 2000. Basiskompetenzen von Schülerinnen und Schülern im internationalen Vergleich.* Opladen: Leske + Budrich.

Biehler, R., Heymann, H. W., & Winkelmann, B. (Eds.). (1995). *Mathematik allgemeinbildend unterrichten: Impulse für Lehrerbildung und Schule.* Köln: Aulis.

Bigalke, H.-G. (1984). Thesen zur Theoriendiskussion in der Mathematikdidaktik. *Journal für Mathematik-Didaktik, 5*(3), 133–165.

Bikner-Ahsbahs, A. (2016). Networking of theories in the tradition of TME. In A. Bikner-Ahsbahs & A. Vohns (Eds.), *Theories of and in mathematics education. Theory strands in German-speaking countries* (pp. 33–42). Springer International Publishing AG Switzerland.

Bikner-Ahsbahs, A., Prediger, S., & The Networking Theories Group. (2014). *Networking of theories as a research practice in mathematics education.* New York: Springer.

Bikner-Ahsbahs, A., & Vohns, A. (2016). Theories in mathematics education as a scientific discipline. In A. Bikner-Ahsbahs & A. Vohns (Eds.), *Theories of and in mathematics Education. Theory strands in German-speaking countries* (pp. 3–12). Springer International Publishing AG Switzerland.

Blum, W., & Kirsch, A. (1991). Preformal proving: Examples and reflections. *Educational Studies in Mathematics, 22*(2), 183–203.

Borromeo Ferri, R. (2006). Theoretical and empirical differentiations of phases in the modelling process. *Zentralblatt für Didaktik der Mathematik, 38*(2), 86–95.

Bruder, R., & Schmitt, O. (2016). Joachim Lompscher and his activity theory approach focusing on the concept of learning activity and how it influences contemporary research in Germany. In A. Bikner-Ahsbahs & A. Vohns (Eds.), *Theories of and in mathematics education. Theory strands in German-speaking countries* (pp. 13–20). Springer International Publishing AG Switzerland.

Burscheid, H. J. (1983). Formen der wissenschaftlichen Organisation in der Mathematik-Didaktik. *Journal fürMathematik-Didaktik, 4*(3), 219–240.

Damerow, P., Elwitz, U., Keitel, C., & Zimmer, J. (1974). Elementarmathematik Lernen für die Praxis? *Ein exemplarischer Versuch zur Bestimmung fachüberschreitender Curriculumziele.* Stuttgart: Klett.

Dörfler, W. (2016). Signs and their use: Peirce and Wittgenstein. In: A. Bikner-Ahsbahs & A. Vohns (Eds.), *Theories of and in mathematics education. Theory strands in German-speaking countries* (pp. 21–33). Springer International Publishing AG Switzerland.

Fischer, R. (1983). Wie groß ist die Gefahr, daß die Mathematikdidaktik bald so ist wie die Physik?—Bemerkungen zu einem Aufsatz von Hans Joachim Burscheid. *Journal für Mathematik-Didaktik, 4*(1983)3, 219–240.

Gellert, U. (2008). Validity and relevance: Comparing and combining two sociological perspectives on mathematics classroom practice. *ZDM—The International Journal on Mathematics Education, 40*(2), 215–224.

Greefrath, G., & Vorhölter, K. (2016). *Teaching and learning mathematical modelling. Approaches and Developments from German-speaking Countries*: Springer International Publishing AG Switzerland.

Hefendehl-Hebeker, L. (2016): Subject-matter didactics in German traditions. Early Historical Developments. *Journal für Mathematik-Didaktik, 37,* Supplement 1, 11–31.

Heymann, H. W. (2003). *Why teach mathematics? A focus on general education.* Dordrecht, Boston: Kluwer Academic Publishers.

Institut für Didaktik der Mathematik (Eds.). (2007). Standards für die mathematischen Fähigkeiten österreichischer Schülerinnen und Schüler am Ende der 8. Schulstufe. Klagenfurt. http://www.uni-klu.ac.at/idm/downloads/Standardkonzept_Version_4-07.pdf

Jahnke, H. N. (1978). *Zum Verhältnis von Wissensentwicklung und Begründung in der Mathematik—Beweisen als didaktisches Problem.* Bielefeld: Universität Bielefeld.

Jahnke, H. N. (1990). *Mathematik und Bildung in der Humboldtschen Reform.* Göttingen: Vandenhoeck & Rupprecht.

Jungwirth, H. (2009). An interplay of theories in the context of computer-based mathematics teaching: How it works and why. In: *Proceedings of CERME 6.* Lyon, France. http://ife.ens-lyon.fr/publications/edition-electronique/cerme6/wg9-07-jungwirth.pdf. Accessed 6 August 2017.

Kirsch, A. (1977). Aspects of simplification in Mathematics teaching. In H. Athen & H. Kunle (Eds.), *Proceedings of the third international congress on Mathematical education* (pp. 98–120). Karlsruhe: Zentralblatt für Didaktik der Mathematik.

Krummheuer, G. (2007). Argumentation and participation in the primary mathematics classroom: Two episodes and related theoretical abductions. *Journal of Mathematical Behavior, 26*(1), 60–82.

Krummheuer, G., & Voigt, J. (1991). Interaktionsanalysen von Mathematikunterricht. Ein Überblick über Bielefelder Arbeiten. In H. Maier & J. Voigt (Eds.), *Interpretative Unterrichtsforschung* (pp. 13–32). Köln: Aulis.

Maaß, K., & Mischo, C. (2011). Implementing modelling into day-to-day teaching practice—The project STRATUM and its framework. *Journal für Mathematik-Didaktik, 32*(1), 103–131.

Neubrand, M. (2003). Mathematical literacy/mathematische Grundbildung. *Zeitschrift für Erziehungswissenschaft, 6*(3), 338–356.

Niss, M., & Højgaard, T. (Eds.). (2011). *Competencies and mathematical learning: Ideas and inspiration for the development of mathematics teaching and learning in Denmark.* Roskilde: Roskilde University, IMFUFA.

Nührenbörger, M., Rösken-Winter, B., Ip Fung, Ch., Schwarzkopf, R., & Wittmann, E. (2016). *Design Science and Its Importance in the German Mathematics Educational Discussion.* (ICME-13 Topical Surveys) Rotterdam: Springer.

Perels, F., Otto, B., Schmitz, B., & Bruder, R. (2007). Evaluation of a training programme to improve mathematical as well as self-regulatory competences. In M. Prenzel (Ed.), *Studies on the educational quality of schools* (pp. 197–219). Münster: Waxmann.

Philipp, K. (2012). *Experimentelles Denken: theoretische und empirische Konkretisierung einer mathematischen Kompetenz.* Wiesbaden: Springer Spektrum.

Prediger, S., & Wessel, L. (2013). Fostering German-language learners' constructions of meanings for fractions—design and effects of a language-and mathematics-integrated intervention. *Mathematics Education Research Journal, 25*(3), 435–456.

Prediger, S., Bosch, M., Kidron, I., Monaghan, J., & Sensevy, G. (2009). Introduction to the working group 9. Different theoretical perspectives and approaches in mathematics education research—strategies and difficulties when connecting theories. In: *Proceedings of CERME 6.* Lyon, France. http://ife.ens-lyon.fr/publications/edition-electronique/cerme6/wg9-00-introduction.pdf. Accessed 18 February 2016.

Prediger, S., Gravemeijer, K., & Confrey, J. (2015). Design research with a focus on learning processes. *ZDM—The International Journal on Mathematics Education, 47*(6), 877–891.

Reiss, K., Heinze, A., Kessler, S., Rudolph-Albert, F., & Renkl, A. (2007). Fostering argumentation and proof competencies in the mathematics classroom. In M. Prenzel (Ed.), *Studies on the educational quality of schools. The final report on the DFG priority programme* (pp. 251–264). Münster: Waxmann.

Reiss, K., Heinze, A., & Pekrun, R. (2007). Mathematische Kompetenz und ihre Entwicklung in der Grundschule. In M. Prenzel, I. Gogolin & H. H. Krüger (Eds.), *Kompetenzdiagnostik. Sonderheft 8 der Zeitschrift für Erziehungswissenschaft* (S. 107–127). Wiesbaden: Verlag für Sozialwissenschaften.

Reiss, K., Roppelt, A., Haag, N., Pant, H. A., & Köller, O. (2012). Kompetenzstufenmodelle im Fach Mathematik. In P. Stanat, H. A. Pant, K. Böhme, & D. Richter (Hrsg.), *Kompetenzen von Schülerinnen und Schülern am Ende der vierten Jahrgangsstufe in den Fächern Deutsch und Mathematik. Ergebnisse des IQB-Ländervergleichs 2011* (pp. 72–84). Münster: Waxmann.

Schukajlow, S., Leiss, D., Pekrun, R., Blum, W., Müller, M., & Messner, R. (2012). Teaching methods for modelling problems and students' task-specific enjoyment, value, interest and self-efficacy expectations. *Educational Studies in Mathematics, 79*(2), 215–237.

Steiner, H.-G. (1983). Zur Diskussion um den Wissenschaftscharakter der Mathematikdidaktik. *Journal für Mathematik-Didaktik, 4* (1983)/3, 245–251.

Steiner, H.-G. (1987). A systems approach to mathematics education. *Journal for Research in Mathematics Education, 18*(1), 46–52.

Van den Akker, J., Gravemeijer, K., McKenney, S., & Nieveen, N. (Eds.). (2006). *Educational design research.* London: Routledge.

Vom Hofe, R., & Blum, W. (2016). "Grundvorstellungen" as a Category of Subject-Matter Didactics. *Journal für Mathematik-Didaktik, 37,* Supplement 1, 225–254.

Winter, H. (1995). Mathematikunterricht und Allgemeinbildung. *Mitteilungen der GDM, Nr., 61,* 37–46.

Wittmann, E. Ch. (1995). Mathematics education as a "design science." *Educational Studies in Mathematics,* 29, 355–374 [repr. in A. Sierpinská, & J. Kilpatrick (Eds.) (1998), Mathematics Education as a Research Domain. A Search for Identity (pp. 87–103). Dordrecht: Kluwer].

Open Access Except where otherwise noted, this chapter is licensed under a Creative Commons Attribution 4.0 International License. To view a copy of this license, visit http://creativecommons.org/licenses/by/4.0/.

What Is and What Might Be the Legacy of Felix Klein?

Hans-Georg Weigand, William McCallum, Marta Menghini, Michael Neubrand, Gert Schubring and Renate Tobies

Abstract Felix Klein always emphasised the great importance of teaching at the university, and he strongly promoted the modernisation of mathematics in the classrooms. The three books "Elementary Mathematics from a higher (advanced) standpoint" from the beginning of the last century gave and still give a model for university lectures especially for student teachers. The "Merano Syllabus" (1905), essentially initiated and influenced by Felix Klein, pleaded for an orientation of mathematics education at the concept of function, an increased emphasis on spatial geometry and an introduction of calculus in high schools. The Thematic Afternoon "The legacy of Felix Klein" will give an overview about the ideas of Felix Klein, it will highlight some developments in university teaching and school mathematics related to Felix Klein in the last century, and it will especially be asked for the meaning, the importance and the legacy of Klein's ideas nowadays and in the future in an international, worldwide context.

H.-G. Weigand (✉)
University of Würzburg, Würzburg, Germany
e-mail: weigand@dmuw.de

W. McCallum
University of Arizona, Tucson, USA
e-mail: wmc@math.arizona.edu

M. Menghini
Sapienza University of Rome, Rome, Italy
e-mail: marta.menghini@uniroma1.it

M. Neubrand
University of Oldenburg, Oldenburg, Germany
e-mail: michael.neubrand@uni-oldenburg.de

G. Schubring
University of Bielefeld, Bielefeld, Germany
e-mail: gert.schubring@uni-bielefeld.de

R. Tobies
University of Jena, Jena, Germany
e-mail: renate.tobies@uni-jena.de

When we talk about the legacy of Felix Klein, we are interested in the significance of Felix Klein's work for mathematics education, for our current theory and practice, and above all, for tomorrow's ideas concerning the teaching and learning of mathematics. We are interested in Felix Klein as a mathematician and as a mathematics teacher, but most of all we are interested in his ideas on teaching and learning mathematics, the problems he saw at university and at high school, and the solutions to these problems that he suggested. We are interested in these solutions because we recognise that we are nowadays confronted with similar or even the same problems as 100 years ago. Speaking about Felix Klein's legacy means hoping to get answers to some of the problems we are struggling with today. Speaking about Felix Klein's legacy today means giving answers to—at least—three basic questions:

1. Which situations and which problems at the end of the 19th and the beginning of the 20th century can be seen in analogy to present situations?
2. How did Felix Klein react to these problems and which solutions did he suggest?
3. What do we know nowadays about the effect of the answers and solutions provided by Felix Klein 100 years ago?

Analogies between the situation 100 years ago and today can immediately be seen if we think about the current discussions concerning the goals and contents of teacher education at university, especially the problems of students, with the transition from high school to college or university and the transition back to high school. The problems with these transitions are expressed in Felix Klein's most famous statement, the "double discontinuity" from the introduction to *Elementary mathematics from a higher standpoint, Volume I*:

> A mathematics freshman at the university is confronted with problems he had not been concerned with at school. After finishing university and becoming a teacher, he/she is expected to teach traditional elementary mathematics, which he was not confronted with at university. Therefore, he teaches mathematics the way he was taught some years ago and his university studies remain only a more or less pleasant memory which do not influence his teaching.

When we hear the lamentations of today's university professors about the decreasing abilities of first-year students, and when we note the negative views of young teachers about the effects of their mathematics studies, you can surely be in doubt whether there has been any change in the last 100 years.

However, we also know that answers to problems in education—not only mathematics education—can only be offered taking full recognition of the current political, social, and scientific situation. Answers are not and will never be general statements, they always have to be newly evaluated in an on-going process of discussion between different social groups. What is or what might be the impact of Felix Klein's ideas on these current discussion processes?

In the following, we start with some short biographical notes about Felix Klein and give an introduction to his comprehensive program. Then we give an overview of the three strands we offered on the "Thematic Afternoon", each concentrated on one important aspect of Felix Klein's work: Functional Thinking, Intuitive Thinking and Visualisation, and Elementary Mathematics from a Higher Standpoint—Conception, Realisation, and Impact on Teacher Education.

Felix Klein: Biographical Notes and His Comprehensive Program

The Starting Point

Having been a full professor at the University of Erlangen (1872), the Technical University in Munich (1875), and the University of Leipzig (1880), Felix Klein (1849–1925) joined the University of Göttingen in 1886. He had gained international recognition with his significant achievements in the fields of geometry, algebra, and the theory of functions. On this basis, he was able to create a centre for mathematical and scientific research in Göttingen (Tobies, 2002). Felix Klein was far ahead of his time in supporting all avenues of mathematics, its applications, and mathematical pedagogy. He organised that the establishment of new lectures, professorships, institutes, and curricula went hand in hand with the creation of new examination requirements for prospective secondary school teachers.

Klein was 16 years old when he passed his German *Abitur* examination, 19 when he completed his doctorate, and 21 when he qualified as university lecturer and he was appointed full professor at the age of 23.

Right from the start, Klein also wanted to improve instruction at secondary schools, because all mathematical instruction at German universities was aimed at training future teachers. In letters to foreign mathematicians, for example, to Gaston Darboux (1842–1917), Klein discussed mathematical problems *and* teaching questions as early as the 1870s. Darboux arranged for the first translation of a paper by Klein: his famous article on non-Euclidean geometry. In a paper on Darboux, David Hilbert mentioned that it was Darboux who had influenced Klein's educational efforts to a great extent (the central role of the concept of function, for example). Later, Darboux became the director of the French Education Commission, and, in 1914, he chaired the last meeting of the IMUK to take place in Paris before World War I. Klein, as president, could not take part because of health reasons, but had prepared the content of the proceedings (Tobies, 2016).

International Perspectives

Klein maintained a network of international relations that included, among others, Cayley from Great Britain, Zeuthen from Denmark, Stolz from Austria, Lie from Norway, Cremona from Italy, and Markov from Russia. Mathematicians of the older generation encouraged students from France (recommended by Darboux) and other countries to go to Germany and attend Klein's courses. Klein wholeheartedly supported young students in achieving their own research results. Later, many of them would in turn support him with his other projects (especially of the pedagogical sort).

If we look, for example, at Klein's seminar on hypergeometric functions during the winter term of 1893–94, we will not only see the first female students enrolled at the University of Göttingen (they would complete their doctorates under Klein's supervision two years later). We will also see Emanuel Beke (1862–1946), who gave three presentations (January 10, 17, and 24, 1894) in the seminar and would publish two article in "Klein's" journal, *Mathematische Annalen* (vol. 45 [1894]). Later, in 1906, Beke became the chairman of the Teaching Reform Commission in Hungary, and, in 1908, he endorsed Klein's appointment as the first president of the IMUK/CIEM. Inspired by Klein, Grace Chisholm Young (1868–1944), one of his female doctoral students, wrote a beginner's textbook on geometry that Klein would praise in the second volume of his *Elementary Mathematics from a Higher Standpoint*. Another participant in the seminar, the American Frederick S. Woods (1864–1950), would earn fame for his mathematical textbooks. Finally, the Italian Gino Fano (1871–1952) not only translated Klein's *Erlangen Programme* into Italian, he also wrote a paper on his experiences in Göttingen. There he made a special point to mention the interaction between the university world and secondary teachers. Along with other Italian mathematicians—including Gino Loria, Corrado Segre, and Federigo Enriques—Fano propagated Klein's educational ideas, and they organised the translation of some of his other works as well (Coen, 2012, pp. 210–45).

This seminar of 1893–94 was held after Klein's participation in the World's Columbian Exposition and in the Mathematicians Congress of 1893 (Parshall & Rowe, 1994). At these events, Klein enhanced his international reputation; Charles Hermite (1822–1901) organised the translation of Klein's papers into French and ultimately orchestrated his appointment as a corresponding member of the French Academy (Tobies, 2016). Klein's US trip also caused him to gain more influence in Germany. Friedrich Althoff (1839–1908), an important official at the Prussian Ministry of Culture, came to accept Klein's ideas about the admission of women to university, the improvement of the teacher training, and the promotion of applied mathematics. With Althoff's support, Klein established, in February 1898, the first scientific association for promoting applied physics and mathematics, which brought scientists from the University of Göttingen into contact with 50 influential businessmen from Germany's chemical, electrical, steel, and iron industries. Remarkably enough, the founding members of this association agreed that the improved training of future teachers (in new mathematical fields, applications, and experimental instruction) should be the most important goal. Thus, Klein drafted new examination requirements for prospective secondary school teachers, and these were ratified as early as September 1898. In May 1900, Althoff invited Klein to state his expert opinion in advance of a pedagogical conference in Berlin. This cleared the way for Klein to develop his agenda (Schubring, 1989), which involved contesting opponents on multiple committees and, as member of the Prussian parliament, pushing through a reform of mathematical instruction from kindergarten to the university level including the education of girls and young women (Tobies, 1989).

As a sign of his increased international standing, Klein was made a board member of *L'Enseignement Mathématique*, which was founded in 1899.

This was the first international journal for mathematical instruction, and as such it became the official organ of the first international body, the *Internationale Mathematische Unterrichtskommission* (IMUK) or *Commission Internationale de l'Enseignement Mathématique* (CIEM), founded during the IV International Congress of Mathematicians (Rome, April 6–11, 1908), where Klein was elected—in absentia—as president. The initiator of this International Commission on Mathematical Instruction (ICMI; this name was adopted in 1954), the American David Eugene Smith (1860–1944), had translated (together with W. W. Beman) Kleins *Famous Problems of Elementary Geometry* (1897). In 1912, Smith became Vice-President of this commission.

Felix Klein and Mathematics Education

Klein also stands out for having established the field of mathematical didactics and for having regarded the history of mathematics as a keystone of higher education. He was always keen to underscore the relation between mathematics and culture (Klein, 1912–14). He never pursued the unilateral interests of his subject but rather kept an eye on the latest developments in science and technology. Because of Darwins theory of evolution, for instance, the instruction of biology had been forbidden at Prussian secondary schools since 1882. Together with the biologist Karl Kraepelin (1848–1915) from Hamburg, Klein overturned this ban in 1904 by initiating a mathematical-scientific instruction committee. Klein was successful because he adopted and respected the interests of other groups and because he promoted mathematicians regardless of their nationality, gender, or religion. To cite one final example of Klein's influence: when Poul Heegaard (1871–1848) took over vice-presidency of IMUK/ICTM in 1932, he publically reminisced about the scientific atmosphere under Klein in the mid-1890s, who had inspired him considerably and had discussed with him the idea that would later form the basis for his doctoral dissertation (O'Connor, Robertson, & Munkholm, 2010).

It is no wonder that the term *Klein's reform* was already in wide use during Felix Klein's own lifetime.

In the following, we highlight two main aspects of Felix Klein's work with regard to his influence on mathematics education—organised in two strands at the Thematic Afternoon—and give some information about the new edition of the *Elementary Mathematics*.

Strand A: Functional Thinking

If you understand *functional thinking* as "a way of thinking that is typical for the working with functions," the knowledge about definitions, properties, related concepts, representations, and examples and counter-examples of functions is crucial for the development of functional thinking in mathematics. The understanding of the function concept is a long-standing process, which starts in kindergarten and primary school and can be seen as an open-ended process even in university mathematics. In Strand A we concentrated on two aspects in the frame of the function concept: basic ideas of the function concept and problems with real numbers represented in a number line.

Definitions of Functions

Consider the following definitions of a function:

> A function of a variable quantity is an analytic expression composed in any way whatsoever of the variable quantity and numbers or constant quantities.

> The general concept of a function requires that a function of x be defined as a number given for each x and varying gradually with x. The value of the function can be given either by an analytic expression or by a condition that provides a means of examining all numbers and choosing one of them, or finally the dependence may exist but remain unknown.

> Let E and F be two sets, which may or may not be distinct. A relation between a variable element x of E and a variable element y of F is called a functional relation in y if, for all x in E, there exists a unique y in F which is in the given relation with x.

These definitions exemplify the historical struggle with two key problems in how we think about functions: conceptualizing the function itself as an object, and conceptualizing the domain and range of a function as objects.

Productive Meanings of Functions

The first presentation at the Thematic Afternoon, by Pat Thompson (Arizona, United States), considered the first problem in the light of US and South Korean teachers' meanings for functions and function notation. He reported on a study of ways that 253 US and 368 South Korean teachers understand the ideas of functions and function notation.

Teachers from both countries were given a 46-item diagnostic instrument (43 items in the Korean version). The instrument focused on variation and co-variation; function properties, modelling, and notation; frames of reference; magnitude and measure; proportionality; rate of change; and structure sense. The focus of this talk was modelling and notation.

The presentation started with a discussion of how teachers convey meaning. For example, a teacher might have a complex understanding of the rule $-x = x$ in algebra, involving using the properties of operations to derive this rule. But the conveyed meaning in the student's mind might be something much simpler, for example, "If you see more than one minus sign, write it without any minus signs."

The survey found that significantly higher percentages of South Korean teachers (as high as 68% on some measures) had productive meanings for functions and function notation, whereas the corresponding percentage for US teachers was typically around 30%. Thompson emphasised that it is not that South Korean teachers know more mathematics than US teachers, but rather that a greater percentage of South Korean teachers have mathematical meanings that are potentially productive for student learning.

Thompson suggested that this problem resulted from a vicious cycle, which could also be called the "double discontinuity." Many students leave high school with poorly formed meanings for ideas of the middle and secondary mathematics curriculum. They take mathematics courses in college from instructors who presume, or do not think about whether, students have basic mathematical meanings that they in fact do not have. They therefore apply the same coping mechanisms (e.g., memorisation) in college mathematics that allowed them to succeed in high school mathematics. As a result, they return to high schools to teach ideas they have understood poorly, have rarely revisited, and for which they still have poorly-formed meanings.

He concluded by proposing that professional development focus on teachers' mathematical meanings for the mathematics they teach and on ways that students create mathematical meaning from instruction, suggesting that mathematics departments must play a central role.

Functions and the Number Line

The second presentation, by Hyman Bass (Michigan, United States), considered the problem of conceptualising the domain of a function through an examination of the number line. A robust understanding of the continuous real number line is a central goal of K-12 education, but the extent to which this is achieved is questionable. Bass argued that the roots of this problem rest to a large extent in the early introduction of numbers in grade 1. He discussed a promising way of approaching this due to a theory by Davydov.

Bass considered the question of a student who, at the end of high school, can meaningfully hear, "Let f(x) be a function of a real variable x." The home of this x is the real number continuum **R**. How did this **R**, with its rich algebraic and geometric structure, make itself progressively known to first-grade student Anne, who can be presumed to know little more than simple cardinal counting? There are two possible narratives that can explain this accomplishment.

The construction narrative starts with counting numbers, gradually builds **R** by accretion of new numbers (negative integers and fractions), and eventually a "hole filling" completion (from **Q** to **R**) whose need and nature are often left tacit. A difficulty with this narrative is that fractions appear, conceptually, notationally, and computationally, as a whole new number planet, and their integration into the eventual number continuum that they cohabit with integers and rational numbers is complex and non-intuitive. The construction narrative, characteristic of much of the US curriculum, risks not achieving the final synthesis, an internalised understanding of the real number continuum.

The occupation narrative (promoted by Davydov & colleagues), in contrast, treats the geometric line as present from the beginning; it is the natural environment for linear measurement. This confers on the geometric line (imagine a thin string, flexible but inelastic) a primordial cognitive status on the same footing as cardinal counting. Measurement is at first of quantities (continuous as well as discrete). Numbers then arise as ratios of two quantities, one taken as a unit against which to measure the other. In this way, once an (oriented) interval on the line is chosen as unit, all intervals acquire, at least conceptually, a numerical measure (or ratio); therefore, the continuum of all real numbers is, at least conceptually, present from the beginning. The progression in the construction narrative above is now replaced by the progressive naming of more and more of these numbers as we locate where they take up residence on the (pre-existing) line.

Strand B: Intuitive Thinking and Visualisation

In addition to the functional thinking aspect, the second central aspect is surely the idea of *Anschauung*, a term which is quite hard to translate since it incorporates many facets. It does not stretch only over the geometry part of Klein's books but roots in the basic thinking of Felix Klein on the tasks of the mathematics teachers: They (as Felix Klein once said) will only succeed if they are able to make things "*anschaulich erfassbar*" ("intuitively comprehensible"). The teacher then has to choose "psychological" ways of presentation, and these are not necessarily the "systematic" ways. For this purpose, a whole bundle of possibilities could and should be used by the teacher: Drawings, pictures, models, experiments, any dynamic representation, as far as it could be realised, etc. This broad range of Felix Klein's central idea of *Anschauung* brought us to call Strand B "Intuitive Thinking and Visualisation."

Anschauung, Abstraction, and Visualisation

In the first contribution, Martin Mattheis (Mainz, Germany) displayed the deeper intentions of Felix Klein behind the central aspect *Anschauung*. In many passages of the books and within the papers on school reform, Felix Klein explains what he means by *Anschauung*, and some examples can illustrate these ideas: How numbers should be associated with concrete representations, why geometry should be connected to a vivid intuition in order to develop a full understanding of abstract concepts, how far functions and especially the infinitesimal calculus rest on graphical representation, and, finally, that mathematical intuition always should precede logical reasoning. Felix Klein also pointed out, however, that one should not cut the higher parts of the concepts, as illustrations are necessary for making mathematics more accessible.

The interplay between abstraction and visualisation was also the starting point of Stefan Halverscheid (Göttingen, Germany) and Oliver Labs (Potsdam and Mainz, Germany), who discussed Felix Klein's mathematical heritage as it can be seen today through 3D models and other tools of visualisation. The possibilities have changed since the time of Felix Klein, but not the basic idea of visualisation. Therefore, we are able to visualise two spaces at the same time on a screen, e.g., making visible the influence of parameters onto the curves, even those of higher degrees. We can draw pictures of functions with two variables, or create dynamic pictures by functions in polar coordinates. Different geometries can be visualised on the computer screen. Nowadays, 3D-printing technologies make it even possible to work with and not just see various 3D models. These possibilities overcome some inherent difficulties with the historical models in the famous Göttingen collection. Working with 3D-printer models was concretely demonstrated when the audience was given a chance to have their own experiences with tessellation of the 3-space with given 3D-printer objects. Labs and Halverscheid also provided a small exhibition of 3D models they created with various printers and showed how they used them in teacher education seminars.

Visualisation and Intuition: Historically and Nowadays

The second hour of the session was devoted to the impact of Klein's ideas on visualisation and intuition into the modern teaching of mathematics and the realisation of his ideas in classrooms of various levels. Flavia Mammana (Catania, Italy) detected the modernity of the *Merano Syllabus* for teaching geometry in Grades 10 and 11. Today, an intuitive approach to geometry is facilitated by the use of information technology. However, the central issue of the development from intuition to the concepts still remains. An example how one can use dynamic geometry software to develop concepts showed especially the transitions from 2D to 3D geometry while "seeing" quadrilaterals and/or tetrahedra in certain figures.

For an introductory linear algebra course, Chris Rasmussen (San Diego, United States) invented some visualisation and intuition ideas. Compatible with Klein's views, he presented an instructional sequence that supports students' reinvention of the concepts of span, linear dependence, and linear independence. The approach, labeled as the "Magic Carpet Ride," focuses on vectors, their algebraic and geometric representations in 2 and 3 dimensions, and their properties as sets. Student solutions showed how the creation of formal definitions can proceed from intuition.

Felix Klein was not alone in creating innovative, e.g., visualisation-oriented, teaching ideas in his time. Other contributors to that field, sometimes with explicit reference to Felix Klein, also played a role. Their work has still influence on the teaching today. In her seminars for prospective teachers, Ysette Weiss-Pidstrygach (Mainz, Germany) used classroom models that Peter Treutlein created with the intention of helping students grasp mathematics. The spirit of Felix Klein in using models for teaching was already being carried on from university to school by Peter Treutlein, a teacher and school headmaster, before the First World War. As Klein also did, he attempted to transform visual perception into logical concepts. Today, his models offer a rich variety of possibilities to relate historical, mathematical, and pedagogical aspects in mathematics teacher education. Examples stretched from paper folding to mathematical experiments.

There were still other figures on the scene, as Sebastian Kitz (Wuppertal, Germany) noted. However, he concentrated on the mathematical animated films that Gymnasium teacher Ludwig Münch used as teaching materials, which have been partly reconstructed in the last years. Produced around 1910, they can be seen as a precursor of modern dynamic geometry environments. Film clips were shown on the Apollonian problem in elementary geometry, the transition of the conic sections into one another, the circle of curvature rolling along a curve and tracking the centre, and planetary motion by Ptolemy and Copernicus. Again, we see that the technological progress of that time could become fruitful in transforming Felix Klein's ideas into realisation. These old mathematical animated films give a good impression of that process.

Strand C: Elementary Mathematics from a Higher Standpoint—Conception, Realisation, and Impact on Teacher Education

Strand C was concerned with the three volumes of *Elementary Mathematics from a Higher Standpoint*: their conception, realisation, and impact on teacher education. These lecture notes from the early 20th century were a seminal contribution by Klein to mathematics teacher education, presenting for the first time a methodological orientation for teacher education, not just a content-oriented course. They constituted a model for many later approaches. Strand C intended to assess the importance of this conception of mathematics teacher education, discussed its international impact and reflected about its impact on present-day teacher education.

The Starting Point: Hand-Written Notes

In a short overview, Gert Schubring (Bielefeld, Germany) showed the transformation from lithography to e-book publication of the three volumes of Klein's

Elementary Mathematics from a Higher Standpoint. The volumes originated from a two-semester lecture course, first given in 1907–08 to 1908–09. They were published in the form of a *Nachschrift* of notes taken by experienced students and revised by the professor. This form had become established by students of Weierstraß, who thus enabled that Weierstraß' novel conceptions of analysis, which he did not publish himself, could be disseminated. In Klein's manner of adaptation, a decisively greater number of copies could be disseminated: they were poly-copied by lithography and distributed by the publisher Teubner in Commission: This means that the text was not typeset, but handwritten—fortunately not in the nowadays unusual *Sütterlin* script. Right at the beginning, on the second page of the first volume, one can find the famous quote about double discontinuity.

Volume II on geometry was distributed in 1909 in the same manner.

What became later, in the general revision for typeset printing, the third volume, had originally, in 1901, been a separate lecture course, on application of differential and integral calculus to geometry: a revision of foundations.

On the Way to a Bestseller

After the publication of the third, complete, and revised edition of the *Elementarmathematik* from 1924 to 1928, now in regular book format, the German series became a bestseller, was reprinted many times, and it was also translated—first into Spanish, as *Matemática elemental desde un punto de vista superior* (1927, 1928), and then into English, as *Elementary Mathematics from an Advanced Standpoint* (1932, 1939).

However, this translation, also many times reprinted, suffers decisive problems:

- The two translators, Earle Raymond Hedrick and Charles Noble, although students of Hilbert and Klein in Göttingen and both later mathematics professors at US universities, apparently never managed their translation to be checked by German native speakers.
- Surely non-Germans had difficulty with the *Nachschrift*: It is based on oral lectures in a colloquial style where the idioms are not as easy to grasp for a non-German.
- They used the word *advanced* in the title instead of *higher*.
- Important parts in Volumes I and II were omitted without comment.
- There are inconsistencies in the mathematical terminology.
- And there were numerous translation errors.

Translations into other languages followed: a four-part Japanese translation (*Takai tachiba kara mita shoto: su-gaku*, 1959–60 and 1961) and a Russian translation (*Elementarnaja matematika s tocki zrenija vyssej*, 1987). The most recent translation was into Portuguese: *Matemática Elementar de um Ponto de Vista Superior*, in five parts (2009–2014).

All translations so far excluded Volume III and were restricted to Volumes I and II. The only complete translation so far has been the Chinese translation, first published in the People's Republic of China in 1989 and then reprinted in the Republic of China in 1996.

On the occasion of this 13th International Congress on Mathematics Education (ICME-13, Hamburg 2016), the International Commission on Mathematical Instruction (ICMI), together with Springer, has intended to close the gap in the English versions. This was especially a suitable date because Felix Klein was the first president of ICMI (see Section "Felix Klein: Biographical Notes and His Comprehensive Program").

Notes to the Three Volumes

It is thanks to the initiatives by Gabriele Kaiser and Ferdinando Arzarello that Springer undertook it to publish the first complete translation in modern English of Felix Klein's seminal series of lecture notes—complete meaning that it now also both includes Volume III and integrates the several extended parts of the original of Volumes I and II that had been omitted.[1]

After this overview, the first of the successive two hours of Strand C concerned the content of the three volumes of *Elementary Mathematics from a Higher Standpoint*.

Gert Schubring analysed Klein's innovative establishment of the word *elementary*: It was not used in the everyday way meaning "simple," but as the result of the process of elementarisation of complex developments in mathematics, effecting a restructuration of mathematics from new conceived elements. Klein does not prescribe schools to adopt the latest developments in science; rather, he allows them to make proper choices according to criteria of the school system, yielding a certain "hysteresis" behind the recent, not yet elementarised state.

Through examples taken from Volume III, *Precision and Approximation Mathematics*, Marta Menghini (Rome, Italy) underlined how the relation between applied and pure mathematics was a subject of utmost concern for Klein. For instance, Klein introduces circular inversion starting from physics then creating point sets with particular properties. This example, which was shown using Geogebra, allows illustration of some features of Klein's text: Starting from an intuitive and sometimes practical approach, Klein develops abstract concepts working in rich "mathematical environments," which form the core of an interesting high school mathematics teacher education. Another example concerns the "continuous" transformation of curves, evidencing the invariant properties.

Klein's mathematical, historical, and didactical perspective was illustrated by Henrike Allmendinger (Lucerne, Switzerland), who took a closer look at original

---

[1]Felix Klein, *Elementary Mathematics from a Higher Standpoint*. 3 volumes. Springer, 2016.

text snippets from the chapter on logarithmic and exponential functions from Klein's Volume I, *Arithmetic, Algebra, and Analysis*. Klein describes the customary approach to logarithms in school and criticises it, presenting an alternative approach led by the historical development and giving an insight into the logarithms from the standpoint of function theory.

The International Impact

Jeremy Kilpatrick (Athens, Georgia, United States) commented on the previous part and introduced the second hour, which concerned the international impact of Klein's conceptions, in particular upon teacher education.

Fréderick Gourdeau (Québec, Canada) discussed the interplay between disciplinary mathematics and school mathematics in teacher education. The emphasis placed upon content in the mathematical education of teachers is still being questioned today, just as Klein questioned it in his seminal work. Recently, the notion of "mathematical habits of mind" has been at the forefront of some discussions. This was exemplified by recent proposals for curriculum reform in Canada.

The impact of Klein's work for today's mathematics education in Asia was shown by Masami Isoda (Tsukuba, Japan): in particular, a Japanese secondary school textbook from 1943 reveals pertinent influences of Klein and his collaborator Horst von Sanden. The textbook was also used in some parts of East Asia. The basic principle of the textbook is mathematisation, and after WW II this provided the roots of the curriculum principle of "Extension and Integration" in Japan.

Along the same lines, Katalin Gosztonyi (Szeged, Hungary) showed the impact of Klein's work on today's mathematics education in Eastern Europe. She analysed the traces of Klein's influence on Hungarian mathematics education, from his student and colleague Emanuel Beke, through mathematicians at the mid-20th century such as Kalmár and Péter, until the reform of mathematics education led by Tamas Varga in the 1970s. Common aspects of their conception about mathematics are the emphasis on intuition, visuality, organic development, or cultural aspects of mathematics.

Final Remark

Felix Klein's life shows us that a sensitised person is always necessary in order to recognise problems, think in a visionary manner, and act effectively. The aim of the Thematic Afternoon, "The Legacy of Felix Klein" was to inform, think about, and discuss the meaning of the work and ideas of Felix Klein both currently and in the future. We wanted to show that many ideas that Felix Klein had can be reinterpreted in the context of the current situation and give some hints and advice for handling current problems in teacher education and teaching mathematics in high school. In this sense, old ideas stay young, but new people are always necessary to bring these ideas to life.

References

Coen, S. (Ed.). (2012). *Mathematicians in Bologna 1861–1960*. Basel: Birkhäuser.
Klein, F. (1893–94). Protocols of his mathematical seminars, Vol. 11. http://www.uni-math.gwdg.de/aufzeichnungen/klein-scans/klein/. Accessed September 21, 2016.
Klein, F. (Ed.). (1912–14). *Die Mathematischen Wissenschaften* (Die Kultur der Gegenwart, Part III, Abt. 1). Leipzig/Berlin: B.G. Teubner.
O'Connor, J. J., Robertson, E. F., & Munkholm, H. J. (2010). http://www-groups.dcs.st-andrews.ac.uk/history/Biographies/Heegaard.html. Accessed September 21, 2016.
Parshall, K. H., & Rowe, D. E. (1994). In *The Emergence of the American Mathematical Research Community 1876–1900: J. J. Sylvester, Felix Klein, and E.H. Moore*. USA/UK: American Math. Society/London Math. Society.
Schubring, G. (1989). Pure and applied mathematics in divergent institutional settings in Germany: The role and impact of Felix Klein. In D. E. Rowe & J. McCleary (Eds.), *The history of modern mathematics* (Vol. II, pp. 171–220)., Institutions and Applications Boston: Academic Press.
Tobies, R. (1989). Felix Klein als Mitglied des preußischen, Herrenhauses'. Wissenschaftlicher Mathematikunterricht für alle Schüler—auch für Mädchen und Frauen. *Der Mathematikunterricht, 35,* 4–12.
Tobies, R. (2002). The development of Göttingen into the prussian centre of mathematics and the exact sciences. In N. Rupke (Ed.), *Göttingen and the development of the natural sciences* (pp. 116–142). Göttingen: Wallstein.
Tobies, R. (2016). Felix Klein und französische Mathematiker. In Th. Krohn & S. Schöneburg (Ed.), *Mathematik von einst für jetzt* (pp. 103–132). Hildesheim: Franzbecker.

Open Access Except where otherwise noted, this chapter is licensed under a Creative Commons Attribution 4.0 International License. To view a copy of this license, visit http://creativecommons.org/licenses/by/4.0/.

Part V
National Presentations

Argentinean National Presentation

Esther Galina and Mónica Villarreal

In this text, we aim to offer a brief overview of education and mathematics education (ME) in Argentina and people and actions that have contributed to ME as it stands at present. It includes four topics: the characteristics of the current education system and some special programs, the popularization of mathematics and sciences, a brief reference to the mathematics teacher education system, and the development of mathematics and ME in the country.

Argentinean Education System

Structure and Statistics

Argentina is the second biggest country in South America; it consists of 23 provinces and an autonomous city, Buenos Aires. It has a population of 42 million people, with largest concentrations at the main cities. The total area of Argentina is 3,761,274 km^2, and the continental area is 2,780,091 km^2. The population according to the national census of 2010 was 40,117,096, with 30% between 0 and 17 years old, and the population density (excluding Antarctica) was 14.6 people per km^2.

In spite of the difficulties that such a large country presented for education, universal first-level education for children was implemented in the 19th century in the entire territory. There were problems related to access in the fulfillment of this initiative in some sparsely populated regions and in certain social sectors,

---

E. Galina (✉)
Unión Matemática Argentina (UMA), Consejo Nacional de Investigaciones Científicas y Técnicas (CONICET), Facultad de Matemática, Astronomía, Física y Computación, Universidad Nacional de Córdoba, Córdoba, Argentina
e-mail: galina@famaf.unc.edu.ar

necessitating implementation of a set of official policies to consolidate a universal thereby free compulsory education.

The current Argentinean education system has two main characteristics: It is universal and compulsory from kindergarten to secondary school. Kindergarten (2 years), primary (6 years), and secondary school (6 years) are compulsory. In Argentina, there are public and private schools. Compulsory education depends on each province and is free at public schools.

Since 1884, primary school has been universal, free, compulsory, and secular. In 1905, rural schools were created and boys and girls started sharing classes. In 1993, kindergarten (1 year) and the first three years of secondary school became also compulsory, for a total of 10 years of compulsory education. In 2006, the six years of secondary school became compulsory. Since 2015, one more year of kindergarten has been compulsory, for a total of 14 years of compulsory education.

Regarding the third and fourth education levels, 80% of the university students attend public (federal or provincial) universities. Argentinean public universities have been free for undergraduate studies since 1949. Their budget depends on the federal or state governments. They enjoy a very broad political autonomy, being self-governed mainly by professors and students. In mathematics, the undergraduate and graduate levels are well established and consolidated.

Although higher education levels are primarily concentrated at public or private universities, there are also many non-university tertiary public and private institutions. Many of them are devoted to teacher education.

Excluding university level, in 2014, 10,988,786 students were enrolled in formal common education (73% in public schools), 128,966 in special education (78% in public schools) and 1,240,496 in adult education (91% in public schools). In the same year, there were 1,871,445 university students, with 78% studying at public universities. The number of graduate students at universities was of 144,229, with 77% enrolled in public institutions.

Curricular Documents and Materials

Regarding the many difficulties associated with a non-centralized education system, since each province is autonomous in relation to education, in 2005, the Federal Council of Education (consisting of the National and all the Provincial Ministries of Education, and three representatives from the Council of Universities) approved the Núcleos de Aprendizajes Prioritarios (NAP; "Priority Learning Cores") for primary school. A NAP is a body of knowledge that should be part of the education for all children because it has subjective and social significance and the potential to provide a common basis to reverse situations of injustice. NAP's objectives are to create equal opportunities for access to knowledge in order to contribute to the full social integration of children and to support values that favor the common benefit, social life, shared work, and respect of differences. More recently, NAP for secondary school were also produced. As complement of the NAP, the National

Ministry of Education created the Cuadernos para el Aula, which include pedagogical orientations, discussions, and examples for teachers to approach the contents. Each provincial jurisdiction developed its curriculum based on NAP, its own policies, and contextual conditions.

The textbooks for compulsory education are predominantly published by private publishers, but the National Ministry of Education offers some orientations through the NAP and the Cuadernos para el aula for the teaching of mathematics at different levels of the compulsory educational system. Nevertheless, there are some official programs for distributing books at schools. Another important resource for teachers is National Ministry of Education's website (https://www.educ.ar/), which is oriented towards teachers, families, and students and contains teaching materials, blogs, news, activities, forums, suggestions, experiences, videos, etc.

Although we can say that there is an interesting set of non-compulsory teaching materials that encourage teachers to reflect on their practices and contribute concrete interesting activities, and although some of those materials emphasize the active role of the students as mathematics producers, this is not enough.

Special Educational Programs

Since 2003, the national government has implemented some special education programs such as the Centros de Actividades Infantiles (CAI; "Children's Activities Centers") and Centros de Actividades Juveniles (CAJ; "Youth Activities Centers"). In such centers, operating in public schools, systematic extracurricular activities of diverse types, such as cultural, artistic, recreational, collaborative, scientific, technological, and sportive activities, have been developed. In 2016, with a new government, the National Ministry of Education and Sports decided not to allocate more funds for these programs. Some provincial jurisdictions continue them using their own budgets.

Another important program was Mathematics for All (2010–2012), a program for training teachers and accompanying them during teaching. It was developed in 1700 primary schools and aimed to analyze teachers' practices and "do mathematics" at school using the NAP.

From our point of view, the most important special program has been Conectar Igualdad, which started in 2010. This national program aims at diminishing the digital gap among different population sectors. It includes the distribution of a netbook for each secondary school student and teacher and each student with special needs and their teachers in every public school in the country. As of July 2016, more than 5.5 million netbooks had been distributed. Some facts that contributed to the implementation of this program were the existence of diverse social, cultural, and economic conditions all over de country and the necessity of guarantying all students access to digital technologies (DT); the fact that the use of computers is a determinant for accessing some jobs; the lack of access to diverse sources of information; the fact that technological education does not only imply the use of computers; and the recognition that secondary education has more deficiencies than primary education.

The aims of the program are: (1) Reduce digital exclusion and marginality, (2) give students and teachers direct access to information without mediators, (3) incorporate DT for the construction of knowledge for different school subjects, and (4) improve the quality of public education.

The achievements of this program as of 2016 can be summarized as follows: improved technological literacy among teenagers and their families; greater access to information without mediators; and implementation of several courses for teachers, despite remaining difficulties in teaching with DT. The program installed discussions related to the use of technology in education and the relations among teachers, students, and families around this knowledge.

Popularization of Mathematics and Sciences

Mathematics competitions organized by the Argentinean Mathematical Olympiad (OMA) have had a strong presence in Argentina for many years but have currently lost some relevance. The popularization of mathematics has improved the last years. The effects of the popularization of mathematics accomplished by a variety of public programs and by Adrián Paenza[1] during the last decade in Argentina have been very important. After him, people that had never been attracted by mathematics started to read mathematics books for fun. Radio and television programs conducted by him introduce discussions of mathematics topics, and mathematics facts are treated as curiosities to be elucidated. He has been an initiator of the popularization of science in Argentina.

Encuentro Channel, a television channel belonging to the National Ministry of Education, has been another important medium for popularizing mathematics and sciences since 2007. Its objectives are: (1) Contribute to equity in access to knowledge for all inhabitants of Argentina and the countries of the region regardless of their place of residence or social status, (2) provide schools with television contents and multimedia to improve the quality of education in Argentina, and (3) offer innovative materials to facilitate and improve the processes of teaching and learning in the context of the current challenges for education in order to collectively construct a more equitable society.

In 2011, an important science, technology, industry, and art mega-exhibition called Tecnópolis was inaugurated by the national government. It offers innovative materials to facilitate and improve the processes of teaching and learning in the context of the current challenges for education in order to collectively construct a more equitable society. Other science fairs have been organized by universities.

---

[1] Winner of the 2014 ICMI Leelavati prize for outstanding contributions to increasing public awareness of mathematics as an intellectual discipline and the crucial role it plays in diverse human endeavors.

Popularization of mathematics has been possible thanks to national government policies designed to encourage the development of the sciences and technology and the democratization of scientific knowledge and to increase the number of scientists and science students. Argentina has started to change the scientific culture of its society.

Teacher Education System

Argentina has a number of non-university tertiary institutes and universities for teacher education. Kindergarten and primary school teachers are trained only in tertiary institutions. Secondary school mathematics teachers can study at tertiary institutions or at universities, but most attend non-university institutions. There are 193 tertiary institutes (149 are public) and 34 universities (29 are public) that train mathematics teachers for secondary school. The average number of graduates from each of the 193 non-university institutes in 2015 was 7.6. There is no data available for the average number of graduates from the 34 universities. There are substantial differences between the syllabuses at universities and tertiary institutions. New standards for mathematics teacher education at universities are currently being discussed.

Argentina has a mathematics teacher deficit, so some schools hire mathematics teachers who are students or people with professional degrees without pedagogical training.

Mathematics and Mathematics Education in Argentina

Figures 1 and 2 show timelines for the development of mathematics and ME in Argentina. They include the influence of great mathematicians such as Rey Pastor, Santaló, Levi, Monteiro, Villamayor, Calderón, and others.

Regarding ME, we can say that a mathematics education movement is developing in Argentina. Some of the evidence of this movement are:

- Creation of a variety of ME groups or associations. The Sociedad Argentina de Educación Matemática (SOAREM; "Argentinean Society of Mathematics Education") was created in 1998. The Argentinean Group of Mathematics Education (GAEM) was created in 2007. In 2015, the Argentinean Mathematical Union (UMA) created a special commission devoted to educational issues. In spite of the existence of these ME-related groups, we still do not have a comprehensive ME organization that includes everyone working in ME to coordinate objectives, actions, and meetings.
- Organization of diverse international and national ME meetings in Argentina, such as the XXVII Reunión Latinoamericana de Matemática Educativa (2013) and the VI Reunión de Didáctica de la Matemática del Cono Sur (2002). In the

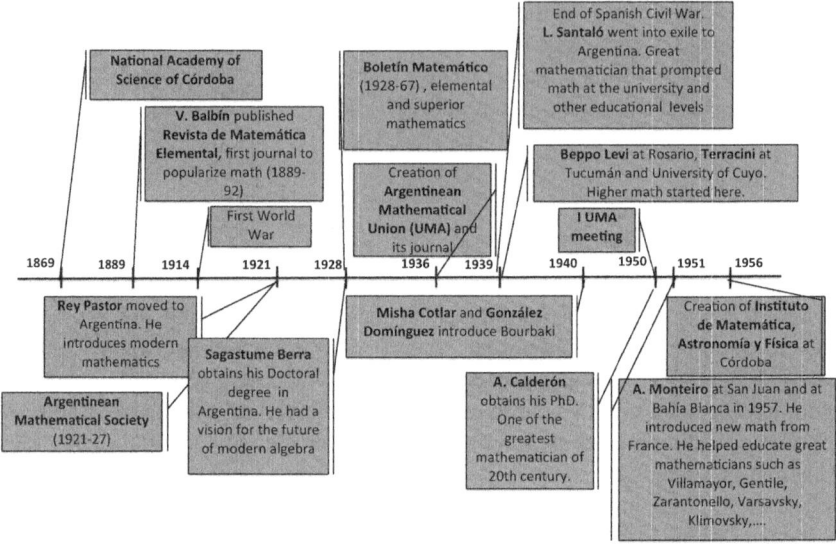

Fig. 1 Mathematics and ME in Argentina 1869–1956

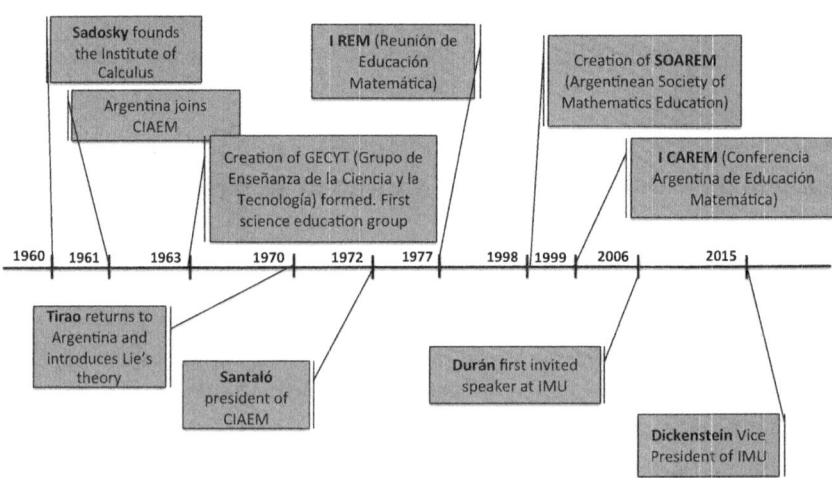

Fig. 2 Mathematics and ME in Argentina 1960–2015

local context, since 1977, UMA has annually organized the Reunión de Educación Matemática; since 1999, SOAREM has organized the Conferencia Argentina de Educación Matemática (12 meetings); since 2003, Jornadas de Educación Matemática by University of the Litoral (5 meetings); since 2000, the EDUMAT Civil Association has organized the Simposio de Educación Matemática (12 meetings); since 2005, the University of San Martín, University

Arturo Jauretche, and UNIPE have organized 7 editions of the Escuela de Didáctica de la Matemática; and since 2006, the University of La Pampa has organized the Reunión Pampeana de Educación Matemática (6 meetings). These are just some examples of local meetings.
- Publication of local journals and books devoted to ME topics. Journals: *Revista de Educación Matemática*, UMA, since 1982; *Revista Premisa*, SOAREM, since 1999; and *Revista Yupana*, University of the Litoral, since 2004. Books: Collection about Mathematics Teacher Education edited by Libros del Zorzal since 2005.
- Creation of many master's degrees and recently some doctoral programs related to ME.
- Visit of foreign researchers.
- Increasing number of research groups related to ME at different universities and tertiary institutes.
- Increasing participation in international meetings with less attendance to congresses in which the official language is not Spanish: Congreso Iberoamericano de Educación Matemática (CIBEM), Reunión Latinoamericana de Matemática Educativa (RELME), and Conferencia Interamericana de Educación Matemática (CIAEM). At ICME-13, there were 38 Argentinean authors for 42 presentations.

Final Remarks

- The idea of a popular, universal, free, and compulsory education is deeply rooted in Argentina.
- Approximately 70% of all educational levels are public.
- Argentina has a high quality set of curricular materials and programs for the popularization of mathematics, but they are still unavailable to certain populations.
- It is necessary to qualitatively and quantitatively improve teacher education.
- Efforts to improve ME to date are still not sufficient.
- There is an increasing community of mathematicians at the international level.
- Mathematicians and mathematics educators are improving their relationships, and universities and non-university institutions in charge of mathematics teacher education are improving their relationships as well.
- Research in ME is increasing both qualitatively and quantitatively.
- Financial support for research in ME mainly comes from public universities. There is very little support from the most important agencies, possibly because in such agencies the evaluating commissions include non-specialists in science and mathematics education.
- It seems likely that a new educational policy started with the new government in December, 2015. There is an uncertainty about the continuity of the educational programs started during the previous government (2001–2015) and implemented for more than 10 years.

Bibliography

Ministerio de Educación y Deportes. Dirección Nacional de Información y Estadística Educativa. Anuarios estadísticos. Argentina. http://portales.educacion.gov.ar/diniece/2014/05/24/anuarios-estadisticos/. Retrieved April 5, 2017.

Ministerio de Educación y Deportes. Instituto Nacional de Formación Docente. Argentina. http://portales.educacion.gov.ar/infd/. Retrieved April 5, 2017.

Ministerio de Educación y Deportes. Portal Educativo. Argentina. https://www.educ.ar/. Retrieved April 5, 2017.

Ministerio de Educación y Deportes. Secretaría de Políticas Universitarias. Departamento de Información Universitaria. Argentina. http://portales.educacion.gov.ar/spu/investigacion-y-estadisticas/. Retrieved April 5, 2017.

Scaglia, S., & Kiener, F. (2013). Aportes sobre el estado actual de la Educación Matemática en Argentina. *Revista Binacional Brasil Argentina, 2*(2), 25–47.

Sistema Federal de Medios y Contenidos Públicos. Encuentro. Canal de TV Educativo. Argentina. http://encuentro.gob.ar/. Retrieved April 5, 2017.

Sociedad Científica Argentina. (1972). In *Evolución de las ciencias en la República Argentina, 1923–1972*. Buenos Aires.

Villarreal, M., & Esteley, C. (2002). Una caracterización de la Educación Matemática en Argentina. *Revista de Educación Matemática, 17*(2), 18–43.

Open Access Except where otherwise noted, this chapter is licensed under a Creative Commons Attribution 4.0 International License. To view a copy of this license, visit http://creativecommons.org/licenses/by/4.0/.

Teachers' Professional Development and Mathematics Education in Brazil

Victor Giraldo

The Aims of the ICME-13 National Presentation

Brazil's National Presentation focused on research in mathematics education in the country and its relationships with projects and initiatives for the improvement of teacher education (especially those involving different communities: mathematicians, mathematics educators, and mathematics teachers), taking into account the challenges imposed by the Brazilian education scenario. The 90-minute presentation was accompanied by printed and interactive multimedia material exhibitions. This presentation was a joint initiative of the Brazilian Society of Mathematics Education (SBEM), the Brazilian Mathematical Society (SBM), the Brazilian Society of Applied and Computational Mathematics (SBMAC), and the Brazilian Statistics Association (ABE).

The Content of the Presentation

In Brazil, in recent decades, efforts have been made aiming at improving the conditions of school education, and teachers' pre-service and in-service professional development. Such efforts involve policies from federal, state, and local governments, as well as initiatives conducted by public institutions and scientific societies.

V. Giraldo (✉)
Federal University of Rio de Janeiro, Rio de Janeiro, Brazil
e-mail: victor.giraldo@ufrj.br

The Structure of the Presentation

We opened the talk with a very brief account of the Brazilian education context. The main focus, however, was on the projects led by the scientific societies that proposed this presentation: The purpose of this presentation was an overview of mathematics education in Argentina and the historic contributions that have converged to reach its present state. The presentation included three principal topics: the characteristics of the education system as it currently exists and some special programs; the teacher education system and its particularities; and the development of mathematics and mathematics education in the country, their historical context, and the current state of mathematics education as a research field.

The Structure of the Presentation (in Detail)

Towards education for all: A brief historical perspective of the Brazilian education system: Public policies, including public elementary and secondary education, inclusive education, multicultural education, and textbook assessment and distribution.

Integrating classroom practice into teacher education (20 min): Mathematics teachers' pre-service and in-service professional development in Brazil, including workshops and other activities and policies especially designed to bring teachers education closer to classroom practice.

Integrating research into teachers' practices: An overview of research and graduate programs in mathematics education, professional graduate programs for teachers, research journals, professional journals, conferences for teachers, and their impact on mathematics teachers' professional development in Brazil.

The Klein Project in Brazil and its outcomes: Workshops for teachers, books, and multimedia instructional materials.

Collaboration between mathematicians and mathematics educators: Development of textbooks and other instructional materials for school mathematics and for pre-service and in-service mathematics teacher education.

An international integration project: The Mathematical Space in Portuguese Language (EMeLP), an international organization of the Portuguese-speaking countries affiliated with ICMI: foundation, initiatives, and meetings.

Closing: Future perspectives and challenges to overcome.

Open Access Except where otherwise noted, this chapter is licensed under a Creative Commons Attribution 4.0 International License. To view a copy of this license, visit http://creativecommons.org/licenses/by/4.0/.

Mathematics Education in Ireland

Maurice OReilly, Thérèse Dooley, Elizabeth Oldham and Gerry Shiel

Introduction

Ireland is a small country on the western edge of Europe, independent since 1922 (while Northern Ireland remained part of the United Kingdom). It has been a member of the European Union (EU/EEC) since 1973.

Although pre-schooling is not compulsory in Ireland, each child is now entitled to two years of education at this level. Primary education normally begins when children are about four or five years old and continues for eight years, while post-primary education follows this for five or six years. Public higher education has two main sectors, namely the seven universities and 14 institutes of technology. Post-primary education is divided into two cycles: the three-year Junior Cycle, and the two- or three-year Senior Cycle, the duration of which depends on whether an optional Transition Year is taken at the beginning of this cycle. High-stakes examinations (the Junior and Leaving Certificates) are offered at the end of the Junior and Senior Cycles. In 2016, 92.7% of all Leaving Certificate students took mathematics at one of three available levels, higher, ordinary or foundation. Accommodation is made for schooling through the medium of Irish (Gaeilge), catering for about 6% of students.

There are two models of initial teacher education (ITE), concurrent and consecutive. The concurrent model requires a student to take a Bachelor of Education degree, while the consecutive model involves a primary degree followed by a Professional Master of Education. Since 2006, the Teaching Council has regulated the teaching profession. The seven universities (either directly or through colleges of education) and one private college deliver ITE. Primary teachers are qualified to teach all curriculum areas, while post-primary teachers qualify (typically) in two subjects. In addition, for mathematics, the University of Limerick delivers a

M. OReilly (✉)
St Patrick's Campus of Dublin City University, Dublin, Ireland
e-mail: maurice.oreilly@dcu.ie

Professional Diploma in Mathematics for Teaching (PMDT) to address the historical problem of 'out-of-field' teachers in the subject.

Transition issues between education sectors have been receiving significant attention in recent years, and in relation to mathematics, in particular. The key transitions have been identified from pre-school to primary, from primary to Junior Cycle in post-primary school, from Junior to Senior Cycle, and from post-primary to higher education. There are active policy innovations and research on all of these transitions, and indeed on identifying the barriers to progression from further education and training to higher education.

Outside the formal education sector, many initiatives support the learning of mathematics and mathematical endeavour generally. Among these is Maths Week Ireland which has been held every year since 2006. This celebration of mathematics involves young and old alike, participants ranging from primary school children to Fields medallists. In 2016 there were over 300,000 participants, making it the largest such event in the world. Mathematics professionals are supported by academic and professional organisations, especially the Royal Irish Academy (RIA, founded in 1785), the Irish Mathematics Teachers' Association (IMTA, 1964), the Irish Mathematical Society (IMS, 1976) and the Irish Applied Mathematics Teachers' Association (IAMTA, 2006).

Policy

There are several key players involved in the development and implementation of policy in mathematics education:

- The Department of Education and Skills (DES, www.education.ie) including its Inspectorate
- Agencies of the DES:
    - The National Council for Curriculum and Assessment (NCCA, www.ncca.ie)
    - The State Examinations Commission (SEC, www.examinations.ie)
    - The Educational Research Centre (ERC, www.erc.ie)
- Stakeholders (teacher unions, subject associations, such as IMTA and IAMTA)
- Support and advocacy agencies:
    - The Professional Development Service for Teachers (PDST, www.pdst.ie)
    - National Adult Literacy Agency (NALA, www.nala.ie).

In the years 2003–2015, the overall performance of Ireland (with OECD average in parentheses) in PISA mathematics was 503 (500), 502 (498), 487 (499), 502 (496) and 504 (490), with performance significantly above the OECD average in 2012 and 2015. A particular area of weakness for students in Ireland in PISA is the Space and Shape content area, where performance has been below OECD average levels.

In 2011, the DES published Literacy and Numeracy for Learning and Life (DES, 2011), its national strategy for improvement in these areas for 2011–2020. It presented a broad set of measures designed to enhance the teaching and learning of mathematics, including an increase in instructional time at primary level and Transition Year, and a stated intention to enhance CPD in mathematics (via the Teaching Council). As part of the strategy, a requirement was introduced for primary schools to submit aggregated standardised test results to the DES at specified grade levels annually. Other actions included: (i) participation in TIMSS in 2015 (the first time since 1995 for second year post-primary); (ii) from 2012, for primary level, an extension of the (concurrent) B.Ed. degree programme from three to four years, and the (consecutive) PME from 1.5 to two years, each with a stronger emphasis on mathematics education; (iii) the introduction from 2012 of school self-evaluation and development planning; and (iv) the revision of the primary mathematics curriculum from 2016. In the five years from 2009 to 2014, there were significant increases in average scores in second and sixth classes (primary) in the National Assessment of Mathematics. Consistent with this, students in fourth class (9–10 year olds) in Ireland in TIMSS 2015 achieved a significantly higher mean score than in 2011 and 1995. Some progress has been made in PISA 2015 in achieving targets for PISA mathematics, but the proportions achieving the highest PISA proficiency levels are still relatively low. Nevertheless, the proportion taking the higher level mathematics at Leaving Certificate continues to rise (perhaps because bonus points for entry to higher education are available to most of these students since 2012).

Data from the 2013 PIAAC study showed that Ireland had a mean score that was significantly below the average for participating countries on numeracy, and that one-in-four adults performed at or below Proficiency Level 1 (the lowest level). NALA supports adults with numeracy (and literacy) difficulties, and published a framework for meeting the professional development needs of tutors of adult numeracy in 2015.

A report by the government-appointed STEM Education Review Group (2016) has made a number of recommendations designed to enhance engagement of students in STEM courses including mathematics, and ultimately, in STEM-related careers. The DES will present its own STEM Strategy in 2017.

Curriculum

Curricula and examinations are ultimately the responsibility of the DES, NCCA and SEC as mentioned above. However, when curricula are revised, various stakeholders are involved in negotiations, with the teacher voice in general strongly represented. There is a free market for textbooks, often written by practising teachers, and schools have relative freedom on timetabling and on the time allocated to individual subjects.

The present curricula (www.ncca.ie/en/Curriculum_and_Assessment/) are best understood in historical context. Two frameworks are helpful here. The first distinguishes levels of curriculum, focusing on how they can differ. These levels are: intended (decreed typically by the state), implemented (taught by teachers), and attained (learnt by students). The second refers to types of curriculum, reflecting different views of mathematics and mathematics education. Relevant types here are empiricist (emphasising moving from the perceived world to that of symbols), structuralist (building up structures within mathematics itself) and mechanist (rules without reasons).

At primary level, a curriculum reflecting the work of Piaget and emphasising discovery learning was introduced in 1971; it can be classified as empiricist with some structuralist features. When it was revised in the 1990s, issues addressed included poor attainment of higher-order objectives; the revised version—reflecting world-wide trends—has greater emphasis on problem-solving and applications in real-world contexts. As regards content, algebra and data are 'strands' throughout the curriculum; probability (or rather 'chance') is included in the higher grades; calculators are introduced in fourth class. A redeveloped mathematics curriculum for 3–8 year-olds (from early years to junior primary) is currently being prepared; the underpinning research reports are discussed below.

At post-primary level, major changes in the 1960s involved adoption of curricula much influenced by 'modern' mathematics, hence archetypally structuralist. They were intended to aid understanding and counteract mechanist teaching. However, implementation was challenging for teachers not attuned to the material; also, the content was too abstract for many students, especially with rapidly increasing retention to Leaving Certificate level. Successive partial revisions through to 2000 —although deeply considered—were largely pragmatic, gradually removing content that was not well implemented or attained, and adding syllabuses for lower attainers. However, mechanist implementation remained a concern, as did teaching and learning over-focused on excessively predictable state examinations.

Dissatisfaction with students' performance, including the moderate PISA 2000 scores, prompted a root-and-branch revision, preceded by surveys of research and practice internationally (for example, Conway & Sloane, 2005). The initiative is known as 'Project Maths'. Revised curricula, phased in from 2008, are more empiricist and less structuralist; the state examination papers now focus strongly on solving problems set in real-world contexts. As regards content, there is increased emphasis on probability and statistics, but less calculus. To support implementation, enhanced professional development was provided, encompassing constructivist pedagogy and, latterly, mathematical content for 'out-of-field' teachers via the PMDT as mentioned above. The Project Maths initiative—incorporating a somewhat altered model of negotiation with stakeholders, a culture revolution with regard to predictable examining, and initial introduction of changes with little lead-in time before the high-stakes Leaving Certificate examinations—has been controversial; its impact will not be evident for some time.

Research

In recent years, STEM research has become conspicuous at a national level, and international exchanges are vibrant. Foci of research in mathematics education in Ireland encompass a broad spectrum. General themes include teaching and learning (at all levels of education), mathematical knowledge for teaching, and exploration of attitude and identity relating to mathematics and its teaching and learning. There are several 'cohort-focused' areas of interest, including mathematics in early childhood, with adult learners, in bilingual contexts, and with diverse international groups. An increasing number of centres are now active in mathematics education research in Ireland. Some have a focus on STEM teaching and learning (for example, EPI-STEM, in Limerick, and CASTeL and STEM-ERC, both in Dublin) while others have a broader remit but include STEM as an area of interest (for example, CRITE, in Dublin). SCoTENS is an all-Ireland (Republic and Northern Ireland) network that has supported many research initiatives in mathematics education. In a European research context, Irish researchers have participated in the Fibonacci Project (2010–2013) which focused on inquiry-based teaching and learning methods in STEM (primary and post-primary). The ERC (mentioned above) has produced national and international studies relating to mathematics education, including reports on PISA and TIMSS, national assessments of mathematics achievement (primary), and evaluation of mathematics programmes.

Since 2004, two series of conferences on mathematics and science education have been held in Dublin, MEI in odd years (2005–2015) and SMEC in even years (2004–2016). Although these series began as national conferences, they have attracted increasing numbers of international participants in recent years, with SMEC 2014 organised jointly with the SAILS FP7 project and MEI 2015 jointly organised with BSRLM. The Irish Mathematics Learning Support Network (IMLSN) has sustained, since 2006, a series of conferences on learning support in mathematics across higher education. The Project Maths Development Team has held three conferences supporting post-primary curricular reform, and several WIMDI (Women in Mathematics Day Ireland) conferences have showcased the work of women in mathematics and in mathematics education. International conferences held in Ireland have included the International Association for Statistical Education (IASE, 2011) and Adults Learning Mathematics (ALM, 2016), while CERME (2017) is forthcoming.

In 2014, the NCCA published two research reports to support the review and redevelopment of the primary school mathematics curriculum for 3–8 year olds. The first report focuses on theoretical aspects underpinning mathematics education for young children while the second is concerned with pedagogical implications. In common with policy documents on mathematics in many countries, mathematical proficiency (NRC, 2001) is identified as a key aim for mathematics education (Dunphy et al., 2014). Attention is given to goals, processes and content, critical transitions and learning paths, as means of achieving proficiency. New research themes emerging in Irish mathematics education include the interplay between outcomes of assessment and practice, dialogic pedagogy of argumentation and discussion, identification of critical ideas for development of key concepts, transitions across education settings, design of rich and challenging mathematical tasks, and equity and access.

References

Conway, P. F., & Sloane, F. C. (2005). *International trends in post-primary mathematics education: Perspectives on learning, teaching and assessment* (NCCA Research Report No. 5). Dublin, Ireland: National Council for Curriculum and Assessment.

DES. (2011). *Literacy and numeracy for learning and life: The national strategy to improve literacy and numeracy among children and young people 2011–2020*. Dublin: Author.

Dunphy, E., Dooley, T., & Shiel, G. with Butler, D., Corcoran, D., Ryan, M., Travers, J., & Perry, B. (International Advisor). (2014). *Mathematics in early childhood and primary education (children aged 3–8 years): Definitions, theories, stages of development and progression* (NCCA Research Report No. 17). Dublin, Ireland: National Council for Curriculum and Assessment.

National Research Council (NRC). (2001). Adding it up: Helping children to learn mathematics. J. Kilpatrick, J. Swafford, & B. Findell (Eds.), Mathematics learning study committee, center for education, division of behavioral and social sciences and education. Washington, DC: The National Academies Press.

STEM Education Review Group. (2016). *STEM education in the Irish school system*. Dublin: DES.

Open Access Except where otherwise noted, this chapter is licensed under a Creative Commons Attribution 4.0 International License. To view a copy of this license, visit http://creativecommons.org/licenses/by/4.0/.

National Presentation of Japan

Toshiakira Fujii, Yoshinori Shimizu, Hanako Senuma
and Toshikazu Ikeda

Japan was honored to be invited to make an ICME-13 national presentation. Japan's National Presentation consisted of a 90-minute National Presentation session and a 5-day Japan Booth. These were organized by the Japan Society of Mathematical Education (JSME) with the help of other organizations. In the National Presentation session, first, a brief history of Japanese mathematics education and the next curriculum were described. Second, Japanese mathematics traditional problems were demonstrated actively using sets of interesting and unique problems involving wasan and origami. Third, two major impacts of TIMSS and PISA on Japanese mathematics education were described. Finally, key factors embedded in the Japanese model of Lesson Study were clarified based on the IMPULS project. At the Japan booth, Japanese mathematics textbooks, curricula, and books were displayed, and 11 posters explaining interesting and unique problems were distributed.

Chapter 1 Structure of Presentation Session

Japan's National Presentation session took place on Saturday, July 30, 2016, from 16:30–18:00. Approximately 120 people attended. The structure of presentation session was as follows:

1. Opening address
   (Toshiakira Fujii, President of the Japan Society of Mathematical Education)
2. Japanese education systems, Japanese mathematics textbooks, and next curriculum
   (Hanako Senuma, Toshikazu Ikeda)

T. Fujii (✉)
Tokyo Gakugei University, Koganei, Japan
e-mail: tfujii@u-gakugei.ac.jp

3. Active demonstrations; Let's give it a try! Interesting and unique Japanese mathematical problems involving *wasan* and *origami*
   (Members of the liaison section of the Japan Society of Mathematical Education: Masakazu Okazaki, Shinya Itoh, Nagisa Nakawa, Akihiko Saeki, and Kensuke Koizumi)
4. Impact of international mathematics achievement tests (TIMSS and PISA) on Japanese mathematics education
   (Keiko Hino)
5. Japanese Lesson Study
   (Toshiakira Fujii)
6. Closing address
   (Masataka Koyama, Vice-President of the Japan Academic Society of Mathematics Education).

Summary of Lectures and Demonstrations in National Presentation Session

Japanese Education Systems and Japanese Mathematics Textbooks and Other Information

Presenter: Hanako Senuma, Tamagawa University, Japan

Summary: The presentation was divided into five parts. The first part discussed the role of the JSME, established in 1919, and the liaison section of JSME, established in 1982. The second part discussed features of Japanese mathematics education from the results of SIMS and TIMSS 1995, 1999 international studies. The third part discussed characteristics of mathematics educational systems and textbooks based on the results of the International Comparison of Mathematics Textbooks in 11 Countries (Nagasaki, 2009). Fourth was a brief history of the Japanese educational system and snapshots of international seminar/conferences held in Tokyo: the United States-Japan seminar on mathematical education in 1971, the ICMI-JSME regional conference in 1983, and the ICME 9 in 2000. Finally, photos of Japan booth were shown and thanks were addressed to all who came to the booth and presentation.

Next Curriculum

Presenter: Toshikazu Ikeda, Yokohama National University, Japan

Summary: The next curriculum will be introduced in 2020 in elementary school, in 2021 in junior high school, and in 2022 in senior high school. Generic skills will be focused on for all subjects and three principles (basic knowledge and skills, mathematical thinking and representation, and meaning and willingness to study) will be set to organize the curriculum. In mathematics, one of the main issues has been how to describe mathematics thinking based on development stages from elementary school to senior high school. In terms of mathematical activity, two types of mathematization will be described as a slogan.

Let's Give It a Try! Japanese Traditional Geometric Tasks: Family Crest Clipping and Origami Crane

Presenter: Masakazu Okazaki, Okayama University, Japan

Summary: The presenter introduced family crest clipping and origami cranes as Japanese traditional geometric tasks (The Association of Mathematical Instruction, 1994). Both tasks include the activities of imagining the completed figure, drawing its design on paper, and checking the geometric relations between the design and the completed figure by folding and unfolding the paper. We believe that these activities contribute to enhancing students' higher-level geometric thinking while enjoying the activity.

Let's Give It a Try! Making Patterns with Triangular Pieces

Presenter: Shinya Itoh, Kanazawa University, Japan

Summary: The presenter introduced a mathematical task from a figure drawn on the cover of a green textbook called *Jinjo Shogaku Sanjutsu* that was edited before World War Two by Naomichi Shiono. He advocated the development of *suri shisoh*, which is an independent-minded approach to cultivating students' ability to independently observe and interpret everyday phenomena mathematically. The task was the final of the textbook for Grade 6, dealing with patterns involving triangular pieces. It involved students determining the sum of areas of figures consisting of smaller triangles arranged in a particular way. The presenter introduced it with *origami* windmills.

Let's Give It a Try! Rearrange a Regular Pentagon

Presenter: Nagisa Nakawa, Tokyo Future University, Japan

Summary: The presenter introduced a mathematical task from a green textbook for Grade 2 published in 1938. This series of textbooks was innovative at that time, stimulating children's scientific and mathematical thinking, and it has also influenced current mathematics education in Japan. The task was to cut a regular pentagon into four parts and rearrange the four pieces of paper to make up figures such as a rectangle, a parallelogram, and two types of trapezoids. The audience used the actual material and solved it during the presentation.

Let's Give It a Try! Tachiawase

Presenter: Akihiko Saeki, Naruto University of Education, Japan
  Summary: The presenter introduced two mathematical problems, *tachiawase*, from *wasan*. *Wasan*, a mathematics native to Japan, was established in the beginning of the Edo period. *Tashiawase* involves making a different geometrical figure by dividing the original geometrical figure into several parts and recombining them. The first problem presented was from Masashige Yamada (1657) involving making a square by dividing a rectangular woolen cloth into two parts and recombining them. The second problem, from Genjyun Nakane (1743), was making a square by dividing two different sizes of squares into three parts and recombining them. The people of the Edo period learned the Pythagorean theorem through *tachiawase*. At the end of presentation, the presenter illustrated that the logo of ICME-13 was constructed using the Pythagorean theorem.

Let's Give It a Try! Sashigane: *Japanese Traditional Ruler*

Presenter: Kensuke Koizumi, Takasaki University of Health and Welfare, Japan
  Summary: The presenter introduced the Japanese traditional ruler called the *sashigane*. The main topic in this presentation was how to use the *sashigane* and how to make use of it in real world. The *sashigane* is a rectangular ruler that includes several scales used by carpenters. For example, it has a *kakume* scale, whose unit is the square root of 2 mm. Carpenters can easily determine the size of a square that can be obtained from a log using *kakume*. For this reason, it is indispensable for carpenters and is said to be one of the carpenters' three treasures.

Impact of International Mathematics Achievement Tests (TIMSS and PISA) on Japanese Mathematics Education

Presenter: Keiko Hino, Utsunomiya University, Japan
  Summary: The two major impacts of TIMSS and PISA on Japanese mathematics education were described (Ginshima & Matsubara, 2012; Nagasaki & Senuma, 2002). The first was the introduction of "mathematical activity" as a core of curriculum design. In the current objectives of mathematics education, "mathematical activity" is placed at the head of the statement. In this presentation, examples of

mathematical activities were given using examples of textbook problems. The second was the implementation of the National Assessment of Academic Ability. Two types of problems are used for this assessment. One type of problem is oriented towards "knowledge" and the other is oriented towards "application." The aim of this test is not only assessing the status quo of students but also giving feedback to schools and teachers. In the presentation, an example of the idea of conducting mathematics lessons using a problem item from the assessment was shown.

Japanese Lesson Study

Presenter: Toshiakira Fujii, Tokyo Gakugei University, Japan

Summary: Lesson Study is an approach to teacher professional development that differs sharply from the professional development practices common in other countries. While the history of Lesson Study in Japan spans more than a century, for Japanese educators, Lesson Study is like air, part of everyday school life. Educators outside Japan, however, having had to learn about Lesson Study less naturally, may sometimes fail to grasp some important aspects of Lesson Study (Fujii, 2016). This presentation tried to clarify key factors embedded in the Japanese model of Lesson Study based on my experience with the IMPULS project (International Math Teacher Professionalization Using Lesson Study) (Fig. 1).

Japan Booth

Japanese mathematics textbooks, curricula, and books were displayed and posters explaining *wasan* and *origami* were distributed from July 25–30, 2016. The Japan booth was mainly conducted by members of the liaison section of the Japan Society of Mathematical Education: Hanako Senuma, Toshikazu Ikeda, Takuya Baba,

Fig. 1 After Japan National Presentation

Akihiko Saeki, Masakazu Okazaki, Shinya Itoh, Takashi Kawakami, Tetsushi Kawasaki, Kensuke Koizumi, Kosuke Mineno, and Nagisa Nakawa, with the help of Satoru Sakanashi and graduate students from Yokohama National University, etc. (Figs. 2 and 3).

Big Four Mathematics Education Organizations in Japan

- The Japan Society of Mathematical Education (JSME; http://www.sme.or.jp/)
- The Japan Academic Society of Mathematics Education (JASME; https://www.jasme.jp/)
- Mathematical Education Society of Japan (http://mes-j.or.jp/)
- The Association of Mathematical Instruction (1994) (AMI; http://www.ne.jp/asahi/math.edu/ami/).

Fig. 2 Mathematics textbooks and posters

Fig. 3 Welcome to the Japan booth!

References

Fujii, T. (2016). Designing and adapting tasks in lesson planning: A critical process of lesson study. *ZDM Mathematics Education*. doi:10.1007/s11858-016-0770-3.

Ginshima, F., & Matsubara, K. (2012). Japan. In I. V. S. Mullis, et al. (Eds.), *TIMSS 2011 encyclopaedia: Education policy and curriculum in mathematics and science*. Vol. 1 (pp. 469–483). Boston College.

Nagasaki, E. (Ed.) (2009). *International Comparison of Mathematics Textbooks in 11 Countries*. National Institute for Educational Policy Research, Japan. (in Japanese) http://www.nier.go.jp/seika_kaihatsu_2/ risu-2-300_suugaku.pdf. Accessed: July 15, 2016.

Nagasaki, E., & Senuma, H. (2002). TIMSS mathematics results: A Japanese perspective. In D. F. Robitaille & Beaton, A. E. (Eds.), *Secondary analysis of the TIMSS data*. (pp. 81–93). Kluwer Academic Publishers.

The Association of Mathematical Instruction (1994). *Origami Sansu and Origami Suugaku*. Kokudosha: Japanese.

Open Access Except where otherwise noted, this chapter is licensed under a Creative Commons Attribution 4.0 International License. To view a copy of this license, visit http://creativecommons.org/licenses/by/4.0/.

National Presentations of Lower Mekong Sub-region Countries

Fidel R. Nemenzo, Masami Isoda, Maitree Inprasitha,
Sampan Thinwiangthong, Narumon Changsri, Nisakorn Boonsena,
Chan Roth, Monkolsery Lin, Souksomphone Anothay,
Phoutsakhone Channgakham, Nguyen Chi Thanh,
Vũ Như Thư Hương and Phương Thảo Nguyễn

Mathematics educators have tried to take responsibility in improving the educational situations in the region. Among the countries in Lower Mekong sub-region, we have been able to gradually create collaboration through a network of a number of collaborative projects and study in mathematics education programs. The mathematics education community in this region was able to be established through the leading role played by Thailand in collaboration among the countries of the region. We should set long-term shared goals to solve the common problems in mathematics education and keep and expand the collaboration for better education in our Lower Mekong sub-region. The presentation consists of three parts: Overviews of (1) mathematics education in each country in the Lower Mekong sub-region (Cambodia, Laos, Myanmar, Thailand, and Vietnam), (2) the establishment of societies of mathematics education and development in each country, and (3) the emergent mathematics education community in the Lower Mekong sub-region.

An Overview of Mathematics Education of Each Country in the Lower Mekong Sub-region

In the first part, an overall summary of mathematics education of each country will be presented.

OECD and UNESCO studies have found that Thai teachers are not being prepared well in teacher education programs and lack nationwide continuing

---

F.R. Nemenzo
University of the Philippines, Manila, Philippines
e-mail: fidel@math.upd.edu.ph

professional development. Moreover, half of the students are not acquiring basic skills in learning (OECD/UNESCO, 2016). In mathematics education in particular, teaching mathematics in Thailand for most teachers means preparing lesson plans by themselves, teaching those lesson plans in their closed classroom, checking the assigned homework, making some quizzes, and prescribing exercises (Inprasitha, 2003, 2015). Inprasitha (2003, 2016) has proposed a paradigm change in teaching approach from the traditional approach to be an open approach incorporating lesson study.

Mathematics education in Cambodia has many problems, for instance, a lack of well-qualified teachers and knowledge in curriculum development, textbook writing, use of Information and Communication Technology (ICT), and methodology of teaching (Roath, 2015). Most classes in schools are large, which are difficult for teachers control; however, the main cause of problems has been teachers' lack of effective methodology in teaching (International Mathematical Union, 2013).

Laos is also struggling in how to improve classroom teaching practice. Classroom observation has shown that the teachers' method involves lecturing and copying the lesson on the blackboard in the front of the classroom and asking questions, while students are passive learners, doing practical exercises (UNESCO, 2011). Laos also lacks continuous professional development and pre-service training (Benveniste et al., 2007). Laos is trying to improve its mathematics education; the goals of the mathematics curriculum education in Laos have been to ensure that students develop mathematical knowledge and skills that they can apply in other subjects and use in higher levels of study and have emphasized knowledge, skills, and attitudes (Thipmany, 2016).

General mathematics lessons in Vietnam involve most of class time for teaching the whole class, 90% for explaining and illustrating methods to the whole class, 8% for working individually, and 2% for working in groups. Reform of teaching methods in Vietnam started in 1992, with a focus on helping teachers to be aware of the need to improve teaching. The focus of the mathematics curriculum in 2000 was to provide students with basic applicable knowledge and skills for living in the community and for future study (Do, n.d.). Among the Lower Mekong sub-region countries, Vietnam does the best in mathematics education. The previous and most recent PISA results shows that Vietnam is in the top 10 of the 72 participating countries.

Establishment of a Society of Mathematics Education and Development in Each Country

This part describes the history of societies of mathematics education and their contributions in each country. This session reveals their roles and impacts on policies in their countries. Transformations of mathematics education will be described.

The Cambodian Mathematical Society (CMS) was established on March 4, 2005, and recognized by the Royal Government of Cambodia. It plays a part in addressing the problems and improving the capacity of mathematics education in Cambodia as well as in the region. In 2013, the Thailand Society of Mathematics Education (TSMEd) was established by mathematics educators in Thailand and has organized an annual conference on mathematics education since 2013. In addition, the Center for Research on International Cooperation in Educational Development (CRICED) of the University of Tsukuba and the Center for Research in Mathematics Education (CRME) of the Faculty of Education, Khon Kaen University, have had a significant collaboration in creating the APEC Lesson Study project from 2006 until the present. This project has addressed the issue of innovations in mathematics teaching in the Asia-Pacific Economic Cooperation (APEC) member economies. The CRICED is a core center of Ministry of Education, Culture, Sports, Science and Technology (MEXT) that is jointly supported by Ministry of Foreign Affairs of Japan, while the CRME is supported and facilitated by Khon Kaen University and other related research funds in Thailand. In 2012, this collaborative work between Japan and Thailand was officially commended at the Fifth APEC Education Ministerial Meeting for supporting human-resource development focusing on developing in-service teachers using Japanese Lesson Study in 19 APEC economies.

Thailand will play an important role in education development in the countries in the Lower Mekong sub-region as a key region in Association of Southeast Asian Nations (ASEAN) community. A long-term collaboration between Thailand and Japan in contribution to mathematics education in this region is an example of education development in the subsequent ASEAN movement.

As the Lower Mekong sub-region is a strategic area for new economic development in the ASEAN community, mathematics is very important to support this development.

Emergent Mathematics Education Communities in the Lower Mekong Sub-region

The third part of the presentation discusses emergent mathematics education communities in the region. Collaboration across the countries is very important in order to improve the educational situations in the region. It provides the possibility of solving educational problems through networking and sharing and learning from each other.

The Faculty of Education at Khon Kaen University (KKU) in Thailand launched the Hoshino Project for training mathematics teachers in the Lao PDR since 2003. In the 2004 academic year, four trainees funded by the Education for Development Fund (EDF) Project studied in the master's degree program in the Department of Mathematics Education and Science Education at KKU. From 2005 to the present, the Department of Mathematics Education in the Faculty of Education at KKU has provided master's degree scholarships for 13 mathematics teachers from the Lao PDR through the EDF Project and the KKU-EDU Partnership Project (7 of the 13 have graduated and are working as network teachers and educators at KKU). From 2012 to the present, three bachelor's degree and three master's degree students from Cambodia received the Her Royal Highness Princess Maha Chakri Sirindhorn Scholarship to study in the mathematics education program at KKU. In this way, the mathematics education network across the countries has gradually been established by study in the mathematics education program at KKU.

In 2003, the Higher Education Commission ordered academic departments to create a cooperative research network (CRN) in Thailand. Mathematics education was separated from the mathematics field and the Cooperative Research Network in Mathematics Education (CRN-MathEd) was created. The Faculty of Education at KKU started a master's degree program in mathematics education program and received a grant from the Project to Support the Competency for Competition of Thailand. This was the starting point for producing young mathematics educators in Thailand. From 2006 until the present, KKU and the University of Tsukuba have run the APEC Lesson Study Project. This project has contributed to the mathematics education community in Thailand. The work of this project since 2006 has contributed to the community through the book entitled *Lesson Study: Challenging in Mathematics Education*, edited by Maitree Inprasitha, Masami Isoda, Patsy Wang-Iverson, and Ban Har Yeap. The first two authors have been the overseers of the APEC Lesson Study since the beginning.

In 2007, the first group of graduate students participated and presented at the fourth East Asia Regional Conference on Mathematics Education (EARCOME 4) in Malaysia. In 2010, KKU graduate students joined EARCOME 5 in Japan. In 2013, Thailand was the host of EARCOME 6 and KKU provided the chance for Laotian and Cambodian teachers to join the conference. During the conference, the meeting for the Capacity and Networking Project (CANP) was organized by Bill Barton and colleagues and Maitree Inprasitha. The CANP is an international development project sponsored by the International Commission on Mathematical Instruction (ICMI) and the International Mathematical Union (IMU). The main aim of CANP3 is for participants from the different countries in Mekong area to work together and form a regional network in mathematics education and mathematics. This meeting led to the first CANP workshop in October 2013 in Cambodia. It engaged young mathematics educators from Cambodia, Laos, Thailand, and Vietnam as the trainees. In 2013, the Thailand Society of Mathematics Education was established by mathematics educators in Thailand, and it organized the first conference on mathematics education in January 2015 at KKU.

In May 2015, Thai mathematics educators and graduate students participated in EARCOME 7 in the Philippines. Graduate students from Cambodia, Laos, and Thailand were supported by a grant from ICMI and Bill Barton and Yeap Ban Har to join the conference and the second workshop of CANP during EARCOME 7.

In June 2015, KKU provided a scholarship for Cambodian and Laotian graduate students to attend the ICMI Study 23 in Macau, China. During the ICMI Study 23, Ferdinando Arzarello, President of ICMI, arranged a meeting for CANP representatives and members from different regions, including some IPC members, to discuss forming a regional network in mathematics education and mathematics. In addition, in November 2015, Thailand had the great opportunity to host two joint international conferences, the World Association of Lesson Study (WALS) and 10th APEC Lesson Study International Symposium at KKU. We made every effort to involve participants from Vietnam and Myanmar in the conference and the third workshop of CANP.

In 2016, Lower Mekong sub-region countries developed and carried out several activities. Khon Kaen University shared ideas from the APEC Lesson Study Project with the Lower Mekong sub-region countries. We conducted the workshops in "Innovations on Teaching for Higher-Order Thinking in Mathematics for Teachers: Lesson Study and Open Approach" that were held June 13–14, 2016, at the National Institute of Education, Phnom Penh, Cambodia, and August 27–29, 2016, at Pakse Teacher Training College (Pakse TTC), Champasak Province, Lao PDR. In addition, the Institute for Research and Development on the Teaching Profession for ASEAN, KKU (IRDTP), and the National Center for Teachers' Development (NCTD) in Japan made an agreement to improve teacher professional development in Lower Mekong sub-region.

With efforts over the last 15 years, the Faculty of Education at KKU in Thailand has produced a number of mathematics educators and young mathematics educators who have to take responsibility for improving the educational situations in the region. Among the countries in Lower Mekong sub-region, we have been able to gradually create collaboration through a network of a number of collaborative projects and study in the mathematics education program at KKU. The mathematics education community in this region was established with Thailand playing the leading role in collaboration in the region. In October 2016, the Department of Teacher Education in the Lao PDR signed an memorandum of understanding (MOU) with Khon Kaen University to send five students to study in PhD programs, including mathematics education, for the next 10 years at KKU with the support of the Australian government through the Basic Education Quality and Access in Lao PDR (BEQUAL) project. We should set long-term shared goals to solve the common problems in mathematics education and keep and expand collaboration for better education in our Lower Mekong sub-region. Finally and hopefully, with the support of ICMI, the group of mathematics educators in this region might take on the challenge of hosting International Congress on Mathematical Education (ICME) in the next 12 or 16 years as Thailand took on the challenge of hosting EARCOME 6 in 2013.

References

Benveniste, L., Marshall, J. & Santibañez, L. (2007). Teaching in Lao PDR. Washington, DC: World Bank. https://openknowledge.worldbank.org/bitstream/handle/10986/7710/429710 ESW0LA0P10Box327342B01PUBLIC1.pdf?sequence=1&isAllowed=y. Accessed January 21, 2017.

Do, Dat. (n.d.). *Mathematics teaching and learning in Vietnam.* http://www.ex.ac.uk/cimt/ijmtl/ddvietmt.pdf. Accessed January 21, 2017.

Inprasitha, M. (2015). New model of teacher education program in mathematics education: Thailand experience. In *Proceeding of 7th ICMI-East Asia Regional Conference on Mathematics Education* (pp. 97–100).

Inprasitha, M. (2016). Research and development of modern mathematics instruction. *KKU: Research, 2*, 2–9.

Inprasitha, M., et al. (2003). *Reforming of the learning processes in school mathematics with emphasizing on mathematical processes.* Khon Kaen Printing.

International Mathematical Union. (2013). *Mathematics in Southeast Asia: Challenges and opportunities (Country Reports).* http://www.mathunion.org/fileadmin/CDC/cdc-uploads/CDC_MENAO/SEA_Reports.pdf. Accessed January 21, 2017.

OECD/ UNESCO. (2016). Education in Thailand: An OECD-UNESCO perspective, reviews of national policies for education. OECD Publishing. Paris. http://dx.doi.org/10.1787/9789264259119-en

Roath, C. (2015). Mathematics education in Cambodia from 1980 to 2012: Challenges and perspectives 2025. *Journal of Modern Education Review, 5*(12), 1147–1153.

Thipmany, O. (2016, February). *National mathematics curriculum in Lao PDR.* Paper presented at SEAMAO RECAM-University joint seminar: Searching for quality mathematics curriculum framework on the Era of Globalization. http://www.criced.tsukuba.ac.jp/math/apec/apec2016/20160216%20Outhit%20Math%20CC%20Development.pdf. Accessed January 21, 2017.

UNESCO. (2011). *World data on education.* http://www.ibe.unesco.org/sites/default/files/Lao_PDR.pdf. Accessed January 21, 2017.

Open Access Except where otherwise noted, this chapter is licensed under a Creative Commons Attribution 4.0 International License. To view a copy of this license, visit http://creativecommons.org/licenses/by/4.0/.

Teaching and Learning Mathematics in Turkey

Huriye Arikan

This presentation discusses mathematics education in Turkey from the perspective of teaching and learning. An overview of mathematics education in Turkey, the contemporary national high school mathematics curriculum, its varied applications, and factors affecting its success are presented. The national exam, which is mandatory for all students to be admitted to a university in Turkey, destructively circumscribes the mathematics curriculum in high schools. In this context, the factors causing unfavorable outcomes and preventive measures are discussed. While a modest selection of national and international attributes of educators and agencies are provided, the presentation mainly focuses on the achievements of the Turkish Mathematical Society in enhancing mathematics education and promoting public interest in the subject. A narration of an innovative school, the Nesin Mathematical Village, which is designed to cultivate deep mathematics appreciation among its participants, is given.

Introduction

In three main parts, this national presentation discusses mathematics education in Turkey from the perspective of teaching and learning. The presentation consists of an overview of mathematics education in Turkey, the achievements of the Turkish Mathematical Society in enhancing mathematics education and promoting public interest in mathematics, and a showcase of a unique revolutionary act in mathematics education, the Nesin Mathematical Village. In this article a very brief

H. Arikan (✉)
Sabancı University, Istanbul, Turkey
e-mail: huriye@sabanciuniv.edu

summary of the talk in the scope of the national presentation along with some highlighted incidents in relation to the promotion of public interest in mathematics is provided. References are provided for further information.

An Overall Summary of Mathematics Education in Turkey

Education in Turkey is a state-supervised national system. Compulsory education, primary, secondary and high school lasts 12 years between the ages of 6 and 18. Pre-primary education includes optional education of children between 36 and 72 months old. Education is financed by the state and free of charge in public schools and in state universities. Secondary or high school education is mandatory but required in order to progress to universities. The student selection and placement system for university is composed of two consecutive exams that are held once a year and are mandatory for all students to be admitted to a university in Turkey. There is severe competition among students who want to be admitted to one of the 146 universities or the open education university in the country. In 2011 the total capacity of universities was 450,000 while the number of candidates was 1.6 million and has been increasing every year. Last year, 2,255,386 high school graduates took the exam. The content of these exams and the subjects that are included or excluded destructively circumscribe the mathematics curriculum in high schools. Some basic subjects, such as volume of revolution, L'Hospital rule, modular arithmetic, matrices and determinants and inverse trigonometric functions are practically omitted from the mathematics curricula in most schools. The multiple-choice nature of the university entrance and placement exams also affects the students' cognition of mathematical concepts and restricts their ability to deduce, prove and express the mathematical results and solutions accurately in written form.

The difference between the curriculum of mathematics education at the high school level designed by the Ministry of National Education and the curricula used in private schools indicates gaps in educational levels among students. During the past years, a great deal of effort has been made, ambitious projects (RTI International & ERI Initiative, 2013) have been utilized, and measures have been taken to improve mathematics education. As a result of the National Education Development Project initiated by Higher Education Council financed by the World Bank, after the 1990s most of the studies in mathematics education have been dominated by "cognitive dimensions" of mathematics, "curriculum studies" and "teaching methods" (Argün, 2008; Ayhan, 1998; Baki, 1997). In the last 10 years the budget allocated for education has dramatically increased. In-service training for all teachers at all levels has been provided by the Ministry of National Education and certificate programs for further degree completion and graduate study opportunities have been provided in cooperation with higher education institutions. Annual nationwide "Good Practice in Education Conferences" organized by the Education Reform Initiative of Sabancı University, brings teachers and instructors together to improve their learning, and the learning environment in class and in school. The Scientific and Technological Council of Turkey has substantially increased the allocated budget for projects related to mathematics education and in 2014, jointly with the Ministry of Education, implemented several financial support programs for successful students who have chosen to study mathematics at the undergraduate level. In spite of these efforts, the international and national reports, PISA (Ceylan, Yetişir, Yıldırım, & Yıldırım, 2013) and TIMSS (Gönen, Parlak, Polat, Özgürlük, & Yıldırım, 2016) results reveal the fact that the governmental incentives and support for maintaining an effective mathematics teaching environment in Turkey have not been enough. In this context the factors causing the unfavorable outcomes and preventive measures to counteract these outcomes (Oral & McGiveney, 2013; Şirin & Vatanartıran, 2014) were discussed further during the oral presentation.

Although higher education institutions had accommodated various degree programs in teacher training and research opportunities in education, until 1982 there were no faculties of education at universities in Turkey. In 1982, the Higher Education Council unified all higher education institutions in the country as universities; after this occurred, several universities established faculties of education that administratively separated the research mathematicians and mathematics educators. The National Educational Development Project restructured the schools of education between 1994 and 1998. In 1999, there were only 46 mathematics educators, including graduate students, located at several universities in the country (Askar & Ubuz, 1999). Research studies in mathematics education were mainly conducted by university instructors. These studies mostly investigated cognitive and affective domains; subjects in geometry were widely studied and the main instruments were data collection using tests followed by a questionnaire. The number of publications in Turkish and in national journals seemed insufficient compared to the number of publications in international journals (Ubuz & Ulutaş, 2008). In time, the number of mathematics educators has dramatically increased, and 63 state and 9 private universities out of 147 offer bachelor of science degree programs in mathematics education. Becoming a school teacher in mathematics requires a degree in education. Becoming a primary school mathematics teacher requires at least a bachelor's degree in mathematics education, while becoming a secondary or a high school teacher requires a minimum of a master's degree in mathematics education.

Contemporary research in mathematics education is advanced and the research environment is cultivated by national and international collaborative projects, platforms and conferences. Significant international conferences such as Psychology of Mathematics Education 2011 (PME35) and the Congress of Research in Mathematics Education 2013 (CERME8) have taken place in Turkey.

The Turkish Mathematical Society, the Mathematicians Association and the Mathematics Education Association are the three foremost non-governmental organizations supporting the enhancement of national mathematics education in Turkey. These institutions all contribute by organizing annual international and national conferences, workshops and project contests.

Achievements of the Turkish Mathematical Society

The Turkish Mathematical Society was founded in 1948 and is a member of the International Mathematical Union and the European Mathematical Society. The society supports the advancement of mathematics, mathematical sciences and related disciplines and promotes mathematics education and the exposition of mathematics at all levels through appropriate venues. Membership is open to all who use mathematics in their vocations. In addition to annual national and international conferences the society provides monthly talks on popular mathematics subjects.

The Friends of Mathematical Research (MAD) campaign society raises money to support young researchers, prominent research activities, international and national workshops, and seminars. The quarterly popular magazine *Matematik Dünyası* (*The World of Mathematics*) has a high circulation. The large-scale International Conference in Teaching of Mathematics at the undergraduate level; ICTM3), which has the aim of bringing mathematicians and math educators together, was organized by the Society in 2006 (Kyle, 2006). The interactive mathematics exhibition "IMAGINARY" visited Turkey as a joint project of the Mathematical Research Institute Oberwolfach, the Turkish Mathematical Society and the Istanbul Center for Mathematical Sciences. There has been a great interest in exhibitions in İstanbul, İzmir, Diyarbakır, Adana, Ankara, and other cities. The profound mathematician Cahit Arf's photograph adorns the 10 Turkish Lira banknotes along with his formula, Arf's invariant, making mathematics a part of the national cultural surplus.

A Dream School: Nesin Mathematical Village

Nesin Mathematical Village is a revolutionary and unique act, forming a mathematical community located at an unexpected place in the world and achieving unexpected success. The Nesin Mathematics Village was established in 2007 by a remarkable mathematics researcher, Ali Nesin (Alladi & Rino Nesin, 2015). The project was inspired by the realization that students needed an outlet that was alternative to traditional methods of mathematics teaching and involved learning mathematics outside the classroom through group work and mutual interaction (Karaali, 2014). Currently the village offers over 140 different courses at high school, undergraduate and graduate levels and has more than 1500 students per year. The village provides community life to help educate and encourage students to research and facilitate the exchange of mathematical information between adults and children. Participants enjoy various mathematics classes, and have the opportunity to attend seminars and lectures by renowned mathematicians, economists and intellectuals. The village is the realization of a dream school where students and teachers cook and clean together and young students learn to take responsibility, to share, and not to fear mathematics. The village has served as a good example and similar projects are under construction.

References

Alladi, K., & Rino Nesin, G. A. (2015). The Nesin mathematics village in Turkey. *Notices of the AMS, 62*(6), 652–685.

Argün, Z. (2008). Lise matematik öğretmenlerin yetiştirilmesinde mevcut yargılar, yeni fikirler. *TÜBAV Bilim Dergisi, 1*(2), 88–94.

Aşkar, P., & Ubuz, B. (1999). Current state of the mathematics education community in Turkey. *Hacettepe Üniversitesi Eğitim Fakültesi Dergisi, 15*(15).

Aydın, A. (1998). Eğitim fakültelerinin yeniden yapılandırılması ve öğretmen yetiştirme sorunu. *Kuram ve Uygulamada Egitim Yönetimi Dergisi, 4*(3), 275–286.

Baki, A. (1997). Çağdaş Gelişmelerin Işığında Matematik Öğretmenliği Eğitim Programları. *Eğitim ve Bilim, 21*(103).

Ceylan, E., Yetişir, M. İ., Yıldırım, H. H., Yıldırım, S. (2013). PISA 2012 Ulusal Ön Raporu. Republic of Turkey Ministry of National Education. http://pisa.meb.gov.tr/wp-content/uploads/2013/12/pisa2012-ulusal-on-raporu.pdf. Accessed: December 12, 2016.

Gönen, E., Parlak, B., Polat, M., Özgürlük, B., & Yıldırım, A. (2016). TIMSS 2015 Ulusal Matematik ve Fen Ön Raporu 4. Ve 8. Sınıflar. Resource document. Republic of Turkey Ministry of National Education. http://timss.meb.gov.tr/wp-content/uploads/Timss_2015_ulusal_fen_mat_raporu.pdf. Accessed: December 12, 2016.

Kyle, J. (2006). The third international conference on the teaching of mathematics. *MSOR Connections, Vol. 6 No 3 August–October 20, 2006*. https://www.heacademy.ac.uk/system/files/msor.6.3r.pdf. Accessed: December 12, 2016.

Karaali, G. (2014). Nesin math village: Mathematics as a revolutionary act. *The Mathematical Intelligencer, 36*(2), 45–49.

Oral, I., & McGiveney, E. (2013). Student performance in math and science in Turkey and determinants of success TIMSS 2011 analysis. Resource Document. Education Reform Initiative (ERG), Istanbul Policy Center (IPC), Sabancı University. http://www.egitimreformugirisimi.org/sites/www.egitimreformugirisimi.org/files/TIMSS%20abstract%20summary.pdf. Accessed: December 12, 2016.

RTI International, & Education Reform Initiative (ERI). (2013). Turkey's FATIH project: A plan to conquer the digital divide or a technological leap of faith? http://www.rti.org/sites/default/files/resources/fatih_report_eri-rti_dec13.pdf. Accessed: December 12, 2016.

Şirin, R. S., & Vatanartıran, S. (2014). PISA 2012 evaluation: Data based education reform proposals for Turkey, Turkish Industry. & Business Association (TUSIAD)—All Private Education Institutions Association (TÖDER), TÜSİAD-T/2014-02/549. ISBN: 978-9944-405-98-0.

Ubuz, B., & Ulutaş, F. (2008). Research and trends in mathematics education: 2000 to 2006. *Elementary Education Online, 7*(3), 614–626.

Open Access Except where otherwise noted, this chapter is licensed under a Creative Commons Attribution 4.0 International License. To view a copy of this license, visit http://creativecommons.org/licenses/by/4.0/.

Part VI
Reports from the Topical Study Groups

Topic Study Group No. 1: Early Childhood Mathematics Education (Up to Age 7)

Elia Iliada, Joanne Mulligan, Ann Anderson, Anna Baccaglini-Frank and Christiane Benz

The Programme

Session 1 papers

- Nathalie Sinclair: *Time, immersion and articulation: Digital technology for early childhood mathematics*
- Iliada Elia: *Gestures and their interrelations with other semiotic resources in the learning of geometrical concepts in kindergarten*
- Jennifer Thom: *Circling three children's spatial-geometric reasonings*

Session 2 papers

- Gina Bojorque: *Ecuadorian kindergartners' SFON development* (presented by Joke Torbeyns)
- Sanne Rathé: *Kindergartners' spontaneous focus on number during picture book reading*
- Christiane Benz: *Measurement makes numbers sensible*

Session 3 papers

- Joanne Mulligan: *Promoting early mathematical structural development through an integrated assessment and pedagogical program*

Co-chairs: Elia Iliada, Joanne Mulligan.

Team members: Ann Anderson, Anna Baccaglini-Frank, Christiane Benz.

E. Iliada (✉)
University of Cyprus, Nicosia, Cyprus
e-mail: elia.iliada@ucy.ac.cy; iliadaelia@gmail.com

J. Mulligan
Macquarie University, Sydney, Australia
e-mail: joanne.mulligan@mq.edu.au

- Miriam M. Lüken: *Repeating patterning competencies in 3- and 4-year old kindergartners*
- Ralf Kampmann: *The influence of fostering children's pattern and structure abilities on their arithmetic skills in Grade 1*
- Ruthi Barkai: *Preschool teachers' responses to repeating patterns tasks*

Session 4 papers

- Ann Anderson: *A study of types of math-in-context that parents and preschoolers share at home*
- Herbert P. Ginsburg: *Interactive mathematics books and their friends*
- Anna Baccaglini-Frank: *Educational multi-touch applications, number sense, and the homogenizing role of the educator*

TSG 1 included research-based contributions on recent trends and developments in early mathematics learning and teaching which stimulated rich discussions and enabled a deep understanding on various important issues in the field of early childhood mathematics education. As indicated in the programme, the wide scope of the studies reported in the presentations were organized into four sessions.

The focus of the first session of TSG 1 was on multimodal, embodied and semiotic aspects of learning with or without technology. The contribution by Sinclair discussed three innovative issues on the use of digital technology in early mathematics education that may transform the learning and teaching of mathematics. The first theme referred to temporalizing early childhood mathematics (time), which suggested that the use of dynamic geometry software and a multi-touch App for counting through embodied actions highlights dynamic aspects of mathematical objects, which promote the learning of sophisticated mathematical ideas. The second aspect concerned children's contact with advanced mathematics (immersion) often beyond the curriculum, because of the less constrained digital environments. The third issue was about the affordance of digital technology to support the articulation of signs in children's mathematical work.

The case study presented by Elia focused on the links between gestures and other semiotic resources in the understanding of geometrical concepts and the changes in these interrelations as learning evolves. Analyzing longitudinal observation data from a kindergarten class, and specifically from a child while interacting with teacher and peers, showed that gestures together with oral language and semiotic inscriptions had a major role in the kindergartner's development of geometric awareness for 2D shapes, their attributes and also in the process of shape dimensional deconstruction. As the child's geometry thinking evolved, his words and gestures were detached from the materiality of the activities and were based on the use of imaginary and general geometrical objects.

The contribution by Thom gave further insight into young children's spatial-geometric reasoning by elucidating the role of embodied actions in children's work. As part of a larger research project, Thom analyzed the forms, acts and processes that constitute children's reasoning while working on a spatial-geometric task. A photograph of a cylinder elicited different mathematical ideas and ways of reasoning in three grade one children. The study showed how these were materialized as gestures, movements, drawings, imagery and verbalizations. Another important finding of the study was the generative co-emergence between the children's reasoning and spatial-geometric conceptions, involving the transition between multiple dimensions, visualisation, decomposition, re-composition, perspective taking, dynamic objects, rotation, symmetry, curved and flat surfaces.

In the second session of TSG 1 different aspects of arithmetical competences of children built the thematic connection of the three presentations. Two of the presentations focused on the SFON effect: the Spontaneous Focusing On Numerosity. The first study was conducted with kindergarten children in Ecuador by Bojorque and presented by Torbeyns. Considering that SFON studies reported a positively associated SFON tendency to the development of early numerical abilities, Bojorque reported on a study where children were tested with SFON tasks at the beginning and end of a kindergarten year and on early numerical abilities at the beginning of the year. Findings showed that there was a limited SFON development and a positive relation between SFON and early numerical abilities. The quality of early mathematics education in Ecuador did not contribute to kindergartners' SFON development.

In the presentation by Rathé the role of picture books and a possible association to the children's SFON was investigated. The study was conducted in Belgium with children at kindergarten age. It examined the association between children's SFON in experimental tasks and number related utterances during everyday picture book reading. No correlation was found between these aspects. Various hypothetical explanations were given for this result.

The relation between measurement and number development was the theme in the presentation by Benz and Pullen which reported on a study with children at ages 5–6 in Australia. A design research was presented where children started formal schooling at a school with a Reggio Emilia and socio-cultural approach. The children did not start with a typical number-focused curriculum but with a measurement-focused curriculum where number concepts were included. Especially the case studies revealed that when young children measure, they use numbers and can acquire number competencies.

During the third session of TSG 1, there were four presentations focusing on patterning and structural competencies for preschoolers through to formal schooling as well as for preschool teachers. Mulligan provided an overview of the Australian Pattern and Structure project comprising a suite of studies with 4–8 year olds. She provided examples of children's Awareness of Mathematical Pattern and Structure (AMPS), a new construct which had been identified and measured, and found to be predictive of general mathematical achievement. An assessment interview, the Pattern and Structure Assessment (PASA) and Learning Pathways of the Pattern and Structure Mathematical Awareness Program (PASMAP) were illustrated.

Similarly, Lüken focused on the development of repeating patterning abilities in early math learning. As part of a longitudinal exploratory study, she analyzed the responses of individual children's patterning from their first to third year of German kindergarten. Three interviews, spanning two years, were presented, suggesting that significant development takes place in children's patterning competencies between the age of three and four. She described two important stages: the abilities to refer to the existing pattern and to alternate two colors.

In a related paper, Kampmann showed that patterning and structural abilities can positively influence arithmetic skills in Grade 1. He described an intervention study with 51 first-graders, showing significant differences between pre- and post-test arithmetic achievement scores for the intervention group. The improvement was particularly beneficial for the lowest achieving children. The intervention lessons included recognizing, describing, explaining and creating patterns with an emphasis on structuring the base 10 system.

The presentation by Barkai presented the work of her team (Pessia Tsamir, Dina Tirosh, Esther Levenson, and Michal Tabach) on a study of preschool teacher knowledge, and the important role of the preschool teachers and their ability to identify and continue repeating patterns. Their study found that preschool teachers were able to identify drawings which represent repeating patterns and find the errors which preclude a drawing from actually being a repeating pattern. However, identifying appropriate continuations was more difficult. They highlighted the use of the Cognitive Affective Mathematics Teachers Education (CAMTE) framework as a research tool and the importance of investigating teachers' knowledge of producing and evaluating solutions.

During the final session of TSG 1, three presenters shared their research into preschoolers' mathematical learning within everyday parent-child mathematical interactions, interactive mathematical books produced for parent or teacher-child shared reading, and teacher-child interactions around two iPad apps, respectively. In the observational study she presented, Anderson and Anderson reported on the types of mathematics that six preschoolers shared with family members during 'naturally occurring' activities that each of their middle class mothers chose to have videotaped. Across the dyads, a range of mathematics concepts were found; with one family sharing mostly number-related activities, while four other families shared more geometry-related activities. Likewise, although activities appeared common across the families, the nature of the specific materials, and/or the specific adult-child interactions oftentimes meant, the mathematics shared within these particular contexts differed.

In his presentation, Ginsburg focused on key features of the digital resource he and his team designed and produced. As Ginsburg indicated, interactive mathematics books, fiction and non-fiction, enveloped in a digital surround of supporting materials—their "Friends"—can delight and educate young children as well as those (e.g., parents, teachers, siblings) who read with them, yet few such books now exist, and little is known about them. In his presentation, he described, and illustrated, the potential of interactive mathematics storybooks (IMS), which entail a special set of affordances that can promote young children's mathematics learning, and the surrounding Friends, which can help the adult understand the mathematics and the child.

The use of digital tools to support and possibly enhance adult-child joint activity was a theme of the final presentation, where Baccaglini-Frank reported on her study of teacher-child interactions. As part of an educational project proposed in Italian preschools, an educator followed a tested protocol proposing two chosen iPad apps to children of ages 5–6, in which interactions with the software occur through multiple-touch gestures on the screen. Though, the educator's interventions were supposedly aimed at strengthening the children's number sense, the result was a homogenization of their schemes, in various cases seemingly inhibiting development of number sense.

Overall TSG 1 drew attention to contemporary and integrated perspectives on mathematics learning and teaching in the early years based on research studies from around the world and contributed to the ongoing discussion on how to advance research, development and practice in early childhood mathematics.

Open Access Except where otherwise noted, this chapter is licensed under a Creative Commons Attribution 4.0 International License. To view a copy of this license, visit http://creativecommons.org/licenses/by/4.0/.

Topic Study Group No. 2: Mathematics Education at Tertiary Level

Victor Giraldo, Chris Rasmussen, Irene Biza, Azimehsadat Khakbaz and Reinhard Hochmuth

The Programme

Research in mathematics education at the tertiary level has experienced tremendous growth over the last decades (Biza, Giraldo, Hochmuth, Khakbaz, & Rasmussen, 2016; Rasmussen & Wawro, 2017). Evidence of this growth includes the continued success of the *Research in Undergraduate Mathematics Education conference in the United States,* now at its 20th annual conference; the university mathematics research contribution to the *Espace Mathématique Francophone* since 2006; and since 2011 the working group on *University Mathematics Education* in the *Congress of European research in Mathematics Education* conference. In 2015 the *Australian Mathematical Society* established a *Special Interest Group in Mathematics Education*, which has among its goals the promotion of inquiry and discussion about tertiary mathematics education. The first biennial conference of the *International Network for Didactic Research in University Mathematics* took place in 2016. Furthermore, in 2015 the new *International Journal of Research in Undergraduate Mathematics Education* published its first issue.

Reflecting this growth, a total of 64 papers and 23 posters were accepted for presentation at TSG 2. There were four main sessions with 14 papers, including two invited talks, one by Elena Nardi and one by Greg Oates. There were 50 short oral

---

Co-chairs: Victor Giraldo, Chris Rasmussen.

Team members: Irene Biza, Azimehsadat Khakbaz, Reinhard Hochmuth.

V. Giraldo
Federal University of Rio de Janeiro, Rio de Janeiro, Brazil
e-mail: victor.giraldo@gmail.com

C. Rasmussen (✉)
San Diego State University, San Diego, SD, USA
e-mail: crasmussen@mail.sdsu.edu

presentations in eight different sessions. Session blocks were organized around more or less coherent themes, including mathematical practices, teaching, professional and curriculum development, connections to engineering, transition to university, preservice teachers, student thinking, and research related to specific courses such as calculus, differential equations, and linear algebra. All talks and authors organized by session type are presented below. As a whole, these talks represent a rich and diverse collection of interests and theoretical approaches characteristic of a maturing discipline.

1st Main Session—Mathematical Practices

- Symbolizing and brokering in fostering inquiry, Megan Wawro, Michelle Zandieh, Chris Rasmussen
- University students' behavior working with newly introduced mathematical definitions, Valeria Aguirre Holguín
- Learning how to axiomatise through paper folding, Dmitri Nedrenco

2nd Main Session—Teaching

- Teaching mathematics to non-mathematicians: what can we learn from research on teaching mathematicians? Elena Nardi
- University mathematics lectures: teaching the same topics but different mathematics, Alon Pinto
- Using a theoretical perspective to teach a proving supplement for an undergraduate real analysis course, Annie Selden, John Selden
- Exploring lecturers' perceptions of using technology to teach mathematics at tertiary level, Jayaluxmi Naidoo

3rd Main Session—Professional and Curriculum Development

- Mathematicians and mathematics education: collaborating for professional development, Greg Oates, Tanya Evans
- Holistic approach to curriculum review of undergraduate mathematics, Pee Choon Toh, Weng Kin Ho, Kok Ming Teo, Khiok Seng Quek, Tin Lam Toh, Eng Guan Tay, Romina Ann S. Yap
- Using technology to develop formative assessment resources for first year undergraduate modules, Ann O'Shea, Sinead Breen, Conor Brennan, Frank Doheny, Fiona Lawless, Christine Kelly, Ciaran Mac an Bhaird, Seamus McLoone, Eabhnat Ni Fhloinn, Caitriona Ni She, Brien Nolan
- The transition from secondary to tertiary mathematics education—a Swedish study, Christer Bergsten, Eva Jablonka, Hoda Ashjari

4th Main Session—Student Thinking

- Students' thinking modes and the emergence of signs in learning linear Algebra, Melih Turgut, Paul Drijvers
- Exploring students' interactions in an online forum that accompanied a course in linear Algebra, Igor' Kontorovich

- Mathematical argumentation of first-year students: the influence of conceptual knowledge, Kathrin Nagel, Kristina Reiss

1st Oral Communication Session Strand A—Linear Algebra

- Difficulties in mathematics experienced by students in a trans-disciplinary engineering study, Evangelia Triantafyllou, Olga Timcenko
- Ideas of mathematical literacy for cultivating students' understanding of concepts of linear Algebra, Ryuichi Mizumachi
- A task design to introduce the concepts of eigenvectors and Eigen values. An embodied approach, María José Beltrán-Meneu, Marina Murillo-Arcila
- Déjà vu in mathematics: what does it look like?, Robyn Pierce, Caroline Bardini

1st Oral Communication Session Strand B—Teaching

- The art of mathematical chatter, Anne D'Arcy-Warmington
- Why students are not motivated to learn mathematics?, Seyed Hadi Afzali Borujeni, Azimehsadat Khakbaz
- "What we need to show is that t is well-defined": gesture and diagram in abstract Algebra, Andrew Francis Hare
- Scripts in mathematics tutorials, Juliane Püschl

2nd Oral Communication Session Strand A—Differential Equations and Calculus

- A case study on the impact of investigating multivariable calculus concepts through geometry and multiple representation, Aaron D Wangberg, Brian Fisher, Elizabeth Gire, Jason Samuels
- Student reasoning about functions, limits, and rate of change in introductory calculus, Caroline Julia Merighi
- Research and practice of college mathematics course assessment in Sichuan University, Jianren Niu, Liang Yang
- How do students of economics understand the concept of marginal cost? Frank Feudel
- About doing geometric approach in differential equations: difficulties and a coherent method, Younes Karimi Fardinpour
- Instrumental action schemes in differential equations using a computer Algebra system, maxima, Fereshteh Zeynivandnezhad

2nd Oral Communication Session Strand B—Teaching

- The practice, guarantee and effect on the second classroom platform in university mathematics teaching, Chen Li, Chen Chaodong
- A comparative study of university students' math achievement of small-class teaching and large-class teaching, Chao dong Chen, Jian ren Niu
- Using journals to support learning: case of number theory and proof, Christina M Starkey, Hiroko Warshauer, Max Warshauer

- Knowledge of rational and irrational numbers of two undergraduate students, Geraldo Claudio Broetto, Vânia Maria Santos-Wagner
- The activity-based learning of mathematics in a technical higher education institution, Elena G. Yevsyeyeva
- Motivating university students to learn mathematics, Azimehsadat Khakbaz, Seyed Hadi Afzali Borujeni

3rd Oral Communication Session Strand A—Professional and Curriculum Development

- The future of mathematics teaching: analysis of the expectations of undergraduates in the federal district, Brazil, Jéssica de Aguiar França, Regina da Silva Pina-Neves, Raquel Carneiro Dörr
- Principles for designing invention tasks for undergraduate mathematics, Ben Davies, Caroline Yoon, John Griffith Moala, Wes Maciejewski
- Meaningful learning in mathematics education for the humanities and social sciences students, Mitsuru Kawazoe, Masahiko Okamoto
- Lecturer education: a course design, Ignasi Florensa, Marianna Bosch, Josep Gascón

3rd Oral Communication Session Strand B—Preservice Teachers

- An investigation into the efficacy of flipped classroom for tertiary mathematics, Weng Kin Ho, Kok Ming Teo, Lu Pien Cheng, Puay San Chan
- Interactive videos: a 21st century necessity for student engagement, Haitham S. Solh
- Systemic integration of programming in undergraduate mathematics: from implementation to theory, Chantal Buteau, Eric Muller
- Undergraduate math students' interactions with programming: developing instruments in institutions, Laura Rose Margaret Broley

4th Oral Communication Session Strand A—Transition to University

- Development of diagnostic self-assessments as a base for individual support for first-year students, Christoph Neugebauer, Sebastian Krusekamp, Kathrin Winter
- Didactical elaboration of multimedia learning materials by recent technological advancements exemplified by computer aid, Tobias Mai, Silvia Becher
- exploration of transfer of first year undergraduate mathematical learning to science, Yoshitaka Nakakoji, Rachel Wilson
- The algebra-to-calculus transition, William Crombie
- What first year university students' recommendations for freshmen reveal about their learning strategies, Robin Göller
- Issues in the transition from secondary to tertiary mathematics, Michael Surman Jennings, Merrilyn Goos, Peter Adams

4th Oral Communication Session Strand B—Connections to Engineering

- Challenges involved when reforming traditional courses in mathematics for engineers, Frode Rønning
- Analysis of typical mathematical competences required to solve tasks in basic engineering courses, Joerg Kortemeyer, Rolf Biehler
- Tree-structured online exercises in mathematics for engineering students: design and evaluation, Robert Ivo Mei
- Learning behaviour, academic success in engineering mathematics, and lecturers' ratings, Birgit Griese, Michael Kallweit
- Mathematical self-efficacy of engineering students at the introductory phase of studies, Ronja Kürten
- A preliminary analysis of the effectiveness of student-produced videos on the relevance of mathematics in engineering, Birgit Loch, Wendy Scott, Michelle Dunn

Posters

- Procedural knowledge as a predictor for success in German math exams for first year engineering students, Mike Altieri
- Mathematics graduate teaching assistants' longitudinal transitions in beliefs about mathematics teaching and learning, Mary Beisiegel
- Algebraic thinking in the understanding and solution of geometric problems amongst year university students, Luis Weng San, Bhangy Cassy
- Results of us national study on calculus, Jess Ellis and Chris Rasmussen
- Studifinder: mathematical e-learning materials for the transition from secondary school to university, Yael Fleischmann, Alexander Börsch, Rolf Biehler, Christoph Colberg, Tobias Mai
- Concept and application of mathematizing to the process of classification, Alfonso J. González-Regaña, Verónica Martín-Molina, José María Gavilán-Izquierdo
- How, when, where and why do students use lecture recordings?, Roland Gunesch
- Digital media as motivating tool for learning descriptive statistics, Mathias Hattermann, Alexander Salle, Stefanie Schumacher, Daniel Heinrich
- A commognitive perspective on students' engagement with the concept of group: the case of students F and M, Marios Ioannou
- Students' perception of group discussions and presentations in a math education course, Seong-A Kim, Jeong-Gyoo Kim, Sunhee Lee
- Perception vs reality: using tutorial videos to aid tutor reflection, Heather Lonsdale, Deborah King
- A comparative analysis of three comprehensive initiatives to redesign developmental mathematics college curriculum, Carolyn Masserang
- Supporting internalisation of mathematical syntax using blocks, Anthony Morphett

- Revision activities of undergraduate mathematics students, Philip Walker, Eabhnat Ní Fhloinn
- Autonomy in mathematics in the secondary-tertiary transition, Pierre-Vincent Quéré, Ghislaine Gueudet,
- Mathematical competencies visible through assessment for engineering students, Kristina Raen
- University teaching assistants' teaching related beliefs, Johanna Rämö, Juulia Lahdenperä, Susanna Oksanen
- Bremath—redesign and implementation of university maths courses for future high school teachers, Ingolf Schäfer
- Assignments and written exams in an ICT learning environment, Karsten Schmidt
- Explicating strategies—planning an intervention to increase the strategic knowledge of university freshmen, Thomas Stenzel
- description and initial results of the preservice teachers seminar "Überpro_wahrscheinlichkeitsrechnung", Gero Stoffels
- Artin's braid group as an introductory example for group theory approaches at the university of hamburg, Sophie Stuhlmann
- Lecturers' pedagogical routines and expectations on students' engagement in closed-book examinations, Athina Thoma

References

Biza, I., Giraldo, V., Hochmuth, R., Khakbaz, A., & Rasmussen, C. (2016). Research on teaching and learning mathematics at the tertiary level: State-of-the-art and looking ahead. In *Research on Teaching and Learning Mathematics at the Tertiary Level*. Berlin: Springer International Publishing.

Rasmussen, C., & Wawro, M. (2017). Post-calculus research in undergraduate mathematics education. In J. Cai (Ed.), *Compendium for Research in Mathematics Education*. National Council of Teachers of Mathematics: Reston, VA.

Open Access Except where otherwise noted, this chapter is licensed under a Creative Commons Attribution 4.0 International License. To view a copy of this license, visit http://creativecommons.org/licenses/by/4.0/.

Topic Study Group No. 3: Mathematics Education in and for Work

Geoff Wake, Diana Coben, Burkhard Alpers, Keith Weeks and Peter Frejd

At ICME-13 TSG 3 aimed to bring together researchers, practitioners and policy makers for the exchange of ideas related to **Mathematics education in and for work**. Our wish was to be inclusive in our endeavours by involving those interested in this area from mathematics education, adult education, workplace education, adult numeracy education, citizenship education, social movement education and other fields.

In pursuit of our work we considered *mathematics* to be inclusive of the formal academic discipline of mathematics and mathematical processes such as modelling and problem solving in addition to many other informal forms of quantitative reasoning that arise in a wide range of work settings and situations.

We took *education* to be inclusive of formal, informal and non-formal learning, that is, in educational settings (e.g. adult community education, vocational and further education) as well as in the community and workplaces; and to involve both individual and collective learning.

We also took a view of *work* to be inclusive of both paid and unpaid work such as work in the home, and activist work in social movements. We recognised that work has very different meanings in the full range of different social and cultural settings and in many cases is evolving rapidly.

Co-chairs: Geoff Wake, Diana Coben.
Team members: Burkhard Alpers, Keith Weeks, Peter Frejd.

G. Wake (✉)
University of Nottingham, Nottingham, UK
e-mail: Geoffrey.Wake@nottingham.ac.uk

D. Coben
University of Waikato, Hamilton, New Zealand
e-mail: dccoben@waikato.ac.nz

The focal topics of the group included empirical, theoretical and methodological issues related to questions such as:

- How is mathematics embedded in work practices; what is this mathematics like and how is it learned?
- How can we define and model competence in work-based mathematics? How can we use diagnostic assessment to understand such competence?
- What mathematics do people currently learn in preparation for work? How could/should this be improved?
- How is mathematics/numeracy valued for and in employment in different societies?
- How does the mathematics learning in and for work meet people's mathematical needs in other domains of their lives?

We planned for and provided a lively forum for debate that involved different modes of exchange, including presentations, posters and discussions and were informed by contributions from a range of different theoretical perspectives and research backgrounds.

On Tuesday we focussed on the Question: What makes for authenticity in mathematics education in and for work?

The following contributions answered this question. Diana Coben & Keith Weeks with the title: "Authenticity in vocational mathematics: Supporting medication dosage calculation problem solving in nursing"; Vincent Jonker, Monica Wijers, Ad Mooldijk, Mieke Abels & Michiel Doorman: "Redesign guidelines to enrich classroom tasks for maths and science"; Kees Hoogland & Birgit Pepin: "The numeracy of vocational students: Exploring the nature of the mathematics used in daily life and work".

The session on Wednesday was chaired by Keith Weeks and focussed on the question: "How do we make sense of mathematics in and for work using different research methodologies and theoretical approaches?" The following contributions were made:

- John Keogh & Theresa Maguire: "Re-contextualising mathematics for the workplace"; Lisa Björklund Boistrop: "Mathematics in the workplace from different perspectives: The case of Anita, a Nursing Aide"; David Pontin: "Vocational mathematics and nursing: Social messiness and complexity".
- "What is the role and place of mathematics in education in and for work?" was the question posed on Friday by Peter Frejd and this was answered by the following presentations by two scholars:
- Phil Kane: "Uncovering estimation and spatial awareness as elements of workplace numeracy" and Karen Reitz-Koncebovski & Katja Maaß: "Dialogue between school and the world of work in teacher professional development (PD)".

On Saturday the question posed by Geoff Wake was "What is the role and place of mathematics in education in and for work?" Nathalie Jennifer van der Wal, Arthur Bakker & Paul Drijvers answered the question with their presentation titled "Techno-mathematical literacies in the workplaces of engineers" and the debate was taken further by Damon Whitten who presented: "Inside a mathematics-for-work lesson on ratio".

Open Access Except where otherwise noted, this chapter is licensed under a Creative Commons Attribution 4.0 International License. To view a copy of this license, visit http://creativecommons.org/licenses/by/4.0/.

Topic Study Group No. 4: Activities for, and Research on, Mathematically Gifted Students

Florence Mihaela Singer, Linda Jensen Sheffield, Matthias Brandl, Viktor Freiman and Kyoko Kakihana

The Programme

TSG 4 from ICME 13 aimed at connecting people from around the world who share interests in recognizing, developing and supporting gifted, talented and promising mathematics students. TSG 4 built on the work of several previous ICME Topic Study Groups. The programme included approximately sixty presenters from twenty different countries in regular presentations, oral communications and poster sessions. The presenters updated colleagues on their most recent work in a relaxed climate where questions and discussions of results were addressed. The TSG 4 co-chairs, team members and IPC liaison collaborated in person as well as electronically before, during and after the conference and all contributed to a smooth management and friendly atmosphere.

To offer a foundation for the discussions in each TSG4 session, a forty-page survey paper, *Research On and Activities For Mathematically Gifted Students* by Florence Mihaela Singer, Linda Jensen Sheffield, Viktor Freiman, and Matthias Brandl (Singer et al. 2016) has been published by Springer as an Open Access book and is available along with survey papers from other Topic Study Groups at http://icme13.org/publications/topical-surveys as well as on the MCG website at www.igmcg.org. The aim of this Topical Survey was to give a brief overview of the

---

Co-chairs: Florence Mihaela Singer, Linda Jensen Sheffield.
Team members: Matthias Brandl, Viktor Freiman, Kyoko Kakihana.

F.M. Singer (✉)
University of Ploiesti, Ploiesti Bucharest, Romania
e-mail: mihaela_singer@yahoo.com; mikisinger@gmail.com

L.J. Sheffield
Northern Kentucky University, Newport, USA
e-mail: Sheffield@nku.edu; lindajsheffield@gmail.com

current state of research on and activities for mathematically gifted students around the world, being of interest to mathematics educators, educational researchers, research mathematicians, mathematics teachers, teacher educators, curriculum designers, doctoral students, and other stakeholders. The focal topics include empirical, theoretical and methodological issues related to the following themes: Nature of Mathematical Giftedness; Mathematical Promise in Students of Various Ages; Pedagogy and Programs that contribute to the development of mathematical talent, gifts and passion; and Teacher Education. Current and historical research and suggestions for new research paths are included in each category.

The four main themes in the survey were also the themes for our TSG 4 sessions at ICME. These are briefly presented below.

1. *Nature of Mathematical Giftedness.* This session was organized around the following questions:

   - What do we know and what do we need to know about mathematical giftedness?
   - Is mathematical giftedness a discovery or a creation?
   - What theoretical frameworks and methodologies are helpful in identifying, creating, valuing, and educating mathematically gifted students in different contexts/societies?

   Answers to these questions were offered by the presenters of the following papers: *Distinguishing Between Gifted and High Achievers at University Level* (authors: Florence Mihaela Singer, Cristian Voica, Ildiko Pelczer); *Characteristics of Mathematical Giftedness: Learning from Extraordinary Minds* (author: Carmel Diezmann); *Characteristics of Mathematical Giftedness in Early Primary School Age* (author: Daniela Assmus), as well as by Matthias Brandl, as chair of the session.

   Discussions during this session included differing definitions of mathematical giftedness and frequently used terms such as mathematically promising, talented, high-achieving, high ability, and precocious as well as questions of nature vs. nurture. Differing answers to these questions often influence how gifted students are identified and served.

2. *Mathematical Promise in Students of Various Ages.* This session addressed questions such as:

   - What does recent research in cognitive science and neuroscience bring to understanding the development of mathematical talent and innovation?
   - In what ways are cognitive, social, and affective aspects connected in gifted students?
   - What are the differences between novices and experts?
   - How are mathematical creativity and giftedness connected?
   - What are new research paths?

Answers to these questions were offered by the presenters of the following papers: *Using Discourse Theory to Analyze Mathematical Giftedness within the South African Education System* (author Michael Mhlolo); *Analysis of the Cognitive Demand of a Gifted Student's Strategies to Solve Geometric Patterns Problems* (authors: Clara Benedicto, Eva Arbona, Adela Jaime, Angel Gutierrez); *Mathematical Problem Solving Techniques Employed by Gifted Students* (author: Andreas Poulos); *Pathways and Dead Ends in the Kingdom of Numbers: Problem Solving Strategies Used by Students in Mathematical Olympiad* (authors: Ingrida Veilande, Liga Ramana, Sandra Krauze), in collaboration with Florence Mihaela Singer, who chaired this session.

One common theme in this session was an emphasis on the problem solving techniques and strategies used by mathematically promising students. Discussions included whether these were innate or teachable as well as their prevalence in students of all ages and from all socio-economic backgrounds.

3. *Pedagogy and Programs.* Moving towards more pragmatic approaches, the questions that drove the discussions were:

- How could teaching best encourage and promote mathematical talents?
- How might classroom interactions and discourse contribute to the development of mathematical reasoning?
- What teaching strategies, curricula, technology, or other in- and out- of school activities might lead students to discover and realize their mathematical promise and talents?
- How is high-level mathematical innovation developed?

Answers to these questions were offered by the presenters of the following papers: *Instructional Models and Pedagogical Tools to Encourage and Promote Mathematical Talents* (author: Ban Har Yeap); *Fostering Talent in Mathematics—a German Perspective* (author: Stephanie Schiemann); *Developing Deductive and Spatial Reasoning with Language-independent Logic Puzzles* (author: Jeffrey J. Wanko); *Enrichment for the Gifted: Generalizing Some Geometrical Theorems & Objects* (author: Michael de Villiers) and by Linda Sheffield, who chaired the session.

Speakers noted that it was important to offer opportunities during the school day as well as extracurricular activities to support and develop mathematical expertise. Several noted the importance of students' collaboration and active involvement in recognizing patterns and constructing their own rules and generalizations to both solve and pose mathematical tasks and problems, with an emphasis on creativity, innovation, depth and complexity rather than speed.

4. *Teacher Education.* This session was focused on the following set of questions:

- What types of mathematics and pedagogy are suitable for educating pre-service and in-service teachers for the gifted?
- How should lessons/units planning be structured in order to address special needs of gifted?

- What types of assessment are most effective for identifying, challenging and nurturing mathematical giftedness and innovation?
- What types of local, regional, national or international co-operation between researchers and educators should be emphasized for the promotion of mathematical talent and giftedness?

Answers to these questions were offered by the presenters of the following papers: *Gifted Students' Expectations and Teachers' Conceptions of Effective Mathematics Teaching* (author: Roza Leikin); *A Cross-country Comparison of Professional Development Programs on Mathematical Promise and Talent* (Elisabet Mellroth, Ralf Benölken); *Examining Ireland's New Second-level Mathematics Syllabus and How it Caters for the High Achiever* (author: Aidan Fitzsimons, Eabhnat Ni Fhloinn); *Addressing the Needs of Gifted Students: Opportunities for Students, Teachers and Researchers* (authors: Hiroko Kawaguchi Warshauer, Max Leon Warshauer, Christina Starkey, Terence McCabe, Christina Zunker), with the contribution of Viktor Freiman, the session's chair.

Pre-service and in-service programs for teachers of mathematically gifted students should be linked to research on best practices for developing mathematical talent and passion. Several presenters noted the importance of teachers themselves persisting in a struggle to solve and pose problems with multiple entry points and multiple methods of solution in a respectful, inspiring, demanding, and joyful atmosphere where it is safe to fail, in order to understand how best to structure similar opportunities for their students. The need for familiarity with resources such as opportunities for students' mathematical competitions, camps, circles, mentors, etc. and sources of rich mathematical tasks and samples of exemplary student work were also cited.

A variety of topics related to the TSG activities were also covered by the short oral communications and posters, among which were the analysis of existing theories in the field of mathematical giftedness; characteristics of motivational factors of mathematically promising students; and metacognitive competencies of mathematically gifted students. Aspects related to the development of mathematical giftedness including identification in primary and secondary school, teachers' characterization of high achieving students in mathematics, and strategies to enhance the teaching of gifted children were other topics addressed during the poster and short oral communications sessions. In addition, several researchers studied features of solving problems by gifted students, and the nature of tasks and enrichment techniques to address the needs of gifted students.

In order to continue to build on the progress made during the TSG, participants were encouraged to join the International Group for Mathematical Creativity and Giftedness (MCG, www.igmcg.org), an International Study Group Affiliated to the International Commission on Mathematical Instruction (ICMI). MCG is free to join and holds biennial conferences in different locations around the world as well as offering a variety of resources such as periodic newsletters and other links to information on current research, problems, and activities related to mathematical creativity and giftedness.

Reference

Singer, F. M., Sheffield, L. J., Freiman, V., & Brandl, M. (2016). *Research on and activities for mathematically gifted students*. New York: Springer.

Open Access Except where otherwise noted, this chapter is licensed under a Creative Commons Attribution 4.0 International License. To view a copy of this license, visit http://creativecommons.org/licenses/by/4.0/.

Topic Study Group No. 5: Classroom Practice and Research for Students with Mathematical Learning Difficulties

Lourdes Figueiras, Rose Griffiths, Karen Karp, Jens Holger Lorenz and Miriam Godoy Penteado

Psychological and clinical research has a long tradition examining cognitive, sensorial and affective difficulties in mathematics, and students with mathematical learning difficulties may require extensive additional teacher resources, curricular adaptation or specific materials. From this starting point, our study group examined a research-based and practice-based agenda focused on the challenges of inclusive mathematics education. The two core ideas we considered were:

1. Teaching mathematics to students with special needs can be a challenging, innovative, and rewarding experience. Importantly, many of the successful strategies identified for working with students with special needs are also useful and effective for all students. We wanted to explore the nature of low-attaining pupils' experience in the classroom, and the ways in which the teacher can support a student, or inadvertently prevent them from making progress. This included examining the development of formative assessments and effective instructional interventions.
2. Inclusion policies and research studies led us to wonder what kind of education we can provide to prospective and practising teachers, to help them meet the diverse needs of a classroom where there is a wide range of student attainment and abilities, including students with special needs.

Co-chairs: Lourdes Figueiras, Rose Griffiths.
Team members: Karen Karp, Jens Holger Lorenz, Miriam Godoy Penteado.

L. Figueiras (✉)
Universitat Autònoma de Barcelona, Barcelona, Spain
e-mail: lourdes.figueiras@uab.cat

R. Griffiths
University of Leicester, Leicester, UK
e-mail: rnag1@le.ac.uk

Our discussions drew upon the work of the ICME Survey Team, "Assistance of students with mathematical learning difficulties—How can research support practice?" Our four sessions were organized in a way that encouraged a healthy debate and included the description of interesting practices, discussion of meaningful initial research projects and presentation of formal research results.

Our topic is a complex one, and we acknowledged that there are many differences of opinion about the causes of difficulties in mathematics, and issues around how we define and use terms such as 'inclusion' and 'special needs'. We wanted to examine the barriers that prevent children from successfully learning mathematics. These may have been identified as sensory and physical difficulties, cognition & learning difficulties, or emotional and behavioural difficulties. There may also be difficulties that have arisen through disrupted or unsatisfactory educational experiences, perhaps linked with illness or trauma. Our interest was in finding ways of improving these situations, to give children the best chance of making greater progress.

Whilst our major focus was on classroom approaches, we were also interested in considering the contribution of home and family to a child's mathematical experience, and wanted to explore ways in which this might be strengthened by positive links with school.

The interest of the organizers was to include as many practical situations as possible. Snapshots of teaching practices (videos or descriptions of real classroom episodes) to reflect on during the sessions and enhance discussion were especially welcomed.

We were pleased to have representation from a wide range of countries with varied experience of work in this field.

The Tuesday TSG session chaired by Lourdes Figueiras and Miriam Penteado, included the following contributions: Mutual learning in an inclusive mathematics classroom from Laura Korten, Germany; The challenge of constructing an inclusive school mathematics from Solange Fernandes and Lulu Healy, Brazil; The delaware longitudinal study of fraction learning: Implications for students with mathematics learning difficulties from Nancy Jordan, Ilyse Resnick, Jessica Carrique and Nicole Hansen, USA; and inclusive practices in the teaching of mathematics: early findings from research including children with downs syndrome from Barbara Anne Clarke and Rhonda Faragher, Australia.

Wednesday's TSG session chaired by Lourdes Figueiras and Jens Holger Lorenz, had the following contributions: Working with children in public care who have difficulties in mathematics: The example of kyle, from Rose Griffiths, United Kingdom; The calcularis learning system: Enhancing individual adaptivity for an inclusive teaching environment, from Michael von Aster, Germany; Response to intervention in mathematics: Research on early prevention of mathematical learning disabilities from Russell Gersten, USA; and PGBM-COMPS math problem-solving program: promoting independent problem solving of students with LD from Yan Ping Xin, Xuan Yang, Ron Tzur, Xiaojun Ma and Joo Young Park, USA.

The Friday TSG session chaired by Karen Karp and Rose Griffiths, covered these contributions: "story-telling tasks on additive relations word problems: The case of MPHO" from Nicky Roberts, South Africa; a novel approach on enabling advanced mathematical communication in absence of sight from Mina Sedaghatjou, Farzad Kooshyar and Stephen Campbell, Canada; and challenging ableism by teaching processes rather than concepts from Rossi DSouza, India.

Useful themed sessions for Oral Communications were held throughout the week. On Tuesday, one group looked at individualized programmes, with contributions from Australia, China and Switzerland. A parallel session contrasted direct instruction with an inquiry approach, led by colleagues from the USA, Germany and France. Later in the day, colleagues from Canada, the USA and Germany began discussions on the teaching of proportional thinking, multiplication, place value and fractions. Friday's parallel sessions had contributions from Italy, Germany, the UK, the USA and Russia, covering a wide range of topics, including work with children with Down Syndrome, those with ADHD, and examining the ways in which provision for children with special needs is organized. The final session of Oral Communications on Friday, led from Brazil, Mexico, the Netherlands and the UK, concentrated on work with children with hearing or visual impairments.

We finished the week with a TSG session on Saturday, with the following contributions: Behavioural difficulties could come from learning difficulties: Why and how to intervene in math class by Lucie DeBlois, Canada; Collaboration between special and common education for inclusive mathematical education in Brazil by Vera Lucia Capellini and Messias Fialho, Brazil; and embodied multimodal mathematics & A reconceptualization of sensory IMPAIRMENTS by Susan Gerofsky, Canada.

Open Access Except where otherwise noted, this chapter is licensed under a Creative Commons Attribution 4.0 International License. To view a copy of this license, visit http://creativecommons.org/licenses/by/4.0/.

Topic Study Group No. 6: Adult Learning

Jürgen Maaß, Pradeep Kumar Misra, Terry Maguire,
Katherine Safford-Ramus, Wolfgang Schlöglmann
and Evelyn Süss-Stepancik

The Program

July 26, 27, 28 and 29, Tuesday to Friday, 12:00–1:30
John O'Donoghue (Ireland): Mathematics education and adult learning in Ireland
Aoife M. Smith (Ireland): An investigation into the concept of Math Eyes with a particular focus on the Math Eyes poster competition
Wolfram Meyerhöfer (Ireland): Mathematics education and adult learning in Ireland
Katherine Safford-Ramus (United States of America): Learning from research, advancing the field
David Kaye (United Kingdom): Defining adult numeracy and mathematics—an academic and political investigation
Pradeep Kumar Misra (India): Open Educational Resources: A potential tool for adult learners, to achieve lifelong learning of mathematics
Maria Elizabete Souza Couto (Brazil): The mathematics in the young people and adult education: The practice in Construction

Co-chairs: Jürgen Maaß, Pradeep Kumar Misra

Team members: Terry Maguire, Katherine Safford-Ramus, Wolfgang Schlöglmann, Evelyn Süss-Stepancik.

J. Maaß (✉)
University of Linz, Linz, Austria
e-mail: juergen.maasz@jku.at

P.K. Misra
CCS University, Meerut, India
e-mail: pradeepkmisra@yahoo.com

Wolfram Meyerhöfer (Ireland): Mathematics education and adult learning in Ireland
Zekiye Morkoyunulu (Turkey): Parent's training in mathematics: A social awareness study
Wolfram Meyerhöfer (Ireland): Mathematics education and adult learning in Ireland
Terry Maquire (Ireland): Math Eyes—A concept with potential
Sonja Beeli-Zimmermann (Switzerland): "I've never cooked with my math teacher"—The duality of mathematics
Shin Watanabe (Japan): Self learning mathematics on lifelong learning
Andrea Maffia (Italy): Adults' conception of multiplication: How does it change along studies?
Jürgen Maaß (Austria): Thinking about relations between adults learning mathematics and reality

The manifold presentations in this topic study group spanned a wide range of issues and fall under the headings: (1) current state of research, (2) numeracy, (3) schooling and lifelong learning, (4) beliefs of adult education teachers, and (5) incorporating technology.

1. Current state of research

In the research of adult mathematic education several areas can be identified. On the one hand a great amount of studies emphases student issues and on the other hand a large number of published research deals with teacher issues. The crucial student issues are math anxiety, self-efficacy and classroom methods. Math anxiety is well documented and it's obvious, that this anxiety looks people out of jobs that require mathematics degrees. The relationship between math anxiety and self-efficacy is inversed. The results of research focused on classroom methods are diverse and show contradictory findings. The explorations cover topics like the effectiveness of Integrated Learning System, the implementation of online, weekend and shot-term courses, the impact of cooperative and collaborative learning methods, etc. The most important teacher issues are teachers' characteristics and the necessity of a professional adult teacher development.

2. Numeracy

The debate around the use and meaning of "numeracy" lasts for a long time and applies to (adult) learners, (adult) teachers, researchers and politicians. Many definitions of numeracy exist and a lot of them mentioned terms like context, solving problems, work-related and empowering in connection with numeracy. All of them agree that numeracy is not less than mathematics but more. Furthermore numeracy is about using mathematics to make sense of the real world and being critical in social and political analysis and also about mathematics itself. Thus numeracy includes personal abilities from basic skills to high-level cognitive abilities. By now numeracy is an important aspect for curriculum development in some countries (e.g. Denmark, England, Ireland). But the perception of numeracy is diverging. Researchers focus on the learners' needs and the relevance of the numeracy to the learners' lives while policy makers' priorities to increase numeracy for economic grow. Over and above this the role of school mathematics in numeracy development is not clarified yet, even if it's confident that numeracy is not automatically developed of schooling.

In Ireland the *Math Eyes* concept was developed as a central component of the professional development for adult numeracy teachers. The idea of Math Eyes is to support individuals to look at familiar things through the lens of mathematics. The hypothesis is, that the development of maths eyes has an impact on the numerate behavior, the motivation and mathematical confidence of the individual and it seems that Math Eyes reengage adult learners in learning mathematics. The evolution of Maths Eyes is proceeding rapidly in Ireland. By now the concept is also relevant for the primary and post-primary secondary schools sector. Despite the widespread use of Math Eyes the output of some projects demonstrate that developing maths eyes is not easy for teachers and learners. Therefore an appropriate professional development especially for adult mathematics teachers is necessary.

The idea of discovering mathematics in one's everyday life is also pursued in many other countries. In Brazil, where Freires pedagogic concept is originated, adult learners were faced with the economy of water in mathematics classes because everyone knows the lack of water. After discussing the significance of this theme the learners studied graphs about the global water consumption, calculated their personal water consumption and debated how mathematics helps to grasp this topic much better. Another interesting example was reported from India, where an adult mathematics class successfully elaborated themes like optimization and multiple representations. Some of the adult learners worked as vegetable sellers and were highly interested to optimize their spending and earning. In the mathematics class they tabulated their data (e.g. vegetables, rates, …), draw graphs and infer functional variations. This work led to a rulebook for the vegetable sellers that helped them to earn more. One conclusion of this experience is that mathematics curriculum should be built on the learners' own lives, especially in their participation in economic activity.

3. Schooling and lifelong learning

Discussing the field of schooling and lifelong learning, it became clear that the duration of compulsory schooling and the duration of schooling also has a big influence on the needs of the adult learners and adult teachers. In Japan for example a lifelong learning strategy is not yet established. This leads to the division of mathematical education into two periods. The first takes place in school and the second much longer period is carried out from everyone without public support. Due to the importance of lifelong learning and a continuing mathematical education the Japanese education system is building up a lifelong learning strategy in which creativity and mathematical thinking plays an important role. From a very different angle a study about adults' conception of multiplication looked at the effect of duration and typology of schooling. In this specific field of arithmetic a qualitative analysis of semi-structured interviews pointed out that people with low mathematical education recall just situations from school while people with medium and high education in mathematics associate their personal life experience to their conception of multiplication. Another aspect in the context of school and lifelong learning is the mathematical education of mathematical analphabets. In Germany within the scope of the "National Strategy for Literacy and Basic Education" the courses for adult learners try to overcome the mathematical autobiography of failure, which mathematical analphabets often have and attempt to regard the mathematical needs of learners.

4. Beliefs of adult education teachers

While mathematics teacher for primary and secondary school run through traditional academic studies many adult educators get their job by circumstances rather than by choice. Usually they attend some kind of training before they start working as an adult mathematics teacher. For research it's interesting to study their views of mathematics and how they relate to their practice. A qualitative study shows that adult mathematics teachers have a broad variety of positively connoted affects and their negative associations are more homogenous and often traced back to discriminating school experience. Positive and negative affective aspects can coexist in the same person and are linked to different strands of mathematics. One consequence could be that further training for adult education teachers in numeracy should result in increasing their awareness of the their mathematical views and how they relate to their practice.

5. Incorporating technology

As the state of research shows several attempts were mad to incorporate technology in adult mathematical education. Today there is great hope that Open Educational Resources (OER) promotes lifelong learning of mathematics, because there are many opportunities to use OER for adult learners. For example OER can be used to design courseware for adult learners without much financial burden. However at least there are two important conditions to speed up the usage of OER The first is the necessity of a special repository of OER for adult learners and the second is the necessity of establishing a mathematical community for adult learners to benefit from their experience.

The scope of this topic study group shows that future research should still concentrate on numeracy, investigate technology as a tool for adult education and should make an effort to develop adult teacher training as well as an advanced degree in adult mathematics education.

Open Access Except where otherwise noted, this chapter is licensed under a Creative Commons Attribution 4.0 International License. To view a copy of this license, visit http://creativecommons.org/licenses/by/4.0/.

Topic Study Group No. 07: Popularization of Mathematics

Christian Mercat, Patrick Vennebush, Chris Budd, Carlota Simões and Jens Struckmeier

The Popularization of Mathematics Study Group (TSG7) gathered for the first time in ICME diverse people using interesting and inspiring mathematics to motivate both young people and the general public.

It brought together those who popularize mathematics through live performance, exhibits, the media and outreach programs. The group of about forty people, included lively article discussions, poster presentations and demonstrations.

Enthusiastic practitioners stepped back to reflect on the impact of their actions. We first discuss their goals.

- Democratize mathematics
- Set a healthier relationship with mathematics
- Raise performance in math education
- Share math beauty, power and pervasiveness
- Justify taxpayer's money in research and education.

The means of expression are diverse:

1. Art and science (theater, films, visual arts)
2. Fixed, itinerant, and virtual exhibitions for museums, science centers or non-dedicated spaces. Science or mathematical festival or forums
3. Competitions in mathematics and computer science

Co-chairs: Christian Mercat, Patrick Vennebush.
Team members: Chris Budd, Carlota Simões, Jens Struckmeier.

C. Mercat (✉)
Université Claude Bernard Lyon 1, Lyon, France
e-mail: christian.mercat@math.univ-lyon1.fr

P. Vennebush
Silver Spring, Discovery Education, Washington, USA
e-mail: patrick.vennebush@verizon.net; patrick@mathjokes4mathyfolks.com

4. Mathematical camps
5. Contact with research mathematics and mathematicians
6. Inquiry/research based projects
7. Math circles/math clubs
8. Recreational mathematics
9. New technologies (apps, websites, …)
10. International exchanges.

It addresses different audiences and target groups tackling unequal access issues, talent, motivation, gender, social, financial or geographical differences, educational opportunities between countries. This TSG resonated very much with the invited lecture of Ricardo Nemirovski about *Informal Mathematics*.[1]

Twenty nine submissions were accepted, with seven posters, five oral presentations and seventeen articles. Their discussion was split in six main sessions.

A popular medium for popularization is printed material, nowadays usually supported by website archives, or totally online articles:

- Frédéric Gourdeau, from Laval university (Canada), presented *Accromath*,[2] a journal of the Institut des Sciences Mathématiques, a consortium of Canadian universities in the Québec region, and CRM, Center for Mathematical Research, founded in 2006, popularizing mathematics, mainly for secondary schoolers, in French.

- Vijayakumar Ambat from Cochin university (India), shared his experiences with massive production of math popularization in India, for example through print media with million copies in daily newspapers.
- Nils Berglund introduced us to *Images des mathématiques*,[3] a French website, stemming from a previous paper journal, dedicated to current *research* topics for the general public, written by a large community of professional mathematicians in France.
- The *Boletín*[4] is an online journal published since nine years by Juan J. Moreno-Balcázar and his 30+ team in the university of Almería (Spain), that identifies that outreach programs have to go beyond the curriculum in order to show usefulness and beauty of mathematics, rooting on extra motivational aspects such as bilingualism, problem posing and students contests.

---

[1] http://informalmathematics.org/.
[2] http://accromath.uqam.ca/.
[3] http://images.math.cnrs.fr/.
[4] http://boletinmatematico.ual.es.

Shows are another popular way to get a message across while entertaining its audience.

- Andrea Oliveira Hall and Sonia Pais, *resp.* from the University of Aveiro and Instituto Politécnico de Leira (Portugal), made us laugh and wonder with a mathemagical show, the *Mathematical Circus*, that since 2012 has pleased more than 10,000 spectators with numerical, topological or combinatorial tricks.
- Nitsa Movshovitz-Hadar, from Technion (Israel), talked about his experience of one-man shows public lectures about mathematics over four years.

Exhibitions, in museums are a traditional form of popularization of knowledge, in many fields but only recently in mathematics.

- Prof. Dr. Albrecht Beutelspacher hold an exhibition excerpt from the *Mathematikum*[5] in Gießen (Germany) that since 2002 welcomes hords of visitors, in a formula taken up by the *MoMath*[6] in New York.

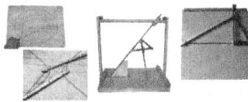

- Michela Maschietto, from the University of Modena (Italy), animates an exhibition of *Mathematical Machines* that are the content of hands-on activities and trainings, based on historical models and backed up by virtual models.[7]
- The *Houses of Mathematics*[8] were presented by Ali Rejali from the University of Isfahan (Iran), as well as similar institutions in Lyon[9] (France), Quaregnon[10] (Belgium), Grenoble[11] (France), Munich[12] (Germany) or *Archimedes Premises* in Belgrade (Serbia).
- Abdulkadir Erdoğan from Anadolu University (Turkey) presented (*in absentia*) examples of popularization activities in the *Math School* project, an interactive exhibition of workshop games based on *Maths à Modeler*,[13] in a specially repurposed classroom.

---

[5]http://www.mathematikum.de/.
[6]http://momath.org/.
[7]http://www.mmlab.unimore.it/.
[8]http://www.mathhouse.org.
[9]http://www.mmi-lyon.fr/.
[10]https://maisondesmaths.be.
[11]http://la-grange-des-maths.fr.
[12]http://www-m10.ma.tum.de/ix-quadrat/.
[13]http://mathsamodeler.ujf-grenoble.fr/.

Computers can be used in many different ways to popularize mathematics: as a modern mean of delivery and establishing a community, a way to coordinate collaboration networks and setting virtual laboratories with interactive material.[14]

- Anna Weltman from the University of California, Berkeley (USA) feeds *Math Munch*,[15] a popular yet sophisticated math blog targeting middle school students, their teachers and parents. The team of young mathematicians acts as ambassadors of the beauty and fun of math to kids, selecting each month items on the internet and putting them into perspective.

- Kerry Cue from Australia, presented her *MathsPigs* blog[16] with lots of down to earth examples of everyday useful mathematics.
- Andreas Daniel Matt and Bianca Violet, from MFO (Germany), invited all contributors to the *Imaginary*[17] international project, gathering software, films, posters and virtual models that can be used in order to build hands-on exhibitions around the globe.

- Ana Cristina Oliveira, from the University of Porto (Portugal), presented *Atractor*,[18] a Portuguese web based platform of realistic computer drawn simulations of mathematical objects, comprising a youtube channel, including stereo 3D movies, photorealistic virtual interactive exhibits and freeware educational software for elementary school and around the notion of symmetry.

---

[14]http://etudes.ru.
[15]http://www.mathmunch.org/.
[16]https://mathspig.wordpress.com/.
[17]https://imaginary.org/.
[18]http://www.atractor.pt/.

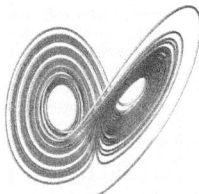

- Sergei Posdniakov from LETI (Russia) introduced computer assisted scientific activities, with virtual laboratories, such as *electronic detective*,[19] as a mean to support and motivate more traditional popularization articles.

Handheld mobile devices have their own niche in popularization of mathematics.

- Donna Ann Dietz, from the American University, writes mathematical apps for mobile devices[20] and evaluates its impact in students' achievements in courses on the content.

- Adi Nur Cahyono, from Univ. Frankfurt (Germany) and Semarang State Univ. (Indonesia), presented the *MathCityMap*[21] project that sets up mathematical modeling tasks that are associated with spots along a city trail and can be explored with a GPS-enabled mobile app.
- Nataly Essonnier from University Claude Bernard Lyon 1 (France), explained how to use *Tetrisquiz*[22] and *EpsilonWriter* in order to gamify easily some math quizz, from simple ones to more sophisticated, usable on a laptop, tablet or handheld device.

Popular culture, especially comics and games might foster motivation.

---

[19]http://edetective.ipo.spb.ru/.
[20]http://www.donnadietz.com/.
[21]https://mathcitymap.eu.
[22]http://tquiz.org/.

- Comics storytelling style is used in the *Magical*[23] project of the National Institute of Education, Nanyang Tech. Uni. (Singapore), presented by Toh Tin Lam for secondary school students. This non academic contextualization of the material aims at changing the motivation of students.
- Andreas Hinz from LMU Munich (Germany) entertained us with popular puzzling enigmas, requiring ingenuity and patience but as well knowledge, pointing out the unifying force of mathematics, especially graph theory, in their resolution.

- Mohammad Bahrami from the Shahid Beheshti University, Tehran (Iran), published a poster about the use of interactive magic tricks and islamic art videogame *Engare*.[24]

Arts and pictures show the beauty of mathematics.

- Jacinto Eloy Puig Portal from the University of Los Andes (Colombia), presented the incredible artwork of students inspired by mathematics, especially mimicking M.C. Escher ideas using nowadays technology and unleashing the creative potential of mathematical thinking.

- Emmanuelle Forgeoux from IREM Rennes (France), reported on *Mathematical Selfies*,[25] a joint program with Western Carolina University where students illustrate mathematical concepts with photographs and produce inspiring galleries.

Math clubs can be close to the classroom, related to competitions, or during holiday camps.

---

[23]http://math.nie.edu.sg/magical/.

[24]http://www.engare.design/.

[25]http://mathematicalselfies.blogspot.fr/.

- Aviva Szpirglas, from Poitiers University (France) described *Math.en.Jeans*,[26] a yearly long initiation to mathematics, with questions introduced by a researcher and followed by a teacher (who is a learner at another level), helping students to go through the frustration and exhilaration of research, until the publication in a national forum, just like professional mathematicians, with poster and lectures in front of hundreds other kids.
- The 180 US *Math Circles*,[27] presented by Brandy S. Wieger from Central Washington University (USA), inspired by Eastern European ones, mixes professional mathematicians, K-12 students and their teachers, in a variety of events formats.

Outreach programs can be events, targeting special groups that require special care, mainly in changing the attitude towards mathematics before actually teaching some specific curricular content.

- Violeta Vasilevska, from Utah Valley university (USA), engages, since 2007, women in mathematics and especially, since 2011, high-school girls in the *Math Girls Rock* project,[28] a math club for young women throughout the school year. She presented the project and survey results showing positive attitude shifts, seeing mathematics as more "likable".
- South African senior high school students benefit from a ten day outreach program tapping into adolescents' positive motivation construction strategies, based on self-determination and identity building, driven by autonomy, connectedness and competence, for a lasting effect on students.
- Veronica Sarungi from the Aga Khan University (Tanzania), showcased the Pre-Pi Days events, exhibitions and hands-on activities in lower secondary schools, a fulfilling experience that, in her opinion, ought to be developed in other countries.

Some higher perspectives on popularizing mathematics were taken:

- R. Athmaraman of the Association of Mathematics Teachers of India in Chennai (India), put macro socio-historical and micro perspectives on the attitude towards mathematics. History of ideas, technology and mathematics, as well as psychological studies might inform us of which socio-cultural background can be mathematically friendly, setting an agenda for further research.
- Alix Boissière from Association *Plaisir-Maths*[29] (France), introduced the didactical and play-based contract as a tool to assess and design popularization activities.

---

[26] http://www.mathenjeans.fr/.
[27] http://mathcircles.org/.
[28] https://www.uvu.edu/math/mgr/.
[29] http://www.plaisir-maths.fr/.

- Martin Andler, from the university of Versailles and president of association Animath[30] (France), told the lessons learnt in the *Cap'Maths* project, uniting most popularization activities in mathematics in France in 2012–2106.
- Hong Zhang from the Sichuan University (China), described how historical turmoils make radical changes in math teaching.

Participating in this Topic Study Group has been a very interesting yet challenging experience. We met interesting people, learnt new ways to share our love and passion for mathematics, but the progress towards establishing Popularization of Mathematics as a new respected field of research in math education is yet to come.

Open Access Except where otherwise noted, this chapter is licensed under a Creative Commons Attribution 4.0 International License. To view a copy of this license, visit http://creativecommons.org/licenses/by/4.0/.

---

[30]http://www.animath.fr/.

Topic Study Group No. 8: Teaching and Learning of Arithmetic and Number Systems (Focus on Primary Education)

Pi-Jen Lin, Terezinha Nunes, Shuhua An, Beatriz Vargas Dorneles and Elisabeth Rathgeb-Schnierer

The purpose of this TSG is to gather congress participants who are interested in research and development in the teaching and the learning of number systems and arithmetic through activities in and out of school. The mathematical domains include whole numbers, integers, ratio and proportion, and rational numbers as well as representations and problem solving using numbers related to each of these domains. We invited submission of research-based proposals for contributions to TSG 8 that could fall into (but are not bounded to) the following issues related to:

- Developing a deep understanding of arithmetic and number systems.
- Developing number sense in children as a foundation for learning arithmetic and flexibility with numbers and operations.
- Assessing conceptual or perceptual knowledge in learning arithmetic and number systems.
- Curriculum development and implementation, for instance, approaches to introducing numbers and comparative analysis of different curricula in one country or across countries.
- Instructional models and strategies for teaching arithmetic and number systems.
- Developing mathematics proficiency in teaching arithmetic and number systems.

Co-chairs: Pi-Jen Lin, Terezinha Nunes.
Team members: Shuhua An, Beatriz Vargas Dorneles, Elisabeth Rathgeb-Schnierer.

P.-J. Lin (✉)
National Hsinchu University of Education, Hsinchu City, Taiwan
e-mail: linpj@mail.nhcue.edu.tw

T. Nunes
University of Oxford, Oxford, UK
e-mail: terezinha.nunes@education.ox.ac.uk

- Using of tools, such as technology, manipulatives, and children literature, etc. in teaching and learning number systems and arithmetic.
- Professional development and teacher education related to teaching arithmetic and number systems.
- Cultural tools and practices for the learning and teaching of arithmetic and number systems.

Three theoretical presentations on teaching and learning arithmetic and number systems provided the basis for discussion in each of the three 120-min slots. These were followed by presentations of empirical research that provide support, challenge or extend the theoretical frameworks. TSG participants were expected to stay with their group throughout the three sessions.

The first day started (chaired by Terezinha Nunes) with presentations by Verschaffel, Lieven; Nunes, Terezinha; Kheu, Natalie Ming Yeng; Torbeyns, Joke.

Our second conference TSG day was split in two rooms and talks were offered by: Rathgeb-Schnierer, Elisabeth; Hickendorff, Marian; Schulz, Andreas; McMullen, Jake; Dorneles, Beatriz Vargas; Chong, Yeog Ok; Koudogbo, Jeanne; Mamede, Ema.

Our third conference day was also split in two slots because there were so many excellent papers. Presentations were offered by Gaidoschik, Michael; An, Shuhua; Mizzi, Angel; Real Ortega, Carolina Rubi; Barros, Rossana; Hadi, Sutarto; Hsu, Wei-Min; Ghosh Hajra, Sayonita.

The last day of your TSG session day ended with papers by Lin, Pi-Jen; Vamvakoussi, Xenia; Rechtsteiner- Merz, Charlotte; Brown, Bruce.

Open Access Except where otherwise noted, this chapter is licensed under a Creative Commons Attribution 4.0 International License. To view a copy of this license, visit http://creativecommons.org/licenses/by/4.0/.

Topic Study Group No. 9: Teaching and Learning of Measurement (Focus on Primary Education)

Christine Chambris, Barbara Dougherty, Kalyanasundaram (Ravi) Subramaniam, Silke Ruwisch and Insook Chung

The Programme

Preparation

Measurement, the topic of this TSG, links not only to everyday contexts and application areas such as engineering, but also with other mathematical topics including number and algebraic thinking. Weak knowledge related to measurement concepts and skills often becomes problematic while studying other subjects. Yet, paradoxically, there is a lack of attention to measurement instruction in primary mathematics education internationally.

The main purpose of the TSG was to better understand conditions and constraints on the teaching and learning of measurement in international contexts, and to consider some possible changes. Specific questions announced as foci of the TSG were the extent to which

(a) measurement could be a topic in and of itself in primary school mathematics;
(b) measurement could be a vehicle for connecting other mathematical topics such as number, operations, algebra, statistics, or geometry;

Co-chairs: Christine Chambris, Barbara Dougherty.
Team members: Kalyanasundaram (Ravi) Subramaniam, Silke Ruwisch, Insook Chung.

C. Chambris (✉)
Université Paris Diderot, Paris, France
and
Université de Cergy-Pontoise, Cergy-Pontoise, France
e-mail: christine.chambris@u-cergy.fr

B. Dougherty
University of Missouri, Missouri, USA
e-mail: doughertyb@missouri.edu; barbdougherty32@icloud.com

(c) measurement learning could be supported by mathematical topics such as number, operations, algebra, statistics, or geometry; and
(d) informal knowledge of and conceptual understanding about measurement (including estimation and knowing how to use some instruments) support or hinder rich teaching and learning of measurement or of other mathematical subjects in school.

The TSG received 18 papers and one poster as submissions including 4 invited presentations from leading researchers, of which nine papers were accepted for long presentations (20 min presentation and 10 min discussion), nine papers for two short oral presentations sessions and one poster. Due to cancellations, the final program of TSG-9 consisted of nine long presentations and eight short presentations.

In relation with the four questions, four major themes emerged and later structured the four sessions: connections between measurement and other mathematical topics, connections between measurement and everyday life, how conceptual understanding of measurement of geometrical quantities develops, connections between math and physics through measurement. From the short oral presentations; a fifth theme emerged: estimation.

Implementation

Day 1 (July 26)

In the main opening session, welcome and introductions were done by Christine Chambris (co-chair) and Silke Ruwisch (team member). In her introductory remarks, Christine Chambris recalled the key issues discussed in the TSG of ICME 12, and referred to the paradox of the lack of attention to measurement in mathematics education referred to above. The two first presentations were chosen for showing how school measurement can on the one hand foster the teaching of other mathematical topics, and on the other hand how it can be connected with everyday life.

First, TSG-invitee Jeffrey E. Barrett (and colleagues) from the United States investigated the intersection of spatial measurement and school mathematics. They featured a research program that connects measurement to other topics in mathematics through tasks involving the coordination of number and space across representations, comparative reasoning about quantity and equivalence, and ideas about units, including grouping and partitioning. Second, Arindam Bose and K. (Ravi) Subramaniam from India showed that school-going children from low income urban households in developing countries often have work related experience through their participation in the informal economy; and how such experience gives rise to measurement related knowledge, which is different from the scientific knowledge of measurement that forms part of the curriculum. They characterized these differences and explored pedagogical implications.

Two sessions consisting of short oral presentations took place. The first session focused on estimation (area, volume and assessment). The second session included presentations on measurement of length, area including the use of "bissemis", and time.

Day 2 (July 27)

The next main session focused on the extent to which measurement can be used as a vehicle for connecting and linking other mathematical topics (such as number, operations, algebra, statistics, or geometry). The converse issue was also addressed how mathematical topics such as number, operations, algebra, statistics, or geometry can support the development of measurement concepts in school.

Three papers were presented in this session followed by insightful discussions. First, TSG-invitee Richard Lehrer (and colleagues) from the United States presented a research project within which geometry and measurement learning was built on informal knowledge through embodied experiences of motion. Based on the teaching of geometrical measurement as a conceptual system, it showed transitions in teachers' pedagogical practices and conceptions of measurement that support children's conceptual change. This presentation suggested that a whole group of teachers can progressively but dramatically change when they are trained and supported with resources within a scientifically coherent project.

Second, Linda Venenciano from the United States presented how measurement can be used as a vehicle for developing algebraic thinking. Based on the works of El'konin and Davydov, mathematics representations (e.g., models and equations) embody quantities and their relationships and scaffold the development of reasoning skills found to be critical to successful preparation for a first course in high school algebra.

Third, Christine Chambris presented a brief history of the relations between numbers and quantities in the primary curriculum in France. She reported the changes that occurred in the French curriculum surrounding the New Math reform which resulted in the elimination of measurement from the theoretical foundations of numbers, and how such changes still impact the teaching of both numbers and measurement. This gave the participants an opportunity to reflect on the role of the New Math in their own countries.

Day 3 (July 29)

Three papers were presented in this session and thought-provoking issues were raised. First, the TSG-invitee Julie Sarama (and colleagues) from the United States presented the development of foundational cognitions and concepts of measurement in the early years framed through the research construct of a learning trajectory for length. They described how conceptual understanding of length measurement develops through a typical curriculum in the USA among children of 4–8 years of age. They focused on a case study of an 8-year-old who could successfully use measurement tools such as a ruler and feet-steps but struggled with strips.

Second, Jeenath Rahaman (and colleague) from India presented two classroom episodes on the construction of the concept of area measurement in India, with students in grades 6 and 7. She first explored the structure of the argumentation and showed that students' and teacher's argumentations were based on different implicit assumptions. Such analysis highlighed tensions when students move between spatial and numerical representations.

Third, Cheryl L. Eames (and colleagues) from the United States presented the evaluation of a hypothetical learning trajectory for length measurement using a partial credit Rasch model with students from grade 4–10. Samples of students' responses were presented, and the analysis highlighted the need for a means to distinguish between density of length, and density of numbers in evaluating students' knowledge.

Much intense discussion linking the ideas contained in the presentations took place in this session: learning trajectories, relations between the development of numbers and quantities, and quantitative equivalences between geometrical shapes.

Day 4 (July 30)

The fourth session consisted of one presentation followed by a review of all the TSG presentations. The TSG-invitee Valérie Munier (and Aurélie Chesnais) from France presented epistemological issues and treatment within French textbooks of measure and measurement in physics and mathematics. The presentation notably raised critical issues on modeling. Indeed, differences in how models are validated in mathematics and physics are often ignored in the teaching. They also demonstrated the need of the improvement of the distinction between empirical and theoretical aspects of measure.

Ravi Subramaniam and Christine Chambris then presented insights gained from the TSG. They noted that viewed historically, everyday measurement has increasingly disappeared from textbooks over a century (Subramaniam & Bose (ICME-12), Chambris). Two possible explanations of this disappearance emerged: first, the process of de-mathematization that characterises the interaction between mathematics and larger societal change, and second, a change within mathematics itself (the grounding of numbers in set theory), leading to a separation of measurement from arithmetic (Chambris). Reconceptualizing the place of measurement in the curriculum may need one to address both these factors.

We might look at counter-trends to demathematization as exemplified in the out-of-school knowledge that children acquire in the informal economy of the community and leverage them for school learning (Bose & Subramaniam). We might give a more central place to estimation: understand more carefully different kinds (or meanings) of estimation, estimation strategies, and build activities around them (Ruwisch; Pizzaro, Gorgorio & Albarracine, Huang). We might introduce activities that build an embodied understanding of quantity through motions of different kinds (Lehrer). We might use measurement as a context to introduce core mathematical topics such as negative integers and algebra (Venenciano: measure up curriculum), or more generally use measurement as a connecting thread (Barrett and others).

Indeed, the teaching of measurement needs to bridge the gap between informal and formal activities. The use of standardized tools and formula often hides the conceptual features of measurement; the most fragile students often struggle when working on representations, not able to see the links with the object represented (ICME 12).

Measurement as a topic is inherently integrative. But integration brings in new theoretical perspectives and theoretical issues even for the way measurement is treated within the mathematics curriculum. Especially, it raises the issue of the theoretical model of quantity (or that of measurement) which is required to build relations between quantities, measurement and numbers (Barrett, Chambris, Eames, Lehrer, Munier & Chesnais, Rahaman, Venenciano).

Reflection

Generally speaking, TSG-9 had regular attendants who were ready to engage in rich discussion throughout the four sessions in a receptive atmosphere. The relatively small number of papers confirms that there is a lack of attention to this domain. Despite this, various and new issues were raised and the necessity of further international studies in the domain of measurement was emphasized by the participants. We hope that the topic study group dealing with measurement continues to serve as a well-recognized group of the congress.

Open Access Except where otherwise noted, this chapter is licensed under a Creative Commons Attribution 4.0 International License. To view a copy of this license, visit http://creativecommons.org/licenses/by/4.0/.

Topic Study Group No. 10: Teaching and Learning of Early Algebra

Carolyn Kieran, JeongSuk Pang, Swee Fong Ng, Deborah Schifter and Anna Susanne Steinweg

The Programme

The full programme for the Topic Study Group on the Teaching and Learning of Early Algebra featured 4 plenary activities, one of which was the opening panel with four presenters and one reactor, 7 research reports, 22 short oral communications, and 9 posters. Close to 85 congress attendees participated in TSG 10 and contributed by their participation to its various activities, all of which allowed brief time for them to pose questions, offer remarks, and engage in some discussion. The four main sessions of TSG 10 were structured as follows:

Session 1, July 26, 2016, 12:00–13:30: Theme of the session: *Epistemological Perspectives on Early Algebra* (a plenary panel involving five contributions—the presenting author's name is underlined).

<u>Maria Blanton</u>, Bárbara M. Brizuela, Ana C. Stephens: *Elementary Children's Algebraic Thinking.*

<u>John Mason</u>: *How Early Is Too Early for Thinking Algebraically?*

<u>Nicolina Malara</u>, Giancarlo Navarra: *Epistemological Issues in Early Algebra: Offering Teachers New Words and Paradigms to Promote Pupils' Algebraic Thinking.*

Co-chairs: Carolyn Kieran, JeongSuk Pang.
Team members: Swee Fong Ng, Deborah Schifter, Anna Susanne Steinweg.

C. Kieran (✉)
Université du Québec à Montréal, Montréal, Québec, Canada
e-mail: kieran.carolyn@uqam.ca

J. Pang
Korea National University of Education, Cheongju, South Korea
e-mail: jeongsuk@knue.ac.kr

David W. Carraher, Analúcia D. Schliemann: *Functional Relations in Early Algebraic Thinking.*

Carolyn Kieran: *Reaction to the Panel Contributions: The Structural Facet of Early Algebraic Thinking.*

Session 2, July 27, 2016, 12:00–13:30: Theme of the session: *Learning Perspectives on Early Algebra* (a plenary presentation followed by two research reports):

JeongSuk Pang: *A Review of Recent Research That Foregrounds the Early Algebra Learner.*

Kathrin Akinwunmi: *On the Development of Variable Concepts by Generalizing Mathematical Patterns in Primary School.*

Aisling Twohill: *The Approaches to Solution of Linear Figural Patterns Adopted by Children Attending Irish Primary Schools.*

Session 3, July 29, 2016, 12:00–13:30: Theme of the session: *Additional Learning Perspectives on Early Algebra and Links to Later Algebra* (a plenary presentation followed by three research reports):

Swee Fong Ng: *A Neuroscience Perspective on Early Algebra: Symbolic and Diagrammatic Approaches to Algebra Problem Solving.*

Yasufumi Kuroda, Naoko Okamoto: *Changes in Brain Activity While Engaging in Number Sequence Questions of Varying Difficulty.*

Catherine Pearn, Max Stephens: *Fraction Tasks and Their Links to Algebraic Thinking.*

Anna Susanne Steinweg: *Algebraic Thinking—Mathematical Key Ideas.*

Session 4, July 30, 2016, 12:00–13:30: Theme of the session: *Teaching Perspectives on Early Algebra* (a plenary presentation followed by two research reports, and then closing remarks by the five members of the TSG 10 organizing team):

Deborah Schifter: *Bringing Early Algebra into Elementary Classrooms.*

Jodie Hunter: *Scaffolding Teacher Practice to Develop Early Algebraic Reasoning.*

Susanne M. Strachota: *Cycles of Generalizing Activities in the Classroom.*

Prior to the unfolding of the ICME-13 activity engaged in by Topic Study Group 10, a pre-conference monograph presenting a topical survey of Early Algebra research (Kieran, Pang, Schifter, & Ng, 2016) was published by Springer and made available on the Internet as an open access eBook (ISBN 978-3-319-32258-2). As described in the monograph, the core of recent research in early algebra has been a focus on mathematical relations, patterns, and arithmetical structures, with detailed attention to the reasoning processes used by young students, aged from about 6 to

12 years, as they come to construct these relations, patterns, and structures—processes such as noticing, conjecturing, generalizing, representing, and justifying. Intertwined with the study of the ways in which these processes are engaged in are the two main mathematical content areas of generalized arithmetic (i.e., number/quantity, operations, properties) and functions. The monograph highlighted how the field of early algebra has gradually come to be more clearly delineated since the early 2000s, bringing with it more comprehensive views and theoretical framings of algebraic thinking. Thus, the contents of the monograph set the stage for TSG 10 contributors to link their newest work to the advances of the fairly recent past, as well as to signal further evolution of the field.

One of the many interesting aspects to emerge during the four main sessions of the ICME-13 Topic Study Group 10 was the attention paid to the key notion of *structure*. For example, Blanton emphasized four essential practices that characterize early algebraic thinking: generalizing mathematical structure and relationships, representing mathematical structure and relationships, justifying mathematical structure and relationships, and reasoning with mathematical structure and relationships. Steinweg pointed to four key ideas of algebraic thinking: pattern structures, property structures, equivalence structures, and functional structures. Mason argued that looking at something structurally is an often-overlooked aspect of algebraic thinking and can be encouraged by offering a partial generalization or a very general statement and then giving students the opportunity to specialize. Malara illustrated how students can learn to gradually represent and express structural aspects of number in transparent, non-canonical ways, aided by collective confrontation in class. Carraher emphasized the affordances of the N-number line representation—a representation that is especially rich for helping young students focus on the structure of numbers and the relation between one number and its neighboring number.

Structure is clearly one of the central pillars in the development of early algebraic thinking. Kaput (2008), one of the pioneers of the Early Algebra movement, included within his three main strands related to algebraic thinking: "algebra as the study of structures and relations arising in arithmetic." But, as some of the TSG 10 presenters argued, the work of noticing underlying structures is not necessarily straightforward. In the language of Radford (2011, p. 23), "the awareness of these structures and their coordination entail a complex relationship between speech, forms of visualization and imagination, gesture, and activity on signs." Furthermore, "the mathematical work of teachers in pressing students, provoking, supporting, pointing, and attending with care" (Bass & Ball, 2003, p. vii) is critical to the development of young students' awareness of structure. As pointed out by Schifter in her TSG presentation, if teachers understand mathematics as procedures for calculating and solving problems, they must widen their view to include looking for and examining structure, and as well, according to Hunter in her TSG presentation, learn to recognize the inherent algebraic structure of number. While much of the pioneering work in early algebra has focused on the process of generalization, several TSG 10 presenters at ICME-13 pointed out that what it is that is generalized in much of early algebraic thinking and activity are the structural aspects of numerical relationships and patterns. Thus, attending to structure is key to the process of generalizing, However, it was also suggested at TSG 10 that more work remains to be done not only in making explicit the various meanings that are attributed to the term *structure*, but also in characterizing the diverse ways in which structure can be expressed by students who are developing algebraic thinking in the different content areas of early algebra. This emphasis will be one of the themes of a planned follow-up volume related to the ICME-13 activity of TSG 10.

References

Bass, H. B., & Ball, D. L. (2003). Foreword. In T. P. Carpenter, M. L. Franke, & L. Levi, *Thinking mathematically: Integrating arithmetic and algebra in elementary school* (pp. v–vii). Portsmouth, NH: Heinemann.

Kaput, J. J. (2008). What is algebra? What is algebraic reasoning? In J. J. Kaput, D. W. Carraher, & M. L. Blanton (Eds.), *Algebra in the early grades* (pp. 5–17). New York: Routledge.

Kieran, C., Pang, J. S., Schifter, D., & Ng, S. F. (2016). *Early algebra: Research into its nature, its learning, its teaching*. New York: Springer (open access eBook).

Radford, L. (2011). Embodiment, perception and symbols in the development of early algebraic thinking. In B. Ubuz (Ed.), *Proceedings of the 35th Conference of the International Group for the Psychology of Mathematics Education* (Vol. 4, pp. 17–24). Ankara, Turkey: PME.

Open Access Except where otherwise noted, this chapter is licensed under a Creative Commons Attribution 4.0 International License. To view a copy of this license, visit http://creativecommons.org/licenses/by/4.0/.

Topic Study Group No. 11: Teaching and Learning of Algebra

Rakhi Banerjee, Amy Ellis, Astrid Fischer, Heidi Strømskag and Helen Chick

The Programme

TSG-11 on Teaching and Learning of Algebra had a small number of presentations in the main session, leaving enough space for discussions and dialogue. The TSG planned to cover the salient themes and ideas in algebra education, including early algebra, algebraic thinking, conjecturing, proving and generalizing and algebra instruction. Each of the sessions had two presentations, one of which was an invited talk by an eminent scholar in the field, focusing on one or more of the themes that were identified in the TSG and another one selected from the papers submitted to the group. The TSG was able to bring forth significant ideas for discussion within the group. The presentations gave theoretical, methodological and empirical insights into students' construction of algebraic knowledge. Below, we give the programme details and brief summary of the sessions.

Co-chairs: Rakhi Banerjee, Amy Ellis.
Team members: Astrid Fischer, Heidi Strømskag, Helen Chick.

R. Banerjee (✉)
Azim Premji University, Bangalore, India
e-mail: rakhi.banerjee@gmail.com; rakhi.banerjee@apu.edu.in

A. Ellis
University of Wisconsin-Madison, Madison, USA
e-mail: aellis1@education.wisc.edu; aellis1@wisc.edu

Day	Speaker	Title
Tuesday, July 26, 2016	Kaye Stacey	Algebra research to guide teaching
	Andrew Izsák, Sybilla Beckmann, Eun Jung, Ibrahim Burak Ölmez	Connecting multiplication, unit fractions, and equations
Wednesday, July 27, 2016	Maria Blanton, Barbara M. Brizuela and Ana C. Stephens	Children's understanding and use of variable notation
	Jan Block	Flexible algebraic action: Solving of algebraic equations
Friday, July 29, 2016	Jinfa Cai	Early algebra learning: Answered and unanswered questions
	Thomas Janßen	Developing algebraic structure sense in linear equations as tuning into a new activity
Saturday, July 30, 2016	Heidi Strømskag	Evolution of the *milieu* for a particular piece of mathematical knowledge
	Erik Tillema and Andrew Gatza	A quantitative approach to establishing cubic identities

Kaye Stacey's talk introduced the participants to the ideas that have emerged through decades of research in the area of algebra education and how they can guide teaching. In the process, she introduced us to the content and structure of the new book The Teaching and Learning of Algebra: Ideas, Insights and Activities, she has co-authored with Abraham Arcavi and Paul Drijvers. It is a useful resource for teachers and researchers. The talk highlighted the knowledge generated about the aims of algebra and its use and the ideas that as students and teachers, one has to deal with. It reiterated the key ideas of algebra, that is, generalizing, exploring properties and relationships, problems solving, and proving theorems. Students' difficulties and challenges with algebra (like, coming to terms with the letter, use it for representing, making sense of it in different contexts and work with symbolic expressions with meaning) and the insights they offer for teaching algebra were discussed. Technology has the potential in supporting teaching and the possibilities need more exploration.

Andrew Izsak's presentation argued for a continuity between arithmetic and algebra and also going beyond the whole numbers to fractions. The presentation talked about an intervention study with pre-service teachers at the middle school, which aimed at helping them connect fractions, proportional relationships and linear equations. The study took a quantitative meaning of multiplication to see fractions as multipliers of unknowns. For instance, in the equation $M \cdot N = P$, the letter M refers to the number of equal-size groups, N refers to the number of units in each of those groups, and P refers to the number of units in M groups. The data from the one-on-one clinical interviews with six pre-service teacher participants was presented. The study revealed that the understanding of unit fraction $(1/b)$ as the number of groups (in $1/b \cdot X$) requires coordination of multiple knowledge elements—unit fraction as a result of splitting a whole, unit fraction as partitioning a group of X units, meanings for multiplication and definitions of fraction and interpreting symbolic expressions.

It was essential for this group also to engage with the early algebra thoughts and literature because it makes us aware of multiple possibilities for introducing algebraic thinking in early years. Maria Blanton presented her group's work on children's understanding and use of variable notation across elementary grades. The talk discussed an intervention study in early algebra for children in grades 3–5 which dealt with generalized arithmetic, equivalence, expressions, equations and inequalities and functional thinking. Children in the intervention group (here only grades 3 and 4 were reported) performed significantly better in the post-test compared to a similar control group, in tasks which required them to make an expression and an equation using the variable and a representation for a functional relationship. It showed the readiness and preference of young children to use the variable notation. She also reported from other intervention studies with children of grades of K-1 showing their capacity to use variable notation for modeling situations as well as in functional relations.

The presentation by Jan Block explored the idea of flexible algebraic action among grade 9 and 10 students in the context of solving quadratic equations. Building on the theory of didactical-cut, it tried to explore what features of a quadratic equation are perceived by students and how they use this information and whether it facilitates or hinders flexible algebraic action. It discussed the wide use of the quadratic formula (pq-formula) and the trial-and-error method to solve equations among these students together with high error rate. It concluded by stating that teaching different strategies for solving quadratic equations is not going to lead to flexible algebraic activity. Rather one has to engage in meta-tasks to identify features that make certain strategies relevant for solving it.

In his talk, Jinfa Cai used statistics from the US National Educational Longitudinal Study (NELS) of 1988 to illustrate why algebra is important. NELS showed that students who take Algebra 1 in high school are much more likely to go to college than those who do not: 83% of students who take Algebra I go to college, whereas 36% of students without Algebra 1 do. Further, Cai showed that students who pass Algebra 2 in high school were 4.15 times more likely to graduate from college than students who have not. Then the LieCal project was presented, which longitudinally investigates the effects of the *Standards*-based Connected Mathematics Program (CMP) curriculum on students' learning of algebra to the effects of more traditional middle-school mathematics curricula. Cai presented example problems from the two curricula: the CMP curriculum represented a functional approach, with emphasis on change, variation, and relationships between variables; and, traditional curricula represented a structural approach, with emphasis on procedures and abstract work with symbols.

Thomas Janßen's presentation was about algebraic structure sense for linear equations, and how it can be developed from structure-seeing. The discussion was based on transcripts and drawings from video-recorded classroom observations of four Grade 8 students working on linear equations. The structure of linear equations had been introduced through a puzzle: On each side of the equal sign there was the same number of matches, some of them in matchboxes, with the same amount of matches in each box. The task was to find a way to determine the number of matches in each box. Janßen showed that the development of algebraic structure sense can be understood as happening in moments of *tuning*—where tuning is a form of social interaction characterized by a common interest and a common understanding of the situation and the goals of the activity, and further, by a common understanding of what actions are necessary to achieve the goals.

In her talk, Heidi Strømskag presented a semiotic analysis of three students teachers' engagement with a generalization task in geometry. She explained how an evolution of the milieu (in Brousseau's sense) enabled the student teachers to create manipulatives (plane geometrical figures) that were instrumental in the generalization process aiming at a relationship between percentage growth of length and area when looking at the enlargement of a square. It was shown how use of different notation systems constrained the interaction among the participants, and how transformation of percentage and fractional notation into geometrical figures—that belong to a different semiotic register—enabled the target mathematical knowledge to be expressed in algebraic notation. Strømskag made a general point about design of milieus for algebraic generalization: the adidactical potential of a situation depends upon a coordination between the particular values that students are asked to work on and the semiotic register(s) expected to the used.

Erik Tillema presented an interview study of eight Grade 10–12 students' generalizations made in the context of solving combinatorics problems about cubic relationships. Students' generalizing actions were characterized by schemes, where a scheme has three parts: an assimilatory mechanism; an activity; and, a result. Tillema showed how two schemes were pre-requisites to establishing the formula that $(x+1)^3 = x^3 + 3 \cdot (x^2 \cdot 1) + 3 \cdot (x \cdot 1^2) + 1^3$. The first was a scheme to quantify the total number of three card hands using multiplication that was coordinated with a systematic way to list all possible outcomes, and the second was a scheme that enabled students to spatially structure 3-D arrays. Further, he showed that images based on quantitative relationships supported student generalizations. Tillema explained that the formula (above) that one student created was a formal statement of generalization (a reflection generalization) that was based on an abstraction in which she connected the activity of her scheme with the results of her schemes (reflective abstraction).

Open Access Except where otherwise noted, this chapter is licensed under a Creative Commons Attribution 4.0 International License. To view a copy of this license, visit http://creativecommons.org/licenses/by/4.0/.

Topic Study Group No. 12: Teaching and Learning of Geometry (Primary Level)

Sinan Olkun, Ewa Swoboda, Paola Vighi, Yuan Yuan and Bernd Wollring

Introduction

The aim of the TSG 12 working group is to promote the sharing of research on early geometrical thinking and understanding with a special focus on the kindergarten and primary education level.

The group provided a forum for discussion of the learning and teaching of geometry, from the historical, epistemological, cognitive, semiotic, and educational points of view, related to students' difficulties and to the design of teaching and curricula.

The topic study groups were organised in four sessions, each of them with 30–40 participants, with 4 invited lectures, 23 papers, and 2 posters.

During the working group, a main important suggestion emerged: The topic of geometry plays a limited role in early mathematical practices, while for young children geometric and spatial thinking could be fundamental in development in and beyond geometry.

Co-chairs: Sinan Olkun, Ewa Swoboda.
Team members: Paola Vighi, Yuan Yuan, Bernd Wollring.

S. Olkun (✉)
TED University, Ankara, Turkey
e-mail: sinanolkun@gmail.com; sinan.olkun@tedu.edu.tr

E. Swoboda
University of Rzeszów, Rzeszów, Poland
e-mail: eswoboda@ur.edu.pl

Description of TSG-12 Sessions

Recognition and Classification of Shapes

The problem of identifying, defining, and classifying geometric shapes was the dominant issue in discussions. This issue is not a new one, but its implementation can be treated very differently. Still, many researchers believe that the ability to recognise shapes is one of the essential elements of early geometrical knowledge. This view does not arouse controversy, but the way it has been implemented in research varies. Much research has determined the current level of students' knowledge in this area, but more and more often didactical proposals have been presented that create an area for collecting of geometric intuition.

Clements and Sarama focused on the 3- to 8-year-old children's knowledge of shapes. They described a large-scale study that mainly dealt with a detailed analysis of children's responses to a set of shape identification tasks (related to typical shapes) using a wide variety of examples and non-examples. The main result obtained is that "pre-school-aged children can learn not only matching and naming shapes but also learn about their components and (some) properties".

However, researchers have devoted more attention to testing various proposals about how to build students' knowledge. Such ideas often take the form of a comprehensive long-term program in which many issues of geometrical education permeate. Often, these proposals are embedded in a rich learning environment. These studies also reveal new aspects of the knowledge of geometric figures in children's minds. The topic of "shape recognition" was the subject of the research presented by Coutat and Vendeira. They developed pre-geometric activities, starting from a collection of 75 shapes to manipulate, using didactical variables such as the number of sides, the convexity, the presence of straight or curved sides, and so on. They analysed the results in terms of "student's perception of the shape," "the use of characteristics," and "the use of a pertinent language," concluding that "manipulation helps the student in the use of characteristics to recognise shapes," promoting a progressive change in their visualisation. Another interesting proposal was presented by Vighi that connects the creation of knowledge of geometric figures with students' own creativity in the artistic environment. Activity was realised with children 5–6 years old, starting from a painting by Kandinsky and reproductions of it made by pupils. The main result involved the problem of manipulation of non-reversible shapes. Jirotková studies the mechanism of birth and development of geometrical schema in a pupil's mind. She presented three manipulative learning environments that contribute to building mental schemas of geometrical objects, relationships, and processes in a pupil's mind.

Definitions of Shapes

Okazaki presented an analytical framework for the paths that students follow in constructing definitions, based on their understanding of inclusion among geometric figures. He found four kinds of understanding of inclusion relations: judgment based on visual characteristics, two common properties, relation between intensions and extensions, and genus–differentia definition.

Iskenderoğlu and Akġan developed a study to analyse the knowledge of definitions of two-dimensional geometrical concepts by prospective teachers, in which the definitions of twelve geometrical concepts was studied (angle, polygon, triangle, rectangle, trapezoid, parallelogram, rhombus, oblong, square, deltoid, and circle). The prospective teachers were struggling with the tests, often the definitions were not fully complete, possibly because of insufficient field knowledge. These results confirm that teacher preparation should be revised. This opens up a new research area on this issue.

Brunheira and Ponte analysed an exploratory task based on the hierarchical classification of quadrilaterals submitted to 30 prospective elementary teachers. The results highlighted the role of the constructions and negotiation of meanings. Gurhan and Zembat investigated the same topic, studying the main tenets of a technology-supported instructional sequence to foster a deep understanding of the hierarchy of quadrilaterals. Based on earlier work, argumentative activities in the study of triangle properties help students detect which properties are important in classifying geometric shapes and, consequently, constructing correct geometric concepts.

Using Language

Starting from the hypothesis that language can reveal some aspects of mathematical knowledge as well as their evolution, Guille-Biel Winder studied the learning of 6- to 7-year-old children during implementations of a situation involving reproduction of figure by folding. She analysed the results obtained from implementations of the PLIOX (a squared paper separated into four square, coloured zones) in two classrooms. An analysis of teachers' activities—language and gestures—with regard to three components (acting, talking, and thinking), in particular the use of language in the process of negotiating meaning, was the focus of the paper presented by Barrera-Curin, Bulf, and Venant. They compared the same mathematical situation in two different contexts (France and Québec).

Building and Representing 3D Shapes

Research has shown that children's perceptions of different components of 3D geometric thinking are variable and complex. In particular, it requires the understanding of the relative locations of 3D objects to each other, the recognition of the properties of 3D objects, the relationship between a geometrical object and its visual representation, and the individuation of its properties starting from its drawings.

Reinhold and Wöller analysed the role of children's (aged 8–9) wooden block building activities in their conceptual knowledge of geometrical solids. In particular, they conducted interviews with children in Germany and Malaysia, and a qualitative analysis of the data shows a wide variety of individual activities. Denizli, Erdoğan, and Olkun developed a test to measure the 3D geometric thinking of first to fourth grade students and to investigate the development of 3D geometric thinking across the grades. In particular, they wanted to evaluate the component of "recognizing the properties of 3D objects". They showed that a student's ability to recognise these properties improves significantly with increasing grade level.

Issues related to 3D geometry reveal the problem of representation of these objects on a piece of paper (or computer screen). Studies have shown that this problem is nontrivial. Kondo presented problems of comparison of segment lengths and of angle amplitudes to students (aged 10–12) using 2D representations of cubes. The segments were edges or diagonals of a cube and the angles were identified by pairs of edges or by an edge and a diagonal. Yuan suggested that the use of virtual manipulatives creates an interactive environment to support multiple representations of geometric objects and to develop the ability of spatial structuring. Her study was focused on counting blocks in a 3D environment, which provided students with an initial understanding of spatial concepts. J.A. Cochran, Z. Cochran, and Hopper proposed activities on the transition from 2D to 3D objects using a new technology, 3D printing. Athias compared tools in the pen-and-pencil environment and dynamic geometry software with young pupils (aged 9–10), showing how a mathematical concept (equal length) could be used in the two environments.

Mental Manipulation

Generally, the problem of dynamic reasoning in geometry has been associated with intuitions of isometric transformations. However, the extension of the theoretical foundations of research in geometry has given another direction to research. It has become important to examine the extent to which children possess the ability to make mental transformations of objects and test didactical proposals, which gives an opportunity to gather experience in a dynamic interpretation of static images.

Swoboda and Zambrowska studied students' mental manipulation of a shape at the early educational level. They analysed the performance of students in one task with attention to dynamic thinking. Ramful and Lowrie also investigated mental manipulation (rotation) with students 11–13 years old through the design of a novel instrument. They concluded that spatial reasoning correlates with performance in mathematics. Jirotková presented manipulative environments in which specially created series of tasks enabled children to develop their competences in dynamic understanding of changes.

Geometrical Relations

Swoboda presented research among 4- to 7-year-old children on development of the "geometrical regularities" script. Additionally, she showed a linkage between the way 6-year-old children use geometric regularities and their later functioning at school. Kim and Kim explored the question "Does the convergent instructional model cultivate core competencies in the field of mathematics?" They concluded that esthetical designs can be thought to produce creative problem-solving ability. Vighi analysed the role of symmetry in a dynamic approach and the difficulties connected with its practice.

Using Non-typical Tools that Support Presentation of Some Geometrical Concepts

Olkun studied both numerical and geometric reasoning promoted by number-line estimation tasks. He concluded that the ability to estimate the relative magnitude of numbers on an empty number line has more to do with geometry achievement than arithmetic. Hassan Mohamed presented special tools that were realised with the aim of helping blind pupils to learn geometrical constructions. The results showed the effectiveness of these tools on the blind pupils' skills in this field. Iwase et al. studied mathematical knots, i.e., closed curves in space. They proposed and analysed some examples of teaching methods in elementary and junior high school.

Open Access Except where otherwise noted, this chapter is licensed under a Creative Commons Attribution 4.0 International License. To view a copy of this license, visit http://creativecommons.org/licenses/by/4.0/.

Topic Study Group No. 13: Teaching and Learning of Geometry—Secondary Level

Ui Hock Cheah, Patricio G. Herbst, Matthias Ludwig, Philippe R. Richard and Sara Scaglia

The Programme

The TSG13 program was organized to focus on the following themes:

- Curricular issues in school geometry
- Technological tools and environments for the study of geometry
- Applications of geometry for modeling real world situations and the study of other disciplines
- Connections between geometry and the study of other branches of mathematics
- Connections between geometry and mathematical practices and processes such as argumentation and proof, visualization, figuration, and instrumentation
- Student conceptions and learning of geometrical ideas and their use in geometric problem solving
- Youth and adult geometrical competencies out of school and at the workplace
- Practices and problems in the teaching of geometry
- Geometry, teacher preparation, and teacher knowledge.

Alain Kuzniak from Université Paris Diderot was the invited speaker for TSG13. His keynote touched on the need of theoretical benchmarks in Geometry Education

---

Co-chairs: Ui Hock Cheah, Patricio G. Herbst.
Team members: Matthias Ludwig, Philippe R. Richard, Sara Scaglia.

U.H. Cheah
Institute of Teacher Education, Cyberjaya, Malaysia
e-mail: uhcrecsam1@gmail.com

P.G. Herbst (✉)
University of Michigan, Ann Arbor, USA
e-mail: pgherbst@umich.edu

research. A total of 16 papers from were reviewed and selected to be presented at the TSG13 sessions.
Invited and selected papers

Tuesday 26 July 2016
Research on geometry education: the need of theoretical benchmarks (Kuzniak, Alain)
The articulation of geometry problems: a major educational challenge (Richard, Philippe R.)

Wednesday 27 July 2016
How to develop spatial ability? results from the research project Geodikon (Maresch, Günter)
Students' use of property knowledge and spatial visualization in reasoning about 2D rotations (Frazee, Leah Michelle; Battista, Michael; Joswick, Candace; Clayton; Emanuel)
Epistemological features of a constructional approach to regular 4-polytopes (Berendonk, Stephan; Sauerwein, Marc)
Symbiosis between specialised and pedagogical knowledge in geometry (Chinnappan, Mohan; White, Bruce; Trenholm, Sven)
Geometry teachers' knowledge: insights from the trapezoid study (Manizade, Agida; Martinovic, Dragana)
Playnig with geometry: a game, an educational inquiry activity or an assessment TASK? (Soldano, Carlotta; Luz, Yael)

Friday 29 July 2016
Exploring models of secondary geometry achievement (Senk, Sharon Louise; Thompson, Denisse Rubilee)
Typical errors in geometry of grade 9 learners in south africa (Steyn, Carine; Morar, Tulsi)
The growth of mathematical understanding: elif's engagement with representations in pirie-kieren levels (Gulkilik, Hilal; Ugurlu, Hasan Hüseyin; Yürük, Nejla; Moyer-Packenham, Patricia)
Enacting functions from geometry to algebra (Steketee, Scott; Scher, Daniel)
Difference in self-reporting implementation of instructional strategies using a dynamic geometry approach (Webre, Brittany April)
The effect of dynamic geometry approach on geometry achievement and conjecture ability (White, Alexander Kevin; Smith, Shawnda; Cuevas, Gilbert)

Saturday 30 July 2016
Designing instruction towards mathematical literacy in geometry: a case study (Cheah, Ui-Hock)
Engaging students with non-routine geometry proof tasks (Cirillo, Michelle)
Is the work of teaching geometry subject specific? (Herbst, Patricio G.)

Oral communications

In addition to the selected papers there were another 20 oral communications and 13 poster presentations:

Tuesday, 26 July 2016

Combinatorial problems in school geometry (Smirnov, Vladimir Alekseevich)
Geometry opportunities for reasoning and proof in secondary school textbooks in trinidad and tobago (Hunte, Andrew Anthony)
Middle school students' (MIS)interpretations in length to volume relationships (Ayan, Rukiye)
Notes for the teaching of geometry in secondary school: a teacher training experience (Villella, José Agustín)
Artifact based geometric constructions (Siopi, Kalliopi)
Mathematics teachers' reflections using instructional design in the teaching of geometry (Jojo, Zingiswa Mybert)
Context integration effects on geometry learning of junior high school students (Chen, Ming-Jang)
Aspects of spatial thinking in problem solving: focusing on viewpoints in constructing internal representation (Arai, Mitsue)
Teachers' proving process in dynamic environment: the inscribed angle theorem (Nagar, Gili Gal)
Inquiry-based learning in geometry teaching (open-ended approach) (Ovsyannikova, Irina)
Center of gravity of various figures (Takayama, Takuma)
Are irrational numbers useful for what? going beyond perimeter, area and volume formulas (Mózer, Graziele Souza)
The use of writing as a metacognitive tool in geometry learning (Orozco Vaca, Luz Graciela)
The interplay between visualization and argumentation in the teaching of geometry (Papadaki, Chrysi)

Friday, 29 July 2016

Prospective teachers' personal and instructional definitions for quadrilaterals (Ulusoy, Fadime)
Irish pre-service teachers' subject matter knowledge of secondary level trigonometry (Walsh, Richard)
Geometry teaching knowledge: a comparison between pre-service and high school geometry teachers (Smith, Shawnda Rae)
Prospective teachers' knowledge about vectors and its applications to algebraic and graphical problems (Bulut, Neslihan)
Enhancing teaching and learning geometry through discovery approach: an example of Iran (Rabbi, Sima)
Is geometric literacy necessary? (Birni, Şeyda)

Poster Presentations

- An analysis of actual conditions of justification to Korean new mathematics textbooks: focus on middle school geometry (Kim, Soocheol)
- Potentially significant teaching units involving 3d geometry and Thales' theorem (Manassés da Silva Batista, Raimundo Nonato Ferreira Tito Filho, Antonio Kennedy Lopes Dantas, Francismar Holanda Holanda)
- Black and light Tangram: learning from fun and interactive way (João Alves da Silva, Manassés da Silva Batista, Antonio Kennedy Lopes Dantas, Francismar Holanda)
- Educational value of the centroid of triangle (Tomohiro Ogihara, Tatsuya Mizogushi)
- Heuristic and inquiry based learning using the seifert graph (Yuki Osawa)
- Cooperative learning as a tool to teach a professional general course in university geometry (Wen-Haw Chen)
- Viewpoints and objects of the observation" in learning space figures (Shinya Ohta, Toshiji Matsubara)
- The concept of center of mass in teaching of geometry (ivko Dimitric)
- Variatio delectat: variation in mathematics (Chris Kooloos, Rainer Kaenders, Gert Heckman, Helma Oolbekkink)
- Doing geometry with 21st century tools and needs (Balvir Singh, Arthur Powell)
- Developing a learning and assessment framework for geometric reasoning to support teaching and learning in years 5-9 (Marj Horne, Rebecca Seah)
- Teaching analytic geometry emphasizing representations and translations (Sunghee Kim)
- School course of geometry: content selection and teaching material distribution (Samvel Haroutunian)

Open Access Except where otherwise noted, this chapter is licensed under a Creative Commons Attribution 4.0 International License. To view a copy of this license, visit http://creativecommons.org/licenses/by/4.0/.

Topic Study Group No. 14: Teaching Learning of Probability

Carmen Batanero, Egan J. Chernoff, Joachim Engel,
Hollylynne Stohl Lee and Ernesto Sánchez

The Programme

We contend that to adequately function in society citizens need to overcome their deterministic thinking and accept the existence of fundamental chance in nature. At the same time, they need to acquire strategies and ways of reasoning that help them in making adequate decisions in everyday and professional situations where chance is present.

By including probability in the curricula at different educational levels and in the education of teachers, educational authorities in many countries have recognized a need for probability literacy. However, including a topic in the curriculum does not automatically assure its correct teaching and learning; the specific characteristics of probability, such as a multifaceted view of probability or the lack of reversibility of random experiments, not usually found in other domains, creates special challenges for teachers, students and citizens.

Research in (what is becoming known as) probability education attempts to respond to the above challenges—as shown by the many papers on this topic presented at conferences such as the European Mathematics Education Conference (CERME), the International Conference on Teaching Statistics (ICOTS), as well as

---

Co-chairs: Carmen Batanero, Egan J. Chernoff.

Team members: Joachim Engel, Hollylynne Stohl Lee, Ernesto Sánchez.

C. Batanero (✉)
University of Granada, Granada, Spain
e-mail: batanero@ugr.es

E.J. Chernoff
University of Saskatchewan, Saskatoon, Canada
e-mail: egan.chernoff@usask.ca

in regional or national conferences such as the Latin-America Mathematics Education Conference (RELME)—is now well established.

The general aim of the Topic Study Group on Teaching and Learning of Probability at the 13th International Congress of Mathematics Education (ICME-13) was to encourage new research in the domain. As such, the organisers welcomed diverse papers, including theoretical analyses and empirical research, in probability education whilst using a variety of research methods. The main topics of the papers were the following:

- *The nature of chance and probability.* This includes different views in the practice of statistics, and in the curricula, as well as philosophical problems and people's personal views throughout history.
- *Statistical versus probabilistic knowledge and reasoning.* Beyond being a tool for inferential statistics, probability is an approach to structure our world and both statistics and probability can connect to mathematical modelling with complementary views. There are also new paradigms in probabilistic reasoning research, once dominated by, for example, intuitions and the heuristics and biases program.
- *Components of probability reasoning and literacy in everyday or professional settings.* This includes dealing with risk and decision-making, and educational programmes to develop the related competences.
- *Probability in school curricula.* Since probability is increasingly being included in world-wide curricula, beginning in primary school in many countries, it is important to reflect on the main ideas that students should acquire at different ages, informal probabilistic reasoning, appropriate teaching methods, suitable teaching situations and successful teaching experiences. Further, the use of technology in teaching and learning probability and analyses of educational resources are also appropriate.
- *Education of teachers.* The field needs suitable models describing the components of teachers' knowledge to teach probability, especially those that take into account the specific features of teaching and learning probability. Research dealing with assessing and developing teacher's knowledge is also expected and encouraged.

The presentations included, as found in other Topic Study Groups, invited papers, contributed papers, and posters. The following invited papers were presented in the Topic Study Group sessions schedule:

Session 1. Theoretical analyses. Chair: Carmen Batanero. Speakers: Manfred Borovcnik and Ramesh Kapadia (Reasoning with risk: a survival guide); Cynthia Langrall (The rise and fall of probability in the k–8 mathematics curriculum in the United States); Hollylynne S. Lee (A framework of probability concepts needed for teaching repeated sampling approaches to inference).

Session 2: Students' reasoning and strategies. Chair: Hollylynne Lee. Speakers: Joachim Engel (Between fear and greed: the six looses); Ernesto Sanchez. (Theoretical dogmatism and empirical commitment in the informal probabilistic reasoning of high school students); Egan J Chernoff (Comparing the relative probabilities of events); Peter Bryant (Teaching 9 and 10 year old children about randomness).

Session 3a. Attitudes and education of teachers. Chair: Joachim Engel. Speakers: Caterina Primi (Statistics anxiety: a mediator in learning probability); Assumpta Estrada (Exploring teachers' attitudes towards probability and its teaching); Emilse Gómez Torres (Prospective teachers' solutions to a probability problem in a sampling context); Robert Adam Molnar (High school mathematics teachers' understanding of independent events); Susanne Podworny (Design of a course for learning probability via simulations with Tinkerplots).

Session 3b. Teaching of probability. Chair: Ernesto Sánchez. Speakers. Pedro Rubén Landín and Jesús Salinas (Probabilistic reasoning in high school students on sample space and probability of compound events); Judah Makonye (Learners' use of probability models in answering probability tasks in South Africa); Roberto Oliveira (The teaching of probability in context through reading and writing strategies at secondary education); Carmen Batanero (Characterizing the probability problems proposed in the entrance to university tests in Andalucia); Haneet Gandhi (Understanding children's conception of ramdomness through explorations with symmetrical polyhedrons).

Session 4. Complementary issues. Chair: Egan J. Chernoff. Speakers: Rolf Biehler (Professional development for teaching probability and inference statistics with digital tools at upper secondary level); Per Nilsson (Interactive experimentation in probability—opportunities, challenges and needs of research); Rink Hoekstra (Risk as an explanatory factor for researchers' inferential interpretations).

In addition to the above, there were also four sessions of contributed short oral communications. The following papers were presented:

Session 1. Teaching resources and experiences. Chair: Egan Chernoff. Speakers: Vincent Martin and Laurent Theis (The teaching of probability to students judged or not with difficulties in mathematics in elementary classes in Quebec); Signe Holm Knudtzon (Pitfalls and surprises in the teaching of probability); Monica Giuliano, Silvia Pérez and Martín García (Teaching probability and statistics with e-status).

Session 2. Teacher education. Co-chairs: Carmen Batanero and Ernesto Sánchez. Speakers: Pedro M. Huerta (Preparing teachers for teaching probability through problem solving); Katharina Böcherer-Linder, Andreas Eichler and Markus Vogel (The impact of visualization on understanding conditional probabilities); Isaias Miranda and Beatriz Rodríguez (Understanding professors' decisions to assess students' learning of probability); Augusta Osorio (Strengthening of elementary teachers in the use of probability in everyday life events); J. Humberto Cuevas and Greivin Ramírez (Performance in stochastic between secondary teachers and teaching students: comparative study in Costa Rica and México); Annarosa Serpe (Mathematization of uncertainty with the aid of computers: a model of activity in high school).

Session 3. Teaching resources and experiences. Chair: Joachim Engel. Speakers: Jorge Soto-Andrade and Daniela Diaz-Rojas (Random walks as learning sprouts in the didactics of probability); Blanca Ruiz (Random variable and its relationship with statistical variable: an educational perspective from a concept analysis); María

Nascimento, Eva Morais and Alexandre Martins (Representations in probability problems).

Session 4. Students' and children's reasoning and strategies Chair: Hollyllynne Lee. Speakers: Ana Serrado-Bayes (Enhancing reasoning on risk management through a decision-making process on a game of chance task); Santiago Inzunsa (Connecting theoretical probability and experimental probability in a modeling environment); He Shengqing and Gong Zikun (Children's learning progressions on probability and suggestions for curriculum improvement); Gong Zikun and He Shengqing (Study on developmental stages and important periods of probability cognition for children aged 6–14).

We were extremely pleased that a number of posters were presented in our Topic Study Group. In particular, we had: Kemal Akoglu (A framework to guide task development for overcoming cognitive issues in learning conditional probability); Roos Blankespoor, Marja van den Heuvel-Panhuizen, Michiel Veldhuis and Anika Dreher (A pilot study on teaching probability in primary school); Melisa Castillo (Achievements and difficulties in learning probability); Eva Morais, María Bascimento and J. Alexander Martins (Representations in probability problems: some examples).

We would be remiss not to mention that the work by the group team, which started about one year before the conference, resulted in the publication of a Topical Survey on Research on Teaching and Learning Probability (Batanero, Chernoff, Engel, Lee, & Sánchez, 2016). Lastly, given that the group sessions were extremely productive, a monograph, with expanded versions of the main papers presented, is being developed and we all look forward to its publication.

Reference

Batanero, C., Chernoff, E., Engel, J., Lee, H., & Sánchez, E. (2016). R*esearch on teaching and learning probability.* ICME-13. Topical Survey series. Springer.

Open Access Except where otherwise noted, this chapter is licensed under a Creative Commons Attribution 4.0 International License. To view a copy of this license, visit http://creativecommons.org/licenses/by/4.0/.

Topic Study Group No. 15: Teaching and Learning of Statistics

Dani Ben-Zvi, Gail Burrill, Dave Pratt, Lucia Zapata-Cardona and Andreas Eichler

The Programme

TSG-15 Rationale

Being able to provide sound evidence-based arguments and critically evaluate data-based claims are important skills that all citizens should have. It is not surprising therefore that the study of statistics at all educational levels is gaining more students and drawing more attention than it has in the past. The study of statistics provides students with tools, ideas and dispositions to use in order to react intelligently to information in the world around them. Reflecting this need to improve students' ability to think statistically, statistical literacy and reasoning are becoming part of the mainstream school and university curriculum in many countries.

As a consequence, statistics education is growing and becoming an exciting field of research and development. Statistics at school level is usually taught in a mathematics classroom in connection with learning probability. To allow for this instructional convention, Topic Study Group 15 (TSG-15) included probabilistic aspects in learning statistics, whereas research with a specific focus on learning probability was discussed in TSG-14 of ICME-13.

Co-chairs: Dani Ben-Zvi, Gail Burrill.
Team members: Dave Pratt, Lucia Zapata-Cardona, Andreas Eichler.

D. Ben-Zvi (✉)
The University of Haifa, LINKS I-CORE, Haifa, Israel
e-mail: dbenzvi@univ.haifa.ac.il

G. Burrill
Michigan State University, East Lansing, USA
e-mail: burrill@msu.edu

TSG-15 Meetings During ICME-13

The growing interest in statistics education was reflected in the popularity of this group and in the more than 60 papers accepted for presentation. The members of TSG-15 came from 34 different countries and varied significantly by experience, background and seniority. The presentations were divided into six themes related to key issues in statistics education research: core areas in statistics education; technology and the teaching of statistics; statistics education at the elementary level; statistics education at the secondary level; statistics education at the tertiary level; teachers' statistical knowledge and statistics education of pre-service/in-service teachers; and future directions in statistics education.

The four meetings of TSG-15 were organized to create a sense of community among all presenters and participants, who shared a common desire and passion to improve statistics education by focusing on conceptual understanding rather than rote learning. To build and support this sense of community we asked participants to prepare for TSG-15 before they arrived in Hamburg by reading all papers in advance, so we could discuss each other's work; the co-chairs kept informal correspondence with all participants before, during and after the conference; and finally, participants were asked to be involved every day of the program so we could get to know one another, develop collegial networks, welcome our emerging scholars and discuss the important work in statistics education research around the world.

Because of the large number of proposals we received, the time available only allowed for relatively short presentations by the authors. However, we felt it critical that all proposals be given time for presentation in some format. The four meetings were therefore organized to capitalize on community-building and discussions around our collective and individual research. Some of the sessions ran in parallel. In addition there was a poster session dedicated to short poster presentations followed by close viewing and discussions, so that the TSG-15 community could engage more directly with the authors and each other in a relaxed setting. Another highlight of the program was a workshop to reflect as a community on the themes, presentations, issues raised and discussions.

The accepted papers were organized in the following ways:

- 13 poster presentations to promote TSG-15 community discussions with diverse and thought-provoking studies;
- 16 short presentations (10 min talk + 5 min discussion) in four "Oral Communication Sessions" organized into four themes to enrich understanding of the themes and extend discussions around common interests;
- 23 long presentations and discussions (15 min talk + 5 min discussion) organized in six thematic sessions (two whole group, and two in parallel sessions) to enhance the overarching themes of the short presentation and poster sessions;

- Short group discussions in almost all sessions to allow for rich interactions and discourse.

TSG-15 had thus nine sessions all together, which were devoted to key issues in statistics education research:

1. Four 90-min sessions of long papers (15 + 5 min), two of which (Sessions 2 and 3) ran in parallel,
2. Four 60- or 90-min sessions of short papers (10 + 5 min), and
3. One poster session.

татры *TSG-15 Beyond the Conference*

Informal feedback received after the conference was extremely positive. We felt at the end that much can be learned by integrating results from such a variety of research and practice in statistics education. Such integration of theories, empirical evidence and instructional methods can eventually help students to develop their statistical thinking. These ongoing efforts to reform statistics instruction and content have the potential to both make the learning of statistics more engaging and prepare a generation of future citizens that deeply understand the rationale, perspective and key ideas of statistics. These are skills and knowledge that are crucial in the current age of information and big data.

An informal set of proceedings was created to allow for immediate distribution of the TSG-15 papers among those within the TSG-15 members. Before the conference an ICME-13 Topical Survey titled "Empirical research in statistics education" was prepared by team members Andreas Eichler and Lucía Zapata-Cardona (Eichler & Zapata-Cardona, 2016). This short book (freely available at http://www.springer.com/gp/book/9783319389677) addresses the current state of research in statistics education. It provides a review of recent research into statistics education, with a focus on empirical research published in established educational journals and on the proceedings of important conferences on statistics education. It identifies and addresses six key research topics: teachers' knowledge; teachers' role in statistics education; teacher preparation; students' knowledge; students' role in statistics education; and how students learn statistics with the help of technology. For each topic, the survey builds upon existing reviews, complementing them with the latest research.

A monograph of the best 20 papers presented in TSG-15 is underway. In TSG-12 (ICME-12) a monograph titled "The teaching and learning of statistics: International perspectives" (Vol I, Ben-Zvi & Makar, 2016) was published. The TSG-15 monograph will be the second volume in this line of publications, and is edited by Burrill and Ben-Zvi (expected publication date: 2016).

References

Ben-Zvi, D., & Makar, K. (Eds.). (2016). *The teaching and learning of statistics: International perspectives (Vol. I)*: Springer International Publishing Switzerland.

Burrill, G., & Ben-Zvi, D. (in preparation). *The teaching and learning of statistics: International perspectives* (Vol. II). Springer International Publishing.

Eichler, A., & Zapata-Cardona, L. (2016). *Empirical research in statistics education.* An ICME-13 Topical Survey, series editor: Gabriele Kaiser. Springer Open.

Open Access Except where otherwise noted, this chapter is licensed under a Creative Commons Attribution 4.0 International License. To view a copy of this license, visit http://creativecommons.org/licenses/by/4.0/.

Topic Study Group No. 16: Teaching and Learning of Calculus

David Bressoud, Victor Martinez-Luaces, Imène Ghedamsi and Günter Törner

Aims

This Topic Study Group was a forum for discussions about research and development in the teaching and learning processes of Calculus, both at upper secondary and tertiary level. Invited and oral presentations, as well as posters, showed advances and new trends.

Organization

TSG-16 had four main sessions, four oral communications presentations (two of them divided in two groups) and a general posters meeting. All the contributions were posted on the website of ICME-13.

The accepted papers were organized as follows:

- 6 invited speakers delivered long presentations
- 9 presentations corresponding to extended papers

Co-chairs: David Bressoud, Victor Martinez-Luaces.
Team members: Imène Ghedamsi, Günter Törner.

D. Bressoud (✉)
Macalester College, Saint Paul, USA
e-mail: bressoud@macalester.edu

V. Martinez-Luaces
FJR-Fing, UdelaR, Montevideo, Uruguay
e-mail: victorml@fing.edu.uy

- 6 sessions devoted to oral presentations
- 6 posters in one general session

The details are described below.

Main Sessions

Tuesday, 26 July.
Session Chair: Imène Ghedamsi.
The invited lecturer David Bressoud (USA) presented a study of university departments of mathematics in the United States, describing efforts being made to improve student success in pre-Calculus and Calculus.

Next, Sarah Mathieu-Soucy (Canada) presented a report on students' perceptions of mathematical theory, suggesting that students see theory as unnecessary for problem solving. She discussed possible remedial strategies.

After that, Young Gon Bae (South Korea-USA) gave an interesting presentation about the flipped classroom as an alternative instructional model. Her team introduced design research for developing a multivariable calculus class.

This first session ended with Günter Törner (Germany) who spoke on the forces that shape the European calculus curriculum and the fractured nature of their influences.

Wednesday, 27 July.
Session Chair: Günter Törner.
The invited speaker Victor Martinez-Luaces (Uruguay) opened the session; he described some experiences with Calculus inverse modeling problems in teacher training courses in Argentina, Guatemala, Mexico and Uruguay.

Next David Webb (USA) delivered a report on the design and use of instructional tasks for active learning at the undergraduate level.

Yuliya Melnikova (USA) divulged a study on what instructors, teaching assistants, and students consider to be the purpose of the lab component in a Calculus I course. She remarked that there is a need for increased communication to improve classroom practice.

To conclude this session, Mike Thomas (New Zealand) spoke on Integrating Digital Technology in the Teaching and Learning of University Mathematics. His examples focused on the accumulation functions and interval perspectives of functions, particularly the idea of average rate of change.

Friday, 29 July.
Session Chair: Victor Martinez-Luaces.
The invited speaker Imène Ghedamsi (Tunisia) delivered a study conducted on the complexity of the cognitive process by which the formal definition of sequence

convergence is conceived. This study was framed within the Theory of Didactical Situation (TDS).

Then, Vilma Mesa and colleagues (USA) presented a broad study which observed lessons taught by various instructors at different institutions. After observing and recording the mathematical tasks used in class, she discussed what we learned about calculus teaching.

Jacqueline Coomes and the co-author (USA) gave a talk about the coordination of the symbolic and graphical meanings of function notation. They employed the notions of procept and actor oriented transfer (AOT) to analyze the data and to further explain the issues.

Kevin Moore and his colleague (USA) talked with regard to reasoning about quantities changing in tandem. They described students' ways of thinking for graph and showed the implications in the context of concepts associated with calculus.

Saturday, 30 July.
Session Chair: David Bressoud.

In the fourth session, a model to analyze the obstacles students have within the transition from calculus to analysis at the entrance to University, was presented by the invited speaker Isabelle Bloch, France.

After that, Claudio Fuentealba and others (Chile-Mexico-Spain) investigated the understanding of the derivative concept in university. The results suggest that the matizing the derivative schema is difficult to achieve.

Tolga Kabaca and colleagues (Turkey) presented research study results that show students' weaknesses in conceptual understanding of integration. In fact, they think that "Integral is a special continuous sum and antiderivative is a genius method to calculate this sum".

Oral Communications

First Session: Tuesday, 26 July.
Group A Session Chair Günter Törner.

Sergiy Klymchuck (New Zealand) described his experiences using counterexamples, puzzles and provocations in calculus classes as an effective pedagogical strategy.

Next, Angie Hodge (USA) talked about the use of active or inquiry-based learning (IBL) and how the teaching of undergraduate mathematics course affects pre-service teachers.

After that, Raquel Carneiro (Brazil) investigated which mathematical concepts freshmen students expect for their further studies to detect any gaps in mathematical training.

At the end, Higinio Ramos (Spain) focused on the difficulties that students have with the concept of inverse function, and presented a theorem to obtain indefinite integrals using that concept.

Group B Session Chair Imène Ghedamsi.

Stefanie Arend (Germany) opened this session and delivered an interesting study centred on the understanding-oriented handling of the epsilon-delta-definition of continuity.

Then, Aggeliki Efstahiou (Greece) presented a teaching sequence focused on building up an alternative definition for the limit of functions of one real variable.

Richard O'Donovan (Switzerland-USA) talked about Calculus using proximities, with an interesting approach in which students can actually prove theorems in a didactical setting.

The last speaker, Analia Berge (Canada), posed a discussion about the transition from Calculus to Analysis, focused on difficulties appearing when systematic theoretical justification is sought.

Second Session: Tuesday, 26 July.
Group A Session Chair Victor Martinez-Luaces.

Ajit Kumar (India) opened the second session by presenting Sage used as an effective pedagogical tool to teach concepts in Calculus.

Hans-Jürgen Elschenbroich (Germany) proposed the use of technology to create a suitable learning environment for the comprehension of the basic ideas of Calculus.

Next, Matti Pauna (Finland) described the evolution of online Calculus courses. He presented an effective advanced online Calculus course covering definitions and theorems.

Igor Subbotin, (USA-Ukraine) presented an approach of introducing elementary functions via linear algorithms, filling the gap that students find, concerning foundational notions of Calculus.

Lastly, Anna Roos (Germany) talked about a study of the detection of mistakes that undergraduates made. Also, she described the methods used and showed the results obtained.

Group B Session Chair Imène Ghedamsi.

That afternoon, the first speaker Laura Conejo, (Spain) presented an alternative supporting textbook to deal with the concept of limit of a function in mathematics lessons.

Laure Barthel (Israel), taking into account the difficulties encountered by students in Calculus courses, proposed useful material related to the local properties of functions.

Then, Christine Herrera (USA) showed a study on the conceptualizations of limits. Findings indicate that covariational reasoning is fundamental to students' understanding of limits.

The following speaker, Rita Desfitri (Indonesia) delivered a research report focused on analyzing in-service teachers' understanding on the concept of limit and derivative.

After that, Behiye Ubuz (Turkey) presented multilevel models developed to explore how mathematical thinking about derivative varies at the student and classroom levels.

The last speaker, Marcel Klinger (Germany) reported about a study that assesses students' understanding of the concept of differentiation and the meaning of parameters of a function.

Third Session: Friday, 29 July.
Session Chair David Bressoud.
The session was opened by Miguel Diaz (Mexico). The presentation documented the understanding process on concepts of Calculus of high school Mexican teachers.

After that, Rebecca Dibbs (USA) presented a course that used post-class reflections to improve students' conceptual understanding of the foundational concepts in calculus.

Monica Panero (France-Italy) shared a study on teachers' practices with the derivative concept and the derivative function. The results show a global perspective on the derivative function.

Finally, Dennis B. Roble (Philippines) talked about a study aimed to determine the levels of students' mathematics comprehension and its impact on their conceptual understanding.

Fourth Session: Friday, 29 July.
Session Chair David Bressoud.
Marcio Vieria (Brazil) opened the session and reported the results of research whose objective is to develop teaching materials for concepts of Differential and Integral Calculus.

André Henning (Germany) shared a linear approximation approach to high school Calculus. The study remarked that classroom implementation of the developed teaching unit is essential.

Then, Jose Fernandez-Plaza (Spain) studied definitions provided by a group of students of Non-Compulsory Secondary Education about the notion of tendency of a function at a point.

At the end, Mario Caballero-Perez (Mexico) delivered a presentation about the development of variational thinking and language for the teaching and learning of Calculus.

Poster Session

Tuesday, 26 July.
In a general session were presented 6 posters of TSG-16. As a key component of communication, the posters allowed the researchers to show a snapshot of their work and interact with colleagues.

Matias Arce from Spain showed how indeterminate forms are perceived by Grade 11 students. Next, Rongrong Cao (USA-China) presented a Calculus course based on arithmetic. Louis Friedler (USA-China) analysed a China-US calculus study. Then, Xuefen Gao (China) made a comparison of calculus in high school Mathematics textbooks between China and USA. Maria Quezada (Mexico) presented on a Calculus laboratory with free software Desmos. Lastly, Marit Hvalsøe Schou (Denmark) described visualisation in upper secondary Calculus teaching.

Conclusions

The main themes addressed in TSG-16 were: Calculus teaching and learning, Calculus understanding, transition from secondary to tertiary level, construction of Calculus concepts, learning theories, technology, visualisation, problem-solving, modeling and applications and teacher training courses, among others.

Most of the papers and posters showed interest in innovative approaches to different topics, in order to help students improve their knowledge and comprehension of Calculus. In several cases, these innovations were directly related to the use of technology, whereas in others, they were more involved in the way of thinking, teaching approaches, courses materials, or specific tasks to be carried out by students of different educative levels and university careers.

It is hoped that this interesting interaction between teachers and researchers from different countries, contribute to stimulate creative ideas that make possible continuous headway in the development of mathematics education, particularly, in Calculus teaching and learning.

Open Access Except where otherwise noted, this chapter is licensed under a Creative Commons Attribution 4.0 International License. To view a copy of this license, visit http://creativecommons.org/licenses/by/4.0/.

Topic Study Group No. 17: Teaching and Learning of Discrete Mathematics

Eric W. Hart, James Sandefur, Cecile O. Buffet,
Hans-Wolfgang Henn and Ahmed Semri

The Programme

Discrete mathematics is a comparatively young branch of mathematics with no agreed-upon definition but with old roots and emblematic problems. It is a robust field with applications to a variety of real world situations, and as such takes on growing importance in contemporary society.

We take discrete mathematics to include a wide range of topics, including logic, game theory, algorithms, graph theory, discrete geometry, number theory, discrete dynamical systems, fair division, cryptography, coding theory, and counting. Cross-cutting themes include discrete mathematical modeling, algorithmic problem solving, optimization, combinatorial reasoning, and recursive thinking.

Discrete mathematics is not always clearly delimited in curricula and can be diffuse. In fact, two separated but linked curricular perspectives emerge: teaching and learning discrete mathematics content and teaching and learning skills of mathematical practice through discrete mathematics problems, both general skills, such as reasoning and modeling, and skills particular to discrete mathematics, such as algorithmic and recursive thinking. Thus, discrete mathematics provides a useful setting in which to pursue the ongoing problem in mathematics education of the didactic transposition of content knowledge and process skill, and it provides an

---

Co-chairs: Eric W. Hart, Cecile O. Buffet.
Team members: Hans-Wolfgang Henn, James Sandefur, Ahmed Semri.

E.W. Hart (✉)
Georgetown University, Washington, D.C., USA
e-mail: ehart@infinitemath.com; ehart@grandview.edu

C.O. Buffet
Université Reims Champagne Ardenne, Reims, France

opportunity to develop and refine models for teaching and learning that develop both.

The main goal of the TSG is to discuss and extend the state-of-the-art about teaching and learning discrete mathematics. The broad focus areas related to this goal are teaching and learning discrete mathematics at all grade levels, research, curriculum development, professional development of teachers, and curricular implementation of discrete mathematics, including policy and standards. Papers from scholars around the world were presented in seven sessions. These papers are briefly summarized below.

Invited Papers

- Margaret Cozzens, Rutgers University—"Food Webs, Graphs, and a 60-Year Old Problem Students Can Help Solve." Food webs describe the flow of energy through an ecosystem. Middle and high school students encounter food webs in their biology classes. The discrete mathematics related to food webs is usually not discussed in these classes, yet it is relatively easy mathematics that teachers can understand. This paper models food webs with directed graphs, and discusses why competition graphs derived from real food webs seem to be interval graphs.
- Robert Devaney, Boston University—"Discrete Dynamical Systems: A Pathway for Students to Become Enchanted with Mathematics." In this paper we show how the topic from discrete dynamical systems known as the chaos game can be used to get students excited about mathematics. In addition, we describe a number of different ways this topic relates to the standard high school mathematics curriculum.
- Susanna Epp, DePaul University—"Discrete Mathematics for Computer Science." This paper explores some of the issues involved in teaching a discrete mathematics course for computer science students. This paper discusses some of the challenges involved in implementing the recommendations for discrete mathematics instruction published by the computer science societies, explores some of the reasons behind them, and suggests considerations that educators should take into account when they prepare instructional materials.
- Solomon Garfunkel, COMAP—"Fairness." For the past 45 years I have worked to bring mathematical modeling and applications of mathematics into the mainstream mathematics curricula at all grade levels. This work has continuously (pun intended) bucked up against those who believe that analysis is mathematics and therefore courseware must be designed to prepare students for continuous mathematics. And even those who give a nod to modeling see it in terms of physics and engineering, reinforcing their belief in the calculus escalator. Discrete mathematics also is important. For example, models of fair division are discussed in this paper.

- Gerald Alan Goldin, Rutgers University—"Discrete Mathematics and the Affective Dimension of Mathematical Learning and Engagement." Discrete mathematics offers some specific affordances for encouraging students to experience mathematics in ways very different from more traditional school subjects such as arithmetic, algebra, geometry, and analysis. Opportunities abound for teachers and curriculum planners to focus on evoking interest and engagement, and on developing powerful affect—emotions, attitudes, beliefs, and values—in relation to mathematics. This paper explores such opportunities in relation to research constructs in the literature on mathematical affect.
- Eric W. Hart, Grand View University—"Discrete Mathematical Modeling in the Secondary Curriculum." In this paper we describe a multistage process of discrete mathematical modeling that is based on many years of curriculum research and development. Five broad problem structures emerge as ways to organize the diversity of discrete mathematics contexts that are important and appropriate for high school—enumeration, sequential change, relationships among a finite number of objects, information processing, and fair decision-making. The process of discrete mathematical modeling is outlined for these five problem structures.

Contributed Papers

- Tom Coenen, University of Twente—"Combinatorial Reasoning to Solve Problems." This study reports on the combinatorial reasoning of students aged 14–16. We study the variation of the students' problem solving strategies in the context of emergent modeling. The results show that the students are tempted to begin the problem solving process on the highest level and otherwise have difficulties transitioning from a lower to a higher level of activities. We advocate matching emergent modeling with teaching combinatorial reasoning, stimulating students to create a relational network of knowledge.
- Aaron Gaio, University of Palermo—"I Like Discrete Mathematics, But I Do Not Know How To Teach It." The paper describes a research project aiming at bringing new mathematical knowledge and competences to students and involving teachers in the activity designing process. In this context, we present an overview of the Italian situation in teaching discrete mathematics in primary and middle school, together with reference to the national teaching guidelines. We then briefly describe the results obtained from our first survey of about 100 teachers.
- Karina Höveler, TU Dortmund—"Children's Combinatorial Counting Strategies and Their Relationship to Mathematical Counting Principles." This paper reports about selected findings from a qualitative study with third graders. The study's main goals were to identify how children solve combinatorial problems and to gain insights into the relationship between their strategies and mathematical ideas.

- Vladimir Igoshin, Saratov State University—"Mathematics and Logic: Their Relationship in the Training of Teachers of Mathematics." In the process of teaching and learning mathematics logic cannot be avoided, mathematics and logic prove to be inseparable and interact closely. This interaction has didactic and training implications. After analyzing different aspects of this interaction, the author identifies and substantiates the principles of logic in mathematics and in the education of teachers of mathematics.
- Elise Lockwood, Oregon State University—"Generalization in Students' Combinatorial Thinking." The purpose of this paper is to characterize and better understand the role of generalizing activity in students' combinatorial problem solving. We do this by drawing upon Lockwood's model of combinatorial thinking. The findings come from a series of interviews in which students solved combinatorial tasks designed to foster generalization.
- Maria Flavia Mammana, University of Catania—"Graph Theory in Primary, Middle, and High School." In this paper we present an experimental teaching activity with topics of graph theory conduced in some primary, middle, and high schools in Sicily. The aim of the whole project is to present a fun, easy approach to mathematics and some connection of mathematics with real life, in order to reach competencies related to the use of mathematical models to solve problems.
- Lisa Rougetet, University of Lille—"Machines Designed to Play NIM Games as Teaching Support for Mathematics, Algorithmics, and Computer Science (1940–1970)." This paper deals with Nim games and machines that were designed to play against a human between the 1940s and the 1970s. We focus on machines that were intended not only to play, but also to explain concepts in mathematics, algorithmics, and computer science.
- James Sandefur, Georgetown University—"Recursion versus Closed Formulas." This paper promotes the use of recursion and difference equations as a means for promoting mathematical understanding and communication through the use of contextual problems. Specifically, three contextual examples are given, with closed form and recursive solutions to the problems being contrasted.
- Ödön Vancsó, Eötvös Loránd Tudományegyetem—"Complex Mathematics Education in the 21st Century: Improving Combinatorial Thinking Based on T. Varga's Heritage." This paper summarizes the ideas and background of a combinatorics research and teaching project including historical reforms in school curriculum in 1978 in Hungary and T. Varga's work. Thereafter we collect the main elements of our project: pretest and developed teaching materials, worksheets with some examples and showing tools for teaching combinatorics such as Poliuniversum.
- Catherine Vistro-Yu, Ateneo de Manila University—"Discrete Mathematics in the General Education Curriculum." This paper describes some key issues that Filipino mathematics educators face as a new mathematics course is developed for inclusion in the new General Education Curriculum for all higher education institutions in the Philippines. The new course is largely a discrete mathematics course, a sample syllabus of which was designed by the first author.

Posters

- Antonio Kennedy Lopes Dantas, Federal Institute of Piauí—"Clothing and the Use of Hanoi Tower: A Learning in Practice." The work presents the construction of the game Tower of Hanoi as a differentiated education episode that allows students to be active in the construction of knowledge, overcoming with motivation and fun, using content like exponential functions, geometric progression, counting, and geometry. It was conducted with 40 students of the 2nd year of high school, in college Zacarias de Gois, in Teresina, Piauí.

TSG Discussion Group

The Discrete Mathematics TSG coordinated with a Graph Theory Discussion Group organized by James Maltas. The TSG concluded with an open discussion forum on: "Discrete Mathematics in Standards, Curricula, Classrooms, and Research around the World: Current Issues and Next Steps."

Open Access Except where otherwise noted, this chapter is licensed under a Creative Commons Attribution 4.0 International License. To view a copy of this license, visit http://creativecommons.org/licenses/by/4.0/.

Topic Study Group No. 18: Reasoning and Proof in Mathematics Education

Guershon Harel, Andreas J. Stylianides, Paolo Boero, Mikio Miyazaki and David Reid

The Programme

There is international recognition of the importance of reasoning and proof in students' learning of mathematics at all levels of education, and of the difficulties met by students and teachers in this area. Indeed, many students face difficulties with reasoning about mathematical ideas and constructing or understanding mathematical arguments that meet the standard of proof. Teachers also face difficulties with reasoning and proof, and existing curriculum materials tend to offer inadequate support for classroom work in this area. All of these paint a picture of reasoning and proof as important but difficult to teach and hard to learn. A rapidly expanding body of research has offered important insights into this area, but there are still many open questions for which theoretical and empirically based responses are sorely needed (for reviews of the literature in this area, see: Harel & Sowder, 2007; Mariotti, 2006; Stylianides, Bieda, & Morselli, 2016; Stylianides, Stylianides, & Weber, 2017).

TSG-18 offered during ICME-13 a forum for an overview of the state of the art, invited contributions from experts in the field (Viviane Durand-Guerrier, Gila Hanna, Eric Knuth, and Maria Alessandra Mariotti), presentation of high-quality research reports from members of the TSG organizing team and other TSG participants, and discussion of directions for future research. Associated with the TSG

Co-chairs: Guershon Harel, Andreas J. Stylianides.
Team members: Paolo Boero, Mikio Miyazaki, David Reid.

G. Harel (✉)
University of California, San Diego, USA
e-mail: harel@math.ucsd.edu

A.J. Stylianides
University of Cambridge, Cambridge, England, UK
e-mail: as899@cam.ac.uk

there were in total 21 regular presentations (8-page papers), 35 oral communications (4-page papers), and 12 posters.

The regular presentations (8-page papers) were organized around four themes as described below. Although several presentations (and associated papers) addressed issues that spanned several themes, practical considerations related to the organization of the TSG sessions during the conference necessitated a best-fit approach.

Theme 1: Epistemological issues related to proof and proving
The following presentations were offered under this theme:

- Reflections on proof as explanation (Gila Hanna);
- Working on proofs as contributing to conceptualization: The case of IR completeness prolegomena to a didactical study (Viviane Durand-Guerrier & Denis Tanguay);
- Types of epistemological justifications (Guershon Harel);
- Reasoning and proof in elementary teacher education: The key role of cultural analysis of the content (Paolo Boero, Giuseppina Fenaroli, & Elda Guala).

Theme 2: Classroom-based issues related to proof and proving
The following presentations were offered under this theme:

- Constructing and validating a mathematical model: The teacher's prompt (Maria Alessandra Mariotti & Manuel Goiuzueta);
- Classroom-based interventions in the area of proof: Addressing key and persistent problems of students' learning (Andreas J. Stylianides & Gabriel J. Stylianides);
- Developing a curriculum for explorative proving in lower secondary school geometry (Mikio Miyazaki, Junichiro Nagata, Kimiho Chino, Taro Fujita, Daisuke Ichikawa, Shizumi Shimizu, & Yasuo Iwanaga);
- Proof validation and modification by example generation: A classroom-based intervention in secondary school geometry (Kotaro Komatsu, Tomoyuki Ishikawa, & Akito Narazaki).

Theme 3: The teaching and learning of proof—issues and dilemmas
The following presentations were offered under this theme:

- Teacher noticing of justifying in the elementary classroom (Mary Kathleen Melhuish & Eva Thanheiser);
- How can a teacher support students in constructing a proof? (Bettina Bedemonte);
- Reasoning-and-proving in school mathematics textbooks: A case study from Hong Kong (Kwong Cheong Wong & Rosamund Sutherland);
- Irish teachers' perceptions of reasoning-and-proving amidst a national educational reform (Jon D. Davis);
- Identifying and using key ideas in proofs (Xiaoheng Yan, Gila Hanna, & John Mason);
- Mathematical argumentation in pupils' written dialogues (Silke Lekaus & Gjert-Anders Askevold);
- What makes a good proof? Students evaluating and providing feedback on student-generated proofs (Tina Kathleen Rapke & Amanda Allan);

- Use of examples of unsuccessful arguments to facilitate students' reflection on their proving processes (Yosuke Tsujiyama & Koki Yui);
- Allowance by experts for a break in "linearity" of deductive logic in the process of proving (Shiv Smith Karunakaran);
- Systematic exploration of examples as proof: Analysis from four theoretical perspectives (Orly Buchbinder).

Theme 4: Issues related to the use of examples in proof and proving
The following presentations were offered under this theme:

- The role of examples in proving related activities (Eric Knuth, Amy Ellis, & Orit Zaslavsky);
- When is a generic argument a proof? (David A. Reid & Estela Aurora Vallejo Vargas);
- How do pre-service teachers rate the conviction, verification and explanatory power of different kinds of proofs (Leander Kempen).

References

Harel, G., & Sowder, L. (2007). Toward comprehensive perspectives on the learning and teaching of proof. In F. K. Lester (Ed.), *Second handbook of research on mathematics teaching and learning* (pp. 805–842). Greenwich, CT: Information Age Publishing.

Mariotti, M. A. (2006). Proof and proving in mathematics education. In A. Gutiérrez & P. Boero (Eds.), *Handbook of research on the psychology of mathematics education: Past, present and future* (pp. 173–204). Rotterdam, The Netherlands: Sense Publishers.

Stylianides, A. J., Bieda, K. N., & Morselli, F. (2016). Proof and argumentation in mathematics education research. In A. Gutiérrez, G. C. Leder, & P. Boero (Eds.), *The second handbook of research on the psychology of mathematics education: The journey continues* (pp. 315–351). Rotterdam, The Netherlands: Sense Publishers.

Stylianides, G. J., Stylianides, A. J., & Weber, K. (2017). Research on the teaching and learning of proof: Taking stock and moving forward. In J. Cai (Ed.), *Compendium for research in mathematics education* (pp. 237–266). Reston, VA: National Council of Teachers of Mathematics.

Open Access Except where otherwise noted, this chapter is licensed under a Creative Commons Attribution 4.0 International License. To view a copy of this license, visit http://creativecommons.org/licenses/by/4.0/.

Topic Study Group No. 19: Problem Solving in Mathematics Education

Peter Liljedahl, Manuel Santos-Trigo, Uldarico Malaspina, Guido Pinkernell and Laurent Vivier

Mathematical problem solving has been an important research and practice domain in mathematics education worldwide. It's agenda focuses not only on analysing the extent to which cognitive, social, and affective factors influence and shape learners' development of problem solving proficiency, but also on the role played as a medium for teaching and learning mathematics and the development of both teachers' and learners' problem solving proficiencies. TSG 19 on Problem Solving in Mathematics Education was dedicated to the furthering and sharing of knowledge on this important topic.

To this end, the mathematics education community was invited to submit contributions that address the aforementioned themes relevant and related to Problem Solving in Mathematics Education. We received 56 submissions from 30 different countries on a wide range of problem solving related topics. From these 56 submissions 15 papers were accepted to be presented as part of our main TSG program (15 min presentation, 5 min discussion) as well as 27 papers to be presented as an oral communication (10 min presentation, 5 min discussion). Within the main TSG program the following 15 papers were presented:

Co-chairs: Peter Liljedahl, Manuel Santos-Trigo.

Team members: Uldarico Malaspina, Guido Pinkernell, Laurent Vivier.

P. Liljedahl (✉)
Simon Fraser University, Burnaby, Canada
e-mail: liljedahl@sfu.ca

M. Santos-Trigo
Center for Research and Advanced Studies, Mexico City, Mexico
e-mail: msantos@cinvestav.mx; manuel.santos.trigo@gmail.com

- *A Framework for Undergraduate Students' Mathematical Foresight*

  Wes Maciejewski, Bill Barton
  University of Auckland, New Zealand

- *Looking Back to Solve Differently: Familiarity, Fluency, and Flexibility*

  Hartono Tjoe
  The Pennsylvania State University, USA

- *Do High- & Low-Achieving Third Graders Benefit in the Same Way from Representational Training when Solving Word Problems?*

  Nina Sturm, Renate Rasch, Wolfgang Schnotz
  University of Koblenz-Landau, Germany

- *Classroom Practices for Supporting Problem Solving*

  Peter Liljedahl
  Simon Fraser University, Canada

- *Concretizing Mathematical Problem Solving with Metaphors*

  Yee, Sean P[1], Thune-Aguayo, Ashley[2]
  [1]University of South Carolina, USA; [2]California State University, USA

- *Problem Solving in Varga's Reform of Hungarian Mathematics Education: The Case of Combinatorics*

  Katalin Gosztonyi
  University of Szeged, Hungary

- *How do Children's Solutions Change when they Solve the same Word Problem in Math and Religion Class?*

  Johansson, Juha Antero
  University of Helsinki, Finland

- *Assessing IBME with Summative and Formative Purpose*

  Maud Chanudet
  Université de Genève, Suisse

- *Beyond the Standardized Assessment of Problem Solving From Products to Processes*

  Pietro Di Martino, Giulia Signorini
  University of Pisa, Italy

- *Toward Developing an Instrument to Assess Mathematical Problem Solving*

  James A. Mendoza Epperson, Kathryn Rhoads, R. Cavender Campbell
  The University of Texas at Arlington, United States of America

- *Pre Service Teachers' Problem Solving Ability in Secondary Level Mathematics (Algebra and Number)*

  Aoife Marie Guerin, Olivia Fitzmaurice, John O'Donoghue
  University of Limerick, Ireland

- *Mathematical Problem Solving With Technology: The Case of Marco Solving-and-Expressing on the Screen*

  Hélia Jacinto[1,2], Susana Carreira[2,3]
  [1]Jorge Peixinho Secondary School; [2]UIDEF, Institute of Education, University of Lisbon; [3]University of Algarve

- *The Spreadsheet Affordances in Solving Intricate Algebraic Problems*

  Nélia Amado[1,2], Susana Carreira[1,2], Sandra Nobre[2,3]
  [1]University of Algarve, Portugal; [2]Research Unit of the Institute of Education, University of Lisbon; [3]Group of Schools Paula Nogueira, Portugal

These presentations, together with the 27 oral communication presentations, were organized into one of 10 themes.

1. problem solving processes
2. problem solving settings
3. problem solving assessment
4. problem posing
5. technology and problem solving
6. meta-cognition and problem solving
7. professional development of problem solving
8. affect in problem solving
9. heuristics and strategies
10. classroom culture and discourse

Much was learned from these 42 presentations about problem solving in general, and about the 10 aforementioned themes in particular. Looking across the corpus of research presented at TSG 19, and across the many diverse and, sometimes contradictory, conclusions a number of key questions began to emerge.

- What is the role of goals in problem solving?
- What is the role/status of heuristics in problem solving?
- What is the utility of Pólya's *look back* stage?
- How to position problem solving in textbooks and curriculum?
- How to assess problem solving so that it is still problem solving?
- What is the role of the extra-logical processes in problem solving?
- How does the availability of tools/technology impact problem solving processes?

Although these questions have been addressed in the literature previously, the research presented at ICME-13 indicated that more work is needed to more adequately understand the answers to, and implications of, these questions.

Open Access Except where otherwise noted, this chapter is licensed under a Creative Commons Attribution 4.0 International License. To view a copy of this license, visit http://creativecommons.org/licenses/by/4.0/.

Topic Study Group No. 20: Visualization in the Teaching and Learning of Mathematics

Michal Yerushalmy, Ferdinand Rivera, Boon Liang Chua, Isabel Vale and Elke Söbbeke

The TSG 20 aimed to focus on issues in visualization in the teaching and learning of mathematics at all levels. The group welcomed studies that tackle wide range of issues including: • What is the role of visualization within and across mathematical knowledge disciplines? • Are there kinds, qualities, and/or hierarchies of visualization and visual skills? • How do learners from different cultural contexts and of varying levels of ability and disability employ visualization in learning mathematics? • Considering recent advances on embodied cognition in mathematics, what theoretical frameworks could link visual and haptic modalities in an effective manner? • What theories on visualization can take into account the specific cognitive nature of mathematical activity and thinking? • What methodological considerations must be accounted for in investigations that focus on visualization? How should tasks, instruments, and measures be designed that will enable investigators to assess changes in students' understanding and learning? • What aspects of mathematics teacher education programs will help teachers understand the affordances and challenges of using visualization as a learning tool in mathematics? • What visual-based tasks can foster creativity leading to meaningful mathematical knowledge? Further, are visual-driven students more creative than nonvisual and other types of learners? In particular, we note how recent and emerging

---

Co-chairs: Michal Yerushalmy, Ferdinand Rivera.
Team members: Boon Liang Chua, Isabel Vale, Elke Söbbeke.

M. Yerushalmy (✉)
University of Haifa, Haifa, Israel
e-mail: michalyr@edu.haifa.ac.il; myerushalmy@univ.haifa.ac.il

F. Rivera
San Jose State University, San Jose, USA
e-mail: ferdinand.rivera@sjsu.edu

technological tools and digital mathematics media enable learners to perform visual engagement and interaction.

The opening session chaired by Boon Liang Chua included two invited lectures. It commenced with the first plenary, *The explanatory value of mathematical visualisations: a philosophical and pragmatic approach*, by Joachim Frans from Germany. This presentation was about the explanatory value of mathematics visualization that lays the foundation for the use and importance of visualization. The next communication was *Means for learning about students' knowledge: automatic assessment of visual examples* presented by Michal Yerushalmy from Israel. This presentation dealt with generating visual examples as a tool for learning about student knowledge throughout feedback from automated formative assessment systems. The two presentations generated a rich discussion about the foundations and the innovations in the field.

The second meeting ran in two parallel meetings and included six presentations. In Session A, Amy Lin from Canada suggested us to look at spatial reasoning of young students in her presentation: *Go figure: can actions promote visual and spatial reasoning?* Lin concentrated on gestural interfaces such as touchscreens that provide a more hands-on experience for the student as a potential support for cognitive processes and mathematical thinking. The research questions addressed in this study concern with the types of gestures (iconic, deictic, metaphoric, rhythmic) emerging when children are solving spatial reasoning problems and the study follow differences between non-spontaneously produced actions through gestural interfaces and learning supported with spontaneous actions. Ulrike Dreher from Germany looks at the multiple representations aspects supported by technology in her presentation titled *Factors that influence representational choice: students' mathematical abilities, self-efficacy and preference* Dreher (with Leuders & Holzäpfel) enquire is there a relationship between students' mathematical skills (translating/working with representations), their self-efficacy beliefs and their preferences for individual representations. The intention was to identify and analyze the relationships between various factors, specifically, the preferences of learners for different representations; the factors of their mathematical and representational self-efficacy as well as their meta-representational competence. The artefact that Natthapoj Vincent Trakulphadetkrai from UK is studying are picture books. In his presentation *Enhancing children's visualisation of multiplication through their self-generated mathematics picture books* he described the extent to which having primary school students create their own mathematics picture books enhance their ability to visualize multiplicative word problems and number sentences. The finding shows that children in the intervention class significantly outperformed their peers in the comparison class in both accuracy and ability to visualize word problems and number sentences.

In session B three presenters made their communications. The first communication *Seeing: an intuitive and creative way to solve a problem* was presented by Teresa Pimentel (with Isabel Vale and Ana Barbosa) from Portugal. This presentation discussed the potentialities of visual solutions and their connections with creativity. The second one, *Visual Patterns: a creative path to generalization*, was presented by Ana Barbosa and Isabel Vale. The presenters reported a study involving 80 pre-service teachers in Portugal solving pattern generalizing tasks. The presentation discussed the teachers' strategies and difficulties, and highlighted the power of visualization to reach a generalization. The last one, *The role of visualization in the mathematical working space of teachers; differentiation of reasoning*, was presented by Carolina Henriquez Rivas from Chile. Using a theoretical framework known as Mathematical Working Space that characterized the epistemological and cognitive aspects activated by teachers when they are engaged in reasoning, the presentation discussed the Chilean mathematics teachers' reasoning when solving a geometrical task which favors a particular way of visualization. The results suggested deepening the theoretical aspect by proposing different categories of tasks and identifying the types of visualization involved.

The first communication of the third meeting was: *Analyzing students' visual thinking in solving selected concepts of mathematical analysis involving the concept of infinity*, was presented by Jonatan Muzangwa (with Ugorjio Ogbonnaya & David Mogari) from South Africa. This presentation analyzed the visual thinking of undergraduates majoring in Mathematics when they solved problems involving the concept of infinity. The presenters claimed that visual thinking was not helpful to the undergraduates in the topic of mathematical analysis. The second one, *Reflex: an educational representation of complex functions*, was presented by Mikael Mayer (with Lucas Willems) from France. In this presentation, the presenters showed the importance of using the REFLEX technology when learning complex functions. With the REFLEX tools students were able to visualize functions through color graphic representations, and connect its graphic representations with mathematical expression. The last one, *The relationship between teacher lens and teacher noticing of students' strategies in figural patterns,* was presented by Rabih El Mouhayar from Lebanon. This presentation discussed how teachers analyze students' strategies when are solving figural growing patterns, focusing on the lens that the teachers used and what they notice in their students' productions. This session saw a lively discussion among the three presenters and all the participants.

The last meeting session had two live presentations and one video presentation from Turkey, Nazan Sezen Yuksel, who was unable to attend ICME due to unforeseen political circumstances in her country. The first presentation, *Eliciting visualisation with techno-modelling tasks*, was presented by Johanna Kotze (with Gerrie Jacobs and Erica Spangenberg) from South Africa. Their study examined the influence of techno-modelling tasks on the visualization of 80 engineering students. The presenter reported positive findings such as the abilities of the students to reverse known processes and to make inferences based on their visual intuition and reasoning. The second presentation, *Onto-semiotic analysis of visualization and diagrammatic reasoning tasks*, was presented by Belen Giacomone (with Goldino, Wilhelmi, Blaanco and Contreras) from Spain. In this presentation, a training framework designed to help mathematics teachers gain competence in epistemic and cognitive analysis of mathematics instruction processes was discussed. The role of visual and analytical languages in establishing mathematical objects was highlighted. In the final presentation via a pre-recorded video, Nazan Sezen Yuksel (with Bulbul) from Turkey presented their paper, *Investigation of development on mental cutting ability by latent growth*. The presenter reported on their study involving over 70 students that examined their mental cutting ability (a component of spatial ability) in three different mathematical activities. Their data analysis showed an increment in the level of students' mental cutting ability.

The group aim was to offer an integrative view on research and practice in the field of visualization, and indeed, participants offered a wide range of approaches and proposals for further development of the field in terms of mathematical foundations, research methods and technology.

Open Access Except where otherwise noted, this chapter is licensed under a Creative Commons Attribution 4.0 International License. To view a copy of this license, visit http://creativecommons.org/licenses/by/4.0/.

Topic Study Group No. 21: Mathematical Applications and Modelling in the Teaching and Learning of Mathematics

Jussara Araújo, Gloria Ann Stillman, Morten Blomhøj, Toshikazu Ikeda and Dominik Leiss

The Programme

TSG21 was well supported by full papers, orals and posters showcasing research into mathematical applications and modelling education. The structure of the main TSG sessions involved invited plenaries and 20 min research and/or practice presentations. As well as these sessions there were parallel oral communication sessions.

The first main session on Tuesday 26 July: 12.00–13.30 was chaired by Jussara Araújo. Following a welcome and overview of sessions by the chair, Gloria Stillman gave a one hour plenary on the *State of the Art on Modelling in Mathematics Education* which was followed by a lively discussion. Later in the day Oral Communications in three parallel sessions took place from 15:00 to 18:00.

The plenary highlighted the many good arguments already presented in the literature as to why real world applications and modelling should be favoured in curricula. The particular goal of modelling and/applications in curricula underpin the approach taken to modelling in many research studies whether these goals be from a mathematical or an informed citizenry perspective. Uptake of curricula goals and implementations vary widely in practice despite advances in many quarters.

Co-chairs: Jussara Araújo, Gloria Ann Stillman.
Team members: Morten Blomhøj, Toshikazu Ikeda, Dominik Leiss.

J. Araújo
Universidade Federal de Minas Gerais, Belo Horizonte, Brazil
e-mail: jussara.loiola@gmail.com

G.A. Stillman (✉)
Australian Catholic University, Ballarat, Australia
e-mail: gloria.stillman@acu.edu.au

A recent curriculum document study by Smith and Morgan (2016) in 11 educational jurisdictions identified three orientations to real world contexts in mathematics: as a tool for everyday life, as a vehicle for learning mathematics and as motivation to learn mathematics. Of concern, though, was the provision of alternative pathways in the majority of these jurisdictions with more mathematically advanced pathways having less emphasis on real-world contexts.

In research into the teaching and learning of mathematical modelling there is strong emphasis on developing "home grown theories" where the focus is on particular "local theories" such as modelling competencies rather than general theories from/with application outside the field. As examples of current theoretical lines of inquiry prescriptive modelling, modelling frameworks (e.g., the dual modelling cycle), and modelling competencies (all local theories) and anticipatory metacognition (a general theory) were overviewed to give a flavour of work being done. Some have been the subject of empirical testing or confirmation whilst others await such work. Focuses of empirical lines of inquiry are just as many and varied so examples focused on student modelers, teachers of modelling and task design. Finally, questions for future theoretical and empirical research based on the fore-going examples were outlined. In all, the emphasis was on doing modelling in the classroom so as to be useful!

On Wednesday 27 July: 12.00–13.30 there were two parallel sessions of full paper presentations. The first session was chaired by Morten Blomhøj. Papers were given by France Caron—*Approaches to investigating complex dynamical systems*, Irit Peled—*Shifts in knowledge and participation of children with mathematical difficulties working on modelling tasks*, Dung Tran—*Authenticity of modelling tasks and students' problem solving*, and Miriam Ortega—*Influence of technology on mathematical modelling of a physical phenomenon.*

The other parallel session was chaired by Toshikazu Ikeda. Papers were given by Takashi Kawakami—*Merging of task contexts and mathematics in dual modelling teaching: Case studies in Japan and Australia*, Jill P Brown—*What do we mean by 'context'?* Andreas Busse—*The negative impact of the new German examination tasks on the modelling classroom in Hamburg* and Corinna Hertleif – *Assessing sub-competencies of mathematical modelling in the LIMO project.*

On Friday 29 July 12.00–13.30 again there were two parallel sessions of full paper presentations. The first session was chaired by Dominik Leiss. Papers were given by Xenia Reit—*The potential of cognitive structures in solution approaches of modelling tasks*, Jennifer Czocher—*Making sense of student-generated conditions and assumptions*, Angles Dominguez—*Model application activity: Integration of concepts and models* and Toshikazu Ikeda—*Organizing mathematical modelling in Japanese mathematics curriculum.*

At the same time a second session was chaired by Jill Brown where papers were given by Juhaina Awawdeh Shahbari—*Adapting a cognitive tool for representing teachers' interpretations of students' modelling activities*, Peter Stender—*Heuristic strategies in modelling problems* and Elizabeth W. Fulton—*Teachers as learners: Understanding and valuing mathematical modelling through professional development.* Again later that day there were Oral Communications in Parallel sessions from 15:00 to 18:00 (see ICME-13 website for details).

The last main session occurred on Saturday 30 August 12:00–13:30 and it was chaired by Gloria Stillman. The session began with two invited half hour plenaries, *Toward a Framework for a Dialectical Relationship between Pedagogical Practice and Research* by Jussara Araújo and *Interplay between Research and Development of Teaching Practices in Mathematical Modelling* by Morten Blomhøj.

Jussara Araújo presented what she saw as the initial steps toward a framework for a dialectical relationship between pedagogical practice and research in the field of modelling in mathematics education. These methodological reflections arose from the development of research on modelling guided by critical mathematics education, and grounded in a socio-political perspective of research. A primary characteristic of the dialectic is that the students|participants are constituted in relation to the teacher|researcher; whereas a second characteristic is that ethical concerns regarding the students|participants help to constitute the methodological rigour of the research that, in turn, is related to the educational quality of the pedagogical practice. Thus, pedagogical practice and research should be seen as part of a single unit, mutually developing and influencing each other; on the other hand, they are different, have different purposes, and may be incompatible, but one presupposes and constitutes the other.

Morten Blomhoej's plenary focussed on the fundamental duality between the teaching aims of (a) developing students' modelling competence and (b) supporting their learning of mathematics through modelling activities. He saw understanding the interplay that results from this duality between research and practice as essential for understanding and furthering the integration of modelling into mathematics teaching. The theories of both teaching and research need to be made concrete and contextualised in relation to teachers' particular modelling projects in order for them to be useful in developmental projects or in-service courses. A learning trajectory through secondary mathematics on the modelling of dynamical phenomena by means of compartment models, difference equations and the use of spreadsheet was used to illustrate how research could and should inform curricula change.

Following a short discussion of these talks, the chair summarised the work of the TSG and closed the session with the expectation that there would be a follow up publication to include at least the main papers and plenaries. Issues related to mathematical applications and modelling in the teaching and learning of mathematics have continued to grow in interest from previous International Congresses on Mathematical Education. This is a very broad field of interest both in terms of educational level range from elementary and primary school to tertiary and from the perspective of mathematics. The TSG thus attracted and catered for a breadth of participants through the individual talks from those interested in the mathematical modelling of primary school students (e.g., Kawakami, Peled), secondary school students (e.g., Busse, Hertleif, Ortega, Tran), and tertiary students (e.g., Dominguez) or both students in schooling and university (e.g., Caron, Czocher,) as well as pre-service and in-service development of their teachers (e.g., Shahbari, Fulton). Others were attracted and catered for by more general issues that surround the teaching and learning of mathematics through mathematical applications and modelling such as task context (e.g., Brown), assessment and difficulty of modeling tasks (e.g., Busse, Reit), intervention strategies when managing modelling by others (e.g., Stender), authenticity in tasks (e.g., Tran) and the curriculum components and sequences involving modelling-related activities for elementary to high school education (e.g., Ikeda). We thank all the contributors to our TSG whether in the audience or presenting.

Reference

Smith, C., & Morgan, C. (2016). Curricular orientations to real-world contexts in mathematics. *The Curriculum, 27*(1), 24–45.

Open Access Except where otherwise noted, this chapter is licensed under a Creative Commons Attribution 4.0 International License. To view a copy of this license, visit http://creativecommons.org/licenses/by/4.0/.

Topic Study Group No. 22: Interdisciplinary Mathematics Education

Susie Groves, Julian Williams, Brian Doig, Rita Borromeo Ferri and Nicholas Mousoulides

The Programme

Topic Study Group 22 *Interdisciplinary mathematics education* included paper presentations and discussion in four main sessions, four oral communication sessions, and one poster presentation session—see below.

The Topic Study Group team also produced the Springer ICME-13 Topical Survey *Interdisciplinary Mathematics Education: A State of the Art* (Williams et al., 2016) for pre-reading.

Co-chairs: Susie Groves, Julian Williams.
Team members: Brian Doig, Rita Borromeo Ferri, Nicholas Mousoulides.

S. Groves (✉)
Deakin University, Burwood, Australia
e-mail: susie.groves@deakin.edu.au

J. Williams
The University of Manchester, Manchester, UK
e-mail: julian.williams@manchester.ac.uk

Main Sessions

Tuesday 12.00–13.30: Chair—Susie Groves	Wednesday 12.00–13.30: Chair—Rita Borromeo Ferri
Introduction to TSG22—Julian Williams	*Introduction to session 2*—Rita Borromeo Ferri
Overview of the TSG22 Topical Survey *Interdisciplinary Mathematics Education: A State of the Art*—Susie Groves	*A modelling perspective in designing interdisciplinary professional learning communities*—Nicholas Mousoulides
Theory of disciplinarity and interdisciplinary activity: Communities, boundaries, voices and hybridity—Julian Williams and Wolff-Michael Roth	*Mathematics in an interdisciplinary STEM course (NLT) in the Netherlands*—Nelleke Susanna den Braber, Jenneke Krüger, Marco Mazereeuw and Wilmad Kuiper
Challenges for mathematics within an interdisciplinary STEM education—Russell Tytler	*Preservice mathematics teachers' interdisciplinary work for STEM education*—Fatma Aslan-Tutak and Sevil Akaygun
Inter-disciplinary mathematics: Old wine in new bottles?—Brian Doig and Wendy Jobling	*Closing discussion*—including possible research agenda between ICME-13 and ICME-14
Closing discussion—including possibilities for a book	
Friday 12.00–13.30: Chair—Nicholas Mousoulides	Saturday 12.00–13.30: Chair—Julian Williams
Introduction to session 3—Nicholas Mousoulides	*Introduction to session 4*—Julian Williams
Scientific inquiry in mathematics and STEM education—Andrzej Sokolowski	*Ratio and proportion in secondary school science*—David Swanson
Using real-life context as an aid for mathematics teaching and learning—Michael Erotoma Omuvwie	*Interdisciplinary communication between music and mathematics: An experience with stochastic music*—María Alicia Venegas Thayer
Quantitative reasoning: Rasch measurement to support QR assessment—Robert Lee Mayes, Kent Rittschof, Jennifer Forrester and Jennifer Christus	*Inspired by Leonardo da Vinci*—STEM *learning for primary and secondary school with the Cross-Link Approach*—Rita Borromeo Ferri, Andreas Meister, Detlef Kuhl and Astrid Hülsmann
Closing discussion—including possible Discussion Group or Working Group at ICME-14	*Closing discussion*—including decision on possible Springer book

Oral Communications

Tuesday 15.00–16.00: Chair—Brian Doig	Tuesday 16.30-18.00: Chair—Rita Borromeo Ferri
Mathematics and sciences teachers collaboratively design interdisciplinary lesson plans: Benefits, limitations, and concerns—Atara Shriki and Ilana Lavy	*Investigating interdisciplinary approaches and commitments through pre-service teachers' use of mathematics and poetry*—Nenad Radakovic, Limin Jao and Susan Jagger
Teaching and applying research methods in a cross-cultural project for students of mathematics education—Mutfried Hartmann, Thomas Borys, Arno Bayer and Tetsushi Kawasaki	*Mathematics and medicine: A study of thinking and variational language* – Gloria Angélica Moreno Durazo and Ricardo Cantoral
Doing inter-disciplinary work in mathematics education: Potentialities and challenges—Sikunder Ali Baber	*Teachers' readiness to mathematics and science integration*—Betul Yeniterzi, Cigdem Haser, and Mine Isiksal-Bostan
Interdiciplinary activities in context—Maite Gorriz, and Santi Vilches	*Incorporating mathematics, creative writing, literature and arts in the classroom*—Frederick Lim Uy
Friday 15.00-16.00: Chair—Susie Groves	Friday 16.30-18.00: Chair—Nicholas Mousoulides
A cloud based performance support system for teaching STEM with hands-on modeling—Roberto Araya	*Integrating mathematics, engineering and technology through mathematics modeling and video representations*—Carlos Alfonso LopezLeiva, Marios Pattichis and Sylvia Celedon-Pattichis
Korean mathematics textbook analysis: Focusing on competence, on contexts and ways of integration—Jong-Eun Moon, Mi-Yeong Park, Jeong Soo-Yong and Mi-Kyung Ju	*An experimental textbook system for financial mathematics for the integration of finance and mathematics*—OhNam Kwon, JungSook Park, JeeHyun Park, Jaehee Park and Changsuk Lee
Mathematics of money dynamics—Francesco Scerbo, Elena Scordo and Laura Vero	*Co-disciplinary mathematics and physics research and study courses (RSC) in the secondary school and the university*—Maria Rita Otero, Vivianna Carolina Llanos, Maria Paz Gazzola and Marcelo Arlego
Transcending the mathematics classroom—Signe E. Kastberg, Rachel Long, Kathleen Lynch-Davis and Beatriz S. D'Ambrosio	*Investigating students' difficulties with differential equations in physics*—Diarmaid Aidan Hyland, Paul van Kampen and Brien Nolan

Posters—Tuesday 18.00–20.00

Assessment of mathematical competencies of biology teacher trainees—Ivana Boboňová and Soňa Čeretková	Relationahips of cognitive domains: Focus on reasoning and applying in mathematics and science—Amanda Meiners, Jihyun Hwang and Kyong Mi Choi
QUBES: Quantitative Undergraduate Biology Education and Synthesis—Carrie Diaz Eaton, Sam Donovan, Stith T. Gower and Kristin Jenkins	Usage of mathematics competency in a new context in science: Experience of Latvia—Ilze France, Līga Čakāne, Uldis Dzērve, Dace Namsone and Jānis Vilciņš
Enacting planets—Emmanuel Rollinde	Geometry from a global perspective—Craig Russell
Students' aspirations for STEM careers—Kathryn Holmes, Adam Lloyd, Jennifer Gore, Max Smith, Leanne Fray and Claire Wallington	Geometry in Slovak blueprint—Soňa Čeretková and Edita Smiešková
Fostering of interdisciplinary competences through basic education in computer science in mathematics in primary school—Peter Ludes	Preparing STEM teachers as researchers: A research experiences for undergraduates project—Jennifer Wilhelm and Molly H. Fisher
An interdisciplinary activity on angiogenesis—Catherine Langman, Judi Zawojewski and Patricia McNicholas	Relationahips of cognitive domains: Focus on reasoning and applying in mathematics and science—Amanda Meiners, Jihyun Hwang and Kyong Mi Choi

Interdisciplinary mathematics education is a relatively new field of research, which has become increasingly prominent because of the political agenda around STEM. However, there are also increasing mathematical demands outside STEM—for example, the need to effectively critique the vast amounts of statistical information evident in all aspects of society—as well as increasing interest in how mathematics inter-relates with other disciplines and contexts.

The level of interest in interdisciplinary mathematics education was evident in the number of presentations and participants at the main sessions and oral communications, and the vibrant discussions that took place. Presentations were complemented by a range of posters that allowed a wide group of attendees to discuss ideas of interdisciplinarity during the poster viewing time.

Disciplinarity is a social phenomenon, marked by increasing specialization and differentiation of practices, professional disciplines—such as nursing, teaching, physiotherapy—often defined by practical competence. It is often difficult for those schooled in one field to relate effectively with others from relatively independent and contradictory fields, with boundaries between disciplines notoriously difficult to cross, which might explain why interdisciplinarity is often praised rhetorically but so difficult to practice.

Interdisciplinarity occurs across a continuum ranging from mono-disciplinarity to meta-disciplinarity. Mono-disciplinarity involves a single discipline only, while multi-disciplinarity involves two or more disciplines, but in both these cases the disciplines themselves may remain intact. Inter-disciplinarity, on the other hand, involves some sort of hybridising of "multi" disciplines—e.g. chemistry and biology becoming biochemistry. Trans-disciplinarity acquires its transcendence due to disciplines being subsumed in joint problem solving enterprises that may perhaps result in a new form of mathematics. While multi-disciplinarity and trans-disciplinarity offering hybridity of disciplines, the disciplines themselves are not displaced, but instead provide the value interdisciplinarity requires. Meta-disciplinarity transpires in an awareness of relationships and differences between disciplines—e.g. the contrasting nature of "using evidence" in history and science may be contrasted and thereby clarified (see Williams et al., 2016).

As can be seen from the programme details, presenters covered a wide range of topics under the umbrella of interdisciplinarity. Presenters provided views of interdisciplinarity from several academic disciplines, including mathematics, physics, medicine and music, as well as across much of the spectrum of interdisciplinarity discussed above. In addition, presenters represented a wide cross-section of countries, which added to the notion that interdisciplinarity is indeed of global interest and importance. Of particular note were presentations and posters outside the expected scientific disciplines, for example, music and poetry. Discussion following each presentation allowed a range of perspectives to be aired. Attendees brought to the discussion the perspectives of many different educational cultures, their affordances and their constraints.

The review of the literature carried out as part of the pre-ICME Topical Survey showed that interdisciplinary mathematics education is relatively under-developed as a research subfield. There is some evidence of beneficial outcomes of interdisciplinary work in integrated curricula that often involves projects, with these benefits typically emphasising motivational, affective and problem-solving learning outcomes. The papers and presentations reinforced these conclusions.

Progress in interdisciplinary mathematics education appears to be hampered by a lack of clarity and consensus about the concept of disciplinarity and how to adequately describe "interdisciplinary" interventions and programs, together with a lack of consistency about relevant learning outcomes and how they can be identified and measured, and a lack of research on which future work can build.

Interdisciplinary mathematics education offers the opportunity to encourage possibly disaffected students to reconsider mathematics. It offers mathematics to the wider world in the form of added value—e.g. in problem solving—and, conversely, it offers the added value of the wider world to mathematics. Therefore interdisciplinary mathematics education should be a major topic for mathematics education and can be expected to become much more prominent in educational research and practice.

Discussion among the large group of session attendees, almost all of whom attended all main sessions, was fruitful. Plans were made to set up a website for continued contact among members of the Topic Study Group between ICME congresses, with a decision to be made closer to ICME-14 as to whether to attempt to continue the Topic Study Group or try to reconvene as a Discussion Group.

All participants were invited to submit abstracts for a proposed Springer monograph, with approximately 25 proposals for chapters currently under review.

Reference

Williams, J. M., Roth, W.-M., Swanson, D., Doig, B., Groves, S., Omuvwie, M., ..., Mousoulides, N. (2016). *Interdisciplinary mathematics education: A state of the art*. Berlin: Springer. doi:10.1007/978-3-319-42267-1

Open Access Except where otherwise noted, this chapter is licensed under a Creative Commons Attribution 4.0 International License. To view a copy of this license, visit http://creativecommons.org/licenses/by/4.0/.

Topic Study Group No. 23: Mathematical Literacy

Hamsa Venkat, Iddo Gal, Eva Jablonka, Vince Geiger
and Markus Helmerich

The Programme

The Mathematical Literacy Topic Study Group 23 at ICME-13 was organized around three key themes that drew from emerging findings in the literature related to discussions about the Mathematical Literacy field:

- Conceptual maps and gaps: What topics, conceptual models, or theories should/could be considered in teaching about or for mathematical literacy? Why? When, for whom, and how should we promote mathematical literacy? What needs exist that notions of mathematical literacy might address?
- The 'place' of mathematical literacy: How are notions of mathematical literacy figured into curricula, and in teachers' identities, beliefs, attitudes and practices, in teacher education, in learning materials, in local/national assessments, etc.? What tensions and challenges emerge? What dis/continuities exist between teaching of mathematical literacy in school—and in tertiary and adult contexts?
- What can large and smaller-scale studies tell us about mathematical literacy?: How do empirical results from studies with differing scales and foci (e.g., on

---

Co-chairs: Hamsa Venkat, Iddo Gal.
Team members: Eva Jablonka, Vince Geiger, Markus Helmerich.

H. Venkat (✉)
University of the Witwatersrand, Johannesburg, South Africa
and

Jönköping University, Jönköping, Sweden
e-mail: hamsa.venkatakrishnan@wits.ac.za

I. Gal
University of Haifa, Haifa, Israel
e-mail: iddo@research.haifa.ac.il

skill transfer, teachers, literacy/language aspects of applied math tasks, inter/national surveys) inform our thinking about the conceptualization, teaching, learning, or assessment of mathematical literacy? Given such results, should mathematical literacy be directly taught, or be integrated across curriculum subjects, and how? Or can it emerge as a by-product of teaching "regular" mathematics? What evidence do we have about possible (re)solutions?

Papers and discussions related to these themes were intended to stimulate discussion about key directions for future research related to mathematical literacy. Overall, our TSG included two invited plenary presentations followed by two discussant reactions, eight regular (invited) papers and six short oral communications, nine posters, and extended general discussions during all sessions which were moderated by the team members.

Invited plenary papers were presented across the first two sessions. The first plenary, presented by Geiger, Goos & Forgasz, followed from their (Geiger, Goos, & Forgasz, 2015) editing of a ZDM Journal issue focused on mathematical literacy. This plenary provided a historical overview of the range of terminology and the capabilities and facets commonly associated with these terms. As such, this provided participants with an introduction related to the first theme, and an anchor for a range of submissions related to this theme. Among these related papers, North used Dowling's theorizing of 'gaze' to consider what a mathematical literacy gaze might be comprised by, and linked his theoretical development to a discussion of possibilities and critiques related to the Mathematical Literacy curriculum in use in the South African post-compulsory phase. Vohns discussed possibilities and avenues for the development of reflective knowledge within mathematical learning, and arguing that this would be important to fulfilling requirements for a mathematically literate orientation. Developing notions of particular topics within the broader notion of mathematical literacy were seen in Engel, Gal & Ridgway's incorporation of citizenship into earlier work focused on statistical literacy with attention to 'civic statistics, and also in Chen's attention to spatial literacy. Askew, in his discussant response to Geiger et al's plenary presentation, presented a range of alternatives for thinking about the relationship between mathematics and mathematical literacy, and raised attention to how tasks and how responses to tasks were evaluated often embodied different conceptualizations of this relationship. His presentations also raised questions about whose responsibility it is to develop some aspects seen as central to conceptualizations of mathematical literacy (e.g., interpretation, critical reflection) that seem to fall outside some conceptualizations of mathematics.

In Session 2, Klieme presented a plenary input drawing from his experience within the PISA Mathematical Literacy international assessment analysis team. He focused particularly on the somewhat under-emphasised aspect of comparisons of learning environments across participating nations. In his presentation he shared both the ways in which learning environments were conceptualized within the PISA data gathering process (based on 'Opportunity to Learn', 'Teaching Practices' and 'Teaching Quality'), and cross-cultural comparisons that were possible to draw from their dataset. Of particular interest in his presentation was evidence of the culturally situated clustering of several factors that made a difference but in particular geographic regions, contrasting with the underpinning view in PISA of the 'universality' of the items selected for international assessments. Jablonka, in her discussant response to Klieme's plenary presentation, emphasized this disjuncture, while focusing also on the nature of the PISA items and the largely mathematical conceptualizations underlying their formulation. Klieme encouraged discussion, in the multi-national participation within the TSG, of whether participants viewed the items he had shared as fitting with national understandings of mathematical literacy and/or mathematics, as well as asking participants to reflect on his learning environment findings with reference to different national contexts. His talk was related to Theme 3, but the focus on learning environments within the presentation allowed for links with Theme 2's attention to teaching practices as well.

A range of papers attended to issues related to mathematical literacy teaching and learning, with paper presentations occurring in Session 3 and in an adjoining Oral Communication session, and supplemented by the poster presentations. Given space restrictions we can discuss only some illustrative examples: Bennison presented examples of the ways in which mathematical literacy could be seen as a cross-curricular enterprise, while noting too, through a focus on 'boundary objects' some of the complexities that needed to be negotiated in taking on mathematical literacy as a multi disciplinary goal. Winter (2016) echoed some of these complexities in his study of pre-service teachers' work with contextual tasks. His data pointed to ways in which the notion of mathematization, developed within Realistic Mathematics Education in the context of mathematics, might need extending to take on board the orientations and goals of mathematical literacy. Curriculum development to take on board mathematical literacy orientations were also foregrounded in presentations from Turkey (Gurbuz) and England (Lee).

In Session 4, the emphasis was on discussion within TSG23 on key issues and points of contention that had emerged across the discussions in the earlier sessions. A key discussion related to a focus on 'mathematical literacy tasks' and how task selections made either by teachers, assessment designers or researchers can provide an important and under-researched avenue for exploring conceptualizations of mathematical literacy in 'bottom-up' ways, rather than the 'top-down' ways that have tended to predominate in approaches that work from definitions and curricular specifications. A potentially productive route that was discussed related to whether descriptors and/or criteria could be developed for mathematical literacy tasks that would help to disaggregate the views of mathematical literacy that they operationalized, and then the potential in this approach for evaluating the extent to which assessments either diverged from, or only partially reflected the mathematical literacy goals and definitions that they purported to cover. An initial list of criteria that emerged from the Session 4 discussion across the participants were as follows:

- Decision making or coming to a judgement/interpretation/understanding the decisions of others/forming an opinion relating to personal, social and political actions or activities (transport/environment/tourism/advertising),
- Use of evidence as the basis for decision making,

- Tasks designed to 'inform' the public about an issue, including consumers,
- Supporting 'literate' reading of text for improving participation in personal and civic life (local and global issues), at work and in lifelong learning,
- Should ML support preparation for employment? Can this be taught in a ' generic' way? (e.g. budgets/allocation of resources),
- Critique/challenge from a basis in evidence and interpretation, and for influence.

The closing discussion included a focus on whether some contexts or situations would or should be privileged over others if mathematical literacy was conceptualized as a 'life-orientated' competence rather than a 'knowledge-oriented' subject. Picking up on the learning environments focus, there was broad agreement on the need for more 'open' pedagogic relationships if the goals of mathematical literacy were to be realized.

In conclusion, our sense of the progress of the Mathematical Literacy TSG23 was that the discussions around the plenary papers and discussant responses, coupled with the paper and poster presentations, highlighted some key under-represented and under-theorised areas in the ongoing debates around notions related to mathematical literacy. We see a need to increase attention in teaching, teacher preparation and professional development, assessment design, and future research to the unique aspects of mathematical literacy (numeracy). For example, more attention is needed to 'use, apply and interpret' situations where some mathematics may be a useful part, but only a part, of the toolkit that can be brought to bear to understand a terrain and/or solve problems within it was central to the discussions. Askew problematized this notion of mathematical literacy with the following diagram and an accompanying question, reproduced below:

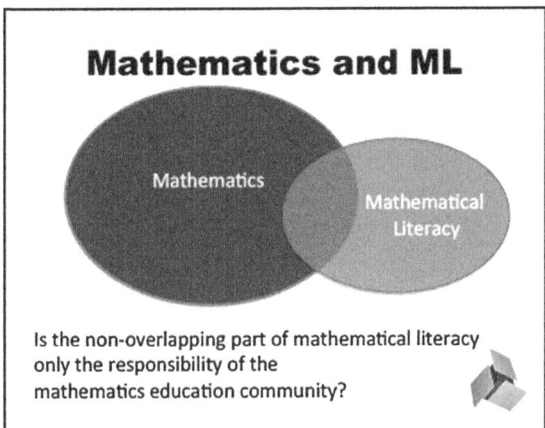

This question and the broader discussion around it and around the three key themes that drove the design of our TSG start to get at the most important (http://dictionary.cambridge.org/dictionary/english/important) or basic (http://dictionary.cambridge.org/dictionary/english/basic) aspect of what it might mean to focus on mathematical literacy in ways that are linked to, but not subsumed by, mathematics.

Our sense, as co-chairs of TSG23 is that this direction, with the strands related to task design and learning environments noted above, indicates some promise for debates and further development in the coming years. As a professional community we should furnish mathematical literacy with a more distinct identity and a gaze that contains some aspects that are independent of mathematics, while retaining some overlaps, in order to better serve learners as well as the needs of society at large.

Reference

Geiger, V., Goos, M., & Forgasz, H. (2015). *ZDM Mathematics Education, 47*, 531. https://doi.org/10.1007/s11858-015-0708-1.

Open Access Except where otherwise noted, this chapter is licensed under a Creative Commons Attribution 4.0 International License. To view a copy of this license, visit http://creativecommons.org/licenses/by/4.0/.

Topic Study Group No. 24: History of the Teaching and Learning of Mathematics

Fulvia Furinghetti, Alexander Karp, Henrike Allmendinger, Johan Prytz and Harm Jan Smid

The Topic Study Group on the history of the teaching and learning of mathematics was formed in 2004 at ICME-10. Its work reflects the growing interest in the history of mathematics education. At ICME-13 in Hamburg, this TSG had more participants than in previous years, but the procedure at the main meetings followed established tradition—short reports were heard, after each of which the speaker could answer some questions. Before turning to a discussion (necessarily very brief) of the contents of the reports, a few words must be said about the beneficial innovations that set apart the 2016 conference.

The first of them was the publication of a booklet, *History of mathematics teaching and learning. Achievements, Problems, Prospects* (Springer, 2016), prepared by Alexander Karp and Fulvia Furinghetti in discussion with the other members of the team. Hopefully, this booklet offers researchers (first and foremost, beginning researchers) useful bibliographic information, and also lists a number of areas in which scholarly work is being conducted in different countries. Another innovation was the introduction of so-called "Oral Communications," that is, additional meetings at which more presentations were heard. Along with poster presentations, they gave participants an opportunity to form a better understanding of what is being studied by historians of different countries' mathematics education. Since these presentations cannot be discussed in detail here, we restrict ourselves to listing the authors of the studies that were presented (names of presenters in

---

Co-chairs: Fulvia Furinghetti, Alexander Karp.
Team members: Henrike Allmendinger, Johan Prytz, Harm Jan Smid.

F. Furinghetti
University of Genoa, Genoa, Italy
e-mail: furinghetti@dima.unige.it

A. Karp (✉)
Teachers College, Columbia University, New York, USA
e-mail: karp@tc.columbia.edu; apk16@columbia.edu

boldface): **Eisso Johannes Atzema**; **Eliete Grasiela Both**, Bruna Carnila Both; **María Teresa González Astudillo**, Myriam Codes Valcarce; **Günter Graumann**; **Zohre Ketabdar**, Maryam Ketabdar; **María José Madrid**; Alexander Maz-Machado, Carmen León-Mantero, Carmen López; **Marvin Roberto Mendoza**, Luis Armando Ramos; **Nicola M.R Oswald**, Nadine Benstein; **Irene Papadaki**; Athanasios Gagatsis, Elena Kiliari; **Emily Timmons Hamilton Redman**; **Ildar Safuanov**.

Finally, it must be noted that plans are underway to publish the results of the work of the TSG as a collection of papers, in which selected studies by the participants will be presented in greater detail than they were either during oral presentations or in electronic publications on the conference website. All of these innovations are very significant.

To turn now to the presentations heard at the main meetings, we would like to give special mention to the report of invited speaker Gert Schubring, *Patterns for Studying the History of Mathematics Education: A Case Study of Germany*. Devoted to the history of German mathematics education, the report went beyond this framework, since it possessed considerable methodological significance. It emphasized that the very subject of the history of mathematics education is multidisciplinary by nature. It is impossible to understand what went on in the teaching of mathematics without taking into account what was taking place in social history. The presenter demonstrated the importance of such an approach, showing how various systems of mathematics education formed in the German states of the nineteenth century, reflecting religious and political differences in their histories.

Probably the most popular period in terms of the number of presentations devoted to it this time was New Math (to use this American term to denote the period of reforms that took place in different countries and usually under different names). This once again indicates the importance of what happened during this period in the sixties and seventies of the last century including those cases about which accepted opinion is that the reforms failed. In a presentation titled *Royamount—Proposals of Arithmetic and Algebra Teaching for the Lower Secondary School Level*, **Kristin Bjarnadóttir** examined the influence of the famous seminar in Royamount on teaching in Icelandic, and more broadly, Scandinavian elementary schools. Changes in instruction in another subject and in another country (Brazil and in part France) were the topic of **Elisabete Zardo Burigo**'s report *Real Numbers in School: 1960s' Experiments*. **Dirk De Bock** and Geert Vanpaemel in their report on *Early Experiments with Modern Mathematics in Belgium* addressed the history of teaching in elementary school and even kindergarten, and the corresponding preparation of teachers, when Belgium became one of the leaders of the reform movement. Finally, **Johan Prytz**, presenting a report titled *New Math for Big Education, Old Math for Small Education. A Study of Different Ways to Reform School Mathematics,* analyzed various teaching materials to investigate broader issues connected with the development of reforms in Sweden (including the question of whether it is correct to believe that the reforms failed).

Also strongly connected to this topic was the presentation by **Gabriella Ambrus**, Andreas Filler, **Ödön Vancsó**, *Functional Reasoning and Working with*

Functions in the Mathematics Teaching Tradition in Hungary and Germany, which naturally devoted attention to the New Math period, when functions once again began receiving greater attention than they had during the preceding period. Comparing the teaching of specific subjects in different countries is a fruitful idea that has been often developed in the past.

The study of teaching materials (first and foremost, textbooks) and the biographies of their authors is one of the principal areas of research in the history of mathematics education. In Hamburg, this topic was addressed, on the basis of Spanish materials, by the reports of **Antonio M. Oller-Marcén**, Vicente Meavilla, *Arithmetic in the Spanish Army at the End of the Nineteenth Century. The Works of Salinas and Benítez,* and in part by **Miguel Picado**, Luis Rico, Bernardo Gómez, *The Metrological Reform in the Spanish Educational System in the Nineteenth Century: Who Were the Authors of the Textbooks?*

Jenneke Krüger's paper *Frans van Schooten Sr: Dutch Mathematics for Engineers, Leiden, 1611–1645* addressed related issues, although its focus was not a textbook, but surviving lecture plans. The most significant aspect of this presentation was probably that the materials studied are not of comparatively recent provenance, but date back to the seventeenth century, from which many fewer documents survive, and which in addition have by no means been everywhere thoroughly studied. It is to be hoped that the seventeenth and eighteenth centuries will attract more attention from researchers in the future.

Textbooks from the period that predates even the seventeenth century are the focus of **Alexei K. Volkov**'s paper *Didactical Function of Images of Counting Devices in Chinese Mathematical Textbooks.* This study can also be said to address the history of using technology in mathematics education (which of course must not be thought of as beginning in the twentieth century).

Harm Jan Smid presented a report on *Becoming a Mathematics Teacher in Times of Change,* devoted to the biography of a Dutch teacher born in 1778, who lived during a time of upheavals, and whose life demonstrates the changes that took place in society and consequently in mathematics education as well. The biographies of teachers—although teachers living in a different country and during a different time—were also the subject of the report *Russian Mathematics Teachers, 1830–1880: Several Examples* by **Alexander Karp**. This paper summed up the author's archival research on the biographies of teachers mainly from St. Petersburg, which shed light on the formation of mathematics teachers as a professional group. Also connected with this topic was the report by **Fulvia Furinghetti** and Annamaria Somaglia *The Professionalization of Italian Primary Teachers through a Journal Issued at the Beginning of the Twentieth Century,* which addressed the journal *Bollettino di Matematiche e di Scienze Fisiche e Naturali* and its role in the formation of Italian elementary mathematics education.

Reforms in mathematics education were not, of course, limited to the New Math period. Probably one of the most successful examples of reform was the Meiji period in Japan, which was the subject of a report by **Marion Cousin**, *The Revolution of Mathematical Teaching during the Meiji Era (1868–1912).* The transfer, or more precisely, the adaptation of foreign achievements in mathematics education may be said to be one of the most recurrent and prominent topics in its history. Successful examples may be considered important not only for historians, but also for practitioners of mathematics education.

We conclude this brief inventory with the presentation of **Sethykar SamAn**, *History of Mathematical Instruction in Colonial Cambodia*, which is of particular interest not least because its subject has received extremely little attention, and because of the methodological difficulty of such a study in view of the almost complete destruction of possible sources during the period of Pol Pot's dictatorship. Meanwhile, the study of the history of mathematics education in Cambodia, and more broadly, the study of the history of education in other colonial or semi-colonial countries, is of enormous interest, if only because it combines many of the issues and topics mentioned above. This kind of research must definitely be continued.

As may be seen from what has been said, the four 90-min meetings that the TSG had at its disposal were quite packed. No less packed is the program for the future development of our field of research, which was the focus of the meetings in Hamburg. The problem of collecting and publishing materials on the history of mathematics education, and even more importantly, the problem of interpreting them as part of the social history of each country and of the whole world, seem especially important today, during our own period of significant changes in education as a result of both technological and social transformations. The work of the TSG at the Hamburg conference was a useful step toward the solution of these problems.

Open Access Except where otherwise noted, this chapter is licensed under a Creative Commons Attribution 4.0 International License. To view a copy of this license, visit http://creativecommons.org/licenses/by/4.0/.

Topic Study Group No. 25: The Role of History of Mathematics in Mathematics Education

Constantinos Tzanakis, Xiaoqin Wang, Kathleen Clark, Tinne Hoff Kjeldsen and Sebastian Schorcht

Aim

TSG 25 aimed to provide a forum for participants to share their research interests and results as well as their teaching ideas and classroom experience in connection with the integration of the history of mathematics (HM) in mathematics education (ME). Special care was taken to present and promote ideas and research results of international interest while still paying due attention to the national aspects of research and teaching experience in this area. Effort was also made to allow researchers to present their work with enough time left to get fruitful feedback from the discussion. This stimulated the interest of all participants and especially the newcomers by giving them the opportunity to get a broad overview on the state of the art in this area.

Co-chairs: Constantinos Tzanakis, Xiaoqin Wang.
Team members: Kathleen Clark, Tinne Hoff Kjeldsen, Sebastian Schorcht.

C. Tzanakis (✉)
University of Crete, Rethymnon, Greece
e-mail: tzanakis@edc.uoc.gr

X. Wang
East China Normal University, Shanghai, China
e-mail: xqwang@math.ecnu.edu.cn

Rationale, Focus, Main Themes

The fruitful and harmonious interplay among history, education, and mathematics as three different but complementary dimensions constitutes what is potentially interesting, stimulating, and beneficial for teaching and learning both mathematics as a subject and about mathematics.

History points to the non-absolute nature of human knowledge: What is acceptable as knowledge is "time dependent" (historicity is a basic characteristic) and is potentially subject to change.

Education stresses the fact that humans are different in several respects depending on age, social conditions, cultural tradition, individual characteristics, etc., and in this way helps to understand these differences and become more tolerant.

Mathematics above all sciences emphasizes most the need for logical/rational/intellectual rigor and consistency in our endeavour to understand the world around us (both its mental and empirical aspects).

Along these lines, contributions to this TSG, explicitly or implicitly attempted to illuminate and provide insights on the following general questions:

- Which history is suitable, pertinent, and relevant to ME?
- Which role can HM play in ME?
- To what extent has HM been integrated in ME (curricula, textbooks, educational aids/resource material, and teacher education)?
- How can this role be evaluated and assessed and to what extent does it contribute to the teaching and learning of mathematics?

The TSG also focused on one or more of the following main themes announced in advance:

1. Theoretical and/or conceptual frameworks for integrating history in ME.
2. History and epistemology implemented in ME, considered from either (a) cognitive or/and affective points of view, including classroom experiments at school and university and teacher pre- and in-service education or (b) teaching material, including textbooks and resource material of any kind.
3. Surveys on (a) research on the HM in ME and (b) the HM as it appears in curriculum and/or textbooks.
4. Original sources in the classroom and their educational effects.
5. History and epistemology as a tool for an interdisciplinary approach in the teaching and learning of mathematics and the sciences by unfolding their fruitful interrelations.
6. Cultures and mathematics fruitfully interwoven.

Programme

Originally, 44 contributions were submitted, of which six were withdrawn. The reviewing process resulted in the remaining 38 contributions, coming from 17 countries on four continents, being distributed as follows: (i) 16 were presented in the four 90-min. regular sessions, allocating a 20-min. time slot for each (a 15-min. presentation followed by a 5-min. discussion); (ii) 17 were presented in the two 90-min. and the two 60-min. oral communication sessions, allocating a 15-min. time slot for each (a 10-min. presentation followed by a 5-min discussion); and (iii) 4 were presented in the poster sessions scheduled for all TSGs. There was a final summary at the closing session.

The TSG 25 sessions were attended by about 50 participants, at least half of whom were newcomers to this TSG. To help participants get a sufficiently clear and comprehensive overview of the research domain covered by the TSG's main themes, its Organizing Team conducted a pre-conference survey on the developments in this domain since 2000, the year of publication of a corresponding comprehensive ICMI study (Fauvel & van Maanen, 2000). This survey, which is available online,[1] was communicated to all contributors in this TSG well before ICME-13 as a useful tool to anyone with interest in this domain show wanted to become informed on the main issues and have a concise guide to the work done in this area. Moreover, participants had been invited to stay in contact with the Organizing Team in order to help keep this document continuously updated by making comments, pointing out omissions, and especially providing further bibliographical references.

Summary of the Presentations

Seen as a whole, research communicated during the TSG regular oral communication or poster sessions had the following two general characteristics:

(i) It covered *all levels of education*; from primary school to tertiary education, with special focus on pre- and in-service teacher education.
(ii) Most of the contributions in one form or another referred to and/or were based on *empirical research* in order to support, illuminate, clarify, or evaluate key issues, main questions, or conjectured theses raised by the authors or in the literature on the basis of historical-epistemological or didactical-cognitive arguments.

More specifically, each contribution's main focus and content fell in one of the five areas as detailed below, though, of course, these areas are closely interrelated:

---

[1]http://www.clab.edc.uoc.gr/HPM/HPMinME-TopicalStudy-18-2-16-NewsletterVersion.pdf.

I. Theoretical and/or conceptual frameworks for integrating history and epistemology of mathematics in ME in connection with relevant empirical/experimental evidence (three regular presentations and the oral communications)
II. Empirical investigations on implementing history and epistemology in ME considered either from a cognitive or an affective perspective (four regular presentations and four oral communications)
III. Original historical sources (of any type, not only written documents) in teaching and learning of and about mathematics (two regular presentations, two oral communication, and one poster presentation)
IV. Design of courses and/or didactical material and its implementation and evaluation (including textbooks and resources) (four regular presentations, six oral communications, and two poster presentations)
V. Interdisciplinary teaching and socio-cultural aspects in the context of empirical investigations (three regular presentations, two oral communications, and one poster presentation)

Items (i) and (ii) above and their more detailed descriptions in Items I–V reflect key issues that were stressed quite early[2] and still remain central to the research in this area and the implementation of its results in educational practice, namely:

- To put emphasis on *pre- and in-service teacher education* as a necessary prerequisite for the integration of the history and epistemology in ME to be possible at all;
- To design, produce, make available, and disseminate a variety of *didactical material* in the form of anthologies of original sources, annotated bibliographies, descriptions of teaching sequences/modules as sources of inspiration and/or as generic examples for classroom implementation, educational aids of various types, appropriate websites, etc.;
- To systematically perform carefully designed and applied *empirical research* in order to examine in detail and evaluate convincingly the effectiveness of the integration of the history and epistemology in ME on improving the teaching and learning of mathematics as well as students' and teachers' awareness of mathematics as a discipline and their disposition towards it; and
- To acquire a deeper understanding of theoretical ideas put forward in integrating history and epistemology in ME and to carefully develop them into coherent *theoretical frameworks* and *methodological schemes* that will serve as a foundation for further research and applications.

It is hoped that the work done before, during, and after this TSG or based on it and still in progress will contribute to these key issues.

---

[2] At least since the influential ICMI study by Fauvel and van Maanen (2000).

Reference

Fauvel, J. G., & van Maanen, J. (Eds.). (2000). *History in mathematics education: The ICMI study, new ICMI study series* (Vol. 6). Dordrecht: Kluwer.

Open Access Except where otherwise noted, this chapter is licensed under a Creative Commons Attribution 4.0 International License. To view a copy of this license, visit http://creativecommons.org/licenses/by/4.0/.

Topic Study Group No. 26: Research on Teaching and Classroom Practice

Yoshinori Shimizu, Mary Kay Stein, Birgit Brandt, Helia Oliveira and Lijun Ye

Focus and Themes

This Topic Study Group aimed to improve understanding of the importance, specific nature of, and challenges associated with research on teaching and classroom practice; to promote exchanges and collaborations around the identification and examination of practices in classroom instruction across different education systems; and to enhance the quality of research and classroom practice. Developing systematic research on classroom practice in school mathematics is a relatively new endeavor. In fact, this Topic Study Group was only the third time in ICME history to take a primary focus on classroom practice.

The focus of TSG 26 was discussion of research related to mathematics teaching and classroom practice. Classroom practice includes the activities of teaching and learning located within the classroom. It requires examination of the interactions among the mathematical content to be taught and learned, the instructional practices of the teacher, and the work and experiences of the students within particular educational settings. In these interaction processes, mathematical content is contextualized through situations, the teacher plays an important instructional role drawing on his/her knowledge, and the students involve themselves in the learning processes. Research aims to understand the nature and extent of the interactions, the

---

Co-chair: Yoshinori Shimizu, Mary Kay Stein.
Team members: Birgit Brandt, Helia Oliveira, Lijun Ye.

Y. Shimizu (✉)
University of Tsukuba, Tsukuba, Japan
e-mail: yshimizu@human.tsukuba.ac.jp

M.K. Stein
University of Pittsburgh, Pittsburgh, USA
e-mail: mkstein@pitt.edu

complexity of the didactic system, the roles of the teacher and students in the interaction processes when the mathematical content is taught and learned, and the complexity of the activities in mathematics classrooms. Papers that explore how factors outside the classroom (e.g., school leadership, policies, organizational dynamics) shape instruction inside classrooms are also welcomed.

The TSG26 was intended to provide an international platform for all interested parties (e.g., mathematics educators, school teachers, educational researchers, etc.) to disseminate findings from their research on teaching and classroom practice with the use of various theoretical perspectives and methodologies, and to exchange ideas about mathematics classroom research, development, and evaluation.

Regular Sessions

Regular sessions of TSG 26 were organized into four 90-minute time sessions. **Session 1** was spent for the invited speaker who provides a state of the art presentation.

Invited Lecture was given by Daniel Chazan and Patricio Herbst with a title of "Reconciling two uses of norm in mathematics education research". They juxtaposed two meanings of norms—norms as sets of expectations deliberately designed by teachers that are co-constructed with students and norms as sets of expectations that come with the canonical uses of curricular tasks. In particular, in the context of a press on mathematics educators to generate instructional improvement, they asked how an understanding of norms associated with canonical uses of tasks might relate to norms that teachers negotiate and co- construct with their students in the context of instructional innovations.

Two theoretical presentations on research on teaching provided the basis for discussion in the following **sessions 2 & 3**.

Session 2 included parallel presentations by the researchers working on research on teacher's questioning on the one hand and on measuring teaching and classroom practice on the other.

- Esther Alice Enright, Lauren Ashley Hickman and Deborah Loewenberg Ball: A typology of questions by instructional function
- Melissa Kemmerle: Questions about questions: How student questions in mathematics classrooms are affected by authority distribution and assessment practices
- Siún NicMhuirí: Using research frameworks to develop practice: Teacher questions in a math talk community
- Jeremy Zelkowsk, Jim Gleason and Stefanie D Livers presented: Measuring mathematics classroom interactions: An observation protocol reinforcing the development of conceptual understanding

- Lidong Wang and Yiming Cao: Using cognitive diagnostic model to build a differential model to measure mathematics teachers' effect on grade 7 students' achievement

Session 3 included parallel presentations on research on focused on teaching and students' learning and describing and comparing teaching.

- Marika Toivola and Harry Silfverberg
- Amanda Allan, Tina Rapke and Lyndon Martin
- Sharon Marianne Calor, Rijkje Dekker, van Drie Jannet Petronella, Bonne Zijlstra and Monique Volman,
- Emily C Kern, Erin C. Henrick, Thomas M. Smith, Paul Cobb and Yiming Cao,
- Yu Bin Lee, Cheong Soo Cho
- Steven Watson, Louis Major and Elizabeth Kimber

Session 4 consisted of two invited presentations of papers and draw conclusions regarding the state of the art in research on classroom teaching.

Mary Kay Stein, Katelynn Kelly, Debra Moore, Richard Correnti, Jennifer Russel presented their paper entitled as "Theorizing and measuring teaching for conceptual understanding". They described and provided initial evidence for a theory of mathematics teaching and learning that can guide efforts to validly and usefully measure teaching in an era of increasingly ambitious standards for student learning. The theory is based on two constructs that past research suggests foster students' conceptual learning: Explicit attention to concepts (EAC) and students' opportunity to struggle (SOS). By crossing high and low values of these constructs, four quadrants are formed that, we argue, represent four discernable profiles of teaching. Fifty classroom videos of Grades 4-8 mathematics teachers were coded using this theory. Results suggest that not only can teaching be reliably placed into one of the quadrants, but also that quandrant-based teaching was characterized by other teaching practices that we had conjectured would be indicative of high or low EAC/SOS. We close with a discussion of the benefits of using a theory to guide investigations into the relationship between mathematics teaching and learning.

Yoshinori Shimizu and Yuka Funahashi presented their paper entitled as "Beyond the labels: Learning from international comparative studies of mathematics classroom". In this talk, they problematized such labels and discussed possibilities of going beyond them by searching for similarities and differences in educational practices by drawing on the data and analyses from the Learner's Perspective Study (LPS). Two studies are used as cases for illustrating possibilities of identifying similarities and differences in the classroom practices: a comparison of lesson event of teacher's summing up during mathematics lessons in China and Japan, and a comparison of teacher's questioning in Germany and Japan. With the recognition that international comparative studies of classroom practices provide researchers and policy makers opportunities for understanding their own implicit theories about how teachers teach and how children learn mathematics in their own context, this paper emphasizes the importance of taking into account the different cultural assumptions underpinning teaching and learning in the international debates on mathematics education.

Four broader categories of research areas were identified, "Teacher's Questioning", "Measuring teaching and classroom practice", "Teaching and Students' learning", "Describing and Comparing teaching".

Open Access Except where otherwise noted, this chapter is licensed under a Creative Commons Attribution 4.0 International License. To view a copy of this license, visit http://creativecommons.org/licenses/by/4.0/.

Topic Study Group No. 27: Learning and Cognition in Mathematics

Gaye Williams, Wim Van Dooren, Pablo Dartnell, Anke Lindmeier and Jérôme Proulx

The Programme

Learning and cognition is a classical and very vital area in research in mathematics education. Different to many other special and related TSGs, such as teaching and learning of algebra, geometry, measurement, statistics, calculus, to mention a few, this TSG has a more general focus.

Originally, research was focused mainly on the cognitive processes taking place in the individual. The past twenty-five to thirty years, however, the research has expanded. Research on learning as well as mathematical cognition are now frequently framed with socio-cultural theories, and closer connections are being made between social and cognitive theories. In addition, influences of materials, classroom contexts, and affective factors such as emotions, beliefs, and attitudes on learning and cognition are foci of interest.

In what follows, we will briefly report on the presentations that took place in the four regular sessions of TSG27. Each session started with an invited plenary speaker, after which a number of papers were presented.

Co-chairs: Gaye Williams, Wim Van Dooren.
Team members: Pablo Dartnell, Anke Lindmeier, Jérôme Proulx.

G. Williams (✉)
Deakin University, Burwood, Australia
e-mail: gaye.williams@deakin.edu.au

W.V. Dooren
University of Leuven, Leuven, Belgium
e-mail: wim.vandooren@kuleuven.be

Session 1

The main focus for Session 1 was on the development of mathematical cognition in individual learners. The focus was strongly, but not exclusively, on cognitive aspects of learning and cognition in mathematics.

Erno Lehtinen (University of Turku, Finland) was the invited keynote speaker. He focused on the development and extensions of number concept before formal schooling and during primary school. He showed how individual level cognitive analysis helps in understanding learning trajectories, crucial changes, and different processes leading to deep and flexible understanding of number systems in some individuals, and superficial and inflexible number knowledge in others. Drawing on research conducted in his group, he described inter-individual differences in children's spontaneous quantitative focusing tendencies, using them to partially explain differences in how children learn to understand natural numbers that prepare students for conceptual changes needed in learning rational numbers. Educational consequences of these findings were also discussed.

The two subsequent papers focused on a sub-aspect of this talk: students' rational number understanding and difficulties experienced in the development of this understanding.

Jo Van Hoof (University of Leuven, Belgium) focused on natural number bias, a tendency in learners to apply natural number properties to rational numbers even when not applicable. She drew attention to three natural number properties that were inappropriately applied in rational number tasks: density, size, and operations. A test constructed to characterize the development of 4th to 12th graders' natural number bias was administered to 1343 elementary and secondary school students. Results showed an overall natural number bias weakest in size tasks, and strongest by far in density tasks. An overall decrease of the natural number bias with grade was found. Educational implications were provided.

David Maximiliano Gomez (Universidad de Chile, Chile) examined learners' understanding of fraction magnitudes. His research extended beyond learners' understanding of fraction magnitudes interpreted through natural number bias to the idea that several qualitative differences may exist among learners regarding their fraction magnitude understanding. The biases possessed and strategies employed by a large group of middle school children in a computerized fraction comparison task were studied. Overall findings suggested the presence of a strong bias for mistaking component magnitude for fraction magnitude. A clustering analysis revealed the coexistence of at least five distinct manners of reasoning. The findings hold promise not only for research purposes, but also as a contribution to teaching practices because of their potential to expose common mistake patterns.

Session 2

In this session, the scope of mathematical cognition was broadened to a situated process distributed over individuals and objects.

Dor Abrahamson (University of Berkeley, USA), the invited keynote speaker, focused on the conceptualization of the mind as embodied, extended, and enactive activity in natural and sociocultural ecologies. He showed how students' immersive hands-on dynamical experiences become formulated within semiotic registers typical of mathematical discourse. He presented analyses of integrated videography, action logging, and eye-tracking data from tutor–student clinical interviews using a technologically enabled embodied-interaction learning environment, the Mathematical Imagery Trainer for Proportion, to describe the emergence of mathematical concepts from the guided discovery of sensorimotor schemes. A central notion was the theoretical construct of attentional anchor. Abrahamson reconciled constructivist and sociocultural models by underscoring the role of artifacts and facilitation in the micro-events of mathematical ontogenesis.

Anke Lindmeier's (University of Kiel, Germany) presentation focused on structured representations for whole numbers. Such representations which have a long tradition in mathematics education are used to foster the formation of mental models and according specific strategies, so that fast, accurate, and flexible solutions for whole numbers tasks can be retrieved. However, evidence of how children actually use these strategies is rare. An eye-tracking experiment explored possibilities to assess strategies when working with structured representations. By comparing the strategies of first-graders and mathematics proficient adults in a basic whole number problem, and the strategies in different representations, she showed characteristic affordances of structurally equivalent representations. The study informs the further use of structured whole number representations and the potential of eye-tracking to infer students' cognitive processes when working with these representations.

Tine Degrande's (University of Leuven, Belgium) studied the extent to which children spontaneously focus on quantitative relations (SFOR), and the nature of this quantitative focus (types of quantitative relations that make up SFOR). Three different variants of a SFOR-task (multiplicative, additive or open task) were offered to second, fourth and sixth graders. Although most children focused on quantitative relations in the task, they focused not only on multiplicative relations but also on additive relations. SFOR. Multiplicative SFOR was found to increase with age whereas the evolution of additive SFOR depended on the task variant. The open SFOR-task was found to be best suited to capture SFOR. These results suggest further research on SFOR requires a broader conceptualization of quantitative relations than only multiplicative conceptualization.

Session 3

This session extends the focus to personal and environmental factors that influence students' learning.

Judy Anderson (University of Sydney, Australia), the invited keynote, and her colleagues captured the essence of various aspects of their multifaceted project. This longitudinal study of over 4000 students (and their teachers) from classrooms in 47 schools employed both quantitative and qualitative research methods including student interviews, teacher interviews, and observations of classrooms with high levels of student engagement. Constructs studied included student motivation and engagement, attitudes, the classroom environment, student achievement, and their 'switching on' and 'switching off' behavior. It was found that: (a) compared with Grade 6 students, those in Grades 7 and 8 significantly declined in mathematics engagement; (b) with regard to future intent, mathematics self-efficacy, valuing, enjoyment, perceived classroom enjoyment, and parent interest were significant predictors; and (c) additional predictors associated with disengagement were mathematics anxiety, perceived classroom disengagement, school ethnic composition, and school socioeconomic status.

The other two papers in this session examined ways in which students process learning.

Miguel Figueirado (Universidad de Lisboa, Portugal) utilised a questionnaire with Year 10 students to study components that build learning styles for mathematics. Her findings confirmed two learning styles previously identified in the literature: meaning-oriented and reproduction-oriented learning. It was found that meaning-oriented learners tended to be more successful than reproductive-oriented learners and more aware of their learning results.

Bishnu Khanal (Tribhuvan University, Nepal) found that students studying mathematics in secondary school in Nepal had difficulties in understanding investigating, and generalizing mathematical situations. Both quantitative and qualitative analyses were employed. The study raises questions about whether these difficulties are due to the ineffectiveness of the learning strategies students employed.

Session 4

The final session focuses on social and contextual influences on learning.

Minoru Ohtani (Kanazawa University, Japan), invited keynote speaker, employed a task designed to engage adolescents which conceived numerical tables, algebraic expressions, and graphs as traces or shadows of a function, and anthropomorphized this function as a Japanese "Ninja" with these representations as shadows of the stealthy and invisible "Ninja" who gives glimpses of its existence. They found students grasped properties of particular functions from bits and pieces of the shadows thus enabling the reification of function as a mathematical object. This activity, designed collaboratively by the researchers and the teacher provided opportunity for dynamic and interactive representations to direct students' attention to features of changes of variables and succeeded in promoting discussions whose main topics were those features of function. Implications for teaching and learning are significant.

P Janelle McFeetors (University of Alberta, Canada) used a constructivist grounded theory study of Grade 12 students to study how they actively shaped their learning processes through the way they approached homework and study for tests. In particular she examined whether students can, through the process of learning to learn mathematics, bring into view how they learn mathematics. She found that students were authoring processes for learning, authoring mathematical ideas, and self-authoring as they began to see themselves as capable mathematical learners (where 'authoring' was conceptualised as a generative activity of making meaning of experiences).

Gaye Williams (Deakin University, Australia) used lesson video and video-stimulated interviews to examine the activity of a group of three Year 5/6 students undertaking an unfamiliar mathematical problem solving task in class. She found that neither a peer-tutoring model nor a collaborative development of new knowledge model fitted the learning that occurred. Instead, 'non-expert others' who were processing new ideas more slowly 'opened out' a new Zone of Proximal Development for a student more expert in relation to the mathematics emerging from the task. This study raises questions about influences of different paces of thinking on learning during problem solving.

Summing Up

These TSG27 Learning and Cognition sessions were well attended as were the other presentations associated with this TSG. The questions asked by participants, and the rich discussions that followed paper presentations enriched these sessions further.

Open Access Except where otherwise noted, this chapter is licensed under a Creative Commons Attribution 4.0 International License. To view a copy of this license, visit http://creativecommons.org/licenses/by/4.0/.

Topic Study Group No. 28: Affect, Beliefs and Identity in Mathematics Education

Markku Hannula, Francesca Morselli, Emine Erktin, Maike Vollstedt and Qiao-Ping Zhang

The Topic Study Group 28 was aimed at addressing all areas of affect, including attitude, anxiety, beliefs, meaning, self-concept, emotion, interest, motivation, needs, goals, identity, norms, values. The different approaches to study affect included psychological, social, and philosophical research perspectives. Moreover, the call for papers explicitly questioned the issue of the mutual relationship between affective constructs and their connection to cognition and other constructs studied in mathematics education, as well as the description of programs for promoting aspects of affect.

The activity of the working group was aimed at:

- Presenting an overview of the state of the art in the research field of affect in mathematics education, both at the students' and the teachers' (pre-service or in-service) level.
- To identify new trends and developments in research and practice in these areas.
- To engage participants in a critical reflection of this research field and generate discussion of an agenda for future research on affect in mathematics education.

The participation to the Topic Study Group highlights a growing interest for affective issues: 86 researchers attended as presenting authors. Due to the high number of proposals, it was necessary to carry out a selection of the contributions

---

Co-chairs: Markku Hannula, Francesca Morselli.
Team members: Emine Erktin, Maike Vollstedt, Qiao-Ping Zhang.

M. Hannula (✉)
University of Helsinki, Helsinki, Finland
e-mail: markku.hannula@helsinki.fi

F. Morselli
University of Genoa, Genoa, Italy
e-mail: morselli@dima.unige.it; fraemme@libero.it

and to organize parallel sessions, so as to provide adequate time for presentation and discussion. One invited lecture and 20 research reports were presented during regular sessions; additionally, there were 44 oral communications and 21 poster presentations. The final part of each session was devoted to a general discussion and synthesis. Below, we synthesize the contents of the regular sessions.

Regular session 1 was held in plenary mode. After a brief introduction and an ice-breaking activity, Leder gave her invited lecture, presenting an overview of the state of the art of research on affect, with a special emphasis on gender issues. The first regular presentation was given by Bofah and Hannula, who reported a quantitative analysis of motivational beliefs as a mediator between perceived social support and mathematics achievement.

Regular session 2 was devoted to the theme of identity, taking into account both student and teacher perspective. A common feature of these presentations was the effort of broadening the construct of identity, linking it to other theoretical constructs and/or adopting innovative theoretical stances. First, Heyd-Metzuyanim proposed a comparison between research on affect and research on discourse, highlighting overlapping and gaps. More specifically, she suggested the construct of identity as a nexus of the study of affect and discourse. Next, Westaway examined the interplay between teachers' identities and mathematics pedagogical practices, adopting the methodological and theoretical framework of social realism. This presupposes a historical analysis that includes teachers' life histories and mathematics histories, and, through this analytic process, the identification of the mechanisms from which teachers' identities emerge. Then Felix gave a presentation focused on the development of the identity of a mathematics teacher, suggesting that the identity is deeply shaped by the struggle for recognition. More specifically, he analyzed autobiographical stories through the theoretical lenses provided by Honneth's three levels of recognition, Kelchtermans' four components of a professional self (or identity) and Heikkinen and Huttunen's circles of recognition. Finally, Karaolis and Philippou presented their instrument for the study of teachers' identity, carried out combining a questionnaire and qualitative interviews. Hierarchical cluster analysis led to single out three clusters of teachers, who differ in terms of self-efficacy, motivation, and task orientation.

Regular session 3 was run parallel with session 2. It was devoted to exploring links between affective factors and mathematical activity and performance. Two presentations concerned student affect. Kohen and Tali explored the impact of learning based visualization, embedded with a tool for promoting self-efficacy, on middle students' achievements and self-efficacy. Fuller and Deshler presented their research on the complex way that anxiety associated to different aspects of the study of mathematics interacts with personality traits. One presentation concerned prospective teachers. Haser investigated pre-service mathematics teachers' feelings of difficulty when faced to problem posing and problem solving tasks. Finally, Hannula and Oksanen presented a large-scale study on the link between teachers' beliefs and their students' affect and achievement, showing that teachers' beliefs may have a small but statistically significant effect on the development of students' affect and achievement. More specifically, student achievement and affect developed more positively, when their teachers emphasized student thinking. Learning outcomes were also positively related with teacher efficacy and student affect was found to develop more positively when their teachers emphasized facts and routines.

Regular session 4 was devoted to the study of teacher affect, with a variety of methods and theoretical lenses. The issue of comparative studies was explicitly addressed. Kahlil and Johnson presented their study concerning novice teachers' "in the moment" affect, as emerging during a mixed-reality simulated classroom. The interpretative lenses they adopted refers to Goldin and colleague's engagement structures. Laschke and Blömeke focused on future teachers' motivation to teach, analyzing data from 15 countries. Their study led to reflections on methodological issues such as the use of the same instrument for countries with different cultural and educational traditions. Adeyemi presented a large quantitative study on the relationship between mathematics anxiety and mathematics teaching anxiety among in-service elementary teachers. The study had a specific focus on the influence of gender. Juwe examined the beliefs of mathematics teachers who have special responsibilities in their schools (Mathematics Curriculum Leaders), proposing a comparison between England and Nigeria.

Organized in parallel to session 4, the regular session 5 was focused on student affect. Two main themes emerged: the interplay between affect and mathematical activity in classroom, and the need for theoretical lenses (and consequent methodological instruments) to better understand phenomena. Elizar presented a study of beliefs and attitudes influencing students' higher order thinking skills in mathematics. Gun and Bulut studied students' attitude towards mathematics, adopting a tripartite model that encompasses cognitive, affective and behavioral components. Branchetti and Morselli studied the interactions during group work from a socio-cultural perspective and networking with two theoretical lenses: the construct of rational behavior and that of identity. Wilkie studied students' affect using an open-response inquiry. In her presentation she discussed the methodological design of her study and used affect-related examples to illustrate the insights that researchers and teachers can gain from eliciting open responses from students.

The last regular session was again held in plenary mode. Pieronkievicz and Goldin addressed belief change combining two theoretical constructs: the concept of affective transgression (consciously crossing emotional boundaries established by prior beliefs) and the concept of meta-affect. They concluded their presentation suggesting teachers to address students' affect explicitly within an emotionally safe teaching environment, and discussed some strategies to this aim. Middleton, Mangu, and Lee presented their study on the impact of motivational variables (Interest, Identity, Self-Efficacy, and Utility) on the career intentions of high schoolers, with particular attention to STEM aspirations. Finally, Achmetli and

Schukajlow presented their study drawing from the project MultiMa, where students were asked to construct multiple solutions while solving real-world problems by applying multiple mathematical procedures. Their results indicated that constructing multiple solutions had a positive influence on students' experience of competence, but no effect on their interest in mathematics.

Open Access Except where otherwise noted, this chapter is licensed under a Creative Commons Attribution 4.0 International License. To view a copy of this license, visit http://creativecommons.org/licenses/by/4.0/.

Topic Study Group No. 29: Mathematics and Creativity

Demetra Pitta-Pantazi, Dace Kūma, Alex Friedlander,
Thorsten Fritzlar and Emiliya Velikova

The Programme

The aim of TSG 29, Mathematics and Creativity, was to bring together mathematics educators, educational researchers, mathematics teachers, and curriculum developers for the international exchange of experiences and ideas related to mathematical creativity. Approximately 50 researchers from 20 countries participated. Eleven articles, seven short oral communications, and one poster were presented during the conference. Participants were given the opportunity to present their work and discuss important aspects of mathematical creativity. The TSG was organised in four sessions with article presentations and two sessions with short oral presentations. The main topics that the TSG addressed were:

1. Definition and measurement of mathematical creativity
2. Tasks, methods, and environments that have the potential to promote mathematical creativity
3. Problem posing and mathematical creativity.

Co-chairs: Demetra Pitta-Pantazi, Dace Kūma.
Team members: Alex Friedlander, Thorsten Fritzlar, Emiliya Velikova.

D. Pitta-Pantazi (✉)
University of Cyprus, Nicosia, Cyprus
e-mail: dpitta@ucy.ac.cy

D. Kūma
Liepaja University, Liepāja, Latvia
e-mail: dace.kuma@liepu.lv

Session 1: Definition and Measurement of Mathematical Creativity

The activities of the TSG started with the co-chairs and team members of the TSG offering a brief overview of the main topics that the group would address. During the first session, three articles were presented. The first two presentations proposed two new tools and methods for the measurement of mathematical creativity. In their article "Creativity-in-progress rubric on proving: Enhancing students creativity," Karakok, El Turkey, Savic, Tang, Naccarato, and Plaxco presented a new formative assessment tool, the Creativity-in-Progress, which can be used to measure individuals' creativity while engaged in mathematical proof. The researchers described the development of this tool and its categories. Joklitschke, Rott, and Schindler, in their article "Revisiting the identification of mathematical creativity: Validity concerns regarding the correctness of solutions," suggested that with the existing methods for the measurement of creativity, students' potential is not sufficiently assessed and valued. Thus, they suggested modifications. One of the modifications they suggested was that students' erroneous or unfinished solutions may also be used for the assessment of mathematical creativity. A third study presented during the first session by Pitta-Pantazi and Sophocleous, entitled "Higher order thinking in mathematics: A theoretical formulation and its empirical validation," went beyond to the identification and measurement of mathematical creativity and extended towards the assessment and measurement of higher order thinking in mathematics. The researchers suggested that higher order thinking is constituted by several subcomponents: basic, critical, and complex mathematical thinking. In their article, the researchers empirically validated a model of higher order thinking and presented tools that they used for its measurement.

Session 2: Tasks, Methods, and Environments that Have the Potential to Promote Mathematical Creativity

In the second session, the articles presented addressed types of tasks, methods, and environments that have the potential to promote students' mathematical creativity. Mathematics educators and researchers presented types of activities as well as qualitative and quantitative data from students' work.

In their presentation "Developing flexibility of problem solving strategies in the classroom," Jesenska and Semanišinová suggested competitions in problem solving for groups of students as a method that could potentially promote students' mathematical creativity. In these competitions, students are prompted to find innovative solutions to given problems. The researchers offered five such problems and corresponding strategies exhibited by their students. In his presentation "Some types of creativity-promoting tasks," Alex Friedlander classified and described various types of tasks that have the potential to promote students' mathematical creativity. He suggested that creativity-promoting tasks may involve some or all of the following actions: planning, implementation of a plan, revision, invention of a new procedure, reflection, and production of new knowledge.

Session 3: Tasks, Methods, and Environments that Have the Potential to Promote Mathematical Creativity

During the third session, researchers continued the discussion on tasks, methods, and environments that may be used to develop mathematical creativity. In their presentation of the article "Ornaments and tessellations: encouraging creativity in the mathematics classroom," Moraová and Novotná presented a number of teaching experiments from the M^3EaL project (Multiculturalism, Migration, Mathematics Education, and Language), which aimed at the development of mathematical creativity in migrant pupils and pupils of different sociocultural backgrounds. They found that when teachers are faced with a cultural heterogeneity in their classrooms and they cannot rely on the traditional textbooks, they tend to become more creative in their planning of the lessons and in some cases this also encourages their students' creativity. El-Sahili, Al-Sharif, and Khanafer investigated mathematical creativity in secondary education. In their article "Mathematical creativity: The unexpected links," they suggested that secondary school students can solve mathematical problems in a non-traditional manner that requires the formation of hidden bridges between different mathematical domains or ideas that at first glance appear unrelated. They suggested that this ability is not restricted only to professional mathematicians or postgraduate mathematics students but that students at the intermediate and secondary school level can also possess it. Assmus and Fritzlar focused on mathematical creativity in primary school and more specifically on primary school students' inventions processes. In their article "Mathematical creativity in primary school," Assmus and Fritzlar also argued that primary school gifted students not only can solve and pose problems but are also able to purposefully or freely create new mathematical objects. The researchers offered examples of such invention processes.

Session 4: Problem Posing and Mathematical Creativity

The topic of the last session was problem posing and mathematical creativity. In this session, three articles were presented. In particular, in his presentation of the article "Remarks on creative posing of problems: Pro et contra," Kasuba discussed the creative posing and design of mathematical problems and what specifically should be included in the formulation of a problem. Sophocleous, in her article "Mathematical problem-posing ability and critical thinking in mathematics," investigated the relationship between primary school students' problem-posing ability and their critical thinking in mathematics. In their presentation of the article "Flexibility of pre-services' teachers in problem posing in different environments," Daher and Anabousy presented the results of the impact of four different environments that aimed at promoting pre-service teachers' flexibility in posing problems. The four different environments were: (1) with technology and with a "what if not" strategy, (2) with technology but without "what if not" strategy, (3) without technology but with a "what if not" strategy, and (4) one without technology and without a "what if not" strategy.

Two Sessions with Short Oral Communications

Two sessions were devoted to short oral communications. Primarily, researchers concentrated on tasks and methods that may promote mathematical creativity and ability in and out of school. Regarding school mathematics, Ferrington proposed that creativity can be developed by asking challenging questions, while Tanaka suggested that asking students to pose problems can be seen as a creative activity. Vilches and Gorriz suggested that open challenges encourage students to create their own innovative product. Furthermore, Mamiy and Mamiy focused on the mathematical circle as a means for developing mathematical creativity. Furthermore, Bártlová claimed that teachers can develop students' mathematical abilities out of school by encouraging students to participate in unconventional environments such as interactive science centres. Moreover, Mikaelian presented an approach to aesthetic education in learning mathematics, while Abdounur highlighted the way in which the relationship between mathematics and theoretical music influenced our understanding of important mathematical concepts. Finally, one poster presented by Choe provided insights into South Korea's education system and its negative impact on students' mathematical creativity.

In the closing session of the TSG, researchers raised questions which are crucial and open to further investigation. They also offered ideas for future research studies and collaborations. Overall, the work of the TSG demonstrated how much research in mathematics creativity has evolved and still how many questions need to be answered. It appears that mathematical creativity is an important topic within the field of mathematics education that is worth pursing and has a lot to offer in our highly demanding and rapidly changing world.

Open Access Except where otherwise noted, this chapter is licensed under a Creative Commons Attribution 4.0 International License. To view a copy of this license, visit http://creativecommons.org/licenses/by/4.0/.

Topic Study Group No. 30: Mathematical Competitions

Maria Falk de Losada, Alexander Soifer, Jaroslav Svrcek and Peter Taylor

Our TSG 30 worked during four of the six days of ICME-13. Sessions were well attended by delegates from all over the world. Each talk was followed by a constructive and productive discussion. New relationships were forged, new collaborative projects envisioned. One of these projects was kindly offered to us by the Convenor of the Congress, Prof. Dr. Gabriele Kaiser: to compose a book of high quality papers on Mathematical Competitions, which may be published by Springer. The program of TSG 30, which follows, may convey the flavor of our study group and its international breadth. The titles of plenary talks are followed by quotations from them that impressed me the most.

Co-chairs: Maria Falk de Losada, Alexander Soifer.
Team members: Jaroslav Svrcek, Peter Taylor.

M. Falk de Losada
Universidad Antonio Narino, Bogota, Colombia
e-mail: mariadelosada@gmail.com

A. Soifer (✉)
University of Colorado at Colorado Springs, Colorado Springs, USA
e-mail: asoifer@uccs.edu

J. Svrcek
Olomouc, Czech Republic

P. Taylor
Canberra, Australia

The Program

July 26, Tuesday, 12:00–1:30
Plenary Talk Alexander Soifer (USA):
Beyond *Lǎozǐ*: The Goals of Mathematics Instruction.

Give a man a fish, and you will feed him for a day.
Teach a man how to fish, and you will feed him for a lifetime.

—老子 (*Lǎozǐ*, VI century BC)

Before we address the purpose of mathematics instruction, it is instructive to ask ourselves, what is the purpose of life itself? It seems to me that the purpose of life is to discover and express ourselves, and in so doing contribute to high culture of our planet. The ultimate purpose of instruction is therefore to aid our students in their quest for self-discovery and self-expression.

Lǎozǐ proposes to teach a man fishing as a method of solving the problem of survival. This does go further than giving a man a fish. However, is it good enough in today's world? Not quite, dear Sage, not in today's rapidly changing world. What if there is no more fish? What if the pond has dried out while your man has only one skill, fishing? A problem solver will not die if the fish disappears in a pond—he'll learn to hunt, grow crop, solve whatever problems life puts in his way. And so, we will go a long way by putting emphasis not on training skills but on creating environment for developing problem solving abilities and attitudes. This is the state-of-the-art. The proverb for today's world ought to be:

Give a man skills, and you will feed him in the short run.
Let a man learn solving problems, and you will feed him for a lifetime.

Every day we confront and solve a myriad of problem. Life *is* about solving problems. And mistakes in solving life's problem can be quite costly. This is where mathematics comes in handy. Mathematics allows us to learn how to think creatively, how to solve problems. And once our student masters problem solving in mathematics, s(he) will be better prepared to confront problems in any human endeavor. Moreover, one *cannot teach* mathematics, or anything else for that matter. State-of-the-art in mathematics instruction is about creating an atmosphere where students can learn mathematics by doing it, with a gentle guidance of a teacher.

1:10 Iliana Ivanova Tsvetkova (Bulgaria):

Mathematics Competitions as a Tool for Development of Gifted Students

July 27, Wednesday, 12:00–1:30
Plenary Talk María Falk De Losada (Colombia):

Are Mathematics Competitions Changing the Way Mathematics Is Being Done and the Mathematics that Is Being Done?

Mathematical problem solving competitions, as a branch of mathematics education, have a feature that distinguishes the work being done from every other initiative in the field. And this has its roots in Hungary at the Eötvös and Kürschák competitions and the journal of problems in mathematics and physics, Középiskolai Matematikai Lapok or KöMaL. With common roots in these pioneering competitions, a school was formed that produced outstanding figures in mathematics, in methodology and in epistemology. Beginning with the work and leadership of Lipót Fejér (Leopold Weiss) who grew up solving problems from KöMaL and who placed second in the Eötvös competition of 1897, a school was formed that came to include, in varying degrees, Paul Erdős, George Pólya and Imre Lakatos, the great mathematician and collaborator with mathematicians around the globe, the influential thinker on problem solving and method, and the philosopher–epistemologist who dared to question formalist mathematics proposing an alternative interpretation of the character, origins, structure and justification of mathematical knowledge and its historic evolution. These three stand out among the many great Hungarian mathematicians whose mathematical formation began in or was intimately related to the competitions, especially because they migrated to England and the United States and worked and published in English, thus opening their ideas and results and bringing them to bear on the worldwide community of mathematicians and mathematics educators…

Timothy Gowers, IMO gold medalist and winner of the Fields Medal, states:

> Loosely speaking, I mean the distinction between mathematicians who regard their central aim as being to solve problems, and those who are more concerned with building and understanding theories…. consider the following two statements.

i. The point of solving problems is to understand mathematics better.
ii. The point of understanding mathematics is to become better able to solve problems.

Most mathematicians would say that there is truth in both (i) and (ii)….

> So when I say that mathematicians can be classified into theory-builders and problem-solvers, I am talking about their priorities, rather than making the ridiculous claim that they are exclusively devoted to only one sort of mathematical activity.

Gowers considers himself to be a mathematician whose priority (in the tradition of Paul Erdős) is problem solving.

1:10 Chen Donglin and Frederick K.S. Leung (Hong Kong)

China Mathematical Olympiad School: A Case Study

July 29, Friday, 12:00–1:30
Plenary Talk Peter J. Taylor (Australia):

Some Reflections, some Suggestions

Competitions have a unique value in the education system. Because the questions are normally set externally, independently of class activity, they can test a student's ability to apply known mathematics in new situations, and as such they can well help equip a student to be more useful in their later career.

Many features of everyday life are competitive, and in particular the development of mathematics and mathematical research have a competitive nature. These have sometimes been driven by tangible challenges such as the 23 problems set by Hilbert in 1900, the offering of million dollar prizes by the Clay Foundation in the 1990s, or necessity, such as the need to tighten financial security and the need to understand genetic structure.

The modern existence of mathematics competitions for school students dates back to 1894 in Hungary, Olympiads started in the early 1930s in the Soviet Union, and large, inclusive competitions commenced in the US in 1950. By 1984 there were many competitions held nationally and internationally and a need had developed for the organisers to form a learned society to enable information exchange...

The World Federation of National Mathematics Competitions (WFNMC) was founded at a meeting attended by about 20 people attending ICME-5 in Adelaide in 1984. It grew into a respectable organisation, an Associated Study Group (ASG) of ICMI, and now has its own refereed Journal, its own web site (http://www.wfnmc. org/), system of awards and conducts Conferences.

The Conferences have been particularly important, not only having allowed people in different countries to meet each other and establish lines of communication, but the first Conference was one of the exciting I have attended. There have been opportunities to meet Paul Erdos and John Conway, and some of the lectures, such as by Erdos, Conway, Robin Wilson, and those by Alexander Soifer on the chromatic numbers, axiom of choice and van der Waerden, have been most memorable. The Journal, offering opportunity to exchange information via refereed papers, has also been useful.

1:10 Luis F. Caceres Duque, Jose H. Nieto Said, and Rafael Sanchez Lamoneda (Puerto Rico, USA):

The Mathematical Olympiad of Central America and The Caribbean: 17 Years Supporting Math Contests in the Region

July 30, Saturday, 12:00–1:30
Plenary Talk Kiril Bankov (Bulgaria):

Numbers on a Circle

The intellectual treasure of every mathematics competition is the set of the problems given to the participants. Competitions present variety of problems: from these that are closely connected to the school curriculum to those that deal with "non-standard" situations. The latter usually stimulate creative thinking and thus remain in the minds for a long time. Many of these problems give rise to numerous mathematical ideas because finding their solutions develop the mathematical abilities. This paper discusses such problems: some are taken from mathematics competitions, others are inspired from competition problems. In both cases, being among the best examples of the beauty of mathematics, they provoke an interest in mathematics that often begins with the consideration of attractive problems.

The life is full of operations. Many times in a day we take decisions about the series of operations to be done in order to obtain particular results. The correctness of these decisions depends on the ability to estimate the final results. Mathematics helps in modeling this reality by tasks using a particular admissible operation to transform a given situation to a different one. These problems lead to interesting generalizations by changing either the admissible operation or the initial/final situations. This part presents such examples taken from mathematics competitions in the context of arrangement of numbers on a circle.

1:10 Borislav Yordanov Lazarov and Albena Vassileva (Bulgaria):

Age Factor in Performance on a Competition Paper

Open Access Except where otherwise noted, this chapter is licensed under a Creative Commons Attribution 4.0 International License. To view a copy of this license, visit http://creativecommons.org/licenses/by/4.0/.

Topic Study Group No. 31: Language and Communication in Mathematics Education

Judit Moschkovich, David Wagner, Arindam Bose, Jackeline Rodrigues Mendes and Marcus Schütte

The Program

Language and communication are recognized to be core components in the teaching and learning of mathematics, but there are many outstanding questions about the nature of interrelationships among language, mathematics, teaching, and learning. Recent research has demonstrated the wide range of theoretical and methodological resources that can contribute to this area of study, including those drawing from cross-disciplinary perspectives influenced by, among others, sociology, psychology, linguistics, and semiotics. In this topic study group participants presented and discussed the latest research in language and communication in mathematics education internationally.

This TSG invited presentation, discussion, and reflection on the latest research on language and communication related to learning and teaching mathematics. We use "language and communication" in its broadest sense to mean the multimodal and multi-semiotic nature of mathematical activity and communication, using not only language but also other sign systems. We thus welcomed contributions focusing on all modes of communication—oral, written, gestural, visual, etc. The TSG built on the strong body of research in mathematics education that addresses these issues and also considered important questions that remain.

Co-chairs: Judit Moschkovich, David Wagner.
Team members: Arindam Bose, Jackeline Rodrigues Mendes, Marcus Schütte.

J. Moschkovich (✉)
University of California, Santa Cruz, USA
e-mail: jmoschko@ucsc.edu

D. Wagner
University of New Brunswick, Fredericton, Canada
e-mail: dwagner@unb.ca

Several themes described in the TSG31 description were addressed during the main sessions: the role of theory in understanding language and communication in mathematics education; multiple methods for researching mathematics education; relationships among language (and other sign systems), mathematical thinking, and learning mathematics; language, communication, and mathematics in classrooms and communities; and using theoretical and methodological tools from other disciplines such as linguistics, semiotics, discourse theory, sociology, etc.

The aim of TSG31 at ICME-13 was to examine and discuss research on mathematics education focused on language and communication. The TSG had 13 presentations in the main TSG sessions and 16 oral communications. Each main session concluded with a period of discussion of cross-cutting themes. A joint session was also organized with TSG32 (Mathematics education in a multilingual and multicultural environment).

Panel: "Trajectories of Research on Language and Communication in Mathematics Education: Where We Have Been, Where We Are Going"

The panel include the following three presentations:

Some sixty years of language data in mathematics education: A brief and skewed history
David Pimm, Simon Fraser University, Canada
Recommendations for research on language and learning mathematics
Judit N. Moschkovich, University of California, Santa Cruz, CA, U.S.A.
Subject specific academic language versus mathematical discourse
Marcus Schütte, Technical University Dresden, Germany

TSG Session 2

This session included the following three presentations:

Identity fostered language communication in a mathematics classroom: An Analysis

- Arindam Bose, University of South Africa, South Africa
- K. Subramaniam, Tata Institute of Fundamental Research, India
- Mamokgethi Phakeng, University of South Africa, South Africa

A teacher's use of revoicing in mathematical discussions

- Kaouthar Boukafri, Universitat Autònoma de Barcelona, Spain
- Marta Civil, University of Arizona, USA

- Núria Planas, Universitat Autònoma de Barcelona, Spain

  The significance of linguistic negotiation in inclusive learning of mathematics in primary school

- Judith Jung &Marcus Schütte, Technical University Dresden

TSG Session 3

This session included the following three presentations:
How can teachers provide learning opportunities for oral explanations?

- Kirstin Erath, TU Dortmund University, Germany

  Four-year old language repertoire in a counting situation

- David Wagner, University of New Brunswick, Canada
- Annica Andersson, Stockholm University, Sweden

  Making student explanations relevant in whole class discussions

- Jenni Ingram, Nick Andrews & Andrea Pitt, University of Oxford

  From a question to questioning within context

- Jin-Woo Cho, Seoul National University
- Eun Jung Lee, Korea Foundation for the Advancement of Science & Creativity
- Min-Sun Park, Seoul National University
- Kyeong-Hwa Lee, Seoul National University

TSG Session 4

This session included the following three presentations:
How learners communicate their mathematics reasoning in a mathematics discourse

- Benadette Aineamani, Pearson Holdings, South Africa

  Authority and politeness: A combined analysis of a teaching episode

- Konstantinos Tatsis University of Ioannina, Greece
- David Wagner, University of New Brunswick, Canada

"I am sorry: I did not understand you": The learning of dialogue by prospective teachers

- Raquel Milani. Federal University of Rio Grande, Brazil

Joint Session with TSG32

The joint session of TSGs 31 and 32 provided the opportunity for participants in the two TSGs to discuss common concerns and significant distinctions in mathematics education research on language considering (or not) multi-lingual and multi-cultural dimensions. The joint session consisted of a panel and discussion focused on the theme: "Intersections and differences in work on language in monolingual and multilingual/multicultural classrooms and settings". The panelists were Richard Barwell, Arindam Bose, Aldo Parra, Jackeline Rodrigues Mendes, Dave Wagner and Lena Wessel. The panel was chaired by Judit Moschkovich and Marcus Schütte. As a prompt for the discussion, the panelists provided a handout of some provocative statements related to the TSG foci (which we do not have the space to include in this report) and participants were invited to discuss the following questions:

- What do you see or experience as points of intersection between these two foci: mono and multilingual/multicultural?
- What do you see or experience as differences between these two foci: mono and multilingual/multicultural?
- Why do you think these two topics are treated as separate?
- How can insights from one focus contribute to the other focus and vice versa?

A productive discussion of these questions involving panel members and the audience then ensued.

Open Access Except where otherwise noted, this chapter is licensed under a Creative Commons Attribution 4.0 International License. To view a copy of this license, visit http://creativecommons.org/licenses/by/4.0/.

Topic Study Group No. 32: Mathematics Education in a Multilingual and Multicultural Environment

Richard Barwell, Anjum Halai, Aldo Parra, Lena Wessel and Guida de Abreu

The Programme

All over the world, mathematics education takes place in multilingual and multicultural environments, including situations affected by historical diversity, colonialism, migration and globalisation. Research on the issues arising in such environments is growing and is of wide relevance. The aim of TSG32 at ICME-13 was to examine issues that arise in conducting research on mathematics education in such environments. The TSG saw 9 presentations in the main TSG sessions and 13 oral communications. Each main session concluded with a period of discussion of cross-cutting themes. A joint session was also organised with TSG31 (Language and communication in mathematics education).

Three themes mentioned in the original TSG description were addressed during the main sessions: the interaction between policy, practice and research; the role of theory in understanding mathematics education in multilingual and multicultural environments; and cross-disciplinary perspectives in researching mathematics education in multilingual and multicultural environments. We organise our summary of the presentations in the main sessions of TSG32 around these themes (the presentations were not, however, presented in this sequence during the TSG ses-

---

Co-chairs: Richard Barwell, Anjum Halai.
Team members: Aldo Parra, Lena Wessel, Guida de Abreu.

R. Barwell (✉)
University of Ottawa, Ottawa, Canada
e-mail: richard.barwell@uOttawa.ca

A. Halai
Aga Khan University, Karachi, Pakistan
e-mail: anjum.halai@aku.edu

sions). Following the list of presentations for each theme, we list some of the questions that arose in relation to the theme.

The Interaction Between Politics, Policy, Practice and Research

Papers responding to this theme addressed questions such as: How can the interaction between politics, policy, practice and research strengthen mathematics education in multilingual and multicultural environments? What challenges arise? What insights can be developed from the careful analysis of practice to inform research, policy and future practice? The following five presentations addressed this theme:

Translanguaging between Maltese and English: the case of value, cost and change in a Grade 3 classroom

Marie Therese Farrugia, University of Malta, Malta

Textbook language accessibility in English medium classes

Lisa Kasmer, Anthony Snyder and Esther Billings, Grand Valley State University, USA

How the choice of artifacts may enhance communication between different communities

Vanessa Sena Tomaz and Maria Manuela David, Universidade Federal de Minas Gerais, Brazil

The culturally rich mathematics class

Sonja Van Putten, Hanlie Botha, Batseba Mofolo-Mbokana, Jeanine Mwambakana and Gerrit Stols, University of Pretoria, South Africa

Is Grade 7 too late to start with bilingual mathematics courses? An intervention study

Lena Wessel, Susanne Prediger, Alexander Meyer and Taha Kuzu, TU Dortmund, Germany

Discussion of these presentations led to the following questions:

- How can teachers translate something to convey an idea in another language? If we wish to avoid direct translations, what kind of strategies can teachers use to convey an idea?
- How is research in on this topic different if the researcher does not speak the language(s) of the informants? Might results be different if the researcher were a member of the language community of the informants?
- In the teacher's view, what is the future of mathematics in their language? Do they want to develop the same mathematics in their own language, in a way that makes sense to the community? Or do they feel they have to change their language to accommodate western mathematics?
- How would we prepare teachers to develop both mathematics and linguistics competence?

- Some presentations appeared to identify problematic practices, but where do we draw the boundaries between what is problematic and what is not?
- How can we deal with populations of traumatised refugees who are told not to use their first languages? How much does mathematics teaching need to address these issues?
- How might emphasising mathematical vocabulary help or hinder learning? How might this differ in different cultural or like linguistic contexts?

The Role of Theory in Understanding Mathematics Education in Multilingual and Multicultural Environments

Papers responding to this theme addressed questions such as: What theories have been used and why? What do current theoretical frameworks not address? How can theory help to challenge normative assumptions? How has theory and research developed in the context of multilingual and multicultural environments contributed to understanding the learning and teaching of mathematics more generally? The following three presentations addressed this theme:

Multiple language resources in an elementary school mathematics class for learners of French in Quebec

Richard Barwell, University of Ottawa, Canada

Epistemic dimension of multilingualism: the bright side of Babel

Aldo Parra, Aalborg University, Denmark

Beyond the "language of instruction": using formal and informal discourse practices in linguistically diverse classrooms

William Zahner, San Diego State University, USA

Discussion of these presentations led to the following questions:

- When re-appropriating theories from outside of mathematics education, are we losing the context of these theories?
- If we take theories from linguistics, how far do we go back to the original context? What does it do to take these theories?
- Where was mathematics in anything we discussed? If we lose sight of the mathematics in what we do, we will be overwhelmed by social issues, to the point where mathematics education (not just education) gets lost. Where do we set the conceptual boundaries so that we can do our work?
- The discussion on the artificial nature of the monolingual/multilingual distinction is interesting. However, what happens if we declare that all classes are multilingual? If everything is multilingual, then nothing is, because the term is vacuous. It might be more helpful to observe a distinction between multilingual and multivocal interaction? All classrooms are multivocal, but not all are multilingual.

Cross-Disciplinary Perspectives in Researching Mathematics Education in Multilingual and Multicultural Environments

This theme included questions like: What are the advantages, challenges and tensions arising from working across disciplines, including psychology, linguistics, sociology, etc.? What has research on mathematics education in multilingual and multicultural environments contributed to these disciplines? One paper responded to this theme:

Descriptive and typological linguistic methodologies in mathematics education research

Cris Edmonds-Wathen, Umeå University, Sweden

Discussion of these presentations led to the following questions:

- What are the reasons for the methodological choices we make when working across languages and cultures in mathematics education research?
- How can you work with an expert informant? How is it possible?
- If research involves mathematics, methodological issues can become more challenging depending on whether the language is codified and written, whether it has technical vocabulary encompassing math-like terminology, and so forth.
- How does someone conduct research involving a language in which they have minimal competence?

Joint Session with TSG31

The joint session of TSGs 31 and 32 provided the opportunity for participants in the two TSGs to discuss common concerns and significant distinctions in mathematics education research on language considering (or not) multi-lingual and multi-cultural dimensions. The joint session consisted of a panel and discussion focused on the theme: "Intersections and differences in work on language in monolingual and multilingual/multicultural classrooms and settings". The panellists were Richard Barwell, Arindam Bose, Aldo Parra, Jackeline Rodrigues Mendes, Dave Wagner and Lena Wessel. The panel was chaired by Judit Moschkovich and Marcus Schütte. As a prompt for the discussion, the panellists provided a handout of some provocative statements related to the TSG foci (which we do not have the space to include in this report) and participants were invited to discuss the following questions:

- What do you see or experience as points of intersection between these two foci: mono and multilingual/multicultural?
- What do you see or experience as differences between these two foci: mono and multilingual/multicultural?
- Why do you think these two topics are treated as separate?
- How can insights from one focus contribute to the other focus and vice versa?

A productive discussion of these questions involving panel members and the audience then ensued.

Concluding Remarks

Prompted by the rich set of presentations, the discussions at the end of each session raised and addressed some important and challenging issues for research in multicultural and multilingual mathematics classrooms. Participants, for example, debated the extent to which research in this area needs to include a mathematical focus. It appeared that for some participants, mathematics classrooms were a context in which questions of social structure, marginalisation and social justice should be addressed. Others countered that such issues are not specific to mathematics and that research in mathematics education should focus on specifically mathematical questions.

A second issue concerned the role of theory and, in particular, the use of theories from outside of mathematics education. Research in this area frequently draws on theories from, among other fields, sociolinguistics, social theory or bilingual education. While participants generally recognised the value of 'importing' theories in this way, concerns were expressed about whether such theories were treated with the depth they would have in their 'home' domain. Some participants also asked whether mathematics educators should do more to develop theory on this topic using ideas from within mathematics education. One approach to address this point would be to conduct longitudinal and cross-national studies, such as the learner perspectives study.

A third general focus for discussion concerned the complexity of multicultural and multilingual classrooms, both in relation to research and practice. The diversity of presentations and, in particular, the diversity of contexts to which the presentations referred highlighted how multilingual and multicultural classrooms vary enormously. Some participants proposed developing classifications of different contexts to avoid being overly simplistic and in order to better situate individual research projects. Relatedly, participants proposed that there needs to be more collaboration with teachers to raise awareness of complexity of multilingualism and to develop strategies to use in classrooms. Finally, participants discussed the need for more convergence in research in this area, in contrast to the current rather fragmentary approach.

Open Access Except where otherwise noted, this chapter is licensed under a Creative Commons Attribution 4.0 International License. To view a copy of this license, visit http://creativecommons.org/licenses/by/4.0/.

Topic Study Group No. 33: Equity in Mathematics Education (Including Gender)

Bill Atweh, Joanne Rossi Becker, Barbro Grevholm, Gelsa Knijnik, Laura Martignon and Jayasree Subramanian

The Topic Study Group 33 at ICME-13 (TSG33) provided a venue for discussion by researchers and practitioners from different countries who are passionate about issues of equity and are working in their particular settings toward achieving the goal of *mathematics for all*. Certainly variations exist among countries in the terms used (e.g. equity, diversity, inclusivity, social justice) and the targeted groups (e.g. based on race, indigeneity, socioeconomic background, physical and cognitive disabilities). Our understanding of the complexity of issues related to opportunity to learn, participation in, and achievement in, mathematics have also changed as new theoretical models have informed our collective work.

The aims of TSG33 sessions included, but were not limited to the following:

- Problematise the equity agenda itself, as increasing and sometimes competing demands for social justice from different groups require attention;
- Examine new theoretical frameworks that help us understand and study equity;
- Consider the prevalence of (in)equity around the world;
- Analyse intervention programs around the world with an eye to identifying characteristics of successful interventions that may transfer to different cultural settings; and,
- Query equity in participation in mathematics education research and international dialogue, with a focus on who is excluded from participation.

Co-chairs: Bill Atweh, Joanne Rossi Becker.
Team members: Barbro Grevholm, Gelsa Knijnik, Laura Martignon, Jayasree Subramanian.

B. Atweh
Philippines Normal University, Manila, Philippines
e-mail: b.atweh@oneworldripples.com

J. Rossi Becker (✉)
San José State University, San José, USA
e-mail: joanne.rossibecker@sjsu.edu

We commence, by making two observations about the equity agenda in research and policy around the world. First, we note that in the past decades, equity has become mainstream in mathematics education in the sense that it is an integral part of curriculum documents and policy in many countries, many research and professional conferences, and professional publications in the field. However, in the ever increasing dominance of educational testing as a springboard for education policy and evaluation that often equates educational outcomes with the results on standardized testing in many countries, Clarke (2014) observed that "equity has been colonised by, and subordinated to, discourses of quality in education, becoming, in a sense, another form of accountability, if one with a conscience" (p. 594).

Second, and perhaps related, is that discussion of equity has been an integral part of other areas of theory and the implied curriculum approaches in the discipline as articulated by critical mathematics education, ethnomathematics, culturally relevant mathematics education, political and social justice approaches, and, sociocultural and sociopolitical perspectives to mathematics education. It is worthwhile to note that some of these lines of research have been reflected in national and international policy formulations and wide adoption in practice more than others.

Here we identify two challenges to understanding equity as access, participation and outcomes that were raised by a variety of authors is more recent literature—from post-structural and from sociopolitical perspectives respectively. On one hand, recent literature in equity and mathematics education provided alternative understanding of the concept of identity as seems to be assumed in traditional approaches of participation and achievement. In the pioneering understanding of equity, identity and group belonging were taken as fixed and given. However, from a postmodern perspective(s) identity, of students and teachers, is seen as "multiple, fluid, or contradictory" (Gutiérrez & Dixon-Román, 2011, p. 21). The authors argue that "while documenting the inequities that marginalized students experience daily in mathematics education could be seen as the first step towards addressing hegemony, most research stops there" (p. 22). However, as Gill and Tanter (2014) noted "such developments were harder to capture in measurable terms and hence less likely to be written into policy" (p. 281).

On the other hand, some authors writing from sociopolitical perspectives have raised questions about the im/possibility of understanding and remedying equity within an intrinsically unequal society. Martin (2015) argued that the equity principles promoted by the high status policy statement reflects white rationality and promotes the participation in a system that has long oppressed African American and Latin@s students. By its silence on critical mathematics that aims at empowerment of marginalized students and their societies, it promotes an educational system that is more colonizing rather than liberating. Although using different social theories of oppression, similar concern is expressed by Pais and Valero (2011) who point out questions that often remain unraised with regards to equity such as: "Why is there inequity? Why is there a gap at all? That is, why does school (mathematics) systematically exclude/include people in/from the network of social positionings?... Why does school perform the selective role that inevitably creates inequity?" (p. 44). The authors go on to add "[a]s far as society remains organized under capitalist tenets, there will always be exclusion because exclusion is not a malfunction of capitalism, but the very same condition that keeps it alive" (p. 44).

The Programme

Invited Papers

- Renato Marcone (Brazil): "I Don't Wanna Teach This Kind of Student": Silence in Mathematics Education and Deficiencialism
- Danny Martin (USA): From Critical to Radical Agendas in Mathematics Education
- Margaret Walshaw (New Zealand): Recent Developments on Gender and Mathematics Education.

Paper Presentations

- Maria Alva Aberin, Ma. Theresa Fernando, Flordeliza Francisco, Angela Fatima Guzon and Catherine Vistro-Yu (Philippines): After-School Mathematics Program
- Bill Atweh (Philippines) & Dalene Swanson (Scotland): Alternative Understandings of Equity and their Relationship to Ethics
- Arindam Bose, Renato Marcone and Varun Kumar (Brazil): Non-typical Learning Sites: A Platform where Foreground Interplays with Background
- Grant Adam Fraser (USA): An Intervention Program to Improve the Success Rate of Disadvantaged Students in Pre-Calculus Courses
- Mellony Holm Graven & Nicky Roberts (South Africa): Focusing Attention on Promoting Learner Agency for Increased Quality and Equity in Mathematics Learning
- Barbro Grevholm (Norway), Ragnhild Johanne Rensaa (Norway) & Joanne Rossi Becker (USA): Interventions for Equality—Their Creation, Life and Death. What Can We Learn from Them?
- Jennifer Hall (Australia): Gender, Mathematics, And Mathematicians: Elementary Students' Views and Experiences
- Gelsa Knijnik & Fernanda Wanderer (Brazil): Mathematics Education. Cultural Differences and Social Inequalities
- Anina Mischau and Katja Eilerts (Germany): Without Gender Competent Math Teachers No Gender Equity in Math Education at School
- Eva Norén & Lisa Björklund Boistrup (Sweden): Gender Stereotypes in Mathematics Textbooks

- Anita Movik Simensen, Anne Berit Fuglestad and Pauline Vos (Sweden): Lower Achieving Students' Contributions in Small Groups—What if a Student Speaks with Two Voices?
- Jayasree Subramanian (India): Gender of The School Mathematics Curriculum.

Oral Communications

- Chang-Hua Chen and Chia-Hui Lin (Taiwan): Developing Differentiated Instruction to Close Learning Achievement Gap in Mathmematics
- Rosie Lopez Conde (Philippines): Pre-Service Teachers' Praxeology in Teaching Mathematics for Social Justice and Equity
- Alice Larue Joy Cook: (USA) Implementation of Social Justice Mathematics: Experiences & Perceptions of Secondary Math Teachers
- Guilherme Henrique Gomes da Silva (Brazil): Equity in the Higher Education: The Role of Mathematics Education Faced with Affirmative Actions
- Jennifer Marie Langer-Osuna and Jennifer Munson (USA): Supporting Elementary Teachers' Capacity to Foster Equitable And Productive Mathematics Classrooms
- Lena Lindenskov, Steffen Overgaard, Pia Tonnesen & Peter Wenig (Denmark): Research on Early Intervention Programs in Denmark as a Means to Equity
- Niamh O'Meara & Mark Prendergast (Ireland): An Investigation into the Inequity Surrounding Mathematics Instruction Time
- Sally-Ann Robertson (South Africa): Teacher's Questioning Practices And Issues Of Learner Agency In Mathematics Classrooms.

Posters

- Suzanne Beth Antink (USA): Contributing Replicable Factors in K-12 Female Student Mathematics Success
- Susan Holloway (USA): Language Learning Adolescent Girl's Math Achievement: The "Ophelia Effect" In Colorado
- Inge Koch, J. McIntosh, M. O'Connor (Australia): Choose Maths: Australian Approach Towards Increasing Participation Of Women
- Ji-Eun Lee, J. Kim, W. Lim, Sang-Mee (USA): A Cross-National Study Of Conceptualizing Equitable Mathematics Classrooms
- Luis Leyva (USA): Blending Academic And Social Support Through Apoyo And Consejos For Undergraduate Mathematics Success Among Latin@s
- Daouda Sangare & Nangui Abrogoua (Ivory Coast): Gender Differences In Mathematics Performance In Sub–Saharan Francophone Colleges And Universities, Through The Pan
- Neila de Toledo e Toledo (Brazil): Agricultural School, Its Mathematics Education And Social Inequalities.

References

Clarke, M. (2014). The sublime objects of education policy: quality, equity and ideology. *Discourse: Studies in the Cultural Politics of Education, 35*(4), 584–598.

Gill, J., & Tranter, D. (2014), Unfinished business: Re-positioning gender on the education equity agenda. *British Journal of Sociology of Education, 35*(2), 278–295.

Gutiérrez, R., & Dixon-Román, E. (2011). Beyond gap gazing: How can thinking about education comprehensively help us (re)envision mathematics education. In B. Atweh, M. Graven, W. Secada, & P. Valero (Eds.), *Mapping equity and quality in mathematics education* (pp. 21–34). New York: Springer.

Martin, D. (2015). The collective black and principles to actions. *Journal of Urban Mathematics Education, 8*(1), 17–23.

Pais, A., & Valero, P. (2011). Beyond disavowing the politics of equity and quality in mathematics education. In B. Atweh, M. Graven, W. Secada, & P. Valero (Eds.), *Mapping equity and quality in mathematics education* (pp. 35–48). New York: Springer.

Open Access Except where otherwise noted, this chapter is licensed under a Creative Commons Attribution 4.0 International License. To view a copy of this license, visit http://creativecommons.org/licenses/by/4.0/.

Topic Study Group No. 34: Social and Political Dimensions of Mathematics Education

Murad Jurdak, Renuka Vithal, Peter Gates, Elizabeth de Freitas and David Kollosche

Pre-Congress Activities

The inclusion of a Topic Study Group (TSG 34) on social and political dimensions of mathematics education was a first under that title in ICME's history. Such recognition weighed on TSG 34 Team members to live-up to the challenge of organizing TSG 34. The Team spent quite some time trying to understand each other's perspectives and to formulate a mission statement that, on one hand, recognized the integrity of the perspectives of its members, and on the other hand was meaningful to the international mathematics education community. The intensive and lengthy deliberations led to a mission statement (reproduced in Section "TSG 34 Mission Statement" below) that was published in the 2nd ICME-13 Announcement.

Also TSG 34 decided to make use of ICME-13/Springer offer to publish a topical survey on the social and political dimensions of mathematics education. The Team collectively prepared a survey (Jurdak, Vithal, de Freitas, Gates, & Kollosche, 2016) entitled '*Social and Political Dimensions of Mathematics Education-Current Thinking*', a description of which is given in Section "TSG 34 Topical survey on 'Social and Political Dimensions of Mathematics Education-Current Thinking'".

Co-chairs: Murad Jurdak, Renuka Vithal.

Team members: Peter Gates, Elizabeth de Freitas, David Kollosche.

M. Jurdak (✉)
American University of Beirut, Beirut, Lebanon
e-mail: jurdak@aub.edu.lb; murad.jurdak@gmail.com

R. Vithal
University KwaZulu-Natal, Durban, South Africa
e-mail: vithalr@ukzn.ac.za

TSG 34 Mission Statement

TSG 34 will critically examine the social and political dimensions of mathematics education scholarship and practice. The Group will examine the different meanings of the constructs of 'social' and 'political' as they relate to mathematics education, attending to a diverse range of scales, from the global to the micro-political, and examining a diverse range of international contexts, particularly contexts characterized by poverty and conflict, 'liberation' movements, and immigration. The Group is preparing the 'Essentials', a pre-ICME13 publication, whose aim is to present an overview of research and open the discussions on concerns in mathematics education, such as issues of equitable access and quality education, the role of economic and historical factors, distributions of power and cultural regimes of truth, dominant and counter discourses around identity and dis/ability, and activism and material conditions of inequality. In addition, the Essentials will examine salient implications of these concerns to domains such as: curriculum and reforms; learning and cognition; nature and measures of student outcomes; teaching and teacher education; media and digital technologies; research practice and impact. A Facebook Page and a Facebook group has been set up for TSG34, both called ICME13 TSG34 Social and Political Dimensions of Mathematics Education. A call for papers dealing with, but not limited to, the themes of the Essentials will be advertised. During the Congress, the Group will provide a variety of interactive formats for conveying and discussing relevant issues including invited lectures, plenary panels, presentations of accepted papers and posters and small group discussions.

TSG 34 Topical Survey on 'Social and Political Dimensions of Mathematics Education-Current Thinking'

The authors of '*Social and Political Dimensions of Mathematics Education-Current Thinking*' (Jurdak et al., 2016) ruled out a conventional survey of literature on the social and political dimensions of mathematics education and opted to focus on what they considered five critical areas of the social and political dimensions of mathematics education:

- Equitable access and participation in quality mathematics education: ideology, policies, and perspectives
- Distributions of power and cultural regimes of truth
- Mathematics identity, subjectivity and embodied dis/ability
- Activism and material conditions of inequality
- Economic factors behind mathematics achievement.

Furthermore, the team opted to focus mainly on current thinking in those five areas and only to go back in history as far as was needed to contextualize the current issues. Each author took primary responsibility for writing one of the sections and for reviewing one section written by another author.

Based on a critical review of current thinking in five selected areas, the survey found that (1) equitable access and participation in mathematics education is achievable in some countries; (2) mathematics is increasingly perceived as a negotiable field of social practices arising from specific needs and serving certain interests; (3) research seems to re-entrench stereotypes about identities that excel at mathematics and tends to assume a binary between structure and agency; (4) the relations between activism, the material conditions of inequality and mathematics education has remained under-developed and under-represented; and, (5) the nature of a society's economic structure influences relations in a classroom and may lead to a marginalisation of mathematics learners, specifically those from poor and working class households.

Implemented Programme

The programme was delivered in four 90-minutes sessions, distributed over four days. The first and last days had one 90-minutes session each, while the second and third days had two parallel 90-minutes sessions each.

First Day

The 90-minutes session was an openinning session, which was chaired by Renuka Vithal and Murad Jurdak and included two events. The first event included a welcome and program overview, self-introductions by Group members, and introduction of the *Topical Survey on Social and Political Dimensions of Mathematics Education-current thinking*. The second event was a panel which was chaired by Elizabeth de Freitas and David Kollosche and entitled (*conceptions of social and political dimensions of mathematics education*). The panel featured four panalists: Paola Valero, Lisa Darragh, Renuka Vithal, Murad Jurdak.

Second Day

The second day included two parallel sessions each of 90 min. The first parallel session was chaired by Renuka Vithal and had the theme of *activism and material conditions of inequality*. The session included four presentations. The first presentation titled '*Teaching Mathematics for Social Justice Here And There: Teacher Candidates' Reactions In The United States and Uruguay*' was given by Paula Patricia Guerra Lombardi on behalf of her co-authors Wooing Limand and

Hyunjung Kang. The second presentation titled '*Nomadic Topologies Change Mathematics Educators' Subjectivities and Hence Their Worlds*' was given by Peter Appelbaum. The third presentation titled '*Financial Education and Mathematics Education: A Critical Approach*' was given by Celso Ribeiro Campos on behalf of his co-author Aurelio Hess. The last session titled '*"Its Influence Taints All": Mathematics Teachers Resisting Performativity through Engagement with the Past*' was given by Gill Adams on behalf of her co-authors Hilary Povey and Rosie Everley.

The second parallel session which was chaired by David Kollosche had the theme of '*distributions of power and cultural regimes of truth*" and included five presentations. The first presentation titled '*Truths and Powers in Mathematics Education*' was given by Alexandre Pais, an invited speaker. The second presentation titled '*The Ethics of Mathematical Application and the Ideology of Solutionism*' was given by Hauke Straehler-Pohl. The third presentation titled '*Outcome of The Market Logic: The Academic-Professional Development of the Mathematics Teacher*' was given by Alex Rodrigo Montecino Muñoz. The fourth presentation titled '*Enacting Hybridity in a Home-school Mathematics Activity*' was given by Laura Black on behalf of her co-authors Sophina Choudry, Kelly Pickard-Smith, Bethany Ryan, and Julian Williams. The fifth presentation titled '*Mathematics, the Axiomatization Movement, and its social Implications*' was given by Sabrina Bobsin Salazar.

Third Day

The third day included two parallel sessions each of 90 min. The first parallel session was chaired by Murad Jurdak and had the theme of '*equitable access and participation in quality mathematics education: ideology, policies, and perspectives.* The session included five presentations. The first presentation titled '*Mathematics Curricula: Issues of Access and Quality*' was given by Tamsin Meaney, an invited speaker. The second presentation titled '*Mathematics Education for Social Justice: A Case Study*' was given by Natalia Ruiz López on behalf of her co-authors Gustavo Bruno, César Sáenz de Castro, and José Bosch Betancor. The third presentation titled '*Social, Political, Personal, and Imagined Constraints on Enacting Change after Professional Development*' was given Lisa Jean Darragh. The fourth presentation titled '*The Production of "Common Sense" in the Media about More Mathematics in Early Childhood Education*' was given by Troels Lange on behalf of his co-author Tamsin Meaney. The fifth presentation titled '*The Influence of Habitual Dispositions according to Pierre Bourdieu in Handling Mathematical Problems*' was given by Belgüzar Kara.

The second parallel session was chaired by Elizabeth de Freitas and had the theme of '*Mathematics identity, subjectivity and embodied dis/ability*'. The session included four presentations. The first presentation titled '*Parody and Power: Producing and Resisting Mathematics 'Ability'*" was given by Yvette Solomon, an invited speaker. The second presentation titled '*Maths Moves Me: The Body as a Political Space for Learning*' was given by Anna Chronaki. The third presentation titled '*Unequal Bodies—Corporeality and Social Inequality in the Context of Mathematics Education*' was given Nina Bohlmann. The fourth presentation titled '*The Biopolitics of Number Sense: Ordinality and Ontology*' was given by Elizabeth de Freitas on behalf of her co-author Nathalie Sinclair.

Fourth Day

The session of the fourth day which was chaired by Murad Jurdak and Renuka Vithal included open whole-group discussion on four topics:

1. Open discussion on economic dimension of mathematics education
2. Reporting back from parallel sessions
3. Open discussion on the implications of social and political dimensions of mathematics education.
4. Group wrap-up.

Concluding Remarks

1. There was an appreciation on the part of the Team members of the collaborative work among them, though they came from different perspectives on social and political dimensions of mathematics education.
2. There was a general feeling on the part of the Team as well as participants that the programme was well planned and efficiently executed.
3. There was a consensus that the economic dimension of mathematics education should be added to the social and political dimensions as a distinct dimension of mathematics education in future ICME's.
4. There was a general agreement to recommend that the TSG on the social and political dimensions of mathematics education be included in future ICME's. The relevance and significance of this Study Group was evidenced by the number, richness, and scholarly and cultural diversity of the presented papers.

Reference

Jurdak, M., Vithal, R., de Freitas, E., Gates, P., & Kollosche, D. (2016). *ICME 13 Topical Surveys series, social and political dimensions of mathematics education-current thinking*. Cham: Springer Open.

Open Access Except where otherwise noted, this chapter is licensed under a Creative Commons Attribution 4.0 International License. To view a copy of this license, visit http://creativecommons.org/licenses/by/4.0/.

Topic Study Group No. 35: Role of Ethnomathematics in Mathematics Education

Milton Rosa, Lawrence Shirley, Maria Elena Gavarrete and Wilfredo V. Alangui

Introduction

Participants of TSG35 addressed numerous themes related to ethnomathematics and its pedagogical action. In order for us to better understand the development of ethnomathematics, members discussed both current and future perspectives of this program. As well, its goals, objectives, and assumptions were analyzed in regards to the encouragement of an ethics of respect, solidarity, and cooperation across cultures. These topics were connected by themes of culturally relevant pedagogy, innovative approaches in ethnomathematics, and the role of this program in mathematics education.

In the ethnomathematics topic study group at ICME-13, there were 28 accepted papers written by 36 researchers from 19 countries: Australia, Belgium, Brasil, China, Costa Rica, Greece, India, Israel, Italy, Mozambique, Nepal, New Zealand, Peru, Philippines, Portugal, South Africa, Spain, Tanzania, and United States of America. From these papers, 24 were presented. Approximately, 30 researchers participated in the discussions conducted in each one of the 11 sessions (7 regular sessions and 4 oral communication sessions available in the congress). The majority of the papers presented in the ethnomathematics study group were by researchers from Brazil (7) and Nepal (3).

Co-Chairs: Milton Rosa, Lawrence Shirley.
Team members: Wilfredo V. Alangui, Maria Elena Gavarrete.

M. Rosa (✉)
Universidade Federal de Ouro Preto, Ouro Preto, Brazil
e-mail: milrosa@hotmail.com; milton@cead.ufop.br

L. Shirley
Towson University, Towson, MD, USA
e-mail: lshirley@towson.edu

Description of the Activities

Ethnomathematics grew out of the history of mathematics, mathematics education, and issues of mathematics in anthropology, sociology, economic, environmental concerns, and political science. It recognizes that the members of cultural groups develop activities that involve mathematical thinking. In order to allow the ethnomathematics community to discuss important issues related to this program, 7 regular sessions of the TSG35 were developed.

(a) **July 26th, 2016**

  Opening Session: Pedagogical Action of the Ethnomathematics Program
  Ubiratan D'Ambrosio and Milton Rosa (Brazil)
  Ethnomathematics and its pedagogical action

(b) **July 27th, 2016**

  Two Parallel Sessions
  Marcos Cherinda (Mozambique)
  From defrosting hidden mathematical knowledge to its formal learning: reviewing Gerde's research approach
  Wilfredo Alangui (Philippines)
  There is a theory behind what we're doing! Ethnomathematics and indigenous peoples' education in the Philippines
  Morane Almeida Oliveira (Brazil)
  Proposal for a methodological approach for the technical course for indigenous agroforestry agents in the state of Acre
  Tony Trinick; Uenuku Fairhall; Tamsin Meaney (New Zealand)
  Cultural and Mathematical Symmetry in Maori meeting houses
  Veronica Albanese; Natividad Adamuz-Povedano; Rafael Bracho-López (Spain)
  Ethnomathematics: two theoretical views and two approaches to education
  Charoula Stathopoulou (Greece)
  Once upon a time ... the Gypsy boy turned 15 while still in the first grade

(c) **July 29th, 2016**

  Symposia: Innovative Approaches in Ethnomathematics
  Daniel Clark Orey (Brazil)
  The Critical-reflective Dimension of Ethnomodeling
  Parallel Session
  Miriam Amit; Fouse Abu-Qouder (Israel)
  Weaving culture and mathematics in the classroom: the case of Bedouin ethnomathematics
  Karen François (Belgium)
  Wittgenstein's late philosophy as a philosophical foundation for ethnomathematics
  Mogege Mosimege (South Africa)

The role of language in ethnomathematical research and implications for mathematical teaching and learning

(d) **July 30th, 2016**

Two parallel sessions
Jaya Bishnu Pradhan (Nepal)
Chundaras' culture and mathematical ideas
Maria Cecília Fantinato; José Ricardo Souza Mafra (Brazil)
Aritapera's craftswomen: informal learning processes in an ethnographic study in ethnomathematics
Tod Shockey; John Bear Mitchel (United States of America)
An ethnomodel of a Penobscot lodge
There were also 4 oral communication sessions attached to the ethnomathematics topic study group with 10 presenters from Australia, Brazil, India, Italy, Nepal, Peru, and Tanzania.

(e) **July 26th and 27th, 2016**

Four Parallel Sessions
Hongshick Jang (Tanzania)
Language, ethnomathematics and technology in mathematics education challenges and pitfalls: the case of Tanzania.
Toyanath Sharma (Nepal)
Meaningful mathematics through cultural artifacts.
Alexandrina Monteiro; Jackeline R. Mendes (Brazil)
Knowledge mobilization in cultural practices: ethnomathematics as a counter-conduct movement.
Kay Owens (Australia)
The role of culture and ecology in visuospatial reasoning: the power of ethnomathematics.
Franco Favilli; Fiorenza Turiano (Italy)
On which finger will the number fall?
José Ricardo Mafra; Maria Cecilia Fantinato (Brazil)
Perceived techniques and processes of craftswomen in Santarém/PA.
María del Carmen Bonilla (Peru)
Tools of history of mathematics and dynamic geometry in the pre-service training in intercultural bilingual education.
Ramesh Neupane (Nepal)
Teaching and learning mathematics in a cultural context: ping as a project.
Sudhakar Agarkar (India)
Understanding the units of length measurement used by tribal people in India.
André Gerstberger; Ieda Maria Giongo (Brazil)
Ethnomathematics look at mobile usage regarding teaching mathematics processes in elementary education final years.

The presentations focused on and discussed ethnomathematics as a line of study and research of mathematics education from the many diverse perspectives and points of view brought to the group from all over the world. Together, presenters are investigating the roots of mathematical ideas and practices, starting from the way individuals see and use mathematical thinking in different cultural groups. In other words, ethnomathematics studies seek to identify mathematical practices that begin with the knowledge of the *others* in their own terms and rationality.

Because ethnomathematics studies the cultural roots of mathematical knowledge beginning with the various ways in which different cultural groups mathematize including academic mathematics, our study of ethnomathematics considers the historical evolution of mathematical knowledge with the acknowledgment of all social and cultural factors that form this ongoing and dynamic development.

Discussions and Reflections

The various presentations recognized how members of distinct cultural groups develop unique techniques, methods, and explanations that allow them alternative understandings, comprehensions, new actions, and a transformation of societal norms. Such historical research field has been a foundational area in ethnomathematics and continues to build a database of examples of mathematical thinking in distinct cultural groups.

It is evident from the discussion from TSG35 that the theoretical basis of an ethnomathematics program offers a valid alternative to traditional studies of history, philosophy, cognition, and pedagogical aspects of mathematics. Therefore, there is a growing sensitivity to the understanding and comprehension of mathematical ideas, procedures, and practices developed by the members of distinct cultural groups. This is due primarily to the expansion of studies related to culture, history, anthropology, linguistics and ethnomathematics.

Because ethnomathematics offers a broader view of mathematics, including its ideas, notions, procedures, processes, methods, and practices rooted in distinct and diverse cultural environments, this aspect leads to increased evidence of cognitive processes, learning capabilities, and attitudes that influence the learning processes occurring in classrooms. In addition to reflecting on social and political dimensions of ethnomathematics, another important aspect of this program is the possibility for the development of innovative approaches for a dynamic and *glocalized society* as outlined by D'Ambrosio.

The results of our discussions within the TSG35 show that it is important to understand the diverse sociocultural representations and concepts of *ethno* developed from distinct ideas, procedures, practices, and dimensions of space and time through the relationships between members of cultural groups. This aspect shows that, currently, a more sensitive understanding of diverse mathematical ideas, procedures, and practices developed by members of diverse cultural groups has become increasingly available through the growth of the fields of ethnology, culture, history, anthropology, linguistics, and ethnomathematics.

The insights gleaned from the presentations and from our discussions demonstrated the breadth and depth of how ethnomathematics influences the teaching and learning of mathematics, how it broadens our understanding of the nature of mathematical knowledge, and how it helps to create a just and inclusive society. Our discussions at the TSG 35 shows that it is necessary to pursue the current agenda of the ethnomathematics program in order to continue its progressive trajectory that contributes to the achievement of social justice and peace with dignity for all.

Final Considerations

From the presentations and discussions conducted during the regular and oral communication sessions of TSG35, it is possible to identify three unique characteristics that are interrelated.

1. It is necessary to continue to support and encourage further investigations in regards to innovative approaches in ethnomathematics programs especially in relation to social justice, civil rights, indigenous education, professional contexts, the playing of games, urban and rural contexts, and ethnomodelling.
2. An important change in mathematical instruction needs to accommodate continuous and ongoing changes in the demographics of students in mathematics classrooms around the world. Since it proposes that educators contextualize their mathematics teaching/learning by relating mathematical content to the sociocultural experiences of their students, it has become necessary to integrate culturally relevant pedagogies and diverse ethnomathematics perspectives into existing teacher education programs. Ethnomathematics intends to make school mathematics relevant to students through a more culturally relevant view of mathematics.
3. It is important to look at the diverse circumstances leading to the formulation of social, historical, cultural, political, and educational imperatives, and to realize that these are linked to one of the main goals of ethnomathematics, which is to broaden our conception of the diverse nature of mathematics. This includes specific examples of mathematical applications and models from an increasing number of diverse cultural groups.

From the discussions provided in this topic study group it is possible to conclude that mathematical knowledge is constructed by developing ideas, procedures, and practices that are common to the members of distinct cultural groups. This mathematical knowledge enables these members to elaborate and use their abilities that include the universal processes of counting, locating, measuring, drawing, representing, playing, understanding, comprehending, explaining, and modeling to solve problems they face daily.

Ethnomathematics provides mathematics educators an important framework to enable the transformation of mathematics so that it can better contribute in realizing the dream of a just and humane society. In this regard, mathematics is considered a powerful tool to help people build a civilization with dignity for all, in which inequity, arrogance, violence, and bigotry have no place, and in which threatening life, in any form, is rejected.

Presenters in this group shared the necessity for further discussions of issues related to mathematics education, classroom practices, and valuing the mathematical knowledge developed in specific cultural groups, and which helps to clarify the nature of mathematical knowledge.

Open Access Except where otherwise noted, this chapter is licensed under a Creative Commons Attribution 4.0 International License. To view a copy of this license, visit http://creativecommons.org/licenses/by/4.0/.

Topic Study Group No. 36: Task Design, Analysis and Learning Environments Programme Summary

Jere Confrey, Jiansheng Bao, Anne Watson, Jonei Barbosa and Helmut Linneweber-Lammerskitten

The Programme

Our TSG36 program included contributions from continents around the world: Africa, Asia, Europe, North and South America, and Oceania. Plenaries were presented by Anne Watson (UK), Koeno Gravemeijer (NED), Kazuhiko Nunokawa (JP), Berta Barquero, Ioannis Papadopoulos, Mario Barajas, and Chronis Kynigos (SP, GR) Angelika Kulberg (SWE), and Celia Hoyles and Richard Noss (UK). Twenty six papers and thirteen posters were presented. Confrey (USA) authored a closing summary. The presentations are listed below (posters omitted due to space constraints); the text references these contributions.

Topic Study Group 36's presentation began with the framework proposed by a prior ICME Topic Study Group (Watson) organized into the categories (or parameters) of Theory, Intentions, Likely Activity and Implementation. Over the course of the conference, additional components of a framework for TSG 36 emerged around (a) tasks, (b) learning environments, and (c) theory. Within the component of tasks, the topics addressed included how tasks are sequenced and structured (Gravemeijer, Brady et al., Goa et al.) what representations and tools were used in tasks (Thiel-Schneider, Johnson), what kinds of activities and actions

---

Co-chairs: Jere Confrey, Jiansheng Bao.
Team members: Anne Watson, Jonei Barbosa, Helmut Linneweber-Lammerskitten.

J. Confrey (✉)
North Carolina State University, Raleigh, NC, USA
e-mail: jconfre@ncsu.edu; jere_confrey@ncsu.edu

J. Bao
East China Normal University, Shanghai, China
e-mail: jsbao@math.ecnu.edu.cn

were taken (Palatnik, Schäfer & Linneweber-Lammerskitten), the meaning of cases and classifications (Kulberg) and their effects on instruction and the type of feedback provided. Within the topic of learning environments, topics included student-to-student and student-to-teacher interactions (Mok et al.), curricular approaches (Goa et al.), assessment (Sharma et al.) and teacher knowledge and teacher roles in instruction (Dietiker et al., Lee) A variety of views of the student and learning were addressed. The use of theories in the presentations ranged from "grand theories" such as Realistic Maths (Gravemeijer), Socio-Epistemological Theory (Montiel et al.), Socio-Constructivism and Constructionism (Hoyles & Noss, Barquero et al.), and intermediate or "bridging theories" such as local instructional theories (Gravemeijer), task variations (Kulberg) and exemplifying (Kim & Park), problem solving approaches (Nunokawa), project-based instruction (Barquero et al.), learning trajectories (Confrey), and place-based design (Zender & Ludwig). Discussions indicated that task design work is evolving towards ever more careful description of local instructional theories to ensure that grand theories are held accountable to their impact on the design and research. Attention to addressing how tasks reside within sequences (Brady et al.), hierarchies (Tan et al.), frameworks of mathematical relations (Gravemeijer), and cases (Kulberg) arose across presentations.

The TSG also had a strong and consistent focus on both the design and research processes. This focus drew attention to iterative and agile design (Confrey), emergent modeling (Gravemeijer), shifting communities of interest (Barquero et al.), mind maps (Barquero et al.), progress maps (Isidro Camac et al.), navigation (Zander et al.), pre- and post-design (Kulberg), studies of classroom practices and implementation (Johnson), clinical interviewing (Tan et al.), and crowd-sourcing (Zander et al.).

From the talks and presentations as a whole. four themes emerged that bear further examination and development. The TSG36 members varied in the extent to which they acknowledged an important role for (1) re-envisioning the purpose and characterizations of doing mathematics, (2) considering how to design from the perspective of the student, (3) articulating, sharing, and strengthening the foundation of professional knowledge, and (4) articulating a role for learning systems in providing feedback, diagnostic/formative assessment, and supporting iterative design.

Attempts to re-envision the purpose and characterization of doing math came from a variety of presentations. Hoyles and Noss stated "in technology-enhanced mathematics classrooms, the use of digital tools can disrupt routine practices in a transformative sense, and ensuing breakdowns can promote further reflection and thinking again." Olsher et al. described how "The design principles of this game setting provide a unique and innovative implementation to the use of online dynamic figures in a game setting that is strongly rooted in meaningful mathematical work of students." Barquero et al. also emphasized in their joint enterprise of building an e-book that "... also, mathematical modelling was considered as a crucial approach to cope with most of the extra-mathematical questions, and progressively build up mathematical models to study the complexity of the questions posed along the unit, linking and articulating the models appearing at each step."

Consideration of how to design from the perspective of the student arose repeatedly in presentations emphasizing the need to connect to student experiences (Johnson, Nunokawa) and as relevant to their current and evolving identities. For instance, Nunokawa addressed how students learn to appreciate what kind of advantages the new ideas have over the previous ones. The importance of leveraging multiple representations (Johnson) arose along with articulating different student models (Kraemer et al., Rojas et al.). Wilkes described how sophisticated mathematics could be recognized in elementary students' activities.

Ways to strengthen the foundation of professional knowledge around the implementation of tasks was a topic that for which numerous scholars offered, as virtually everyone rejected the notion of materials that are "teacher proofed." Kulberg and Montiel et al. discussed how teaching can be improved through the way that teachers understand and approach cases. Stories were used to reach teachers and strengthen their involvement in task activities (Noruzi et al., Dietiker et al.). Joubert & Mostert and Lee involved teachers as co-designers and partners. Viewing teachers as central to the process of identifying and fostering emergent thinking and supporting discourse (Olsher et al.) was critical to many participants (learning to listen and support discourse and argumentation) There was also an emphasis on teacher community in this process including comparing novice and experienced teachers (Gao & Zhang) and studying collaborative teacher tasks (Lee et al.).

The final theme emerged around the idea of learning systems, systems in which tasks are embedded but allow students to proceed in personalized ways and at times, include assessments and feedback (Basila, Confrey). Sometimes this resided in Microworlds and dynamic software environments such as a similarity game (Olsher et al.) and for others the learning system actually included the learning systems that permitted a multi-site, multi-theory construction of an e-textbook. (Barquero et al.) with two geographically disparate communities designed an e-book. MathCity Math (Zender & Ludwig) even described a system that leveraged GPS and permitted mathematics activities to be constructed in real time at different locations around a city.

Overall the TSG 36 showed evidence that the field of task design, analysis and learning environments is continuing to grow and offer new forms of insights.

Tuesday, 26 July 2016, 12.00–13.30, 1st Session
• Confrey, Jere: Welcome/overview
• Watson, Anne: Parameters for practice and research in task design in mathematics education
• Gravemeijer, Koeno: A personal take on instructional design

Tuesday 26 July 2016

- Paoletti, Teo; Moore, Kevin C.; Stevens, Irma E.: Task-design principles for covariational reasoning
- Palatnik, Alik: Towards a typology of students' mathematical research projects
- Mok, Ida Ah Chee: Experiencing meaningful school mathematics: rich tasks for inequality
- Dietiker, Leslie; Brakoniecki, Aaron; Miller, Elyssa R.; Richman, Andrew S.: Enacted task design: tasks as written in the classroom

Tuesday 26 July 2016

- Gao, Xiang; Zhang, Bo: A comparison of novice and experienced teachers' design of a question sequence
- Schäfer, Marc; Linneweber-Lammerskitten, Helmut: Enhancing mathematical curiosity through Vitalmaths video clip tasks
- Lee, Arthur Man Sang: Developing collaborative rich tasks with teachers in Hong Kong classrooms
- Tan, Phei Ling; Kor, Liew Kee; Sam, Prof. Dr. Lim Chap: Applying attribute hierarchy method in task design and item analysis for the topic "time" in primary mathematics
- Udinkaew, Choosak; Saengpun, Jensamut: Designing mathematical tasks for developing mathematical thinking in classroom taught through open approach
- Wilkes II, Charles E.: Sophisticated mathematics: What does it look like for fifth graders

Wednesday 27 July 2016

- Nunokawa, Kazu: Bridging students' ideas and lessons' goals
- Kulberg, Angelika: Variation within sets of examples
- Barquero, Berta; Papadopoulos, Ioannis; Barajas, Mario; Kynigos, Chronis: Cross-case design in using digital technologies: Two communities of interest designing a c-book unit

Friday 29 July 2016

- Thiel-Schneider, Alexandra: How does the connection of different perspectives on exponential growth succeed?
- Brady, Corey; Eames, Cheryl; Jung, Hyunyi: Design principles for curricular sequences focused on models and modeling
- Albersmann, Natascha: Construction of mathematical tasks for parents and their children on secondary school level
- Cheng, Jing; An, Shuhua; Bao, Jiangsheng: Cognitive demand of mathematics opening problems exhibited by expert secondary mathematics teachers in Shanghai-China

Friday 29 July 2016

- Kraemer, Jean Marie; Brocardo, Joana Maria; Mendes, Fatima; Delgado, Catarina: Designing tasks for adaptive/flexible multiplicative reasoning
- Isidro Camac, Lilian Edelmira; Ordoñez Montañez, Candy Clara; Paz Huaman, Gina Patricia: Authentic tasks to assess math competence in learning progress maps
- Kim, Dong-Won; Park, JinHyeong: Building mathematical statements through exemplifying
- Montiel, Gisela; López-Acosta, Luis; Cantoral, Ricardo; Scholz, Olivia: Design-based socioepistemological research

Friday 29 July 2016

- Forsythe, Susan Kathleen: Analysis of students' work with a dynamic figure through the lens of Duval
- Olsher, Shai; Shternberg, Beba; Yerushalmy, Michal: Guess who: Addressing meaningful characteristics as means to discover which is the chosen dynamic figure
- Posov, Ilya Aleksandrovich; Mantserov, Dmitry Irikovich: Using free software to implement verification problems with parameters
- Zender, Joerg; Ludwig, Matthias: Mathcitymap (MCM): from paper to smartphone—a new approach of an old concept

Friday 29 July 2016

- Noruzi, Sepideh; Mehrmohammadi, Mahmoud: Teaching mathematics through different genres of stories
- Johnson, Heather Lynn: Designing technology-rich tasks to foster secondary students' covariational reasoning
- Rasila, Antti; Sangwin, Christopher J.: Development of stack assessments to underpin mastery learning
- Joubert, Marie; Mostert, Ingrid: Using 'learning experiences' in South African classrooms: Implications for a teacher toolkit
- Sharma, Bibhya; Kumar, Bijeta; Bali, Akeshnil: Online mathematics diagnostic test and remediation for new entrants in higher education in the pacific region

Saturday 30 July 2016

- Hoyles, Celia & Noss, Richard: Mathematics and digital technology: Challenges and examples from design research
- Confrey, Jere: A reflection on the evolving agenda of the TSG 36 on task design, analysis and learning environments

Open Access Except where otherwise noted, this chapter is licensed under a Creative Commons Attribution 4.0 International License. To view a copy of this license, visit http://creativecommons.org/licenses/by/4.0/.

Topic Study Group No. 37: Mathematics Curriculum Development

Anita Rampal, Zalman Usiskin, Andreas Büchter, Jeremy Hodgen and Iman Osta

The Programme

- Christian Hirsch. Print and digital curriculum design in the U.S.: The case of 'Transition to college mathematics and statistics'.
- Jerry Lipka. The emergence of the 'center of everything': Insiders and outsiders working together developing mathematics curricula from indigenous knowledge.
- Mark Prendergast, Cormac Breen, Michael Carr & Fiona Faulkner. Investigating third level lecturers' awareness of second level curriculum reform.
- Anita Rampal. What math for all? For and from life?
- Teresa Rojano & Armando Solares. The mathematics curriculum design from an international perspective: methodological elements for a comparative analysis.
- Dawn Teuscher, Lisa Kasmer, Travis Olson, and Shannon Dingman. Isometries in new U.S. middle grades textbooks: How are isometries and congruence related?
- Vivien Townsend. The 'mastery' curriculum in England: A battle with authoritative discourses of ability and accountability.
- Zalman Usiskin. Paradigms of curriculum development in school mathematics.
- Yan Guorui & Frederick K.S. Leung. A comparative case study of teachers' use of mathematics textbooks in Beijing and Hong Kong.

Co-chairs: Anita Rampal, Zalman Usiskin.

Team members: Andreas Büchter, Jeremy Hodgen, Iman Osta.

A. Rampal
Delhi University, New Delhi, Delhi, India
e-mail: anita.rampal@gmail.com

Z. Usiskin (✉)
University of Chicago, Winnetka, Chicago, IL, USA
e-mail: z-usiskin@uchicago.edu

Summary

Curriculum concerns what we want students to learn, and when we want them to learn what we want them to learn; it is the heart of mathematics education. *Curriculum development* involves the processes that influence policy, the development of curriculum materials and pedagogical practices, and planning and implementation.

Other TSGs at ICME-13 discussed curriculum development in specific topic areas or with specific populations or age groups. TSG 37 sought papers of broader concern: the status of curriculum development in countries or regions; policies and influences shaping the development process; implementation, learnings, and future visions for curriculum development.

Usiskin identified five kinds of curriculum: ideal (intended), textbook (materials), implemented (taught), tested, and learned. The first four of these are created by people called policy makers, curriculum developers, teachers, and item writers, respectively. He offered three broad paradigms of curriculum development: traditional, test-influenced, and innovative. In the traditional paradigm, these five kinds are developed in the order shown above. In the test-influenced paradigm, tested curriculum specifications occur before materials are written. Discussions during TSG-37 sessions indicated that test-influenced curricula are exceedingly common throughout the world, with PISA and TIMSS results having particular impact in some countries.

For the innovative paradigm, Usiskin offered a sequence of events: (1) work by pioneers, the individuals who design the curriculum and assume a role comparable to policy makers in traditional curricula; (2) proselytizing of and by first users, who become apostles for the curriculum; (3) use by those disenchanted with the old curriculum; (4) acceptance by government agencies and teacher trainers if there is evidence of success; (5) if accepted; a forcing of the enchanted, those people who were content with the old curriculum and need persuasion or coersion to change. At this point, test creators take notice and, with enough support, the tests change to reflect the new ideas. But simultaneously, there can be pushback by those who question the need for the changes that the new curriculum embodies.

Rampal and Lipka each emphasized the need for curriculum to be developed from the cultures and lives of all students, rather than developing content from the hierarchy of mathematics. Rampal discussed efforts in the national primary mathematics curriculum of India to rethink mathematics, for and from life, for all children, through pedagogies of empathy that enable democratic participation. She noted the daunting challenges of curriculum development for public education in the 'gatekeeping' subjects of science and math in socially and culturally diverse poor countries, in the light of national pressures to reserve professional opportunities for the 'talented' privileged, and increasing global pressures for curriculum homogenization and standardization, tied to aggressive formulations of learning outcomes, with increasing transnational corporate interests in promoting 'low fee' private schools for the poor.

The curricular materials and textbooks developed by Rampal et al. follow a constructivist approach and also involve thematic issues of work, entrepreneurship, heritage, craft knowledge, history of monuments and pre-historic cave paintings, etc. using contexts that invoke and integrate concepts already learnt. In a similar vein, Lipka engaged the TSG participants in activities involving the "center of everything" from a curriculum he and individuals from the Yupiaq community in Alaska designed together. He noted that indigenous knowledge is rarely part of a core academic subject and that in the USA and elsewhere, indigenous people and their knowledge have been suppressed. His work uses indigenous knowledge to create an accessible curriculum for teaching the foundations of mathematical thinking for elementary school students. This work has included scholars and knowledge holders at Saami University, at the University of Greenland, and in Micronesia with Carolinian knowledge holders from Yap State and Chuuk who are associated with their respective Departments of Education. This work fits squarely into the role of ethnomathematics in mathematics curriculum.

Rojano presented a study to evaluate, from an international perspective, the current mathematics curriculum for compulsory education in Mexico. The Mexican curriculum was compared with those of the United Kingdom, Chile, and South Korea. Four influences on conceptions of school mathematics were noted: the international context of mathematics education; mathematics education research; the role of mathematics in the broader school curriculum; and distinctive traits in the quality of curriculum design. Differences were found concerning how school mathematics is conceived in the four curricula. Differences also exist in the presentation of the curriculum—e.g., how much detail, the presence or absence of discussions of relevance and of problem-solving. Notable in all four curricula was the lack of explicit references to research literature, as well as their inclusion of innovations in terms of content or teaching approaches (use of ICT or early algebra, for instance). However, all four programmes showed signs of the influence of education research. Suggestions are given for a redesign of the manner in which the Mexican curriculum is presented.

In England in 2014 a new national mathematics curriculum was introduced. This curriculum introduces some content in earlier grades than before and is accompanied by rhetoric identifying "mastery" as an overriding goal for all students—not just a few. Mastery is specified as mathematical fluency and conceptual understanding, and the ability to reason mathematically and solve problems. Townsend presented a case study of a primary school teacher (whose job entails teaching all subjects) who struggles with bringing all students to mastery, against longstanding beliefs that mastery is possible only for some students. Townsend concluded that teachers like this one "will be left to mediate tensions between adopting new ways of working alongside ingrained practices resulting from historical authoritative discourses."

Prendergast reported on his study of the knowledge of tertiary mathematics lecturers about Project Maths, a recent change in the secondary mathematics course content, teaching, and assessment in Ireland. Most lecturers were aware that there was more emphasis on understanding, more mathematics in context, and more exploration and activities. Those aware of changes in assessment pointed out the fact that it is now more difficult to predict the kinds of questions. Yet in all these domains, the majority of the lecturers indicated they had made no change and did not plan to make any change.

Hirsch described the development of a new course, *Transition to College Mathematics and Statistics* designed for the last year of secondary school, primarily for college-intending students whose planned programs of study do not require calculus. The course is notable for the United States for including of concepts from algebra and functions, statistics and probability, discrete mathematics, and geometric visualization. These branches of mathematics are connected by the central themes of the course, modeling of our world and the nurturing of mathematical habits of mind. A suite of curriculum-embedded technological tools has been developed for the course. The course was developed with the aid of extensive field-testing and has been tested in a variety of settings in the past few years, and the testing suggests that students are able to solve more complex and more open-ended problems. In so doing, they seem to be more able to work in teams and assess their own work and the work of the team.

Teuscher compared the ideal curriculum as found in the U.S. Common Core State Standards in Mathematics (CCSSM) with the approach to isometries and congruence in six current middle school textbook series. Geometric transformations have only been universally found in books at this level since 2010. The researchers found that three series defined congruence without regard to isometries even though the CCSSM ask for a definition in terms of isometries. Very little work was done on properties of isometries, and errors were found in some of the materials with regard to the discussion of orientation. The researchers concluded that teachers need to be aware of these difficulties, but this is unlikely because so many teachers themselves have never previously encountered this content.

Guorui examined how two teachers in each of two schools in Beijing and Hong Kong utilized mathematics textbooks to develop the taught curriculum. The Pythagorean Theorem was chosen so as to have uniform content, though the teachers were purposely selected so that in each school one teacher had at least 18 years and the other had less than 5 years experience. The experienced teachers tended to go beyond the textbook more than the inexperienced. The researchers pointed out the variety of factors that influence textbook use beyond the textbook itself, including the education system, school policies and practices, and teacher personal traits, themselves all influenced by socio-cultural factors.

In all, the papers demonstrate the complexity of developing mathematics curriculum reflecting the broad range of desires and needs of mathematics in our world while simultaneously taking into account the correspondingly broad socio-cultural backgrounds of students and the variety of teachers and schools in any system. Animated discussion in each of the sessions and participants enriched the deliberations by relating some if these issues to participants' own specific contexts.

Open Access Except where otherwise noted, this chapter is licensed under a Creative Commons Attribution 4.0 International License. To view a copy of this license, visit http://creativecommons. org/licenses/by/4.0/.

Topic Study Group No. 38: Research on Resources (Textbooks, Learning Materials etc.)

Lianghuo Fan, Luc Trouche, Chunxia Qi, Sebastian Rezat and Jana Visnovska

TSG38 focuses on issues related to mathematics teaching and learning resources, which mainly refer to school mathematics textbooks but also include other resources such as teacher manuals, student learning and assessment materials, and online resources. It is the hope of the organising team that TSG 38 would bring to foreground and examine various theoretical and methodological approaches used to study teaching and learning resources.

In the pre-congress call for contribution, it was stated that TSG38 sought contributions addressing broadly the areas of resources, teachers, and students, as outlined in the list of possible questions below, with a particular interest in analyses of the evolution of interactions between resources, teachers and students in a time of transition. The following three aspects were particularly highlighted in the call for contribution.

About the resources themselves: Among the learning materials available in mathematics classrooms in different countries, what role do textbooks and other curricular or learning resources play in mathematics teaching, learning, and assessment? How does the digitalization of information and communication affect this role? Is it possible to have a common definition of e-textbooks, and how could we characterize the differences between the traditional textbooks and e-textbooks?

Co-chairs: Lianghuo Fan, Luc Trouche.
Team members: Chunxia Qi, Sebastian Rezat, Jana Visnovska.

L. Fan (✉)
University of Southampton, Southampton, UK
e-mail: L.Fan@southampton.ac.uk

L. Trouche
École Normale Supérieure de Lyon, Lyon, France
e-mail: luc.trouche@ens-lyon.fr

About the teachers: What are the main features of the teacher resource systems in different countries, their crucial resources, stability, flexibility and evolutions? What are the relationships between their individual and collective resources, and how could we model such relationships? What about the relationships between resource designers and users? What are the consequences of evolutions at stake for the teaching of mathematics, and for teacher knowledge and professional development?

About the students: what are the main features of the student resource systems in different countries? What is the effect of modern ICT (particularly internet) on their use and the design of resources? How do these evolutions affect their behavior, learning and relationships concerning the subject of mathematics?

TSG38 received a larger-than-expected number of submissions. According to the statistics released, it was one of the five largest TSGs in terms of the number of submissions received. After the process of review, 67 contributions were accepted, 61 for oral presentation and 6 for poster presentation.

Given the large number of contributions, the programme of TSG38 was organised into four 90-min regular sessions, plus 10 concurrent parallel sessions which took place in 4 time slots. As Sebastian Rezat was unable to attend the congress due to some unforeseen reason, these sessions were chaired by the two chairs, the two team members and the IPC liaison person Birgit Pepin. The four regular sessions are as follows:

Session 1 (26 July 2016). Setting the scene: What role do textbooks and other curricular or learning resources play in mathematics teaching, learning, and assessment? (Chair: Birgit Pepin)

> 1.1 Remillard, J.: Understanding teacher-resource interactions: Perceiving curriculum resources (Contribution invited).
> 1.2 Leshota, M. J., and Adler, J.: Disaggreating a mathematics teacher's pedagogical design capacity.
> 1.3 Fan, L., Mailizar, M., Alafaleq, M., and Wang, Y.: How proof is presented in selected secondary maths textbooks in China, Indonesia and Saudi Arabia.
> 1.4 Qi, C., Zhang, X., and Huang, D.: Research on textbooks used in teaching transformation for secondary school.

Session 2 (27 July 2016). How does the digitalization of information and communication affect the role of resources? (Chair: Jana Visnovska)

> 2.1 Trouche, L., Gueudet, G., and Pepin, B.: Open educational resources: A chance for enriching mathematics teacher's resource systems?
> 2.2 Kynigos, C., and Kolovou, A.: Teachers as designers of digital educational resources for creative mathematical thinking.
> 2.3 Pu, S., Song, N.: Research on international development trends of primary mathematics textbooks in the 21st century.
> 2.4 Rocha, K.: Uses of online resources and documentational trajectories: The case of Sésamath.

Session 3 (29 July 2016). Teachers' collective work through resources (Chair: Chunxia Qi)

3.1 Van Steenbrugge, H., Larsson, M., Ryve, A., Insulander, E., & Brehmer, D.: Curriculum support for teachers: A collective perspective.
3.2 Faughn, A. P., and Borchelt, N.: Mathematics teachers' circles: a resource perspective on classroom transfer.
3.3 Wang, C.: Analyzing teachers' expertise, resources and collective work throughout Chinese and French windows.
3.4 Essonnier, N., Kynigos, C., Trgalova, J., and Daskolia, M.: Studying the role of context in social creativity for the design of digital resources.

Session 4 (30 July 2016). Teachers' and students' interactions through resources (Chair: Lianghuo Fan and Luc Trouche)

4.1 Ruthven, K.: Researching instructional activity and student interaction with digital resources (Contribution invited).
4.2 Visnovska, J., and Cortina, J. L.: Resources as a means of supporting teachers in planning for interactions with students' ideas.
4.3 Naftaliev, E.: Engagements of prospective teachers with e-textbook.
4.4 Kim, O. K.: Teacher decisions on lesson sequence and their impact on opportunities for students to learn.

The 10 parallel sessions in 4 time slots (TS), the first two on 26 July and the last two 29 July, were organised with focuses on different areas concerning the research of mathematics textbooks and resources.

On 26 July, TS1 (60 min) consisted of 3 parallel sessions: the first parallel session focusing on "**Textbook analysis and comparison**" (Chair: C. Qi; presenters: X. Yang, M. S. Aguilar, S. Y. Jeong, and K. Oh), the second on "**Text evaluation and digital tools**" (Chair: L. Trouche; presenters: H.-D. Janetzko, A. Pu, and A. M. Bijura), and the third on "**Teachers' work and learning with tools**" (Chair: J. Visnovska; presenters: R. Lucena, R. Yap, and R. M. A. Filho). TS2 (90 min) consisted of 2 parallel sessions: the first parallel session focusing on "**Problem solving**" (Chair: L. Fan; presenters: E. Bingobali*, E. Santaolalla, S. Walter, and C. A. Fuentes) and the second on "**Representation and illustration in textbooks**" (Chair: B. Pepin; presenters: X. Liu, P. Pausigere, V. Sarveswary, R. E. Borba, V. C. Lianos, and G. Glasnovic).

On 29 July 2016, TS3 (60 min) consisted of 3 parallel sessions with all focusing on "**Students and teachers**": the first parallel session had three presentations (Chair: J. Visnovska; presenters: N. Podevano, X. Jia and L. Zhao), the second parallel sessions had four presentations (Chair: L. Trouche; presenters: V. Gitirana, I. Ercan, E. Benitez, and X. Shao), and the third also four presentations (Chair: L. Fan; presenters: Z. Zhu, C. C. Assis, M. Ribeiro, and M. A. Huntley). TS4 (90 min) consisted of 2 parallel sessions: the first parallel session focusing on "**Collective work**" (Chair: C. Qi; presenters: C. Qi, A. Bapat, S. Basturk, J. Slisko, and D. Wijayanti) and the second on "**Teacher use and interaction with resources**"

(Chair: L. Trouche; presenters: H. Siedel, F. Bifano, D. Paez, L. Jaber, L. Ahl, and E. Aydin*).

The two presentations marked with * by E. Bingobali and E. Aydin were made through video presentations due to the fact that they were unable to attend the congress due to unforeseen reasons taking place in Turkey.

To conclude, it is worth noting that ICME-10 was the first congress in the history of the ICMEs that a group with specific focus on mathematics textbooks (including learning and teaching materials), Discussion Group 14, was programmed. According to the organisers, DG14 received a much smaller number of submissions and accepted only 9 of them for presentation (Fan, Turnau, Dole, Gelfman, & Li, 2008). Compared with about nearly 70 presentations, it signals to us a clear and rapid growth of the interest of the international mathematics education community in the area of mathematics textbooks and resources (also see Fan, Zhu, & Miao, 2013), taking into account digital evolutions (Pepin, Gueudet, Yerushalmy, Trouche, & Chazan, 2015). On the other hand, it should be also noted that among all the submissions, a great majority of studies reported are on textbooks (compared to other resources), on teaching resources (compared to learning resources) and on printed text (compared to digital form), while methodologically, most are on textbook analysis, textbook comparison and teachers' use and interaction with textbooks and resources.

Overall, we are encouraged by the fact that there were so many contributions to TSG38 and wish to thanks all the contributors, presenters, reviewers, and the participants to help TSG38 a success. A post-congress monograph, based on contributions and presentations at TSG38, will be published by Springer.

References

Fan, L., Turnau, S., Dole, S., Gelfman, E., & Li, Y. (2004). DG 14: Focus on the development and research of mathematics textbooks. In M. Niss (Ed.), *Proceedings of the 10th International Congress on Mathematical Education* (pp. 485–489). Roskilde, DK: Roskilde University.

Fan, L., Zhu, Y., & Miao, Z. (2013). Textbook research in mathematics education: Development status and directions. *ZDM-International Journal on Mathematics Education, 45*(5), 633–646.

Pepin, B., Gueudet, G., Yerushalmy, M., Trouche, L., & Chazan, D. (2015). E-textbooks in/for teaching and learning mathematics: A potentially transformative educational technology. In L. English & D. Kirschner (Eds.), *Third handbook of research in mathematics education* (pp. 636–661). New York: Taylor & Francis.

Open Access Except where otherwise noted, this chapter is licensed under a Creative Commons Attribution 4.0 International License. To view a copy of this license, visit http://creativecommons.org/licenses/by/4.0/.

Topic Study Group No. 39: Large Scale Assessment and Testing in Mathematics Education

Rae Young Kim, Christine Suurtamm, Edward Silver, Stefan Ufer and Pauline Vos

Introduction

Topic Study Group 39 aimed to address issues related to large-scale assessment, evaluation and testing in mathematics at all levels. Sound large-scale assessment (LSA) has the potential to provide important feedback about students' mathematical thinking, about classroom mathematical culture, or about a country's curriculum emphasis. Furthermore, LSA can have a strong influence in mathematics education as it often defines the mathematics that is mediated, valued and worth knowing.

Our TSG sought contributions of research in and new perspectives on LSA in mathematics education. We saw these issues as falling into three main strands: purposes and use, design and development, and teacher-related issues. Prospective contributors were requested to address one or more of the following topics:

Purposes and Use

- Purposes and use of LSA in mathematics at the international, national, school, classroom, or individual level
- The use of assessment for learning, as learning, and of learning in mathematics as they relate to LSA

Co-chairs: Rae Young Kim, Christine Suurtamm.
Team members: Edward Silver, Stefan Ufer, Pauline Vos.

R.Y. Kim (✉)
Ewha Womans University, Seoul, Republic of South Korea
e-mail: kimrae@ewha.ac.kr

C. Suurtamm
University of Ottawa, Ottawa, Canada
e-mail: Christine.Suurtamm@uottawa.ca

- Policy issues such as how LSAs frame political discussions and decisions
- The communication and use of results from LSA in mathematics

Design and Development

- The development of LSAs which might include the conceptual foundations of such assessments
- Task design that values mathematical power including problem solving, modeling, and reasoning across disciplines, and that addresses the diversity of learners
- The design and implementation of alternative modes of LSA in mathematics (e.g., online, student investigations)

Teacher-related issues

- The design and development of LSA of teachers' mathematical and pedagogical content knowledge
- The impact of LSA on teachers' knowledge and practice

We initially received over 40 papers for the TSG covering a wide range of areas of interest from all over the world. We discussed how to organize the sessions and participated in reviewing the papers. Each paper was evaluated by two reviewers including co-chairs, team members, and the authors of the papers submitted to TSG 39. Based on the reviews of these papers, 12 of these contributions were chosen for extended papers, 14 were chosen for oral communication, and 12 were recommended for poster presentations. Considering the topics and issues of the papers, we categorized the papers into three extended paper sessions, three oral communication sessions, and one poster session (at general exhibition) facilitated by co-chairs and team members as chair. In addition, we had a joint session with *Topic Study Group 40: Classroom Assessment for Mathematics Learning* to share mutually interesting issues, ideas, and practices around assessment through intensive discussion. We collaboratively produced a pre-conference publication with the classroom assessment group as well. Since some papers were withdrawn, 11 papers were presented in extended paper sessions, 1 paper was presented in the joint session (along with 2 from TSG 40), 11 were presented in oral communication sessions, and 8 were shown in the poster session in the end.

All the sessions of TSG 39 were organized to create a sense of community among all the presenters and participants who share common interests and ideas about large-scale assessment to improve mathematics education. The participants contributed greatly to the sessions and brought in perspectives from a wide range of knowledge, experiences, and practices. They were asked to read all of the papers before coming to the TSG 39 sessions and to bring some questions and comments on the papers. We also generated online space to facilitate further discussion out of sessions. The following are the leading questions in the discussion:

- How do we ensure that we are assessing what is important to assess?
- What framework do people use in task design or assessment evaluation?

- What should be considered in task design?
- How do MKT items developed in one country transfer to other countries?
- What do we need to take into consideration when examining student achievement on LSA?
- In what ways can technology interact with assessment?
- How can LSA assessment be designed and used to improve student learning and equity?

Main Ideas and Discussions in Each Session

Each session consisted of three or four 15-min presentations, short questions and comments after each presentation, and a 20-min whole group discussion at the end. Although each session was originally organized by the main themes, various issues and questions related to several themes came up together in the sessions. Thus, we summarized what was presented and discussed by the main themes shown above: Purposes and use, design and development, and teacher-related issues.

Purposes and Use

More than 17 papers were presented regarding this main theme with various perspectives throughout the sessions. The presentations showed that large-scale assessments have been implemented for multiple purposes and uses in mathematics education. One group of papers focused on the use of large-scale assessments to evaluate systems and to make student placements. For instance, there are analyzing issues in specific regions such as gender and socioeconomic status (SES) in Brazil (e.g., Chagas and Kleinke) and the case of bonus points in Ireland (e.g., Treacy). Some papers presented the use of assessments to make student placements (e.g., Reddy) or to predict student performance by finding some factors or determinants (e.g., Alagoz and Ekici; Seifert, Eilerts, and Rinkens; Weitz and Venkat).

Another group of presentations showed that large-scale assessments could be used to reveal the features of student achievement and affective characteristics in certain contexts or across national contexts. Many papers focused on the analysis of student achievement in specific regions such as Taiwan (e.g., Tam and Leung), Belgium (e.g., Deprez, Nijlen, Ameel, and Janssen), and Thailand (e.g., Jaikla, Changsri, and Inprasitha) or across countries in terms of cognitive domains or levels (e.g., Kanageswari). While discussing several issues and concerns in each context, we also found commonalities across contexts.

The results from large-scale assessments contribute to analysis of factors related to student achievement. For instance, the relationship between self-efficacy and student achievement by their cognitive levels (e.g., Zhou, Liu, Q., and Liu, J.), the effects of socioeconomic status (SES) and opportunity to learn (OTL) at classroom and country levels on student achievement (e.g., Bokhove), the relationship between the use of ICT and mathematics achievement (e.g., Kanoh), didactic contract (e.g., Ferretti, Gambini, and Giorgio), and the factors influencing affective characteristics (e.g., Hwang, Kim, H., and Kim, W.). Some papers suggested a natural model of analysis of student abilities (e.g., Dimitric) or items measuring students' geometric intuition (e.g., Bai, Huang, and Zhang). We discussed pedagogical and political issues from the results of studies as well as methodological concerns around data analysis and interpretation.

Design and Development

Many presentations brought up methodological issues around the design and development of tasks in large-scale assessments. For instance, the validity of the assessment (e.g., Bansilal; Grapin; Kasoka, Jakobsen, and Kazima), cross-cultural adaptations of measures (e.g., Marcinek and Patrová), cultural sensitivity and validity (e.g., Philpot), perceived task difficulty different from empirical one (e.g., Beitlich, Lehner, Strohmaier, and Reiss), and equivalent assessment design (e.g., Inekwe). In addition, many studies showed that individual or cultural differences in solving problems, especially word problem (e.g., Strohmaier, Beitlich, Lehner, and Reiss) or problems with realistic situations (e.g., Chen, Liu, Zhao, Song, and Li), could influence the reliability and validity of large-scale assessment.

Another group of presentations pointed out that large-scale assessments have often measured low level of cognitive demands (e.g., Dogbey and Dogbey; Drüke-Noe and Kühn), which could not reflect current goals in mathematics education. In order to enhance student learning through large-scale assessment, some presentations suggested new ways of evaluating student abilities by developing new items to measure geometric intuition (e.g., Bai, Huang, and Zhang) or providing a new guideline and prescription for interpreting problem situations with multicultural values (e.g., Djepaxhija, Vos, and Fuglestad).

Teacher-Related Issues

Although a relatively small number of papers focused on this theme, we discussed how the results from large-scale assessments could be used for improving teaching practice and teacher knowledge. Since teaching is a cultural activity in a situated context, we also discussed cross-cultural adaptation issues of using measures of Mathematical Knowledge for Teaching (MKT) from a certain context to another (e.g., Marcinek and Patrová) and considered qualitative approaches such as using video clips (e.g., Bruckmaier and Krauss).

We learned from the joint session with the classroom assessment group that large-scale assessment and classroom assessment could complement each other to improve mathematics teaching and learning. In particular, Burkhardt argued that high-stakes assessment could be "a tool for improvement" by playing the roles not only in assuring accountability of systems but also in "measuring student performance", "defin(ing) performance goals for teaching and learning", and "largely determin(ing) the balance of classroom activities in most classrooms." This implies that large-scale assessment and classroom assessment can inform each other and enhance student learning in constructive ways.

Concluding Remarks

All the participants actively participated in the sessions and brought up interesting and important issues around large-scale assessments. We finally found that there were both decontextualized commonalities and contextualized differences across different contexts. In this sense, it was productive to collaborate with TSG 40, the classroom assessment group, to elaborate our discussions around assessments and improve assessments for student learning. We also came to the conclusion that further discussion needs to be continued to develop the emerging ideas from this topic study group.

Acknowledgements The contribution of all the authors and participants in TSG 39 are deeply acknowledged.

Open Access Except where otherwise noted, this chapter is licensed under a Creative Commons Attribution 4.0 International License. To view a copy of this license, visit http://creativecommons.org/licenses/by/4.0/.

Topic Study Group No. 40: Classroom Assessment for Mathematics Learning

Denisse R. Thompson, Karin Brodie, Leonora Diaz Moreno, Nathalie Sayac and Stanislaw Schukajlow

The Programme

Prior to the conference, members of the two topic study groups on assessment (TSG 39 on Large Scale Assessment and TSG 40 on Classroom Assessment) collaborated to develop a topical survey on *Assessment in Mathematics Education* (Suurtamm et al., 2016). The survey addressed five main issues related to the current state of assessment:

- Purposes, traditions, and principles of mathematics assessment
- Design of assessment tasks in mathematics education
- Mathematics classroom assessment in action
- Interactions of large-scale and classroom assessment in mathematics education
- Enhancing sound mathematics assessment knowledge and practices.

Twelve papers were presented in the four main sessions of TSG 40, with the third session held jointly with TSG 39: *Large-Scale Assessment and Testing in Mathematics Education*. In addition, twelve papers were presented in the oral communication sessions and fourteen posters were also presented. Papers addressed topics related to all but the first issue in the pre-conference topical survey. The focus of the main topic study group sessions is indicated below.

Co-chairs: Denisse R. Thompson, Karin Brodie.

Team members: Leonora Diaz Moreno, Nathalie Sayac, Stanislaw Schukajlow.

D.R. Thompson (✉)
University of South Florida, Tampa, USA
e-mail: denisse@usf.edu

K. Brodie
University of Witswatersrand, Johannesburg, South Africa
e-mail: karin.brodie@wits.ac.za

Session 1 included three papers focusing on assessment tools used by teachers, with three different models to assess student learning. Swan and Foster focused on design approaches to formative assessment with tasks developed that incorporate pre-assessment activities, formative feedback questions, and sample work for students to critique. Sia and Lim approached assessment *for* learning by developing a cognitive model for the concept of time and identifying attributes describing the knowledge and skills needed to answer tasks related to that concept; by associating the attributes with items and then finding students' patterns of success with the attributes, teachers have detailed feedback to make inferences to enhance student progress. Krieger, Platz, Winter, and Niehaus focused on the use of the open source web application, Internet Mathematics Assessment System (IMathAS), to help students in writing proofs; building the system for a proof requires identifying segments or phrases that would be appropriate as well as an acceptable sequence so that grading can be semi-automatic.

- Swan, M., & Foster, C. *Formative assessment lessons for concept development and problem solving.*
- Sia, C. J. L., & Lim, C. S. *Using cognitive diagnostic assessments (CDA) as an alternative mode of assessment for learning.*
- Krieger, M., Platz, M., Winter, K., & Niehaus, E. *Classroom assessment and learning support for logical reasoning in mathematics education—suggestion of an e-proof environment.*

Session 2 included four papers focusing on teacher judgments and teacher learnings that occur as part of their assessment process. Sayac collected assessment tasks from primary teachers of varying experience levels and then analyzed the tasks for the level of complexity and nature of the competencies; she found most tasks were low level on both complexity and competency level. Hardie researched the sources that New Zealand teachers reference in making overall teacher judgments about student progress for parents or to document teaching and learning; teachers tended to use formal and informal methods in assessing their students' progress and had to learn to be skilled users of evidence from various assessment methods. Marynowski provided results from a case study of two secondary teachers in which she investigated both their beliefs about assessment as well as their actual practices, including students' perceptions of their teachers' practices; she found that neither teacher fully recognized how various practices related to their beliefs. In the final paper from this group, Pai described the beginning analysis of a phenomenological study investigating three phases of an ephemeral assessment cycle, namely eliciting, interpreting, and acting; he also explored factors that influence these phases.

- Sayac, N. *How are pupils in French primary school assessed in mathematics? A didactical approach to explore this question.*
- Hardie, C. P. *Making overall teacher judgments in mathematics.*

- Marynowski, R. M. *Secondary mathematics teacher assessment beliefs and practices.*
- Pai, J. *In-the-moment decisions: A preliminary investigation on observations and conversations as assessment in secondary mathematics classrooms.*

Session 3 was held jointly with TSG 39, with a focus on psychometric models and other issues at the boundary of classroom and large-scale assessment. Two papers from TSG 40 were presented in this session. Bostic and Sondergeld described a process by which they developed and validated measures for assessing problem-solving ability of middle grades students, including linking of performance between the seventh and eighth-grade measures using Rasch IRT analysis. Ariza-Hernández, Rodríguez-Vásquez, and Arciga-Alejandre focused on assessing undergraduates' understanding of real functions of real variables using Sierpinska's four categories of understanding, namely identification, discrimination, generalization, and synthesis; students' achievement was analyzed using Bayesian IRT analysis to determine which categories of understanding students had achieved.

- Bostic, J. D., & Sondergeld, T. *Validating and vertically equating problem-solving measures.*
- Ariza-Hernández, F. J., Rodríguez-Vásquez, F. M., & Arciga-Alejandre, M. P. *Analysis of the understanding of a mathematical concept using a Bayesian IRT model.*

The three papers presented in the final session focused in some way on aspects of self-assessment within the mathematics classroom. O'Shea described assessment *for* learning through student autonomy, describing actions by an expert teacher to develop an environment in which grade 5 children engaged in critical dialogue to take control of their own learning. Veldhuis, van den Heuvel-Panhuizen, and Zhao compared the effects in the Netherlands and China of professional development support for teachers around issues of assessment on their students' achievement; teachers in the two countries varied in the length of time they participated in the professional development, with resultant differential effects on students' achievement. Straumberger described the use of self-diagnosis sheets as a means to help students assess confidence at using various competencies, and relating achievement to the accuracy of their self-diagnosis.

- O'Shea, A. *Exemplifying the expert primary mathematics classroom: The case of Alex and assessment for learning.*
- Veldhuis, M., van den Heuvel-Panhuizen, M., & Zhao, X. *Supporting primary teachers' assessment practice in mathematics: Effects on student learning in the Netherlands and Nanjing, China.*
- Straumberger, W. *Using self-assessment for individual practice in math classes.*

Three short papers were presented in each of the four oral communication sessions, with the focus of the sessions providing an opportunity to expand on issues raised during the regular topic study group sessions. Most papers focused on issues related to formative assessment, including the design of tasks, use of rubrics, specific formative assessment strategies, or the use of digital or electronic environments to create tasks. In addition, various papers provided cultural perspectives on the use of formative assessment in different countries.

The papers in the topic study group sessions and oral communications, as well as the poster presentations, indicate that assessment at the classroom level is a pressing issue around the globe. Teachers and researchers are engaged in collaborative efforts investigating assessment *for* learning as well as assessment *as* learning, all in the service of enhancing mathematics learning of all students. The various papers also indicate that much work remains to be done within the overall assessment sphere as the mathematics education community attempts to better understand the interplay of teachers' assessment practices and students' learning.

Authors of several of the papers, oral communications, and posters will be expanding their contributions as part of a post-monograph publication related to the work of the TSG.

Reference

Suurtamm, C., Thompson, D. R., Kim, R. Y., Moreno, L. D., Sayac, N., Schukajlow, S., et al. (2016). *Assessment in mathematics education: Large-scale assessment and classroom assessment* (ICME-13 Topical Surveys). SpringerOpen.

Open Access Except where otherwise noted, this chapter is licensed under a Creative Commons Attribution 4.0 International License. To view a copy of this license, visit http://creativecommons.org/licenses/by/4.0/.

Topic Study Group No. 41: Uses of Technology in Primary Mathematics Education (Up to Age 10)

Sophie Soury-Lavergne, Colleen Vale, Francesca Ferrara, Krongthong Khairiree and Silke Ladel

TSG 41 at ICME-13 will explore these issues:

- How do school and teachers around the world, and in differently advantaged communities, use technology to enrich mathematics learning at primary level?
- Which factors contribute to successful and sustained use of technology in primary settings?
- Which innovations in digital technology for education do enable primary children to inquire, problem solve and think mathematically and to share their learning?

Many types of digital technology and environments are available for primary education since before the turn of the century. Yet, individual drill and practice software and interactive tools for exposition still appear to dominate practice in primary classrooms where technology is used. Around the world today, young children bring their experience with hand-held and other technology into the classroom. Moreover, teachers are normally more comfortable using digital tools in the classroom that they use in their personal life. In recent years, these have included tools to communicate in the cloud. Are primary teachers keeping up with digital natives? And which types of technology use are emerging to enrich and foster mathematics learning at primary school?

Co-chairs: Sophie Soury-Lavergne, Colleen Vale.
Team members: Francesca Ferrara, Krongthong Khairiree, Silke Ladel.

S. Soury-Lavergne
École Normale Supérieure de Lyon, Lyon, France
e-mail: sophie.soury-lavergne@ens-lyon.fr

C. Vale (✉)
Deakin University, Burwood, Australia
e-mail: colleen.vale@deakin.edu.au

Regarding the first two issues, we want to learn more about factors and practices that enable teachers to efficiently embed technology use in the classroom, including contexts of differently advantaged communities. These factors might involve the design of the technology, curriculum innovation, instructional leadership, collaborative teacher inquiry, or other interventions. Contributions concerning this theme will need to identify specific technology for primary mathematics teaching and learning and the context. Regarding the third question for this TSG we want to know about innovations in the design of digital technology and tasks. Contributions for the last issue will focus on the impact of innovative technology and environments on children's mathematical inquiry, problem solving and reasoning.

The sessions chaired by Chairs Colleen Vale and Sophie Soury-Lavergne the following contributions were presented: Kevin Larkin, "Enhancing student learning using geometry apps: utilising the homogeneity and heterogeneity of clusters of apps"; Annie Savard, "Robotic tasks: affordances for mathematics learning?"; Anne Voltolini, "Duo of digital and material artifacts dedicated to the learning of geometry at primary school".

The Wednesday session chaired by Silke Ladel the following scholars presented the titles: Patricia Moyer-Packenham, "Using virtual manipulatives on iPads: how app alignment promotes young children's mathematics learning"; Sophie Soury-Lavergne, "Duos of artefacts to enhance mathematical learning"; Sean Chorney, "Exploring the social dimension of using TouchCounts".

On Friday Krongthong Khairiree chaired the session with: Catherine Attard, "Is current research assisting the implementation of contemporary ICT in the primary mathematics classroom?"; Nigel Calder, "Reshaping the learning experience through apps: affordances"; Shannon Larsen, "Using 1-1 mobile technology to support student discourse".

And on Saturday Krongthong Khairiree, "Enhancing students' visualize skills in solving word problems using bar model and the Geometer's Sketchpad"; Stéphane Cyr, "Impact of a video game on fractions concept learning in elementary school students"; Piata Allen, "He puawaitanga harakeke—using technology to accelerate learning in indigenous language schools".

Open Access Except where otherwise noted, this chapter is licensed under a Creative Commons Attribution 4.0 International License. To view a copy of this license, visit http://creativecommons.org/licenses/by/4.0/.

Topic Study Group No. 42: Uses of Technology in Lower Secondary Mathematics Education (Age 10–14)

Lynda Ball, Paul Drijvers, Bärbel Barzel, Yiming Cao and Michela Maschietto

There is no doubt that digital technology nowadays has a tremendous impact on society. As a consequence, the question arises as to what impact this has on the teaching and learning of mathematics. The ICME-13 Topic Study Group «Uses of technology in lower secondary mathematics education» addressed this topic for education to 10–14 year olds and aimed to:

- establish an overview of the current state of the art in technology use in mathematics education, including both practice-oriented experiences and research-based evidence, as seen from an international perspective;
- suggest important trends for technology-rich mathematics education in the future, including a research agenda and school level implementation strategies.

This focus is related to the topic of other ICME-13 TSGs. TSGs 41 and 43, respectively, focused on primary or upper secondary; TSG 44 addressed e-learning and blended learning. As there was one general TSG on in-service and professional development of secondary mathematics teachers (TSG 50), the TSG 42 described here included both a learner's and a teacher's perspective on digital technology in lower secondary mathematics education.

The following themes were core in the work of TSG 42.

Co-Chairs: Lynda Ball, Paul Drijvers.
Team members: Bärbel Barzel, Yiming Cao, Michela Maschietto.

L. Ball
The University of Melbourne, Melbourne, Australia
e-mail: lball@unimelb.edu.au

P. Drijvers (✉)
Freudenthal Institute Utrecht University, Utrecht, The Netherlands
e-mail: p.drijvers@uu.nl

- Evidence for effect:
  What are the research findings about the benefits for student learning of the integration of digital tools in lower secondary mathematics education?
- Mathematics education in 2025:
  What will lower secondary mathematics education look like in 2025, with respect to the place of digital tools in curricula, teaching and learning? How can teachers integrate physical and virtual experiences to promote deep understanding of mathematics?
- Digital assessment:
  What are features of appropriate online assessment of, for and as learning?
- Communication and collaboration:
  How can digital technology be used to promote communication and collaborative work between students, between teachers, and between students and teachers? What are the potential professional development needs of teachers integrating digital tools into their teaching, and how can technology act as a vehicle for such professional development activities?

Even if these themes are not exhaustive for the topics of the TSG, most contributions in these proceedings focus on one of them, and as such offer an excellent overview of the advancements in this field.

Open Access Except where otherwise noted, this chapter is licensed under a Creative Commons Attribution 4.0 International License. To view a copy of this license, visit http://creativecommons.org/licenses/by/4.0/.

Topic Study Group No. 43: Uses of Technology in Upper Secondary Education (Age 14–19)

Stephen Hegedus, Colette Laborde, Luis Moreno Armella, Hans-Stefan Siller and Michal Tabach

The Programme

TSG 43 addressed the use of technology in upper secondary mathematics education from four points of view:

- theoretical analysis of epistemological and cognitive aspects of activity in new technology mediated learning environments;
- the changes brought by technology in the interactions between environment, students and teachers;
- the interrelations between mathematical activities and technology;
- skills and competencies that must be developed in teacher education.

The group received 42 submissions for a presentation and 5 posters coming from 23 different countries. From these submissions, 12 were selected for a long presentation during the sessions of the Topic Study Group. The other submissions gave rise to short oral communications in slots external to the sessions of the Topic Group. In order to stimulate and structure the discussion, four additional presentations were planned by the Topic Group for introducing the four main themes of the group: they were done by L. Moreno Armella and C. Brady, S. Hegedus and

---

Co-chairs: Stephen Hegedus, Colette Laborde.
Team members: Luis Moreno Armella, Hans-Stefan Siller, Michal Tabach.

S. Hegedus
Southern Connecticut State University, New Haven, CT, USA
e-mail: heguduss1@southernct.edu

C. Laborde (✉)
Cabrilog, Fontaine, France
e-mail: colette.laborde@cabri.com

S. Dalton, by H.-S. Siller, by M. Tabach and J. Trgalova. These presentations gave rise to an ICME 13—Topical Survey (Hegedus et al., 2016).

The programme of the TSG was organized with the intention of meeting three aims: bringing information, supporting discussion and formulating critical questions for the future.

The first session was devoted to these four presentations providing a state of art presentation of the four main issues mentioned above and formulating critical questions related to those issues. The long presentations took place in the second and third sessions in parallel. The fourth session was a collective session divided into three parts:

– showcase examples of uses of technology by participants;
– discussion of issues raised during the previous sessions in small groups;
– reports summarizing the group discussion and conclusion.

A large diversity of themes was addressed by the presentations and during the discussion slots. The following section summarizes the main issues and questions of the group sessions.

Main Issues and Questions

Technologies at use

Although the group called for presentations about new emerging technologies, the presentations mainly dealt with "classical" technology like Dynamic Geometry Environments or CAS, but with a stronger focus on Dynamic Geometry Environments in the long presentations as well as in short oral communications.

Three presentations only reported about the use of emerging technologies: multimodality involving various sensory modalities (like sight, touch, sound), 3D augmented realities, 3D pen, Wii graph a software application modeling the movement of two controllers of the game console Nintendo Wii. It is interesting to note that the two studies about 3D pen and Wii console shared a common theoretical framework in which diagrams and gestures are strongly linked and in which the meaning of mathematical objects is to be found in the interactions between diagrams and gestures. This framework can also be linked to the notion of co-action (Moreno-Armella & Hegedus, 2009): the learner is co-acting with the representational systems of mathematics offered by dynamic technologies.

Topics and questions

The most mentioned technological environments were Dynamic Geometry Environments or Dynamic Mathematics Environments. In the presented studies, they were used

- for teaching specific notions;
- for modeling; in particular, a study shows that technology is used at different stages in the modeling cycle;
- for fostering exploration.

Some participants mentioned that it would be nice to see more applications/investigations in other areas than geometry, algebra, in particular. There is an innate appeal to the visual nature of DGE, so that CAS teaching questions are challenging. Of course, it must be kept in mind that algebra is not only accessible through CAS.

However, some presentations addressed issues related to the use of CAS. A focus was made in particular on justification and proof. There may be changes in the type of justification brought in textbooks when using a CAS. The example of Denmark was given. In this country, more than one textbook make use of CAS assisted proof, i.e. a proof of a theorem or a statement when steps are outsourced to a CAS. The number of steps may vary from one or few steps to the whole proof, the CAS playing the role of an authority. A connected issue was addressed about the norms for evaluating the work of students solving problems by means of a CAS. In absence of verbalization, it may be difficult to reconstruct the work of the students.

This discussion about CAS and justification raised the questions of the emergence of new norms when using technology and not only for CAS. A study devoted to the use of graphing calculators showed that students using calculators were asking less critical and less why questions than those not using calculators. There is presently an absence of policy that may make the situation difficult for teachers.

Problems as well as contribution brought by the use of technologies for teachers in their teaching and teacher education were addressed by several presentations and gave rise to discussion. As shown in some presentations, new opportunities of interaction between the teacher and the students may emerge due to the presence of technology providing feedback to students' actions. The nature of these opportunities and the role of the teacher within these new forms of interactions constitute a wide field to investigate. For example, what are the new ways of interaction among multiple students or between students and their teacher that can occur as a result of these digital technology tasks?

It seems that there is a need of investigating the impact of research on teacher development: in what ways can research provide or has research provided teachers with knowledge, recommendations, and professional development around an optimal use of student work with digital technologies to take advantage of what the digital tools have to offer?

Teachers' choices, views and norms seem to play a crucial role for a successful integration. A study reported how new socio and socio-mathematical norms emerge in technology enhanced lessons done by preservice teachers. In particular, more justifications are requested from students by the teacher and discussion among students about the "reasons" was encouraged. This led to a more general question: Are there any interesting ways of taking advantage of technology for working with proofs in math classrooms?

Teachers' beliefs regarding discovering learning and time constraints are most influential when determining frequency of technology use. Teachers supporting discovery learning are more inclined to resort to the use of technology whereas paying much attention to time constraints may prevent them from integrating often technology often into their teaching.

Conclusion

From the whole set of presentations and discussions emerged the complexity of the questions of the use of technology in the teaching and learning of mathematics. The studies shed light on many interactions between the mathematical structures and their representations in technology environments, between the learners, the environments and the task they are faced with, between the teacher interventions and the students' work on technology, between teachers' beliefs and the frequency and type of use of technology.

Research developed several theoretical tools for analyzing these interactions but the question of the dissemination of these studies and their possible use by teachers is still remaining.

How to disseminate research-based tasks with technology (designed by researchers or teacher educators) in a way to help teachers use them "properly"? In the case of collaborative design of tasks (by groups of teachers and researchers/teacher educators), how should researchers/teacher educators act in a way not to impose their research view but leading towards tasks with required educational quality?

"How to motivate teachers to adopt technology?" was among the final questions of the group …

References

Hegedus, S., Laborde, C., Brady, C., Dalton, S., Siller, H.-S., Tabach, M., et al. (2016). Uses of technology in upper secondary mathematics education. *ICME-13 topical surveys*. Cham: Springer Open.
Moreno-Armella, L., & Hegedus, S. J. (2009). Coaction with digital technologies. *ZDM Mathematics Education 41*(4), 505–519.

Open Access Except where otherwise noted, this chapter is licensed under a Creative Commons Attribution 4.0 International License. To view a copy of this license, visit http://creativecommons.org/licenses/by/4.0/.

Topic Study Group No. 44: Distance Learning, e-Learning, and Blended Learning

Rúbia Barcelos Amaral, Veronica Hoyos, Els de Geest, Jason Silverman and Rose Vogel

In Topic Study Group 44 we built on current and emerging research in distance learning, e-learning, and blended learning. Specifically, we pushed the boundaries of what is known through an examination and discussion of recent research and development in teaching and learning through these modalities, with a focus on primary, secondary, and higher education. Some of the subtopics considered were utilization of both Web 2.0 and Web 3.0 resources in e-learning, blended learning, and distance education modalities (for example, how are OER utilized as a resource by users); MOOC (what are the affordances and constraints of this approach through specific cases); emerging work on the usage of mobile technologies (such as cell phones and tablets) for distance learning; transitioning traditional classroom practices to use online affordances and constructing bi-learning environments; enabling mathematical collaboration in online mathematics education; online distance education and blended learning in the professional development of mathematics teachers; e-portfolio for reflected mathematics teaching and learning; orchestrating productive mathematical conversations in an online or blended learning setting; the role of the faculty/moderator in online mathematics education; emergence and sustainability of communities of practice in online environments of collaboration and co-construction of resources; research methodologies and paradigms for studying online and blended mathematics education; and evaluation and effectiveness of distance education, e-learning, and bi-learning.

Co-chairs: Rúbia Barcelos Amaral, Veronica Hoyos.
Team members: Els de Geest, Jason Silverman, Rose Vogel.

R.B. Amaral
São Paulo State University, São Paulo, Brazil
e-mail: rubiaba@rc.unesp.br; rubiaba@gmail.com

V. Hoyos (✉)
National Pedagogical University, Mexico City, Mexico
e-mail: vhoyosa@upn.mx

The TSG served to disseminate significant contributions as seen from international perspectives by providing an overview of the current state-of-the-art research, sharing and discussing emerging work (trends, ideas, methodologies, and results), and a calling for the development of a canon of research for online, blended, and distance math education. As a part of TSG 44 activities, research presentations by distinguished people in the field, posters, and sessions of collective discussion and reflection around previously accepted research contributions were considered.

On the first day during the conference, two sessions of work were accomplished, one on the topic of cases and perspectives of distance learning, e-learning, and blended learning and the other on online student learning, both chaired by Rose Vogel. The scholars who presented the talks in the first session were Fabian Mundt and Mutfried Hartmann; Kar Fu Yeung, Rachel Ka Wai Lui, William Man Yin Cheung, Eddy Kwok Fai Lam, and Nam Kiu Tsing; Karin Landenfeld, Martin Göbbels, and Antonia Hintze (invited paper); and Tatjana Hrubik-Vulanovic. The scholars in the second session were Jonathan T. Lee; Bijeta Kumar and Bibhya Sharma; and Yasuyuki Nakamura, Tetsuya Taniguchi, Kentaro Yoshitomi, Shizuka Shirai, Tetsuo Fukui, and Takahiro Nakahara.

There were also two sessions on the second day covering the topics of online collaborative learning and teacher PD through online tools and learning, both chaired by Rubia Barcelos. The first topic's presentations were given by Arthur Powell (invited talk); Kadian M. Callahan and Anne-Marie S. Marshall; and Mandy Lo, Julie-Ann Edwards, Christian Bokhove, and Hugh Davis. The talks on the second topic in the afternoon were given by Maman Fathurroman, Hepsi Nindiasari, Nurul Anriani, and Aan Subhan; Tatyana A. Oleinik; Andrey I. Prokopenko and Sergeevich S. Zub; Maria E. Navarro, Veronica Hoyos, Victor Raggi, and Sergio Vazquez; and Eugenia Taranto, Virginia Alberti, and Sara Labasin.

The third session, was chaired by Jason Silverman and covered the topic of mathematics teacher education at a distance and mediated by ICT. The talks were by Elizabeth Fleming, Daniel Chazan, Pat Herbst, and Dana Grosser-Clarkson (Invited Paper); Cosette Crisan; Tamar Avineri, Hollylynne S. Lee, Dung Tran, Jennifer N. Lovett, and Theresa Gibson; and Yaniv Biton and Osnat Fellus. During the afternoon, a third session was chaired by Veronica Hoyos on the topic of distance learning and quantitative assessment, and talks were given by Tajana Hrubik-Vulanovic, Ferlisa Bundalian Lavador, Mary J. Castilla, and Richard Vinluan and Maxima Joyosa Acelajado.

The last session on the topic of communities of learning at a distance and enhanced by ICT, was chaired by Veronica Hoyos. The following scholars presented their work on that topic: Marcelo Borba (Invited talk); Giovannina Albano, Maria Polo, and Pier Luigi Ferrari; and Angela María Restrepo.

Open Access Except where otherwise noted, this chapter is licensed under a Creative Commons Attribution 4.0 International License. To view a copy of this license, visit http://creativecommons.org/licenses/by/4.0/.

Topic Study Group No. 45: Knowledge in/for Teaching Mathematics at Primary Level

Carolyn A. Maher, Peter Sullivan, Hedwig Gasteiger and Soo Jin Lee

As the teaching of primary level mathematics (ages 5–13) is complex, it requires teachers to master a variety of types of knowledge. These are outlined as follows:

Understanding of important mathematical concepts that underpin meaningful student learning of the main strands of the mathematics curriculum;

- Appreciation of the mathematical processes (conceptual understanding, problem solving, and reasoning) in which students engage in doing mathematics, building mathematical arguments, and their justifications of solutions to problems;
- Selecting and building into lessons tasks that engage students in meaningful mathematics and numeracy learning;
- Awareness and knowledge of activities, tasks, and interventions that engage and develop persistence in students while exploring mathematical investigations;
- Awareness and appreciation of the value of tools (manipulatives) and technology in students' building multiple representations of mathematical ideas;
- Awareness of children's development in their learning of mathematical ideas (e.g. place value, number sense, operations) from informal to formal understandings;

Co-chairs: Carolyn A. Maher, Peter Sullivan.
Team members: Hedwig Gasteiger, Soo Jin Lee.

C.A. Maher
Rutgers University, New Brunswick, USA
e-mail: carolyn.maher@gse.rutgers.edu

P. Sullivan (✉)
Monash University, Melbourne, Australia
e-mail: Peter.sullivan@monash.edu

- Knowledge of pedagogies that are appropriate with heterogeneous classes including specific actions to support students' learning, such as collaborative group activities;
- Knowledge of resources (collaborative communities, lessons, activities, video collections) to support teacher learning.

TSG 45 participants explored the types of knowledge represented by these various challenges, and how teachers can be supported in their learning.

The major presentations for the group were:

- Professional knowledge for early mathematics education by Hedwig Gasteiger and Christiane Benz
- Supporting teachers in improving their knowledge of mathematics by Peter Sullivan
- Teacher learning about mathematical reasoning: An instructional model by Robert Sigley and Carolyn A. Maher
- What is required for teachers to reorganize math textbooks?—Textbook analysis based on key developmental understandings by Soo Jin Lee and Jaehong Shin
- Using task design to build teacher knowledge by Brenda Bicknell and Jenny Young-Loveridge
- Using tasks from contexts to engage students in meaningful and worthwhile mathematics learning by Doug Clarke and Anna Roche
- Teaching the language of mathematics: What teachers need to know and do by Louise C. Wilkinson
- Structure and development of primary teacher's professional competencies by Dennis Meyer, Andreas Busse, Jessica Hoth, Martina Dohrmann
- Pupils as knowledge agents and monitors in the construction of mathematical ideas by Therese Dooley.

The presentations covered a broad range of topics for supporting the improvement in knowledge for teachers of mathematics. In particular topics addressed included knowledge for:

- teaching mathematics in the early years
- diagnosis and support for students with special needs
- incorporating reasoning in teaching and assessment
- text analysis with a focus on fractions
- the design and use of contextual tasks
- supporting pupil construction of mathematics
- processes and purpose of task design and adaptation
- development and use of appropriate language of mathematics
- structure and algebra in all years
- effective incorporation of measurement into teaching and assessment
- catering for the needs of gifted students
- using the study of patterns as a prompt to abstraction and generalization
- analysis of misconceptions associated with learning decimals

- the methods of teaching proportional reasoning
- approaches to teaching equivalent fractions
- characteristics of an equitable and balanced curriculum
- providing corrective feedback and analysis of incorrect answers
- listening to and interpreting student thinking
- establishing classrooms as communities of learners and inquiry
- effective representations of mathematics concepts
- processes of teaching students to solve problems
- the connections between culture and classroom processes
- the drama of teaching.

Open Access Except where otherwise noted, this chapter is licensed under a Creative Commons Attribution 4.0 International License. To view a copy of this license, visit http://creativecommons.org/licenses/by/4.0/.

Topic Study Group No. 46: Knowledge in/for Teaching Mathematics at the Secondary Level

Ruhama Even, Xinrong Yang, Nils Buchholtz, Charalambos Charalambous and Tim Rowland

The program of TSG 46 focused on three themes:

1. Conceptualization and theorization of knowledge in/for teaching mathematics at the secondary level.
2. Methods for measuring, assessing, evaluating and comparing knowledge in/for teaching mathematics at the secondary level.
3. Connections between knowledge and practice of teaching mathematics at the secondary level.

The first three sessions centred on the three themes, while the fourth was devoted to summary, discussion and reflections. Below we describe the activities that took place during the four sessions.

Session 1: Conceptualization and Theorization of Knowledge in/for Teaching Mathematics at the Secondary Level (Chairs: Nils Buchholtz and Tim Rowland)

The problem: A number of international studies investigate the professional knowledge for teaching mathematics at the secondary level and for this purpose draw back on various different theoretical conceptualizations. Within the session the similarities and differences of different conceptualizations were analyzed and discussed, but also the current challenges of these conceptualizations were faced, especially with regard to the interaction between theoretically-assumed knowledge

---

Co-chairs: Ruhama Even, Xinrong Yang.
Team members: Nils Buchholtz, Charalambos Charalambous, Tim Rowland.

R. Even (✉)
Weizmann Institute of Science, Rehovot, Israel
e-mail: Ruhama.Even@weizmann.ac.il

X. Yang
Southwest University, Chongqing, China and University of Hamburg, Hamburg, Germany
e-mail: xinrong.yang@yahoo.com

facets and their visible manifestation in the practice of teaching. The current challenge seems to be to differentiate rather better current conceptualizations for teaching mathematics at the secondary level according to the theoretically sound and empirically-based integration of action-based knowledge facets. Orienting theoretical conceptualizations more to practical school-based contexts offers a basis for empirical research that is oriented more to the realities of mathematics teaching in school. The invited contributors of the session presented promising perspectives in this field, with a valuable overview from the first presenter:

Presentation 1: *Conceptualization and Theorization of knowledge in/for teaching mathematics at secondary level*, by Michael Neubrand from the University of Oldenburg in Germany.

Presentation 2: *Academic mathematics or school mathematics? What kind of content knowledge do mathematics teachers need?*, by Aiso Heinze and Anika Dreher from IPN—Leibniz Institute for Science and Mathematics Education in Germany (together with Anke Lindmeier, IPN, Germany).

Presentation 3: *Analysing secondary mathematics teaching with the knowledge quartet*, by Tim Rowland from the Universities of Cambridge and East Anglia in the UK (together with Anne Thwaites and Libby Jared from the University of Cambridge, UK).

Session 2: Methods for Measuring, Assessing, Evaluating and Comparing Knowledge in/for Teaching Mathematics at the Secondary Level (Chairs: Charalambos Y. Charalambous and Xinrong Yang)

The problem: The last two decades have seen considerable work not only in theorizing the knowledge needed for the work of teaching mathematics, but also in operationalizing and measuring this knowledge. These last two facets pose significant challenges to scholars working on exploring teacher knowledge and its effects on instructional quality and student learning, since at least two critical questions need to be addressed when it comes to considering these issues: (a) what (aspects of teacher knowledge) to measure—especially given the multifaceted nature of (recent) teacher knowledge conceptualizations—and (b) how best to measure them to ensure that valid and reliable data are collected, and legitimate inferences are drawn. Although these questions have attracted significant scholarly interest for elementary school grades, the field of measuring teacher knowledge at the secondary school is still developing. These issues were taken up by both Session-2 presentations. The first of these pointed to the importance of focusing on the knowledge entailments of key mathematical teaching tasks as opposed to simply attending to different types of knowledge; further capitalizing on videos to measure teacher knowledge as embedded in practice; and measuring teacher knowledge in cost-efficient ways. The second presentation made a case about the importance of measuring both generic teaching tasks as well as content-specific tasks, and the knowledge entailments associated with them.

Presentation 1: *Measuring Secondary Teachers' Knowledge of Teaching Mathematics: Developing a Field,* by Heather C. Hill of the Harvard Graduate School of Education, the USA.

Presentation 2: *Measuring Instructional Quality in Mathematics Education,* by Lena Schlesinger and Armin Jentsch of the University of Hamburg, Germany.

Session 3: Connections Between Knowledge and Practice of Teaching Mathematics at the Secondary Level (Chair: Ruhama Even)

The problem: That expertise in mathematics teaching requires adequate mathematical knowledge is a trivial statement, but what "adequate" means is not clear. In many countries, the education of secondary school mathematics teachers traditionally includes a strong emphasis on advanced mathematics courses at the college or university level, taught by mathematicians, assuming that it would contribute to the quality of classroom instruction. This tradition, however, has been reconsidered in recent years, and the relevance of advanced mathematics courses to the quality of secondary school mathematics teaching is being debated. Is there a need for advanced mathematics studies in the professional education and development of secondary school mathematics teachers? What might be the relevance of advanced mathematics courses taught by research mathematicians to teaching secondary school mathematics? This issue was the focus of the three presentations in session 3, all of which reported on studies that addressed the overarching question: *What are the relevance and the contribution of advanced mathematics studies to secondary school mathematics teaching?*

Presentation 1: *Accommodation of teachers' knowledge of inverse functions with the group of invertible functions,* by Nicholas H. Wasserman from Teachers College in the USA.

Presentation 2: *Senior secondary school teachers' advanced mathematics knowledge and their teaching in china,* by Haode Zuo and Frederick K.S. Leung from the University of Hong Kong in China.

Presentation 3: *Teachers' views on the relevance of advanced mathematics studies to secondary school teaching,* by Ruhama Even from the Weizmann Institute of Science in Israel.

Session 4: Summary, Discussion and Reflections (Chairs: Xinrong Yang and Ruhama Even)

In this session, Xinrong Yang from Southwest University in China, Nils Buchholtz from the University of Hamburg in Germany, and Charalambos Charalambous from the University of Cyprus in Cyprus reflected on the first three sessions. Below is a summary of their reflections.

Current theoretical conceptualizations of knowledge in/for teaching mathematics at the secondary level primarily focus on knowledge as a personal disposition that can be tapped for empirical surveys. At the theoretical level, drawing on the seminal work of Shulman (1986), various dimensions of knowledge are often distinguished and segregated depending on assumed content-related aspects, or on aspects of practical teaching. When such knowledge is operationalized in empirical studies, it becomes more possible to separate these different facets empirically. Michael Neubrand and Tim Rowland pointed out that the more context-oriented knowledge gets analyzed in such studies, the harder it gets to empirically differentiate the knowledge in actu from other factors such as the teacher's personality or the affective level, which leads us to look more at the performance of mathematics teachers and at classifications of situations in which mathematical knowledge surfaces in teaching.

In retrospect, thirty years after Shulman's (1986) pioneering work, we can now claim that much has been accomplished on different fronts. Reflecting on this rapidly accumulating work, Charalambos Charalambous argued that the polyphony in the different theoretical frameworks and conceptualizations advanced thus far seems to be productive; he nevertheless voiced concern as to whether this polyphony will eventually be turned into cacophony, in the sense that we might run the risk of creating a Tower of Babel when it comes to talking about, studying, and measuring teacher knowledge. He thus suggested that scholars invest more in *exploring synergies* between different conceptualizations. Given that a shift seems to be observed from studying components of teacher knowledge to investigating tasks of teaching and the knowledge requirements these tasks impose on teachers (cf. Gitomer & Zisk, 2015), the need to develop a comprehensive framework encompassing such tasks and detailing their knowledge requirements for teachers was also underlined. Finally, the merit of employing the different approaches pursued so far to study teacher knowledge was highlighted. At the same time, a series of open issues was also outlined. For example, at what level of granularity should teacher knowledge be measured to ensure both its predictive validity and generalizability? To what extent might certain measures be culturally specific, and what might the implications of this specificity be? To what extent might certain items used in teacher-knowledge measures function differently when used in different contexts? To what extent does the knowledge measured actually impact teachers' teaching practice and students' mathematics achievement? This indicative list of questions indicates that there remains significant uncharted terrain to explore when working on studying teacher knowledge, and its effects on instruction and student learning.

References

Gitomer, D. H., & Zisk, R. C. (2015). Knowing what teachers know. *Review of Research in Education, 39*(1), 1–53.

Shulman, L. S. (1986). Those who understand: Knowledge growth in teaching. *Educational Researcher, 15*(2), 4–14.

Open Access Except where otherwise noted, this chapter is licensed under a Creative Commons Attribution 4.0 International License. To view a copy of this license, visit http://creativecommons.org/licenses/by/4.0/.

Topic Study Group No. 47: Pre-service Mathematics Education of Primary Teachers

Keiko Hino, Gabriel J. Stylianides, Katja Eilerts, Caroline Lajoie and David Pugalee

The Programme

Topic Study Group 47 (TSG 47) included paper presentations on significant new trends and developments in research, theory, and practice about all different aspects that relate to the mathematics education of pre-service primary teachers. The phrase "different aspects" was interpreted broadly to include (among others) the following:

- pre-service teachers' mathematics-content preparation as well as their mathematics-specific pedagogical preparation;
- pre-service teachers' mathematical knowledge for teaching as well as their beliefs about mathematics or mathematics teaching and learning;
- textbooks and other curriculum materials as well as assessment tools used in mathematics teacher education programs for pre-service teachers;
- pre-service teachers' experiences in mathematics classrooms and issues related to their school placements; and
- teacher educators' knowledge for teaching pre-service teachers.

TSG 47 offered a forum for an overview of the current state-of-the-art, invited contributions from experts in the field (Fou-Lai Lin and Skip Fennel), presentation of high-quality research reports from TSG participants, and discussion of directions

---

Co-chairs: Keiko Hino, Gabriel J. Stylianides.
Team members: Katja Eilerts, Caroline Lajoie, David Pugalee.

K. Hino (✉)
Utsunomiya University, Utsunomiya, Japan
e-mail: khino@cc.utsunomiya-u.ac.jp

G.J. Stylianides
University of Oxford, Oxford, UK
e-mail: gabriel.stylianides@education.ox.ac.uk

for future research. In discussing the findings of research studies that took place in different countries, the TSG participants also had an opportunity to learn about practices used around the world in relation to the mathematics education of pre-service primary teachers such as similarities and differences in the formal mathematics education of teachers, types and routes of teacher education, and factors that can influence similarities or differences.

Associated with the TSG there were in total 19 regular presentations (8-page papers), 29 oral communications (4-page papers), and 18 posters. The regular presentations (8-page papers) were organized around five themes as described below. Although several presentations (and associated papers) addressed issues that spanned several themes, practical considerations related to the organization of the TSG sessions during the conference necessitated a best-fit approach.

Theme 1: Mathematics-Content and Mathematics-Specific Pedagogical Preparation

This theme is about the mathematical and pedagogical aspects of teachers' preparation in teacher education. The following presentations were offered under this theme:

- Using mathematics-pedagogy tasks to facilitate professional growth of elementary pre-service teachers (Fou-Lai Lin and Hui-Yu Hsu)
- Investigating the relationship between prospective elementary teachers' math-specific knowledge domains (Roland Pilous, Timo Leuders, and Christian Rüede)
- A self-study of integrating computer technology in a geometry course for prospective elementary teachers (Jane-Jane Lo)
- Pre-service elementary teachers generation of multiple representations to word problems involving proportions (Ryan Fox)

Papers emphasized the importance of pedagogy focused tasks to promote professional growth (Lin and Hsu). Papers focusing on content explored math-specific domains (Pilous et al.), specific areas such as proportional reasoning (Fox), and the role of computer technology in geometry (Lo).

Theme 2: Activities and Assessment Tools Used in Mathematics Teacher Education Programs

This theme is about activities and tools for assessing prospective teachers' knowledge or skills used in mathematics teacher education programs. The following presentations were offered under this theme:

- Preparing elementary school teachers of mathematics: A continuing challenge (Skip Fennell)
- Designing non-routine mathematical problems as a challenge for high-performing prospective teachers (Marjolein Kool and Ronald Keijzer)
- Preservice teachers' procedural and conceptual understanding of fractions (Eda Vula and Jeta Kingji-Kastrati)
- Appraising the skills for eliciting student thinking that preservice teachers bring to teacher education (Meghan Shaughnessy and Timothy Boerst)

Fennell presented current and emerging challenges related to elementary education programs in the United States. Effective characteristics of the learning environments were found through prospective teachers' activities of designing non-routine mathematical problems (Kool & Keijzer). Through various assessment tools, certain aspects of teachers' content knowledge or teaching skills were found to be in need of more stimuli or to be built in teacher preparation programs (Vula & Kingji-Kastrati, and Shaughnessy & Boerst).

Theme 3: Mathematical Knowledge for Teaching and Beliefs

This theme is about the mathematical knowledge that teachers need for their work and about teachers' beliefs and how the might affect teaching practice. The following presentations were offered under this theme:

- A study of prospective primary teachers' argumentation in terms of mathematical knowledge for teaching and evaluation (Yusuke Shinno, Tomoko Yanagimoto, Katsuhiro Uno)'
- Image vignettes to measure prospective teachers' beliefs about mathematics teaching and learning (Stephanie Schuler, Gerald Wittman)
- The mathematics background and mathematics self-efficacy perceptions of pre-service primary school teachers (Gonul Gunes)
- Developing together: measuring prospective teachers' intertwined, topic specific knowledge and beliefs (Erik Jacobson, Fetiye Aydeniz, Mark Creager, Michael Daiga, Erol Uzan)

The sessions provided multiple perspectives on the mathematical knowledge and beliefs for teaching at the primary level. Participants explored self-efficacy as it relates to pre-service teachers' mathematics background. Measures of knowledge and beliefs were the focus of several sessions including the use of image vignettes and teachers' beliefs for topic specific knowledge. Mathematical knowledge for teaching and evaluation was explored in terms of argumentation.

Theme 4: Experiences in Mathematics Classrooms/Teacher Educators' Knowledge for Teaching

This theme is about prospective teachers' experiences in mathematics classrooms or opinions about the learning opportunities for teaching to diverse students in the teacher education programs, and the work of mathematics teacher educators. The following presentations were offered under this theme:

- Preservice mathematics teachers' gains for teaching diverse students (Derya Çelik, Serhat Aydın, Zeynep Medine Özmen, Kadir Gürsoy, Duygu Taşkın, Mustafa Güler, Gökay Açıkyıldız, Gönül Güneş, Ramazan Gürbüz, and Osman Birgin)
- The day will come when I will think this is fun: First-year pre-service teachers' reflections on becoming mathematics teachers (Elisabeta Eriksen, Yvette Solomon, Camilla Rodal, Bjørn Smestad, and Annette Hessen Bjerke)
- Learning and teaching with teacher candidates: An action research for modeling and building faculty school cooperation (Oğuzhan Doğan and Hülya Kılıç)
- Understanding the work of mathematics teacher educators: A knowledge in practice perspective (Wenjuan Li and Alison Castro Superfine)

Faculty cooperation and near peers were shown as playing vital roles to make rich field experience (Doğan & Kılıç, and Eriksen et al.). The prospective teachers' learning opportunities for teaching to diverse students were found not homogenous even within a country (Çelik et al.). Four practices by the teacher educators were identified as they connect preservice teachers' learning to the practice of teaching mathematics to students (Li & Superfine).

Theme 5: Developing Ability to Notice

This theme is about prospective teachers' developing their ability to notice. The following presentations were offered under this theme:

- Learning to act in-the-moment: prospective elementary teachers' roleplaying on numbers (Caroline Lajoie)
- The role of writing narratives in developing pre-service primary teachers noticing (Pere Ivars and Ceneida Fernández)
- Noticing and deciding the "next steps" for teaching: a cross-university study with elementary pre-service teachers (Dittika Gupta, Melissa Soto, Lara Dick, Shawn Broderick and Mollie Appelgate)

The sessions provided multiple perspectives on the ability to notice and the development of that ability amongst prospective teachers. The analysis of a role-play with pre-service primary school teachers involving the use of a calculator has been used to illustrate the complexity of learning to notice and learning to act in-the-moment (Lajoie). Writing narratives have been used as a successful way to help pre-service teachers develop their skill of noticing pupils' mathematical thinking (Ivars and Fernández). Pre-service teachers' skills to recognize, identify and make instructional decisions have been examined in a context in which they were provided with opportunities to engage in noticing practices (Gupta et al.).

Open Access Except where otherwise noted, this chapter is licensed under a Creative Commons Attribution 4.0 International License. To view a copy of this license, visit http://creativecommons.org/licenses/by/4.0/.

Topic Study Group No. 48: Pre-service Mathematics Education of Secondary Teachers

Marilyn Strutchens, Rongjin Huang, Leticia Losano, Despina Potari and Björn Schwarz

The Programme

During Topic Study Group 48 regular sessions, significant new trends and developments in research and practice on the mathematics education of prospective secondary teachers were discussed. An overview of the current state-of-the-art and recent research reports from an international perspective were provided. In keeping with the call for papers, presentations focused on similarities and differences related to the development of mathematics content and pedagogical content knowledge of teachers; models and routes of teacher education, curricula of mathematics teacher education; the development of professional identities as prospective mathematics teachers and a variety of factors that influence these different aspects; field experiences and their impact on prospective secondary mathematics teachers' development of the craft of teaching; the impact of the increasing availability of various technological devices and resources on preparing prospective secondary mathematics teachers; and others.

We received fifty-four 4-page submissions to TSG 48. From the submissions, 20 papers were selected for presentations during the regular meetings. Each of these 20 papers was scheduled in one of the following TSG sessions based on the topic of the paper: (1) Field Experiences (two sessions); (2) Prospective Teachers

---

Co-chairs: Marilyn Strutchens, Rongjin Huang.
Team members: Leticia Losano, Despina Potari, Björn Schwarz.

M. Strutchens (✉)
Auburn University, Auburn, USA
e-mail: strutme@auburn.edu

R. Huang
Middle Tennessee State University, Murfreesboro, USA
e-mail: Rongjin.Huang@mtsu.edu

Knowledge (two sessions); (3) Technologies, Tools and Resources (one session); and (4) Prospective Teachers Professional Identities (one session). Also, we invited four speakers to submit an extended article, one for each of the major themes: Blake E. Peterson and Keith R. Leatham (field experiences), João Pedro da Ponte (teachers' knowledge), Rose Zbiek (tools, technologies, and resources), and Márcia Cristina de Costa Trindade Cyrino (teachers' identities). Thus, 24 papers were scheduled for the TSG 48 regular meetings. We had 3 cancellations. Below is a list of the regular session presenters and the titles of their presentations:

Field Experiences

- Peterson, Blake E. & Leatham, Keith R.; The Structure of Student Teaching Can Change the Focus to Students' Mathematical Thinking.
- Martin, W. Gary & Strutchens, Marilyn E.; Transforming Secondary Mathematics Teacher Preparation via a Networked Improvement Community.
- Akcay, Ahmet Oguz; Boston, Melissa; An Examination of Pre-Service Mathematics Teachers' Integration of Technology into Instructional Activities.
- Potari, Despina & Psycharis, Giorgos; Prospective Mathematics Teachers' Argumentation While Interpreting Classroom Incidents.
- Kilic, Hulya; Pre-Service Teachers' Reflection on Their Teaching.
- Heinrich, Matthias; Consequences from the Learning Level of Students for the Lesson Planning in Mathematics.
- Jackson, Christa DeAnn & Mohr-Schroeder, Magaret; Increasing Stem Literacy via an Informal Learning Environment.
- Losano, Leticia & Villarreal, Mónica: Prospective Teachers Working Together Before and During Their First Teaching Practices.

Teachers knowledge

- da Ponte, Joao Pedro; Lesson Studies in Preservice Teacher Education.
- Arnal-Bailera, Alberto; Cid, Eva; Muñoz-Escolano, José M.; & Oller-Marcén, Antonio M.; Marking Mathematics Exams as a Tool for Secondary Teacher Training.
- Juhász, Péter; Kiss, Anna; Matsuura, Ryota; & Szász, Réka Judit; Developing Teacher Knowledge in Preservice Teachers through Problem Solving and Reflection.
- Lin, Fou-Lai; Yang, Kai-Lin; & Chang, Yu-Ping; Designing a Competence-Based Entry Course for Prospective Secondary Mathematics Teachers.
- Manouchehri, Azita; Infusing Mathematical Modeling in Teacher Preparation: Challenges and Outcomes.
- Chang, Yu-Ping & Yang, Kai-Lin; Apos Theory Applied to Identify Key Challenges for Improving Prospective Mathematics Teachers' Teaching.
- Park, Jung Sook; Oh, Kukhwan; & Kwon, Oh Nam; An Exploratory Study on the Prospective Teachers' Lesson of Analyzing Math Textbooks.
- Olmez, Ibrahim Burak; Izsak, Andrew; & Beckmann, Sybilla; Future Teachers' Use of Multiplication and Fractions When Expressing Proportional Relationships.

Technologies, tools and resources

- Zbiek, Rose Mary; Framing Secondary Mathematics Teacher Understanding.
- Moreno, Mar; & Llinares, Salvador; Prospective Secondary Mathematics Teachers' Perspectives about the Use of Technology for Supporting the Maths Learning.
- Wu, Yingkang, Promoting Pre-Service Secondary Mathematics Teachers' Learning to Teach Mathematics: A Video-Based Approach.

Teachers professional identities

- Cyrino, Márcia Cristina de Costa Trindade, Teacher Professional Identity Construction in Pre-Service Mathematics Teacher Education: Analyzing a Multimedia Case.
- Hine, Gregory Stephen Colin, Exploring Pre-Service Teachers' Self-Perceptions of Readiness to Teach Mathematics.
- Durandt, Rina; & Jacobs, Gerrie J, Pre-Service Teachers' Attitudes towards Mathematical Modelling.

Next highlights from the regular TSG 48 sessions are presented. A major theme discussed in the field experiences sessions is the notion of creating opportunities for prospective secondary mathematics teachers (PSMTs) to learn effective strategies for teaching students and simultaneously ensure that their students are developing the mathematical skills and knowledge that they need. Three presentations (Peterson & Leatham; Martin & Strutchens; Losano & Villarreal) focused on structuring field experiences in such a manner that more than one teacher candidate is placed with the same mentor teacher to capitalize on collective planning, teaching, and monitoring of student growth, as well as fostering student-focused teaching strategies for the teacher candidates. Other papers (Potari & Psycharis; Heinrich) focused on teacher candidates noticing students' actions and asking the appropriate questions or redirecting instruction to better meet students' needs. Jackson and Mohr-Schroeder helped participants to think about field experiences that go beyond the regular mathematics classroom and that help teacher candidates to connect mathematics to other STEM areas. Akcay and Boston presented a study that determined pre-service teachers' ability to integrate technological tools into instructional activities and showcase portfolios in mathematics in ways that support students' high-level thinking and reasoning. Participants were also happy to support Hulya Kilic who had to present her talk virtually, since academic travels abroad were forbidden by the Turkish government during the time of the conference.

Within the PSMTs' teacher knowledge sessions, 'structures in mathematics teacher education that support the development of PSMTs' knowledge' was a major theme. Ponte reviewed studies which focused on lesson study in teacher education and addressed emerging challenges. He described strategies that have been used in teacher education, such as microteaching, reflection in oral and written-form, and face-to-face and digital context for planning and reflecting. He also discussed the constrains that exist in using lesson study in initial teacher education. Arnal-Bailera et al. discussed how grading mathematics exams can be embedded in PSMTs' programs in ways that promote reflection and provide opportunities for professional learning. Park, Oh, and Kwon addressed the analysis of textbooks as an approach in teacher education to bring PSMTs closer to the curriculum and to their future work as teachers. Lin, Yang, and Chang discussed a course in which PSMTs had opportunities to understand students' mathematical thinking, cultivate the competencies of exploration and practice, and develop positive beliefs towards mathematics teaching and learning.

In the second session on PSMTs' knowledge, four presentations focused on teacher education practices related to specific mathematical content areas and processes. Manouchechri's study prepared PSMTs to become familiar with mathematical modelling. The study shows that teachers experienced difficulty enacting the mathematical practice of mathematical modelling. Lin and his colleagues examined PSMTs' teaching of mathematical induction in the context of their field experiences. Juhász et al. focused on the development of PSMTs content knowledge and pedagogical content knowledge in a teacher education course based on problem solving. Finally, Ölmez, Izsák, and Beckmann pointed out that the quantitative definition for multiplication is linked to PSMTs capacity to visualize the relationships between the multiplier and multiplicand in strip diagrams. Coincidentally, the research threads discussed in PSMTs' knowledge reflected the same emphases as some of the studies found in the topical survey that the group published (Strutchens et al., 2016).

Three presentations were given in the technology focused session. Zbiek discussed the conceptual tools for framing secondary mathematics teacher preparation and technology use. She argued that TPACK (Technological, Pedagogical, And Content Knowledge), long used as a framework for knowledge and recently proposed as an orientation towards technology use, is productively enriched by elaboration. Within this frame, she further illustrated conceptual tools for framing technology, content, and pedagogy. She suggested that PSMTs should encounter multiple forms of technology in all venues of their preparation: content courses, pedagogy course, and practical experiences. Wu examined how PSMTs learn to teach mathematical concepts via a video-based approach. The participants' reflection reports documented how they learned from an expert teacher, their peers, and self-reflections. Moreno and Llinares shared PSMTs' perspectives on the use of technology for supporting mathematics learning. They found that PSMTs perspectives on the use of technology were defined by the way in which technological resources were used and the nature of the mathematical activity.

Two papers were presented at the session devoted to PSMTs' professional identity. They revolved around two notions related to identity. The first one is PSMTs' *self-perceptions* and was employed by Hine in his analysis of how PSMTs understand and perceive their "readiness" to teach mathematics based on their pre-service education. The second one, employed by Durandt and Jacobs, is PSMTs' *attitudes*. The authors investigated PSMTs' attitudes towards modeling based on their initial exposure to a model-eliciting task. Although invited speaker, Márcia Cyrino, was unable to attend the congress due to force majeure circumstances, her article[1] introduces new aspects for the notion of identity. Particularly, her work highlights that identity is related to *agency* and *vulnerability*. One of the issues discussed during the session was the following: What are the theoretical links between the notions of belief, attitude, conception, emotion, agency and identity? Furthermore, participants agreed on the importance of reflecting on methodological tools for collecting data—interviews, narratives, surveys, field notes, etc.—best suited for capturing not only the PSMTs' self-perceptions, attitudes and professional identity but also its development.

Overall, TSG 48 sessions were well received and attended. Participants were intrigued that the countries had so many issues in common around prospective secondary mathematics teacher education.

Reference

Strutchens, M., Huang, R., Locano, L., Potari, D., Ponte, J. P., Cyrino, M. C., et al. (2016). *The mathematics education of prospective secondary teachers around the world*. ICME-13 Topical Surveys. Berlin: Springer.

Open Access Except where otherwise noted, this chapter is licensed under a Creative Commons Attribution 4.0 International License. To view a copy of this license, visit http://creativecommons.org/licenses/by/4.0/.

---

[1]Cyrino, M. C. T. Teacher professional identity construction in pre-service mathematics teacher education: analysing a multimedia case.

Topic Study Group No. 49: In-Service Education and Professional Development of Primary Mathematics Teachers

Akihiko Takahashi, Leonor Varas, Toshiakira Fujii, Kim Ramatlapana and Christoph Selter

The Program

Professional development is a never-ending pursuit for a teacher. This is why teacher and teacher educators are involved in the learning process throughout their entire professional lives. The complexity of mathematical teaching practices means in-service teachers face many challenges, such as the demands of new curricula, the introduction of new technologies in the classroom, and the adaptation of teaching practices for students with different abilities and in varying contexts. To address these challenges in each country's agenda for 21st century education, teachers' professional development should link the intended curriculum with students' success.

Topic Study Group (TSG) 49 discussed not only the experience and approaches of effective in-service teacher education and professional development of primary school mathematics teachers, but also contributed to building up a comprehensive overview of the state of the art, the impact of extended and lasting policies (e.g., accountability systems, standardized testing of educational outcomes, etc.), new categories and emphasis introduced by educational researchers (e.g., pedagogical

---

Co-chairs: Akihiko Takahashi, Leonor Varas.
Team members: Toshiakira Fujii, Kim Ramatlapana, Christoph Selter.

A. Takahashi (✉)
DePaul University, Chicago, USA
e-mail: atakahas@depaul.edu

L. Varas
Universidad de Chile, Santiago, Chile
e-mail: mlvaras@dim.uchile.cl

content knowledge, mathematical knowledge for teaching, mathematical competencies, etc.), and current discussions on the available evidence for effective professional development programs and the need for related specific research.

To achieve this goal, researchers were invited to contribute to TSG 49. 50 papers were kindly submitted. The authors of 20 papers were invited to present at TSG 49s themed sessions. The authors of 16 papers were invited to present at the oral communication sessions. In addition, the authors of 14 papers were invited to present at the poster sessions. The following is a list of the papers presented at TSG 49s themed sessions and oral communication sessions:

TSG Session 1: Lesson Studies

- Clivaz, S., & Ní Shúilleabháin, A. Developing Mathematical Knowledge for Teaching in Lesson Study: Propositions for a Theoretical Framework.
- Fujii, T. Lesson Planning in Japanese Elementary School Lesson Study.
- Takahashi, A. Collaborative Lesson Research (CLR).

TSG Session 2: Developing Models of Efficient In-Service Teacher Training

- Cobbs, G. A., Chamblee, G., Luebeck, J. Enhancing In-Service Elementary Mathematics Teachers' Content Knowledge: A Discussion of Two U.S. MSP Projects.
- Kimmins, D., Huang, R., Winters, J., Hartland, K. In-Service Teachers' Perceptions and Interpretations of a Learning Trajectory: Division of Fractions.
- Loh, M. Y., & Seto, C. Mentoring and Mathematics Teacher Noticing: Enhancing Teacher Knowledge.
- Panorkou, N., & Kobrin, J. L. Enhancing Teachers' Formative Assessment Practices: Using Learning Trajectories in Professional Development.
- Peri, A., Espinoza, C. G., Darragh, L. Questions and Quality of Classroom Instruction of Math After a Professional Development.
- Venkat, H., Askew, M., Abdulhamid, L., Morrison, S., Ramatlapana, K. A Mediational Approach to Expanding In-Service Primary Teachers' Mathematical Discourse in Instruction.

TSG Session 3: New Relations and Partnership Experiences in Teachers' Professional Development

- Ader, E. Investigating Classroom Teachers' Development of Quality of Implementation of Mathematical Tasks.

- Amador, J. M., Bennett, C. A., Avila, C. Understanding Rural Teachers' Perceived Needs and Challenges in Creating Rich Learning Environments.
- Glanfield, F. A., Mgombelo, J., Simmt, E., Binde, A. Primary Mathematics Teacher Development in Rural Communities: Lessons Learned From an International Research Partnership.
- Guiñez, F., & Martínez, S. A B-Learning Approach to Developing Mathematical Knowledge for Teaching for In-Service Primary School Teachers.
- Martínez, M. V., & Varas, L. Identifying Elements of Teachers' Change in a Professional Development Experience.
- Martinez, S., & Varas, L. On The Development of a Collaborative Partnership Model Involving In-Service Teachers and Researchers.
- Nutov, L., & Sriki, A. Teacher and Students as a Collaborative Inquiry Learning Community: A Means for Teachers' Professional Development.

TSG Session 4: Scaling Up Sustainable Interventions

- Kristinsdóttir, J.V. Co-Learning Partnership in Mathematics Teacher In-Service Education.
- Morgan, D. Teaching For Mastery: A Strategy for Improving Attainment in Mathematics in English Primary Schools.
- Schliemann, A. D., Carraher, D. W., Teixidor i Bigas, M. Teacher Development and Student Learning.
- Selter, C. The Pikas Project—Using Knowledge Gained From Implementation, School Development & In-Service Teacher Training Research.

Oral Communication Session 1

- Kaplan, H. A., & Argun, Z. Knowledge for Diagnosing Student Thinking: How it Affects Diagnostic Competence?
- McCoy, L. J. An Experiential Learning Approach to Developing In-Service Elementary Teachers' Content Knowledge for Teaching Mathematics.
- Putra, Z. H., Evaluation of Elementary Teachers' Knowledge on Fraction Multiplication Using Anthropological Theory of the Didactic.
- Wang, D. Probing Into the Ways Teachers Learn Mathematics and its Teaching.

Oral Communication Session 2

- Abdullah, N. A., & Leung, F. Highlighting Teacher's Values in Teaching Primary School Mathematics During Lesson Study Process.
- Hlam, T. L. A Teacher Collective as a Professional Development Approach to Promote Foundation Phase Mathematics Teaching.

- Shanmugam, K., Lim, C. S., Razhi, M. Insights of Lesson Study Process From Malaysian Mathematics Teachers: A Case Study.
- Strom, A. D., Kimani, P., Watkins, L. Amping Up Professional Development Through a Collaborative Community of Learners (CCOL).

Oral Communication Session 3

- Eichholz, L. "Mathe Kompakt"—Design and Evaluation of an In-Service Course for Out-Of-Field Mathematic Teachers.
- Dogbey, J. Reforming Elementary School Mathematics Instruction Through Classroom Discourse and Cooperative Learning.
- Pasquali, G. G. The Impact of the Mathematic Olympiads in Paraguayan Teachers.
- Swai, C. Z., & Binde, A. L. A Study of Primary School Teachers' Beliefs of Pedagogical Strategies in Mathematics Lessons in Tanzania.

Oral Communication Session 4

- Husband, M., Rapke, T., Ruttenberg-Rozen, R. "Yes, And ...": Conceptualizing and Characterizing Authority as Fluid in Professional Learning Communities.
- Justo, J. C. R., Da Silva Rebelo, K., Borga, M. F., Dos Santos, J. F., Echeveste, S. S. In-Service Education of Primary Mathematics Teachers With Focus on Problem Solving.
- Karunakaran, M. S., Adams, A. E., Wnek, B., Blackham, V., Klosterman, P., Knott, L., Ely, R. Making Mathematical Reasoning Explicit: Responsive PD.
- Quiroz Rivera, S., Castañeda, E., Rodríguez, R. Less Theory and More Practice: How to Design a Lesson Based in Mathematical Modeling?
- Based on the above paper presentations during the ICME-13, TSG 49, we had a fruitful discussion about important issues for effective in-service professional development for primary mathematics teachers. Although the sessions were filled with innovative ideas, the discussion did not end with a consensus on what would be the essentials to make a vital impact on both student learning and teacher development in the field of mathematics. We are all looking forward to continue our discussion to improve the quality of in-service professional development programs.

Open Access Except where otherwise noted, this chapter is licensed under a Creative Commons Attribution 4.0 International License. To view a copy of this license, visit http://creativecommons.org/licenses/by/4.0/.

Topic Study Group No. 50: In-Service Education, and Professional Development of Secondary Mathematics Teachers

Jill Adler, Yudong Yang, Hilda Borko, Konrad Krainer and Sitti Patahuddin

The Programme

The aim of TSG 50 at ICME-13 was to share, discuss and advance knowledge and understanding of key aspects of research, policy and practice in the in-service education and professional development of secondary mathematics teachers. In TSG 50, there were a total of 97 contributions (19 Paper presentations, 49 Oral presentations and 29 Poster presentations) and well over 100 participants given this total does not include the TSG organizing team, and some paper co-authors.

During ICME-13 conference, the work of TSG 50 was organized into four main TSG Sessions, and these were supplemented by oral communication and poster sessions. The four main sessions each focused on key questions. **Session 1** reported and reflected on A Survey of the Field of Research on Mathematics Professional Development. The session was chaired by Jill Adler, and focused on two key questions: **Q1**. What key research questions, theories and methods are used to study in-service education and professional development of secondary mathematics teachers? **Q2**. What are the accumulating results of research? What do we still not know or understand? What are the implications for policy and practice and across contexts?

Co-chairs: Jill Adler, Yudong Yang.
Team members: Hilda Borko, Konrad Krainer, Sitti Patahuddin.

J. Adler (✉)
University of the Witwatersrand, Johannesburg, South Africa
e-mail: jill.adler@wits.ac.za

Y. Yang
Shanghai Academy of Educational Sciences, Shanghai, China
e-mail: mathedu@163.com

Hilda Borko, Stanford University, USA, presented a "stimulus paper" entitled *Research on Mathematics Professional Development.* This paper is a review of research in the field, and Hilda's presentation was followed by reflective comments from each of the remaining TSG 50 Team Members. Konrad Krainer, reflected on the format of the study, in particular focusing on the research process, its strengths and absences. Yudong Yang, drew attention to the Asian perspective and specifically to important work that has been done on Lesson Study, an aspect of the field not dealt with directly in the review. Sitti Patahuddin emphasized the ICT perspective of professional development for secondary mathematics teachers. Jill Adler pointed out that research in the field of education development was not incorporated in the review, and was important to consider as there was evidence of failure in many developing contexts of professional development initiatives focused on reform, and as Chair of the session concluded with a short summary of TSG 50's first session, and introduction to the remaining sessions of the TSG for all participants.

The second main session focused on the question: **Q3**. What have we learned about secondary mathematics teachers' learning through participation in PD, related to beliefs, knowledge and practice? There were two parallel sessions, one chaired by Sitti Patahuddin, focused on teacher learning in relation to their knowledge, beliefs and practice; the other was chaired by Yudong Yang, and focused on teacher educator/facilitator/coach learning. The presentations' titles and their authors are the following, the first four in the teacher learning session and the remaining four in the teacher educator learning session:

- Enhancing reflective skills of secondary mathematics teachers via video-based peer discussions: a cross-cultural story (Karsenty, Ronnie; Schwarts, Gil).
- The challenges of upgrading mathematics teachers: a case study from one developing country (Kazunga, Cathrine; Bansilal, Sarah).
- Teachers noticing students' potentials while analysing video clips (Schnell, Susanne).
- Development of mathematical knowledge for teaching of mathematics teachers in lesson analysis process (Baki, Müjgan).
- Fostering an intimate interplay between research and practice: Danish "maths counsellors" for upper secondary school (Jankvist, Uffe Thomas; Niss, Mogens).
- Opportunities for learning of secondary math teacher leaders in the context of a video club (Kobiela, Marta; Savard, Annie; Merovitz, Scosha; Chandrasekhar, Vandana).
- District coaches facilitating teachers' use of inquiry-oriented math textbooks: a professional development design study (Boufi, Ada).
- The role of facilitator feedback in shaping teacher attention and response to student thinking (Glennie, Corinne Rose; Brizuela, Bárbara).

The third main session focused on the question: **Q4**. What innovative professional development programs with/without ICT for in-service mathematics teachers at secondary level have been developed and implemented in different cultural contexts? What is the role of professional learning communities in these programs, including online communities? Here too there were two parallel sessions, chaired respectively by Jill Adler and Hilda Borko. One focused on a range of Professional Development programs including a specific focus on Learning and Lesson Study initiatives; and the other focused on those with a specific ICT focus. The presentations' titles and their authors are as following, with the last three on ICT:

- Learning study and the idea of variation and critical aspects of learning (Runesson, Ulla).
- Supporting teachers in ambitious mathematics teaching (Ronda, Erlina).
- (In)visible theory in mathematics teacher education (Österling, Lisa).
- Developing mathematical identity and 'understanding mathematics in depth': conceptions of secondary mathematics teachers (Stevenson, Mary).
- Pedagogical explorations integrated with practical experiences transforming teachers' knowledge (Niess, Margaret Louise; Gillow-Wiles, Henry).
- An online course for inservice mathematics teachers at secondary level about mathematical modelling (Bosch, Marianna; Barquero, Berta; Romo, Avenilde).
- Virtual ethnographic intervention through facebook group: a case study in a disadvantaged context (Patahuddin, Sitti; Lowrie, Tom).

The fourth main session (chaired by Konrad Krainer and concluded by Jill Adler) focused on the question: **Q5**. What do we know about the impact, sustainability and scalability of PD programs? What are the various ways in which sustainability, scalability and effectiveness are defined and then assessed? The presentations' titles and their authors are as following:

- Attending to context when designing mathematics professional development with scale in mind (Smith, Thomas M.; Borko, Hilda; Sztajn, Paola).
- Researching the sustainability of professional development programmes (Zehetmeier, Stefan).
- Improving teachers' mathematical content knowledge and the impact on learner attainment (Pournara, Craig).
- Transformative cascade model for mathematics teacher professional development (Lin, Fou-Lai; Yang, Kai-Lin; Wang, Ting-Ying).

Supplementing these main TSG sessions, were six parallel Oral Communications sessions where 49 papers were presented and discussed; and two sessions for 29 Poster papers (due to the pages' limitation, the titles and presenters of Oral Communication and Poster will not be listed here). It is also not possible to communicate the richness of research and ideas presented and discussed. We nevertheless mention some exemplary aspects that were highlighted.

In the review of the field presented in session 1 most studies examined primarily utilized qualitative methods, which can offer detailed accounts of PD models and rich descriptions of teachers' experiences in the PD while providing proof of concept that the PD design can effectively support teacher learning. Opportunities to analyze excerpts of classroom video or mathematical tasks were provided as examples of contexts and practices promoting learning in professional communities. Knowledge of student mathematical thinking was drawn attention to as critical for teacher learning. In addition, PD facilitators' learning was identified as a new focus of research in the field. These studies demonstrated the continued value of foundational, exploratory, development, and design research for growing the knowledge base on mathematics PD. Still, very few studies investigated causal links between participation in PD and teacher or student outcomes. In addition research that examines what it takes to bring these programs to scale on the one hand, and that compares the effectiveness of different mathematics professional development programs on the other, needs further development.

Secondary mathematics teachers' professional development programs in specific contexts were introduced and discussed in TSG 50. The analysis of sustainable effects is crucial for short-term and long-term programs. The way in which sustainable impacts were researched, the theoretical models and empirical findings were discussed. More evaluation and analysis of sustainable impacts are needed in the future, along with sustainability in teacher education.

The use of technology in mathematics professional development was also a popular topic in TSG 50. Here foci were on the use of technology as interaction tools in and for mathematics teachers' professional development, as it provides resources and platforms for teachers at scale; and on digital technology as an interactive way to support mathematics teachers' learning that has great potential prospect. However, research on the effectiveness of using technology as a mode of interactions is limited. Further research is also needed to understand the possible challenges of the transformation in technology-based programs, such as the design of digital PD to match with the context and needs of teachers, the roles of the facilitators, the specialized pedagogy to conduct digital mathematics professional development, and the utilization of big data in understanding digital PD effectiveness.

In conclusion, TSG 50 was a learning space where mathematics professional development programs were overviewed theoretically, using different analytical frameworks, as well as empirically. New programs for secondary mathematics teachers' professional development in specific context were shared, including some attention to Lesson study, though less so in the Asian context; and all of these with or without information and communication technology. Presentations explored teacher educators' role and their knowledge and skills promotion within diverse backgrounds; and the impact of sustainability of professional development. Others introduced technology-based professional development programs. Some presenters put forward more research questions, including the prospective of using big data to better understand the effectiveness of teachers' PD. However, we are still short of evidence-based studies and of tools for measuring and evaluating teachers' learning. These areas remain a challenge for researchers in this field.

Open Access Except where otherwise noted, this chapter is licensed under a Creative Commons Attribution 4.0 International License. To view a copy of this license, visit http://creativecommons.org/licenses/by/4.0/.

Topic Study Group No. 51: Diversity of Theories in Mathematics Education

Tommy Dreyfus, Anna Sierpinska, Stefan Halverscheid, Steve Lerman and Takeshi Miyakawa

The Programme

Session I

- Presentation 1: Anna Sfard: On the need of theory of mathematics learning and the promise of 'commognition'.
- Presentation 2: Cristina Frade: The social construction of mathematics teachers' identity: Rorty's pragmatistic perspective.

Session II

- Presentation 3: Ricardo Cantoral: Origins and evolution of the socioepistemological program in mathematics education.
- Presentation 4: Carolina Tamayo Osorio and Antonio Miguel: Wittgensteinian 'therapeutic couch' and indigenous experience in (mathematics) education.
- Presentation 5: Higinio Dominguez: Reciprocal noticing in mathematics classrooms with non-dominant students.

Session III

- Presentation 6: Yasuhiro Sekiguchi: Theories and traditions: Tensions between mathematics teaching practices and a recent school reform in Japan.

Co-chairs: Tommy Dreyfus, Anna Sierpinska.
Team members: Stefan Halverscheid, Stephen Lerman, Takeshi Miyakawa.

T. Dreyfus (✉)
Tel Aviv University, Tel Aviv, Israel
e-mail: tommyd@post.tau.ac.il

A. Sierpinska
Concordia University, Montreal, Canada
e-mail: Anna.sierpinska@concordia.ca

- Presentation 7: Verena Rembowski: Semiotic and philosophical-psychological aspects of concept formation.
- Presentation 8: Stefan Halverscheid: An example for interdisciplinary networking of theories for the design of modeling tasks: A case study on ethical dilemmas.

Session IV

- Presentation 9: Michèle Artigue: The challenging diversity of theories in mathematics education.
- Discussion: What have we learned?

Summary

In the closing discussion we have asked what were the implied or explicit ideas about some of the fundamental questions about mathematics, its teaching and its learning, mentioned in the abstract of TSG51? We look at some of these questions in turn.

What is Mathematics?

Anna Sfard's response was that mathematics is a collection of collections of stories about different things (shapes, numbers, sets, functions, …) told using specialized discourses characterized by (1) keywords and their uses (e.g., number, function, limit, derivative, …); (2) visual mediators (sign systems) intended to clarify what the particular story is about (e.g., positional number systems, functional notation, graphs, …); (3) routine actions of the storytellers (abstraction, generalization, deduction, induction, reasoning by contradiction to prove the impossibility of some hypothetical story, testing stories for internal consistency and coherence with other stories, …), and meta-rules (explicit definitions of technical terms, laws of logic, …) (Sfard, 2007).

Ricardo Cantoral said that this is a philosophical question and therefore best left to philosophers. A question for mathematics educators is what are the differences between mathematics as a body of theoretical knowledge and mathematics as a school subject?

In relation to the nature of mathematics, both Anna Sfard and Ricardo Cantoral addressed a question that was not explicitly posed in the abstract of our group: Where do mathematical concepts come from? For Anna Sfard, mathematical concepts come from a feedback loop between practical activities and discourses; for Ricardo Cantoral—from cultural practices, techniques, traditions. It is an important question in mathematics education for, whenever we plan to teach a mathematical concept, we seek the sources of its meaning so that we can construct instructional situations that will help students to construct these meanings for themselves. This question underlies Davydov's concept of "object sources" and Brousseau's concept of "fundamental situations" for particular mathematical notions.

What is Mathematics as a School Subject?

Regarding mathematics as a school subject, there were differences of opinion between Anna Sfard and Ricardo Cantoral. Anna Sfard described school mathematics as a discourse obtained from mathematicians' mathematics by "customization" to the needs and capacities of young learners. She said that it differs from mathematicians' mathematics mostly by the meta-rules, which are less strict and also different in nature.

Ricardo Cantoral, on the other hand, proposed that school mathematics is not only a transposition of scholarly mathematical knowledge. For him, school mathematics is part of the culture of a given society and place; it contains cultural traditions, riddles and games known from popular culture, technology, and other things used in that culture to construct mathematical knowledge in the classroom. School mathematics is culturally situated. There is popular mathematical knowledge, technical mathematical knowledge and scientific mathematical knowledge, and all have to be taken into account when building theories of teaching and learning mathematics at school. There is no hierarchy among the three kinds of knowledge. They are all part of human wisdom (Cantoral, 2013).

What Does it Mean to Teach Mathematics in General or a Particular Mathematical Concept or Process? What Does it Mean to Teach it Well?

According to Ricardo Cantoral, to teach a fundamental mathematical concept (e.g., derivative), we need to find a "cultural basis" for it and help students anchor their understanding of the concept in this cultural basis (e.g., the idea of taste—sweetness, salinity—can be used as a cultural basis for the concept of rate and hence of slope of a linear function).

Higinio Dominguez looked at a more social-interactive aspect of teaching and proposed that to teach well it is necessary to engage in "reciprocal noticing" with students: "In reciprocal noticing, what is noticed is not individual reasoning but rather the emerging and continuous influence of people's reasoning WITH (not *for*) one another ...."

Cristina Frade's perspective was focused on the person of the teacher, but it was not psychological: the question was how a person constructs her or his identity as a teacher. This construction is an important element of becoming a teacher—and therefore, a condition of "teaching well". It is social in nature, by means of language: the individual develops a vocabulary with which to justify their actions, compare their past and present behaviors, and generally narrate their live stories.

A theory of teaching well was the main concern of Yasuhiro Sekiguchi's presentation. The theory of "School as Learning Community" (SLC) was proposed in the 1980s by Manabu Sato and has become the foundation of a school reform in Japan (Saito et al., 2015). Preparation for teaching well is sought through a well-developed system of "lesson study" conferences and developing collaboration between children, teachers, parents, and people in the region.

What Does it Mean to Know Something or a Given Particular Thing in Mathematics (e.g., Fractions)?

According to Anna Sfard, to know something in mathematics means to "extend one's discursive repertoire by individualizing" the discourse in which a particular mathematical story or a collection of such stories is told in school mathematics. "To individualize a discourse means to be able to communicate according to its rules" with others and with oneself. Carolina Tamayo Osorio pointed to a subtle transitional aspect of the process of individualization of a discourse: based on her observations of the Gunadule indigenous Community of Alto Caiman (Colombia) and Wittgenstein's notion of grammar, she proposed that to learn something in mathematics is to extend one's repertoire of language games by constructing a third "border" grammar between one's native grammar and the grammar of school mathematics.

Ricardo Cantoral's take on "knowing something in mathematics" can be summarized as follows: To be able to participate in a certain social (cultural) practice, not only to be able to perform a certain individual intellectual act. This practice may take place in school, but also out of school (e.g., building a log cottage requires knowledge that is technical mathematical knowledge).

Yasuhiro Sekiguchi stressed that in the SLC theory, learning is a collaborative endeavor. For an individual to know something, a whole community must know it; school must be a learning community.

The question, *What does it mean to know something **well***? did not raise much discussion, but it is worth mentioning Anna Sfard's thesis that success and failure in mathematics are elements of a discourse. An individual who participates in that discourse constructs his or her identity as a "success" or a "failure" in mathematics.

Meta-Questions

In her presentation in the last session, Michèle Artigue criticized the questions asked in the abstract from the perspective of the French approach to mathematics education. Some of the questions contained the verb "should" and thus suggested value judgements and a normative point of view that the French didactic culture strongly rejects. Such questions are not scientific: answers to them are not verifiable by scientific means. A scientific question would be: what are the consequences (for the practice of teaching and learning, for example) of such and such normative perspective? Other participants pointed out that in other cultures the construction of mathematics education research as "science" is not necessarily a priority, and that choices made in posing research questions and selecting aspects to consider are inevitably value-laden and guided by more or less explicit answers to the fundamental questions asked in the abstract for our group. So it is useful, for understanding the motives and aims of research, to know the researchers' position on these questions.

The last question debated in the group was,

Why Do We Have So Many Theories in Mathematics Education?

Several hypotheses appeared.

A theory is created or borrowed because it is useful. But usefulness is relative to values and needs (problems?). Since the latter are diverse in mathematics education, a diversity of theories is needed to respond to them. (Anna Sfard)

Teaching and learning of mathematics is culturally situated. Cultures of mathematics education in different populations must be studied empirically. To explain differences between these cultures, existing theories may not be enough. (Ricardo Cantoral)

In Japan, "theory" is always a theory of some practice. A practice develops, somebody notices it, reflects upon it and constructs a theory of this practice. Practices evolve, change; new practices emerge. Hence many theories. (Takeshi Miyakawa)

References

Cantoral, R. (2013). *Teoria Socioepistemológica de la Matemática Educativa. Estudios sobre construccion social del conocimiento*. Barcelona, España: Gedisa.

Saito, E., Murase, M., Tsukui, A., & Yeo, J. (2015). *Lesson study for learning community: A guide to sustainable school reform*. London: Routledge.

Sfard, A. (2007). When the rules of discourse change, but nobody tells you: Making sense of mathematics learning from a commognitive standpoint. *Journal for Learning Sciences, 16*(4), 567–615.

Open Access Except where otherwise noted, this chapter is licensed under a Creative Commons Attribution 4.0 International License. To view a copy of this license, visit http://creativecommons.org/licenses/by/4.0/.

Topic Study Group 52: Empirical Methods and Methodologies

David Clarke, Alan Schoenfeld, Bagele Chilisa, Paul Cobb and Christine Knipping

TSG 52, "Empirical methods and methodologies," was devoted to explorations of common themes and underlying issues in the use of empirical methods and methodologies. The challenge for the group was defined as follows:

> Research in mathematics education employs an extensive range of Methods, Methodologies, and Paradigms (M/M/Ps) in the service of key goals. But which M/M/P combinations help us understand which phenomena, in robust and reliable ways?

The group met for four sessions at ICME-13. The first three sessions were devoted to parallel sessions in which the assembled groups explored methodological issues related to the following six goals central to ongoing research in mathematics education:

1. Improving Mathematics Instruction (instructional materials, strategies, organisation, assessment).
2. Understanding the Learning of Mathematics.
3. Understanding the Teaching of Mathematics (teacher beliefs, knowledge, decision-making and professional development).
4. Classroom Processes and Interactions.
5. Mathematics Education and Social Justice.

Co-chairs: David Clarke, Alan Schoenfeld.
Team members: Bagele Chilisa, Paul Cobb, Christine Knipping.

D. Clarke (✉)
University of Melbourne, Melbourne, Australia
e-mail: d.clarke@unimelb.edu.au

A. Schoenfeld
University of California, Berkeley, CA, USA
e-mail: alans@berkeley.edu

6. Understanding the Role of Culture and Language in Shaping the Teaching and Learning of Mathematics.

These are diverse goals, which might be addressed using research designs that integrate different M/M/P combinations. For each goal, one might be ask: "Suppose you have an hypothesis about this goal. How do you set about evaluating it?" Alternatively, "Suppose you are trying to explain some aspect of individual or group behavior relevant to that goal. How would you characterize and then theorize that behavior?" Or, "How might cultural, historical and political perspectives shape one's understandings of the contingencies related to realizing this particular goal?" One could imagine radically different approaches not only across the six goals, but within each goal. The question, then, was whether the group could, in the final collective discussion, find some degree of coherence in the varied approaches, and produce a larger frame within which the discussions of the six goals could be situated.

That frame emerged as the group considered a broad framing of empirical work in education, as an act of modelling—the idea being that empirical methods and methodologies are, in effect, ways to characterize "real world" phenomena. The challenge, then, was to problematize the enterprise. The challenge was taken up with enthusiasm, which led to some discussions that pushed the boundaries of previous framings.

The group rejected the standard version of the modelling process,

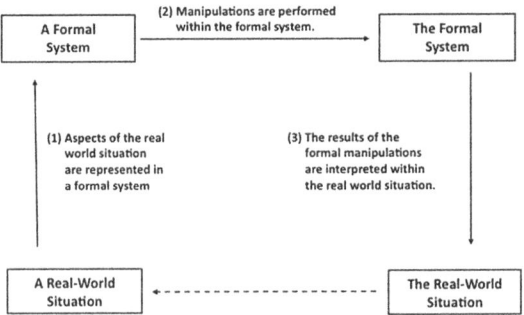

for a more complex one,

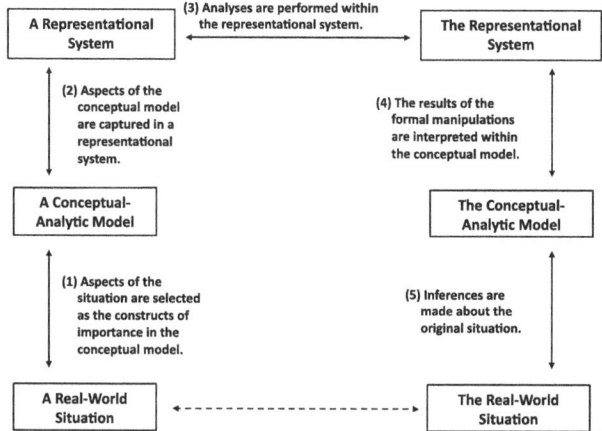

and began to problematize this representation by noting all the ways that empirical work could go wrong—e.g., (arrow 1) for decades of medical research, the "human population" was represented by male experimental subjects, and the results of many studies later turned out not to apply to women. Similarly, (arrow 2) what one decides to consider important in learning makes a big difference—we noted that nations rank differently on TIMSS and PISA, because the two mathematics tests capture different aspects of what might be considered mathematically important. Group members similarly (arrow 3) discussed alternative interpretations of statistical findings—looked at one way, certain results seemed significant, but from another perspective, they were not. Nor were challenges limited to the quantitative realm; questions of how to be confident about interpretations of discourse analyses received attention, as did issues of triangulation and the use of multiple methods. Likewise mapping back from analyses to the "real world" was an issue: the fact that many "significant" laboratory studies did not translate to meaningful learning gains in real classrooms was raised as an issue of concern.

Perhaps the most interesting, and passionate, part of the conversation dealt with what has been called the "political turn" or "socio-political turn" in mathematics education. Members of the group noted that the conceptual analytic models in the second figure often were shaped by tacit social biases—that the "clean" analytic descriptions in the figures could obscure various forms of racism, sexism, gender bias, and more. There was a general concern among the group that the field needs to attend in more explicit ways to the possibility of such bias, and address it in our work. At the same time, there was appreciation for the fact that, although there is a lot more to be done, the field has made tremendous progress since the first ICME in 1969.

Open Access Except where otherwise noted, this chapter is licensed under a Creative Commons Attribution 4.0 International License. To view a copy of this license, visit http://creativecommons.org/licenses/by/4.0/.

Topic Study Group No. 53: Philosophy of Mathematics Education

Paul Ernest, Ladislav Kvasz, Maria Bicudo, Regina Möller and Ole Skovsmose

What is the philosophy of mathematics education? It can be an explicit position that is formulated, reformulated, criticized, refined, etc. But it can also refer to implicit assumptions and priorities, including paradigmatic assumptions that one need not be aware of, but which might be identified through, let us call it, a philosophical archaeology.

The philosophy of any activity comprises its aims or rationale. Thus we ask: what is the purpose of teaching and learning mathematics? An answer explains *why* we engage in these practices and *what* we hope will be achieved. But just considering such purposes quickly leads to seeing the divergence in aims and values of different groups.

A broader view of the philosophy to mathematics education looks at the applications of topics such as epistemology, philosophy of mathematics, ethics and aesthetics. It applies philosophical methods to a critical examination of the assumptions, reasoning and conclusions of mathematics education, systematically enquiring into fundamental questions:

- What is mathematics?
- How does mathematics relate to society?
- Why teach mathematics?
- What is the nature of learning (mathematics)?

Co-chairs: Paul Ernest, Ladislav Kvasz.
Team members: Maria Bicudo, Regina Möller, Ole Skovsmose.

P. Ernest (✉)
University of Exeter, Exeter, UK
e-mail: p.ernest@ex.ac.uk

L. Kvasz
Charles University Prague, Prague, Czech Republic
e-mail: ladislavkvasz@gmail.com

- What is the nature of mathematics teaching?
- What is the significance of information and communication technology in the teaching and learning of mathematics?
- What is the status of mathematics education as knowledge field?

The philosophy of mathematics education matters because it gives people new 'glasses' through which to see the world. It enables people to see beyond official stories about the society, mathematics, and education. It provides thinking tools for questioning the status quo, for seeing 'what is' is not what 'has to be'; enabling us to imagine alternatives possibilities. A preconference overview of the field was published: Ernest et al. (2016), available free from Springer Open.

At the conference, itself the sessions included expert presentations on key questions and issues of the field and was time made available for questions, discussion and participation.

The Tuesday session began with a welcome and an introduction from Paul Ernest and Ladislav Kvasz. Chaired by Ladislav Kvasz and Ole Skovsmose two main presentations were given:

Paul Ernest　An overview philosophy of mathematics education.
　　　　　　Due to personal circumstances the second speaker was unable to attend so the second presentation was:
Paul Ernest　The collateral damage of learning mathematics.

The Wednesday session was chaired by Paul Ernest and Ladislav Kvasz and consisted of two presentations:

Ole Skovsmose　The politics of meaning in mathematics education
Maria Bicudo　Developments in philosophy in/of mathematical education: ontological questions posed by the presence of computers etc.

On the Friday, there were two parallel sessions chaired by Paul Ernest and Ladislav Kvasz. The presentations were as follows:

Jeff Evans and Keiko Yasukawa	Researchers as policy actors? examining the interaction between mathematics education research and PIAAC
Jörn Schnieder and Ingrid Scharlau	Reading mathematical texts with philosophical methods.
Iskra Nunez	Theoretical incompleteness and mathematics education.
Uwe Schürmann	the order of the discourse on modelling.
Nadia Stoyanova Kennedy	Opening a philosophical space in the mathematics curriculum.
Michael Meyer	Concept formation as a rule-based use of words.

Cintia Aparecida Bento Santos & Fernanda Aparecida Ferreira	Possibilities of the phenomenological approach and of philosophical hermeneutics in type search state of art.
Filipe Santos Fernandes	History of scientific and academic production in mathematics education—representation, institution and policy.

On Saturday, the session was chaired by Paul Ernest and Maria Bicudo, and the following talks were given.

Ladislav Kvasz	The language of mathematics in a historical, epistemological, and educational perspective.
Regina Möller	The teaching of velocity in mathematics classes—chances for philosophical ideas.

There was a closing panel discussion with the main presenters Paul Ernest, Ladislav Kvasz, Maria Bicudo, Regina Möller, Ole Skovsmose and the audience. Lastly, the publication strategy of the group was discussed, with two elements. First, there is the publication of all papers presented at TSG 53 including poster presentations, in *Philosophy of Mathematics Education Journal* (2016) number 31, a special issue dedicated to TSG 53 papers. Second, the publication of the best papers rewritten and expanded, plus invited outside expert contributions in a special monograph (Ernest, forthcoming).

References

Ernest, P. (Ed.). (forthcoming). *The philosophy of mathematics education today*. Switzerland: Springer.
Ernest, P., Skovsmose, O., Van Bendegem, J. P., Bicudo, M., Miarka, R., Kvasz, L., et al. (2016). *The philosophy of mathematics education, ICME-13 surveys*. Switzerland: Springer Open. Consulted on 23/01/2017 via URL http://link.springer.com/book/10.1007%2F978-3-319-40569-8
(2016) *The Philosophy of Mathematics Education Journal, 31* (November 2016) Special issue— The philosophy of mathematics education at ICME 13. Consulted on 23/01/2017 via URL http://socialsciences.exeter.ac.uk/education/research/centres/stem/publications/pmej/pome31/index.html

Open Access Except where otherwise noted, this chapter is licensed under a Creative Commons Attribution 4.0 International License. To view a copy of this license, visit http://creativecommons.org/licenses/by/4.0/.

Topic Study Group No. 54: Semiotics in Mathematics Education

Norma Presmeg, Luis Radford, Gert Kadunz, Luis Puig and Wolff-Michael Roth

Aims of the Topical Study Group on Semiotics

The general aim of TSG 54 at ICME-13 was to explore the significance—for research and practice—of semiotics for understanding issues in the teaching and learning of mathematics at all levels. This aim resulted in a pre-conference Topical Survey (Springer, 2016) that addressed the following aspects:

- Nature of semiotics, and its significance for mathematics education;
- Influential theories of semiotics;
- Applications of semiotics in mathematics education;
- Various types of signs in mathematics education;
- Other dimensions of semiotics in mathematics education.

The four regular sessions of the TSG, and the two sessions of the associated Oral Communications, drew both experienced researchers already using semiotics of various forms in their work, and interested participants who were new to these theoretical orientations. Thus the first regular session (with 48 participants) started with a 10-minute introduction by Presmeg and Radford, in order to summarize the structure of the sessions, and the general nature of semiotic theories (the science of signs), especially as these pertain to mathematics education research.

Co-chairs: Norma Presmeg, Luis Radford.
Team members: Gert Kadunz, Luis Puig, Wolff-Michael Roth.

N. Presmeg (✉)
Illinois State University, Bloomington, IL, USA
e-mail: npresmeg@msn.com

L. Radford
Laurentian University, Greater Sudbury, ON, Canada
e-mail: lradford@laurentian.ca

Structure of the Regular 90-Minute Sessions: Day 1

After the introduction, three plenary papers were presented, as follows:

- Luis Radford: *The ethic of semiosis and the classroom constitution of mathematical objects.*
- Adalira Sáenz-Ludlow: *Geometry examples of diagrammatic reasoning.*
- Gert Kadunz: *A matter of translation.*

Radford explored the production of subjectivities in the mathematics classroom as a semiotic problem. To do so, he discussed an example with pre-school children involved in an arithmetic game. Drawing on the late Vygotsky's semiotics, he focused on the manner in which children, through a vast array of embodied and material signs, position themselves as mathematical subjects and make sense of the mathematics and the rules of the arithmetic game.

Drawing on Peirce's semiotics, Sáenz-Ludlow discussed the concept of diagrammatic reasoning. She argued that diagrammatic reasoning is based on abductive, inductive, and deductive reasoning, leading to a deeper understanding of the objects that signs represent. Then mathematical diagrams appear as epistemological tools that, through systematic observation, can lead the students to better understand the structure and interrelation of mathematical concepts in mathematics problem solving and proving.

Kadunz focused on the nature of mathematical objects. To make his ideas explicit, he organized his considerations on the relation between mathematics in school and at university, and on means of translation between these two realms. He argued that a successful translation does not need the assumption of an objectively existing mathematical instance.

Day 2. Again there were three plenary presentations:

- Wolff-Michael Roth: *Birth of signs: From triangular semiotics to communicative fields.*
- Candia Morgan: *Use of social semiotics to explore institutional assumptions.*
- Michael Otte: *Semiotics, epistemology, and mathematical generalization.*

Roth's paper followed a line of theoretical development of the late Vygotsky, who, near the end of his life, envisioned a radical revision of his work on sign operations. Roth showed that a focus on the developing communicative (semiotic) field that is common to participants gives rise to a fruitful rethinking of traditional approaches to the sign and sign operations.

Morgan's presentation revolved around the general principles and use of social semiotics in mathematics education. Through social semiotic lexical analyses of official documents she pointed out some textual patterns that make visible some assumptions about mathematics, teachers, and students. These patterns configure the kinds of agency that are made available for teachers and students (e.g., students are portrayed as "having needs," and as entities in "need of receiving appropriate support").

Otte's presentation started with an overview of the difficulties that semiotics faced before coming to be seriously considered in mathematics education research in the early 2000s. He discussed several key differences between Saussure's and Peirce's semiotics, and, referring to some omnipresent semiotic problems in mathematics (e.g., intension/extension; function/predication, and syntax/semantics), he articulated questions about mathematics generalization from a semiotic viewpoint.

Day 3. The 8 presentations on day 3 were arranged in two *parallel sessions*. The first 4 of these, with a broad focus on gestures and multimodality, were as follows:

- Ulises Alfonso Salinas and Isaías Miranda: *Artifact mediation in the process of objectification.*
- Osama Swidan, Naomi Prusak, and Baruch Schwarz: *Objectifying the hierarchical classification of quadrilaterals in a synchronic-interactive computer environment.*
- Debbie Stott: *Gesturing: A key aspect of mediation for young learners in a South African context?*
- Alexander Salle and Christina Krause: *On the role of gestures for the descriptive analysis of 'Grundvorstellungen': A case of linear functions.*

In this session, the theory of knowledge objectification and (artifact, gesture) mediation constituted a common thread. Salinas et al. reported on the role of artifacts and gestures when high school physics students attempted to solve problems concerning the frame of reference when objects moved across an inclined plane. The results show some of the difficulties that students face when trying to make sense of the Cartesian origin, on the one hand, and the mathematics graph as a co-variational concept, on the other. Using the framework of the multimodality of communication, Stott analyzed the gestures that third-grade students in an after-school mathematics club used in the context of solving a mathematics puzzle. Stott's interest was in understanding the subtle semiotic means to which students resort in order to catch other participants' attention. She examined in fine detail the dynamic structures of various semiotic nodes, the multi-modal form of doing/saying, and what is said and what is done. In both studies, artifacts and gestures provided means for knowledge objectification and, thus, for making sense of difficult situations. Guided by the theory of knowledge objectification, Swidan et al. investigated patterns emerging from student investigations of quadrilaterals within the Virtual Math Teams environment. The study illustrates the students' processes of objectification that underpin the production of a hierarchical classification of different types of quadrilaterals. Based on the idea of semiotic bundles, Salle and Krause analyzed gesture–word relations in contexts where students solved problems involving linear relations. Gestures constituted Petra Menz and Nathalie Sinclair objectifications of *Grundvorstellungen*, basic representations from which mental models are built.

The *second parallel session* had a broad focus of semiotic chaining and semiotic resources:

- José Francisco Gutiérrez: *Exploring tensions in the 'object-subject' dialectic.*
- Nejla Gürefe: *Analysis of semiotic resources used in process of hearing-impaired students' triangle concept explanation.* (Presented in absentia: Turkish presenters could not attend.)

- Barbara Kinach: *Digital visualization tasks for mathematics teacher development: A semiotic chaining analysis.*
- Édith Petitfour: *Teaching geometry to visual-spatial dyspraxic pupils.*

Gutiérrez concentrated on the dialectic of objectification-subjectification when learning mathematics in a social context. He argued that students' semiotic means of objectification sometimes also functions as means of subjectification. In her talk, Kinach observed teacher education through the lens of visualization tasks. She used semiotic chaining as means to analyze innovative visual learning of preservice elementary school teachers. Petitfour presented an approach to the teaching of geometry to students facing difficulties using geometrical instruments with the required precision. She was able to identify and categorize these, and some means of overcoming them.

Day 4. The session started with three presentations, followed by general discussion of issues:

- Petra Menz and Nathalie Sinclair: *Diagramming and gesturing during mathematizing.*
- Yasmine Abtahi: *Semiotic: Signs, tools, and meaning-making.*
- Corin Mathews: *Division means less: Chains of signification in a South African classroom.*

Drawing on new materialism, Menz discussed a classroom episode in which an instructor interacts with a mathematical diagram and makes sense of the mathematics. Abtahi focused on Vygotsky's idea of the ratio object/meaning and its transformation into meaning/object. She illustrated the research question through a mathematical problem in which students resort to concrete materials and the meaning of these objects changes as the classroom activity unfolds. Mathews drew on the idea of chains of signification to show the evolution of the meaning of division in a primary school mathematics classroom.

Oral Communications Associated with TSG 54

- There were 6 of these, arranged in two sessions.
- Uta Priss: *A semiotic-conceptual analysis of conceptual development in learning mathematics.*
- José Luis López and José Guzmán: *Artifacts and gestures in the process of objectification of the concept of variation.*
- Daniela Behrens: *Bundling and de-bundling by dragging: From acting to gesturing.*

In this session, the three presenters invoked theories of semiotic-conceptual analysis based on Peirce's writings (Priss), objectification based on Radford's work (López & Guzmán), and semiotic bundling in a study involving gestures (Behrens). A common thread in the papers was students' evolving change in use and conception of notation and other signs in communication.

- Gloria Inés Sanabria: *Translations between semiotic systems.*
- Nicole Engelke Infante: *Highlighting key links through gesture: A case study of the second derivative test.*
- Anna Shvarts: *Dual eye-tracking as a method to investigate the acquiring of theoretical perception of visual representations.*

Sanabria's presentation featured the problem of translation between semiotic systems from Duval's perspective, invoking discussion about the role of signs and semiotic activity in the production of mathematical objects. Engelke Infante presented a study the goal of which was to determine the effect of the instructor's gestures in the students' understanding of derivatives. Shvarts's presentation dealt with the forms and dynamics of child/adult joint activity through a fine-grained study of perception, oral interventions, and gestures. The use of dual eye-tracking technology facilitated a detailed qualitative analysis of synchronized data. The child not only followed the adult's cultural intervention, but he or she actively coordinated several semiotic registers; understanding appeared to be a result of this coordination at the moments of joint attention at the micro level.

Conclusion

Following the rich discussion of issues that arose from the presentations, a Springer Monograph is being planned that will allow most of the authors to strengthen and expand their papers in the coming months.

Open Access Except where otherwise noted, this chapter is licensed under a Creative Commons Attribution 4.0 International License. To view a copy of this license, visit http://creativecommons.org/licenses/by/4.0/.

Part VII
Reports from the Discussion Groups

Classroom Teaching Research for All Students

Shuhua An, Steklács János and Zhonghe Wu

The theme of the Discussion Group was promoting classroom teaching research through the exploration of effective instruction strategies to support mathematics learning for all students, including struggling, special needs, and excelling students, as well as between different educational systems and cultures.

The presentations of the Discussion Group focused on three aspects: (1) Framework of classroom teaching research; (2) Examples of classroom teaching research by scholars and teacher researchers; (3) Technology in classroom teaching research. The Discussion Group discussed the classroom teaching research framework that focuses on linking research to classroom practice, making classroom research applicable and making sense to classroom teachers (Wu, 2016). All presentations have shown classroom teaching research plays a key role in understanding how to support effective mathematics classroom teaching and learning. A case study by Cheng (2016) addressed the features of expert secondary mathematics teachers' verbal presentation in Chinese mathematics classroom as using precise language, transition and variation strategies to help students well understand the clue and the purpose of teaching.

An (2016) introduced a Model-Strategy-Application-Reasoning (MSAR) approach in teaching and learning mathematics, addressed the conceptual framework of the MSAR, demonstrated examples of the effects of the MSAR approach on the third graders' mathematics learning in US, and discussed the benefits of using the MSAR for diverse student groups. Chen, An, Cheng, and Sun's presentation (2016) demonstrated five types of modeling by Chinese 3rd Graders resulting from use of the MSAR for two-digit number multiplication in China: repeated model, area model, set model, and tree diagram model.

The teacher-researcher has a pivotal role in classroom teaching research. A teacher-researcher, Dunia Zeineddine (2016) demonstrated a video lesson that

S. An (✉)
California State University, Long Beach, CA, USA
e-mail: shuhua.an@csulb.edu

showed how to teach geometry to struggling students with limited language skills by incorporating NCTM's Orchestrating Discussion, which promotes understanding of concepts, engages students in mathematical discourse, and teaches the geometric vocabulary in a supportive classroom environment. Siemssen and Sahr (2016) demonstrated how the use of technology that students already own, such as cellphones, can be used to motivate students to learn rigorous mathematics. Middle school students from an economically disadvantaged district in Texas were challenged to investigate new topics in mathematics and use free or low cost technology to create presentations. For this project, the teacher served as a mentor while the students took full control over both the topic of study and the method of presentation. Topics chosen by the students included statistics, trigonometry, calculus, matrices, forensic analysis, and code breaking. Visuals included board games, graphic novels, 2D and 3D models, and videos. Several videos were presented, including a Sci-Fi movie trailer depicting exponential growth, a super hero using geometry to teach others how to dance, and a music video that explored calculus and trigonometry. Using their own technology, the students were able to produce projects that were both creative and showed deep understanding of complex mathematics.

Siemssen and Sahr's example shows that technology is an increasingly important area in classroom teaching research. Lee (2016) discussed a study on using an online teaching approach for college student mathematics learning in US. János (2016) demonstrated how to using Eye Tracking technology and other biometrical tools in learning process observation and classroom teaching research in Hungary. Such international perspectives in classroom teaching research was also demonstrated by Cao (2016) using examples from Chinese mathematics classrooms.

Open Access Except where otherwise noted, this chapter is licensed under a Creative Commons Attribution 4.0 International License. To view a copy of this license, visit http://creativecommons.org/licenses/by/4.0/.

Mathematical Discourse in Instruction in Large Classes

Mike Askew, Ravi K. Subramaniam, Anjum Halai, Erlina Ronda, Hamsa Venkat, Jill Adler and Steve Lerman

Aims and Rationale for the Topic

This discussion group attracted congress participants interested in exchanging ideas and discussing issues and challenges related to mathematics education in schooling systems where large classes (40+ students) are the norm. Large classes are sometimes viewed as limiting students' opportunities to learn, but the fact that being in a large class is the reality for many students, particularly in developing nations, led the organisers of this discussion group to consider the study of teaching large classes, at both primary and secondary levels, to be a worthy object of research and inquiry.

In contexts where not only are class sizes large, but also teaching resources are frequently limited, it is largely the teachers' instructional practices that provide the main point of access to mathematics for the learners. Thus this discussion group team have been investigating large class pedagogies (and in particular pedagogies which are predominantly teacher-centred) with a view to how such pedagogies bring mathematical objects of learning into being in the classroom space. To this end we have been working with an analytical framework for studying *mathematical discourse in instruction*, MDI. MDI is characterised by four interacting components in the teaching of a mathematics lesson: *exemplification, explanatory talk, learner participation* and the *object of learning* (goal).

Aims of the DG sessions include identifying, sharing and discussing common key issues in teaching and learning in large classes and exploring the potential of the MDI framework to examine such issues. Through sharing cross-national and cross-phase experiences we aimed both to broaden the base of lessons that the framework might be applied to and to explore ways in which the framework might be developed. Through participants sharing experiences and research interests we hope to explore the potential for future collaborations.

Key Questions and Issues

Key questions to discussed included:

(1) What are the key issues that need to be researched in teaching large classes?
(2) What mathematics do teachers of large classes make available to learn?
(3) What is the role of examples and representation in large class teaching?
(4) What forms of learner participation are made possible in large classes?
(5) How does the quality and nature of teacher explanatory talk vary and how does this variation affect the mathematics made available to learners?
(6) How helpful is the framework of Mathematics Discourse of Instruction in study pedagogies of large class teaching?
(7) How might the framework be developed?
(8) What is the potential for future research and development and cross-national collaborations?

Open Access Except where otherwise noted, this chapter is licensed under a Creative Commons Attribution 4.0 International License. To view a copy of this license, visit http://creativecommons.org/licenses/by/4.0/.

Sharing Experiences About the Capacity and Network Projects Initiated by ICMI

Angelina Matinde Bijura, Alphonse Uworwabayeho, Veronica Sarungi, Peter Kajoro and Anjum Halai

The Capacity and Network Project (CANP) is a development project of the International Commission of Mathematical Instruction (ICMI) supported by the International Mathematical Union (IMU), UNESCO and the International Council of Scientific Unions (ICSU) as well as regional governments and institutions. The project is a response to Current Challenges in Basic Mathematics Education (UNESCO, 2011), which includes a call not just for mathematics education for all but also for a mathematics education of quality for all. Five CANPs have come into existence between 2011 and 2016: French West Africa (started in 2011), Central America and the Caribbean (started in 2012), South East Asia (started in 2013), East Africa (started in 2014) and the Andean Region and Paraguay (started in 2016). Each program comprises 4-6 countries. Each CANP workshop has combined plenary sessions (courses and synthesis) and group work (tutorials, workshops and discussion groups). Satellite activities to a wider audience, such as public lectures, were organised. The Discussion Group was an opportunity for all organizers and participants in the five CANPs and ICMI officers to share their experiences with challenges and opportunities in preparing for a CANP event so that they could suggest directions for future CANPs. The Discussions were based around the key questions: What further steps can be taken to support mathematics education in

A.M. Bijura (✉)
Institute for Educational Development, Aga Khan University, Dar es Salaam, Tanzania
e-mail: angelina.bijura@aku.edu

developing countries? How can the five CANP regions and the five CANP networks as well as possible new CANP regions build synergies, be strengthened and get support?

Structure:

Chair: Bill Barton **Co-chair**: Angelina Matinde Bijura

Day 1: Tuesday, 26th July 2016

Time	Topic	Material/working format/presenter
16:30–16:35	Aims and organization of the DG	Angelina Matinde Bijura
16:40–16:50	Welcome remarks	ICMI President Ferdinando Arzarello, Bill Barton and Angelina Matinde Bijura
16:50–17:15	Findings from survey on CANPs	Lena Koch
17:15–17:50	Comments from representatives of each CANP	Participants of CANPs 1–5
17:50–18:00	Closing Day 1	Bill Barton

Day 2: Friday, 29th July 2016

Time	Topic	Material/Working format/Presenter
16:30–16:35	Aim of the session	Angelina Matinde Bijura
16:35–17:20	Interventions guided by the key question	All participants
17:55–18:00	Closing	Bill Barton

Some of the interventions discussed on the second day were: strengthening the cooperation between ICMI/CANP networks with existing cooperation partners and networks, reaching out to policy makers, involving CANP participants in ICMI activities and in particular in ICMI research activities, supporting research activities in developing countries, creating a database about existing networks and organisations in mathematics education worldwide and holding summer schools as a tool for regional development.

Open Access Except where otherwise noted, this chapter is licensed under a Creative Commons Attribution 4.0 International License. To view a copy of this license, visit http://creativecommons.org/licenses/by/4.0/.

Mathematics Teacher Noticing: Expanding the Terrains of This Hidden Skill of Teaching

Ban Heng Choy, Jaguthsing Dindyal, Mi Yeon Lee and Edna O. Schack

Aims and Key Questions

Research on what and how mathematics teachers notice in the classrooms has gathered momentum, with an emphasis on developing noticing expertise in mathematics teachers. This Discussion Group aimed to explore and expand the terrains of research on teacher noticing in three aspects: conceptualizations of noticing, methodologies for studying noticing, and the study of noticing in different contexts. There are three key areas for discussion: (1) conceptualizations of noticing, (2) methodological challenges of noticing, and (3) contexts in which the study of noticing can be situated. During our discussion group, our three invited speakers, John Mason, Sergiy Klymchuk, and Julie Amador, shared their perspectives on the topic to provide an overview of our current understanding of the three set of questions. Their sharing generated many questions for us to think about. In this short report, we will highlight two pertinent questions that may be useful for guiding future research.

Is It Noticing or Is It …?

Most researchers view noticing as a form of professional vision, consisting of three interrelated component skills, attending to, interpreting, and responding (Sherin, Jacobs, & Philipp, 2011), while others see it as a set of practices that work together to improve teachers' sensitivities to enacting a shift in attention to recognize possibilities (Mason, 2002). Each of these conceptualizations can inform us about

---

B.H. Choy
National Institute of Education, Nanyang Technological University,
Singapore, Singapore
e-mail: banheng.choy@nie.edu.sg

various aspects of noticing. However, our discussion uncovered an even wider spectrum of these different conceptualizations, from formative assessment to meta-cognition. However, it is not always helpful to have different conceptualizations of noticing, and more work is needed to achieve better clarity in our conceptualization of noticing.

What Is Studied Regarding Teacher Noticing, and How Is It Studied?

Investigating noticing poses considerable methodological challenges. Current research uses either an in-the-moment approach through utilizing a wearable camera or a retrospective approach through analyzing reflections after the lessons are conducted (van Es, 2011). A few studies, such as Choy (2016), have also adopted a pre-spective look at noticing during task design. Our discussion focused on the grain size or detail of the analysis of teachers' noticing. Even though we could not come to any consensus at the end of the discussion, it was clear that unpacking the "black box" of noticing (Scheiner, 2016, p. 229) would be critical if for advancing the study of teacher noticing. It remains to be seen how future technological advances in capturing what teachers see and think may mitigate the methodological challenges we are facing in this field.

References

Choy, B. H. (2016). Snapshots of mathematics teacher noticing during task design. *Mathematics Education Research Journal, 28*(3), 421–440. doi:10.1007/s13394-016-0173-3
Mason, J. (2002). *Researching your own practice: The discipline of noticing.* London: RoutledgeFalmer.
Scheiner, T. (2016). Teacher noticing: enlightening or blinding? *ZDM Mathematics Education, 48* (1), 227–238.
Sherin, M. G., Jacobs, V. R., & Philipp, R. A. (Eds.). (2011). *Mathematics teacher noticing: Seeing through teachers' eyes.* New York: Routledge.
van Es, E. (2011). A framework for learning to notice students' thinking. In M. G. Sherin, V. R. Jacobs, & R. A. Philipp (Eds.), *Mathematics teacher noticing: Seeing through teachers' eyes* (pp. 134–151). New York: Routledge.

Open Access Except where otherwise noted, this chapter is licensed under a Creative Commons Attribution 4.0 International License. To view a copy of this license, visit http://creativecommons.org/licenses/by/4.0/.

Connections Between Valuing and Values: Exploring Experiences and Rethinking Data Generating Methods

Philip Clarkson, Annica Andersson, Alan Bishop, Penelope Kalogeropoulos and Wee Tiong Seah

What do teacher colleagues learn when they read our research? Do they wonder what it might be like to teach values that they are not sure of? Do our research colleagues wonder whether role-play could be a set of new data collection methods we could use?

In the first session of this Discussion Group we explored the background to values and valuing research in mathematics education with short presentations leading to discussion among participants. Presentations were headed 'Enacting values' [review of foundational literature (Clarkson, Bishop, & Seah, 2010)], 'Projects exploring values and valuing' [describing our work and that of participants (Seah, Andersson, Bishop, & Clarkson, 2016)], and finally 'Methodology challenges' (discussion of affordances and difficulties encountered with present methods). The last part of the session introduced participants to the notion of role-play (Belova, Eilks, & Feierabend, 2013). We self divided into a small group of 'students' and a second larger group of 'research observers'. At the core of the role-play was to experience what it was like to act out a given valuing role ('students'), or observing players who do so ('research observers'), and ascertaining whether identifiable behaviours are more likely to be associated with specific values. To this end we discussed with the 'students' how they might play out a role that showed they were valuing one of the six values that Bishop (1988) had identified, and with the 'research observers' we discussed what behaviour they might expect to be associated with each of these values (Clarkson, 2015).

P. Clarkson (✉)
Australian Catholic University, Melbourne, Australia
e-mail: Philip.Clarkson@acu.edu.au

Session 2 started with the role-play. We then explored the experiences of individuals and the groups of the different 'players'. Interestingly most 'research observers' were able to correctly identify the valuing acted out by the 'students'. However in the following group discussion the different nuances associated with the values, the overlap between them, and the difficulty finding the language to express oneself clearly concerning values and valuing all became evident. We also wondered together whether such an approach would be useful for both teachers and research students in coming to understand more deeply what it feels like to experience valuing a given value, and deciphering what behaviours point to particular values. We concluded that this experiencing did bring a sharper understanding of the role that values play in the teaching and learning of mathematics.

References

Belova, N., Eilks, I., & Feierabend, T. (2013). The evaluation of role-playing in the context of teaching climate. *International Journal of Science and Mathematics Education, 13,* S165–S190.

Bishop, A. (1988). *Mathematical enculturation.* Dordrecht, Holland: Kluwer.

Clarkson, P. C. (2015). Discussion group report: Connections between valuing and values: Rethinking data generating methods. *PME Newsletter,* November/December, pp. 7–10.

Clarkson, P. C., Bishop, A., & Seah, W. T. (2010). Mathematics education and student values: The cultivation of mathematical wellbeing. In T. Lovat, R. Toomey, & N. Clement (Eds.), *International research handbook on values education and student wellbeing* (pp. 111–136). Dordrecht: Springer.

Seah, W. T., Andersson, A., Bishop, A., & Clarkson, P. (2016). What would the mathematics curriculum look like if values were the focus? *For the Learning of Mathematics, 36*(1), 14–20.

Open Access Except where otherwise noted, this chapter is licensed under a Creative Commons Attribution 4.0 International License. To view a copy of this license, visit http://creativecommons.org/licenses/by/4.0/.

Developing New Teacher Learning in Schools and the STEM Agenda

Pat Drake, Jeanne Carroll, Barbara Black, Lin Phillips and Celia Hoyles

The purpose of the Group was to make explicit ways of supporting the preparation of teachers in schools where the reality is that expertise may be thin on the ground. There is a paradox, schools are short of mathematics and science teachers whilst at the same time required to provide more of them.

This was explored by drilling into experience of teaching and learning by teachers or beginner teachers of mathematics and science who make no pretence at being mathematics specialists. Experience came from practice drawn from policy, from research and from scholarship. Short stimulus presentations from the team in the first session were followed in the second session by short presentations from participants who bid for time in between session 1 and session 2. There was a summary of key issues by the respondent to conclude each session.

Discussion illustrated the importance of paying attention to the power of international collaborations; that the problem is recurring and so political networks are very important. Experience in the UK suggests the need for strategy to maintain unerring focus on mathematics; evidence-based interventions; developing excellence in teaching mathematics across all phases of education by sharing knowledge and practice; commitment to placing teachers' needs and goals at the core; commitment to working in partnership to influence policy and practice. These are developed through interlocking networks: research networks, teacher networks and head teacher networks. Outcomes of the group included a sense of direction and purpose in the reality of preparing new mathematics teachers through school-centred approaches; an international network to sustain over time and lead to collaborative intervention projects; ways for mathematics teacher educators to position their work alongside policy and practice in schools. A systematic and politically aware strategy is an evident next step for Australia.

P. Drake (✉)
Victoria University, Melbourne, Australia
e-mail: pat.drake@vu.edu.au

Professional development for teachers needs be sustained over time; is collaborative; and knowledge creation integral. To date there is no explicitly agreed definition of STEM but STEM-ness includes authentic problem-solving; working collaboratively; and builds on scientific, mathematical and design principles and reasoning.

The first session (Tuesday 26th July) was attended by 15 people from Australia, Colombia, Denmark, England, Hong Kong, Mexico, Spain, USA. The second session (Friday 29th July) was attended by 11 people from Australia, England, Spain, Taiwan, USA.

Presentations were offered
Barbara Black, Lesson Study; Dr Lin Phillips, Action learning; Dr Jeanne Carroll, Mindset and performance; Professor Pat Drake, Out-of-field as a resource; Associate Professor Colleen Vale (Deakin University, Australia), Out-of-field as a contested concept; Dr Mary Stevenson (Liverpool Hope University, UK), Mathematics subject enhancement and in-depth mathematical understanding; Dr Jeanne Carroll (Evidence to support growth mindset and difference); Associate Professor Inge Koch (Australian Mathematical Sciences Institute, University of Melbourne, Australia) 'Choose Maths' initiative and scaling up; Professor Dame Celia Hoyles key questions and factors to take forwards. A longer report summarising the input and discussion is available from the Organiser until December 2017.

International network
Barbara Black, Jeanne Carroll, Pat Drake, Ignasi Florensa, Celia Hoyles, Inge Koch, Jian Liu, Lin Phillips, Mary Stevenson, Colleen Vale, Kai-Lin Yang.

Open Access Except where otherwise noted, this chapter is licensed under a Creative Commons Attribution 4.0 International License. To view a copy of this license, visit http://creativecommons.org/licenses/by/4.0/.

Videos in Teacher Professional Development

Tanya Evans, Leong Yew Hoong and Ho Weng Kin

The aim of this Discussion Group was to propose and discuss models of video-based professional development (PD) programmes that are strongly grounded theoretically.

The following presentations were given:

- Kristin Lesseig (Washington State University), "*Using videos to support teacher inquiry and noticing*".
- Mary Beisiegal (Oregon State University), "*To know or not to know? Exploring effects of viewing known and unknown mathematics teachers' instruction.*"
- Leong Yew Hoong (Nanyang Technological University), "*Video-based Unit Study*".
- Tanya Evans (University of Auckland) and Greg Oates (University of Tasmania), "*The use of videos in professional development of academic staff teaching mathematics at university*".
- Heather Lonsdale (Curtin University) and Deborah King (University of Melbourne), "*Perception vs reality—using tutorials videos to aid tutor reflection*".
- Ho Weng Kin (Nanyang Technological University), "*The Impact of Online Video Suite on the Singapore Pre-service Teachers Buying-Into Innovate Teaching of Factorisation* via *Algecards*".

The discussions were framed using the following questions:

T. Evans (✉)
University of Auckland, Auckland, New Zealand
e-mail: t.evans@auckland.ac.nz

1. Discuss the design of a successful model of video use in PD for mathematics teachers. Provide evidence of its "success". Explicate the role of videos in the PD model.
2. What are existing design principles for successful use of video in PD? What is the connection between these principles and existing theories of teacher learning and of video as teacher learning tool? What cross-countries and cultural differences exist?
3. How can we calibrate a video-based PD model in a way that addresses different emphases of knowledge needs of mathematics teachers along relevant knowledge strands (such as the now well-known domains of Mathematics content knowledge, mathematics pedagogical content knowledge, and knowledge of student learning)?

The discussion group was attended by participants from all around the world which afforded an international perspective on the range of successful video-based PD models for mathematics teachers covering all levels of school and university teaching. During the discussions a request was made by I.K. Rana (Indian Institute of Technology) for access to resources that were mentioned and that can be shared. To that end, a list of contact emails of participants was collected and will be used for sharing resources and future notifications. Contact Tanya Evans t.evans@auckland.ac.nz if you want to be added to this list of researchers with a focus on teacher PD.

Open Access Except where otherwise noted, this chapter is licensed under a Creative Commons Attribution 4.0 International License. To view a copy of this license, visit http://creativecommons.org/licenses/by/4.0/.

National and International Investment Strategies for Mathematics Education

Joan Ferrini-Mundy, Marcelo C. Borba, Fumi Ginshima, Manfred Prenzel and Thierry Zomahoun

Mathematics education is an essential pathway for economic security and technological advancement at the national, community, and individual levels. In developed and developing countries, there is an impetus for innovation and improvement of mathematics curriculum and pedagogical practices that meet local and practical needs. Partnership and input from practitioners is essential for policy makers and research funding organizations to navigate the path forward. This ICME13 Discussion Group explored the following global issue: What is the appropriate role of funding agencies, ministries, and related institutions in influencing and advancing improvements in mathematics education research and policy, as well as in facilitating international research in mathematics learning?

The discussion addressed a number of key issues where there are differences internationally. highlighting how much we can learn from examples and organizational structures in various countries. It became clear that different types of organizations—a federal funding agency to "promote the progress of science" (Ferrini-Mundy, U.S. National Science Foundation), a non-profit institute directed toward specific goals that are of economic importance to a country [Zomahoun, African Institute for Mathematical Sciences (AIMS)], a council designed to provide advice to federal and state governments (Prenzel, German Council of Science and Humanities), and a government agency concerned with developing evidence to help guide policy (Ginshima, Japan's National Institute for Educational Policy Research) —have significant roles in the ecosystem for mathematics education.

Examples were provided of ways that government policy affecting mathematics education is informed, shaped, and furthered through that interconnected system. AIMS is focused on increasing the pipeline of women in mathematics on the basis

---

J. Ferrini-Mundy (✉)
National Science Foundation, Arlington, USA
e-mail: jferrini@nsf.gov; barussel@nsf.gov

of data and needs for job creation in Africa to counter high unemployment rates. Germany's "Excellence Initiative", which is promoting young researchers, top-level research, and capacity-building in institutions, provides an example of how mathematics education can benefit from being situated within wider government initiatives that have been recommended by distinguished leaders. We learned how assessments undertaken within the Ministry of Education in Japan are helping to shape curricular emphases in schools in mathematics.

By sharing information about this range of intuitions and organizations, arranged differently within different countries, it became clear that several features seem critical in ensuring coordinated impact on mathematics education policy and resources. Those may include: capacity for analyzing and sharing data and evidence in forms that are useful to practitioners, can inform policy makers, and help funders set priorities. It is also clear that when mechanisms for convening respected experts in education and research in structures charged with advising government exist, focus and strategic direction are possible.

Several issues emerged that would clearly benefit from continued international discussion. Those included the role of "big data" in mathematics education research, with concerns expressed about implications for the nature and amount of time devoted to assessments in schools; the need for active attention to translational research that can bring findings to practitioners and policy makers, in meaningful ways; the potential of making progress on shared challenges world-wide through new means of research collaboration, including infrastructures for data sharing; and the importance of making excellent mathematics education available to diverse groups of students, supported by arguments about economic benefits to a country.

Open Access Except where otherwise noted, this chapter is licensed under a Creative Commons Attribution 4.0 International License. To view a copy of this license, visit http://creativecommons.org/licenses/by/4.0/.

Transition from Secondary to Tertiary Education

Gregory D. Foley, Sergio Celis, Hala M. Alshawa, Sidika Nihan,
Heba Bakr Khoshaim and Jane D. Tanner

Aims and Ideas

The modern world thrives on quantitative information. Consequently, many university majors are becoming increasingly mathematical. Many secondary school graduates, however, are not ready for tertiary course work in mathematics and statistics. For example, 35.1% of U.S. college mathematics enrollments are in pre-college remedial courses: 1.4 million out of 3.9 million in autumn 2010 (Blair, Kirkman, & Maxwell, 2012). Such deficiency in mathematical knowledge and skills can influence students' decisions to abandon their intended major and transfer to a less mathematically demanding major, or even to quit tertiary education. This discussion group examined the difficulties that students encounter in making the mathematical transition from secondary to tertiary education. The group explored methods used to assess student readiness in mathematics and programs to help beginning tertiary students when they face mathematical struggles. The discussion considered both students who seek mathematically intensive majors at the tertiary level and those who pursue less mathematically intensive degrees.

G.D. Foley (✉)
Ohio University, Athens, Ohio, USA
e-mail: foleyg@ohio.edu

Structure

The organizing team (co-chairs and team members) provided introductions on the theme of the *Transition from Secondary to Tertiary Education* to provide a framework for the discussions and spent most of the sessions facilitating the discussion among the participants. The sessions were structured (a) to provide contexts for discussion, (b) to simulate discussion, and (c) to get the participants involved and draw out their ideas. The participants represented a wide variety of nations. The following tables list topics that were presented and discussed.

Tuesday timeline	Topic	Format and presenter
16.30–16.35	Welcome, introductions, and overview	5-min led by Foley
16.35–16.55	Issues in the transition from secondary to tertiary mathematics education: North American perspectives	16-min presentation by Foley and tanner + 4 min of questions and answers
16.55–17.10	Are University freshmen mathematically ready?	12-min presentation by Khoshaim for Er + 3 min of questions and answers
17.10–18.00	Issues in the transition from secondary to tertiary mathematics education	Small-group discussion followed by whole-group discussion, facilitated by team members

Friday timeline	Topic	Format and presenter
16.30–16.35	Welcome, introductions, and overview	5-min led by Foley
16.35–16.50	Issues in the transition from secondary to tertiary mathematics education: a South American perspective	12-min presentation by Celis + 3 min of questions and answers
16.50–17.05	How Saudi Arabia addresses the challenges of transitions from secondary to tertiary education	12-min presentation by Khoshaim + 3 min of questions and answers
17.05–17.15	The role of Universities in the transition from secondary to tertiary mathematics education in Jordan	8-min presentation by Foley for Alshawa + 2 min of questions and answers
17.15–18.00	Issues in the transition from secondary to tertiary mathematics education	Small-group discussion followed by whole-group discussion, facilitated by team members

Reference

Blair, R., Kirkman, E. E., & Maxwell, J. W. (2012). *Statistical abstract of undergraduate programs in the mathematical sciences in the United States: Fall 2010 Conference Board of the Mathematical Sciences survey.* Providence, RI: American Mathematical Society.

Open Access Except where otherwise noted, this chapter is licensed under a Creative Commons Attribution 4.0 International License. To view a copy of this license, visit http://creativecommons.org/licenses/by/4.0/.

Teachers Teaching with Technology

Ian Galloway, Bärbel Barzel and Andreas Eichler

The two sessions of the T^3-discussion group covered the main aspects of the work of T^3 Europe, particularly in Germany. Each session was organized along two sub-topics beginning with a brief input on a concrete example and ending with a specific question or questions. A discussant then opened a plenary discussion by making some comment on the questions or the input. This resulted in four lively debates which of course did not provide answers to the questions but did provoke the participants to think metaphysically about them.

Using technology: best practice in using technology for teachers' professional development.

Oliver Wagner used some best practice examples from his own work, and raised the following questions:

Is it possible to construct a PD session without "button pressing"?

What are the main aims for PD sessions and how do you reach them?

What are the wishes of teachers on PD sessions?

Using technology: the use of analysis and experimental work to bridge mathematics and science.

Daniel Thurm talked about the use of technology in mathematical modelling to link mathematics and science. He discussed the modelling cycle and the role that technology can play at different stages of the modelling process. Using a cooling cup of coffee as a concrete example illustrated the potential problems that can arise. He raised the question:

How do we avoid falling into the trap of finding a mathematical model and then failing to encourage students to explore the underlying reasons as to why the model fits?

Changing knowledge and beliefs of teachers as they begin to use technology in mathematics teaching

Angela Schmitz illustrated the way that the use of technology can have a significant influence on the learning of mathematics. But teachers have divergent opinions on its use and for every change in instructional methods, their beliefs play a decisive role. After a brief look at the state of research on the beliefs of teachers in secondary schools on the use of technology in mathematics instruction the question was asked:

How can teachers' beliefs about the use of technology be changed?

Changing knowledge and beliefs of teachers using formative assessment

Hana Ruchniewicz described a digital self-assessment tool. FaSMEd is concerned with raising achievement through formative assessment and has partners in 8 European countries. She raised the following questions:

How can one assess whether or not a student can

- model real situations mathematically?
- use mathematical representations?
- translate a description of a process or situation into a graph?

Open Access Except where otherwise noted, this chapter is licensed under a Creative Commons Attribution 4.0 International License. To view a copy of this license, visit http://creativecommons.org/licenses/by/4.0/.

Mathematics Education and Neuroscience

Roland H. Grabner, Andreas Obersteiner, Bert De Smedt,
Stephan Vogel, Michael von Aster, Roza Leikin
and Hans-Christoph Nuerk

The interdisciplinary research field of educational neuroscience—linking neuroscience, psychology, and education—has witnessed a tremendous growth in the past five to ten years. By combining behavioral and neuroscientific methods, its general aim is to achieve a broader understanding of the neurocognitive mechanisms underlying learning and to support the development of effective instruction. It has been repeatedly questioned whether the obtained neuroscientific evidence has implications for education (including research and practice) or whether the connection between neuroscience and education is a bridge too far (e.g., Bowers, 2016; Verschaffel, Lehtinen, & Van Dooren, 2016). Has the inclusion of the neuroscientific level of analysis furthered our understanding of successful mathematics learning and how to support it? The aim of this discussion group was to bring together neuroscientists, psychologists, and math educators, and to discuss the chances and limitations of educational neuroscience research on selected topics of mathematics education.

The session began with brief statements by each of the three presenters about their view on the emerging research field of educational neuroscience. These statements were followed by an initial discussion with the audience, in which controversial arguments from different perspectives were raised.

After that, each of the presenters introduced a more specific research area of educational neuroscience. Michael von Aster focused on children with severe numerical difficulties. Neuroscience research was able to show that brain activation patterns in these children differ from those of typically developing children. Based on such findings, von Aster presented a computer game that was developed specifically to enhance dyscalculic children's understanding of numerical magnitudes.

R.H. Grabner (✉)
Institute of Psychology, University of Graz, Graz, Austria
e-mail: roland.grabner@uni-graz.at

Hans-Christoph Nuerk presented studies on typical early mathematical development. Brain imaging studies found that there is a neural link between number representations and finger gnosis, suggesting that finger-based numerical representations are beneficial for numerical development. This finding has potential implications for mathematics education, where many researchers and teachers do not support children in using their fingers to solve arithmetic problems.

Roza Leikin presented research on the cognitive mechanisms of higher mathematics. Using different types of complex mathematical problems, she explored the brain activation patterns in students with distinct mathematical abilities and in generally gifted students. The results seem to support the neural efficiency hypothesis, stating that efficient problem solving is related to a general decrease rather than increase of brain activation.

The discussion group ended with a general discussion of the potentials and limitations of integrating neuroscience into mathematics education research. In conclusion, while the aim of neuroscience is not to provide immediate suggestions for classroom practice, neuroscience might help in better understanding the cognitive mechanisms that underlie mathematical problem solving, as the presentations in this discussion group have shown.

References

Bowers, J. S. (2016). The practical and principled problems with educational neuroscience. *Psychological Review*. doi:10.1037/rev0000025

Verschaffel, L., Lehtinen, E., & Van Dooren, W. (2016). Neuroscientic studies of mathematical thinking and learning: A critical look from a mathematics education viewpoint. *ZDM Mathematics Education, 48*, 385–391.

Open Access Except where otherwise noted, this chapter is licensed under a Creative Commons Attribution 4.0 International License. To view a copy of this license, visit http://creativecommons.org/licenses/by/4.0/.

Reconsidering Mathematics Education for the Future

Koeno Gravemeijer, Fou-Lai Lin, Michelle Stephan, Cyril Julie and Minoru Ohtani

There is a broad consensus that we should foster 21st century skills. Mathematics education seems a perfect place to work on those skills, yet, which of the competencies that are assembled under the label 21st century skills can, and should, be fostered is still an open question. In addition to introducing 21st century skills in the school curriculum, changes will have to be made in content goals. This is especially the case for mathematics education. Increasingly, machines are doing all kinds of mathematical calculations; at the same time mathematics becomes invisible while disappearing in black boxes. Consequently, it will become important to understand the mathematics in those black boxes, and to know how to work with it. This touches on the content of mathematics education, which has to be adapted accordingly. Much research has already been done on getting a handle on mathematics at today's workplace—which proves to be idiosyncratic and interwoven with tools and practices. Those elements will of course stay relevant for the times to come. If we look ahead, however, more fare-reaching issues emerge. We will have to attune mathematics education to the fact that computers and computerized machines do most mathematical work outside school.

In order to begin a conversation on what mathematics education looks like in the future, the Working Group team assembled a variety of research articles on this issue and invited interested ICMI participants to read them prior to attending the Working Sessions. We intended for these articles to provide a frame for discussing what mathematics content and processes will be relevant in our society in 20 years.

In the first session, Gravemeijer presented a short talk to orient participants to the aims of the sessions: *How can mathematics education prepare students for meaningful participation in the future, digital society?* Participants then met in self-formed groups to brainstorm answers to the question: What mathematics will citizens need, and what is the role of general abilities known as 21st century skills?

K. Gravemeijer (✉)
Eindhoven University of Technology, Eindhoven, The Netherlands
e-mail: koeno@gravemeijer.nl

Afterwards, participants shared out and together created an inventory of topics for consideration: Statistics and probability, variable and function, 3D geometry, measurement, basic mathematics/some advanced, interpreting results of data, quantitative literacy, visualization, critical thinking, problem solving and posing, modeling, deduction/logic/proof, argumentation and communication, collaboration, representation, financial literacy and digital mathematics.

In the second session, the team displayed the Inventory of Topics and asked participants to choose an area to develop more thoroughly with a small group. Results of these discussions yielded several implications: (1) it will be important to know the meaning of basic operations, and to work with number systems, properties and relations; (2) three-dimensional geometry should be emphasized, especially visualization and the relation between the digital world and geometry; (3) modeling processes need to have greater priority, and should involve more open problems; (4) data analysis processes—using technology—are critical, especially in contexts from other disciplines; including thinking backwards when interpreting results; (5) special attention will have to be given to mathematical argumentation, and communication.

Open Access Except where otherwise noted, this chapter is licensed under a Creative Commons Attribution 4.0 International License. To view a copy of this license, visit http://creativecommons.org/licenses/by/4.0/.

Challenges in Teaching Praxis When CAS Is Used in Upper Secondary Mathematics

Niels Groenbaek, Claus Larsen, Henrik Bang, Hans-Georg Weigand, Zsolt Lavicza, John Monaghan, M. Kathleen Heid, Mike Thomas and Paul Drijvers

The DG focused on the relationship between CAS and mathematics in teaching and learning, educational design and the qualification of teachers' choices concerning the use and non-use of CAS.

Headword Summary

The first DG session was on challenges and opportunities for design, with main themes:

- Reasons for CAS not being more widespread. Reports as far back as the first ICMI studies in the mid 80s were quite optimistic, and although there have been many investigations and modes of development, the situation today is that CAS-use is supported by the same basic ideas. But optimism has faded, mainly because the technical difficulties with CAS are underestimated and CAS is not just an isolated tool but represents an entire environment.
- Documentation and assessment with CAS has proved to be difficult.
- Gainful CAS-use requires an overarching concept of teaching and learning as CAS changes mathematics in several directions—content, style and activities, exemplified by: Content—CAS's enlarged look-up access to knowledge has changed the focus on what is important to teach and what is not; style—CAS shifts the view on what to prove and how; student activities—a shift from more or less demanding problem solving techniques to the logic of solution methods and checking modes.
- Teachers' knowledge about aspects of the underlying mathematics that CAS may reveal, e.g. to what extent, if any, should one master complex numbers and

---

N. Groenbaek (✉)
University of Copenhagen, Copenhagen, Denmark
e-mail: gronbaek@math.ku.dk

deal with or dismiss them didactically, because students encounter them solving quadratic equations. Similarly, the probability that students produce unexpected results increases with CAS, so the teacher decision between giving immediate response or suspending answers allowing for more reflection on the issue in an increased challenge.

The second DG day focused on in-service training, didactical discourse and changes in teacher role:

- The big diversity in teachers' implementation of CAS on one hand makes it difficult to generate common standards but, on the other hand, offers a range of experiences to draw on. Instigation of teacher reflection on CAS and mathematics and the choices made. Coaching provides one option; another is the development of communities where you can share experiences.
- CAS changes and creates situations in the classroom. Teachers are faced with a much wider range of on-spot choices and solutions. As a consequence, the importance of the teacher's exemplary work has increased.
- Examples of how a comprehensive pool of teaching ideas with CAS are available for teachers, combined systematically with teachers' training.
- Examples of how to develop tools used for electronic marking of CAS assignments, taking differing correct answer options, differing notation etc. into account.

Open Access Except where otherwise noted, this chapter is licensed under a Creative Commons Attribution 4.0 International License. To view a copy of this license, visit http://creativecommons.org/licenses/by/4.0/.

Mathematics in Contemporary Art and Design as a Tool for Math-Education in School

Dietmar Guderian

The first session should find out possibilities (in school and in integration courses) for refugees how to use art and design in the surroundings to open doors to elementary mathematics, to culture and especially to the culture of the countries of origin of the refugees entering Europe (from Syria, Iraq, former UDSSR, …). One idea along the way was to show to adult refugees, their children and indigenous children that the countries of origin of refugees hold high culture like the European countries, too. Thus might help refugees and their children to keep little proudness for themselves. Members and the organiser of the Group presented examples (Russian puppets—enlargement in area and space, Islamic mosaics—tessellations, African clothes—shifting; folkloric dances—algorithms and plane geometrical figures, …).

The second session dealed with Mathematics in international contemporary art and had the following background: Contents of contemporary art and design often resemble applied mathematics: they use mathematics, especially 'prescientific mathematics' like: basic geometric forms, series of numbers, plane mappings, … Often they are more interesting than the same mathematical content brought by traditional math education. On the other hand: Today new results in math are known per internet by artists nearly simultaneously to the specialists. And they introduce them into their artwork immediately (news in tessellation; dynamic processes, big data …).

Nearby effect: Subjects also are new and interesting for teachers, the general public and especially for those with low cultural background because of their social, economical or geographical origin. That's why a big part of the work of the Group run outside the Congress sector: Three important commercial galleries in Hamburg showed exhibitions accompanying the discussion group:

D. Guderian (✉)
University of Education Freiburg, Freiburg, Germany
e-mail: guderian@ph-freiburg.de

Galerie Dr: Nanna Preussners presented artworks dealing with the hazard: art pieces of herman de vries (NL) showed twice, the human inability to create hazard and his ability for that by using aleatoric methods. The artworks of Werner Dorsch (GER) and Jo Schöpfer (GER) led the visitor via "Superzeichen" (Max Bense) catastrophically to a wrong interpretation. Michel Jouet (FR) used aleatory to build an incomplete edge-model of an cube and Werner Hotter integrated the Langford-series to simulate hazard-builded artpieces. Galerie Renate Kammer presented the use of elementary plane and three-dimensional geometric figures in contemporary art: Max H. Mahlmann (GER+), Martin Vosswinkel (GER) and Laszlo Otto (HU) applicated grids. Ingo Glass (GER, HU) and Ludwig Wilding (GER+) integrated basic geometrical elements like square, circle, cube und sphere. Gudrun Piper (D+) and Ilse Aberer (AUT) introduced series and proportions. Galerie multiple box showed irritation by moving and pseudo-moving geometrical forms in artpieces of Vera Kovacic (SLO) and Rolf Schneebeli (CH).

Four public vernissages and three longer than the congress running exhibitions gave ICME-members and citizens a good overview. The work will be continued by four personal exhibitions (Aberer, Hotter, Otto, Vosswinkel) in the participating galleries next year and further activities especially of the group of artists.

Open Access Except where otherwise noted, this chapter is licensed under a Creative Commons Attribution 4.0 International License. To view a copy of this license, visit http://creativecommons.org/licenses/by/4.0/.

Exploring the Development of a Mathematics Curriculum Framework: Cambridge Mathematics

Ellen Jameson, Rachael Horsman and Lynne McClure

This discussion group aimed to use a work-in-progress project as an example to fuel discussion of curriculum coherence and the importance of the relationships between curriculum, context, and implementation. These are major considerations influencing curriculum development at national and international levels. Our example was the framework being developed by Cambridge Mathematics for presenting and organising the domain of school mathematics in a form that emphasises connections and interdependencies between learners' mathematical experiences, and the different routes that can successfully facilitate learners' development mathematical understanding.

Two themes stood out strongly in both sessions. The first had to do with the importance of finding ways of communicating design, design methods, and research methods that can drive productive collaboration among researchers, administrators, policy makers, and teachers during framework development. Focal points for communication with one group of stakeholders might not provide critical information needed for another to engage. Consideration of the priorities and needs of each group in the collaborative process can help to make the final result more useful for all groups, and consequently more likely to be put to use and refined.

Some specific features of a curriculum framework were identified as having the potential to benefit collaboration around emerging curriculum frameworks, and the subsequent quality of those frameworks. Framework design and documentation should be able to:

E. Jameson (✉)
University of Cambridge, Cambridge, UK
e-mail: Ellen.Jameson@Cambridgemaths.org

- make the decision process around structural and content-specific choices explicit, including the balance of research, consultation, and experience involved
- demonstrate which decisions involved stronger or weaker support from research and collaboration, allowing the framework to be more directly evaluated by curriculum designers making decisions according to their own sets of criteria
- allow users of the framework to independently evaluate the framework authors' evaluation criteria for the quality of evidence and collaboration
- allow users of the framework to adapt their use based on an understanding of the designers' goals and intended audience.

The second theme that emerged involved the ways in which frameworks represent areas of mathematical understanding that cut across specific topics throughout the entire curriculum, including what are sometimes called habits of mind, a set of broadly defined skills including problem-solving, reasoning, generalising, and critical thinking. This was linked to discussion of curriculum coherence and the role of framework design in setting out structures for communicating the potential for progression in these areas, so that support for the development of higher order thinking can be supported throughout a learner's entire journey through the curriculum.

This discussion took place during an early, exploratory stage of the Cambridge Mathematics framework project, and we felt that it had occurred at a very helpful time to influence the work going forward. A third theme for us therefore became the importance of being able to discuss early-stage work with a diverse group. This is a challenge in many areas of design, since early work often involves unresolved fundamental questions and prototypical outputs that can be difficult to communicate and evaluate. Nevertheless, there was benefit in doing so for all involved. We are grateful to all of the participants in each session for their thoughtful discussions. The ideas raised here have broad application to other curriculum design efforts, and we look forward to being able to share the way our work has been influenced by these themes at the next ICME.

Open Access Except where otherwise noted, this chapter is licensed under a Creative Commons Attribution 4.0 International License. To view a copy of this license, visit http://creativecommons.org/licenses/by/4.0/.

Theoretical Frameworks and Ways of Assessment of Teachers' Professional Competencies

Johannes König, Sigrid Blömeke and Gabriele Kaiser

For the past decades, research on the measurement of cognitive elements of teacher competence has been growing significantly. Research on teacher expertise underlines the importance of teachers' professional knowledge for the successful mastering of tasks that are typical for their profession.

A substantial number of studies have contributed to this by developing tests to assess teacher knowledge, predominantly in the domain of mathematics. Following the influential work by Shulman (1987), they differentiate mathematical content knowledge, mathematical pedagogical content knowledge, and general pedagogical knowledge.

Besides differentiation into such knowledge categories, methodological approaches vary across the different studies. Researchers have developed different conceptualizations of teacher knowledge and use different methods to access teacher knowledge as part of their professional competence.

While for a number of relevant studies the classical paper-and-pencil assessment represents the dominating paradigm, because it enables an efficient and reliable way to measure declarative-conceptual knowledge in large samples, others shift from paper-and-pencil tests to the implementation of instruments using video clips of classroom instruction as item prompts. Video-based assessment instruments are used to address the contextual nature and the complexity of the classroom situation. Thus they try to go beyond the limited scope of classical paper-and-pencil assessments (Blömeke, Gustafsson, & Shavelson, 2015).

Such a shift reflects the need for instruments that allow an investigation of teachers' situational cognition and the impact of individual differences in teaching experience and in-school opportunities to learn during teacher education. Although knowledge acquired during teacher education and represented as declarative knowledge is probably of great significance, especially the research on teacher

---

J. König (✉)
University of Cologne, Cologne, Germany
e-mail: johannes.koenig@uni-koeln.de

expertise has worked out that both declarative and procedural knowledge contributes to the expert's performance in the classroom.

Several studies adopted this approach to provide a more ecologically valid measurement of the knowledge of mathematics teachers thus intending to measure knowledge that is more of a situated nature. With the growing popularity of video-based measurements in the field of mathematical teacher knowledge research, the necessity becomes visible that different approaches applied by various research teams should be brought together to facilitate dissemination and enrich in-depth discussion.

This was the aim of the Discussion Group at ICME 2016, which brought together the major projects on mathematics teachers' professional competencies and thus forwards the exchange of research teams' approaches and new findings brought about by their current research activities.

The following projects were presented in the discussion group:

1. Learning Mathematics for Teaching and Follow-up-projects (LMT): Deborah Loewenberg Ball and Heather Hill
2. Professional teachers' knowledge and cognitive activation (COACTIV): Stefan Krauss and Werner Blum
3. Teacher Education and Development Study in Mathematics (TEDS-M): Sharon Senk
4. Teacher Education and Development Study in Mathematics-Follow-Up (TEDS-FU): Johannes König, Sigrid Blömeke, Gabriele Kaiser
5. Knowledge Quartet: Tim Rowland
6. Professional competencies of educators in the field of mathematics (KomMa and ProKomMa): Katja Eilerts, Sigrid Blömeke
7. Subject-specific action-oriented teachers' competencies: Anke Lindmeier
8. Learning to Learn from Mathematics Teaching: Rossella Santagata.

Hilda Borko closed the session by a summarising commentary.

References

Blömeke, S., Gustafsson, J.-E., & Shavelson, R. (2015). Beyond dichotomies: Competence viewed as a continuum. *Zeitschrift für Psychologie, 223*, 3–13.
Shulman, L. S. (1987). Knowledge and teaching: Foundations of the new reform. *Harvard Educational Research, 57*, 1–22.

Open Access Except where otherwise noted, this chapter is licensed under a Creative Commons Attribution 4.0 International License. To view a copy of this license, visit http://creativecommons.org/licenses/by/4.0/.

Using Representations of Practice for Teacher Education and Research—Opportunities and Challenges

Sebastian Kuntze, Orly Buchbinder, Corey Webel, Anika Dreher and Marita Friesen

Representations of classroom practice offer the chance of referring to the teachers' professional environment both when conceiving opportunities of professional development and when investigating aspects of teacher expertise. Representations of practice can stimulate teachers' criteria-based analysis in environments that do not bring the full pressure and action constraints of the actual classroom. In professional development and in research, approaches which use representations of practice offer prospects and encounter challenges which can be explored along the following key questions: How can representations of practice encourage pre-service and in-service teacher professional development, e.g. through stimuli for reflection, criteria-based analysis, or structured observation? How can representations of practice help to investigate aspects of teacher expertise, such as e.g. criteria-based aspects of noticing or analyzing? What kinds of methodological challenges emerge when designing opportunities for professional learning, which make use of representations of practice? How can these challenges be addressed? What methodological challenges emerge when designing research settings based on representations of practice? How can these challenges be addressed?

By aspects of teacher expertise which can be in the scope of empirical research, we mean—an understanding which is as open and inclusive as possible—teacher characteristics which may be meaningful for supporting students' learning, such as, components of professional knowledge of mathematics teachers, views and convictions (e.g. Shulman, 1986; Ball, Thames, & Phelps, 2008; Kuntze, 2012), competence facets such as "professional vision" (Sherin & van Es, 2009), "noticing" in the sense of "selective attention" (cf. e.g. Seidel, Blomberg, & Renkl, 2013) or in the sense of "knowledge-based reasoning" (Sherin, Jacobs, & Philipp, 2011),

---

S. Kuntze (✉)
Ludwigsburg University of Education, Ludwigsburg, Germany
e-mail: kuntze@ph-ludwigsburg.de

as well as the notion of "awareness" (Mason, 2002), and specific competences of analyzing classroom situations (e.g. Kuntze, Dreher, & Friesen, 2015). Opportunities and challenges related to the use of representations of practice for supporting aspects of teacher expertise have been explored in an ICME discussion group with contributions from Orly Buchbinder, Dan Chazan, Anika Dreher, Marita Friesen, Jessica Hoth, Sebastian Kuntze, Nanette Seago, Karen Skilling & Gabriel Stylianides, Corey Webel, Bill Zahner, and Rina Zazkis. Joint publications related to the theme of this discussion group are planned for the near future.

References

Ball, D., Thames, M. H., & Phelps, G. (2008). Content knowledge for teaching: What makes it special?. *Journal of Teacher Education, 59*(5), 389–407.
Kuntze, S. (2012). Pedagogical content beliefs: Global, content domain-related and situation-specific components. *Educational Studies in Mathematics, 79*(2), 273–292.
Kuntze, S., Dreher, A., & Friesen, M. (2015). Teachers' resources in analysing mathematical content and classroom situations. *CERME 9*.
Mason, J. (2002). Minding your Qs and Rs: Effective questioning and responding in the mathematics classroom. In L. Haggarty (Ed.), *Aspects of teaching secondary mathematics: Perspectives on practice* (pp. 248–258). New York, NY: Routledge Falmer.
Seidel, T., Blomberg, G., & Renkl, A. (2013). Instructional strategies for using video in teacher education. *Teaching and Teacher Education, 34,* 56–65.
Sherin, M., Jacobs, V., & Philipp, R. (2011). *Mathematics teacher noticing*. New York: Routledge.
Sherin, M. G., & van Es, E. A. (2009). Effects of video club participation on teachers' professional vision. *Journal of Teacher Education, 60*(1), 20–37.
Shulman, L. (1986). Those who understand: Knowledge growth in teaching. *Educational Researcher, 15*(2), 4–14.

Open Access Except where otherwise noted, this chapter is licensed under a Creative Commons Attribution 4.0 International License. To view a copy of this license, visit http://creativecommons.org/licenses/by/4.0/.

How Does Mathematics Education Evolve in the Digital Era?

Dragana Martinovic and Viktor Freiman

In preparation for this work, we have looked into some recent curricula ideas, the international assessments (e.g., PISA), as well as the literature on the 21st century skills. For example, The New Vision for Education: Unlocking the Potential of Technology, 2015 report of the World Economic Forum lists 16 most critical 21st century skills (e.g., literacy, numeracy, ICT literacy, scientific literacy, financial literacy, cultural and civic literacy, critical thinking/problem-solving, creativity, communication, collaboration, curiosity, persistence/grit, adaptability, leadership, social and cultural awareness, and initiative), and proposes that technology is used to help people achieve these skills. How do these skills interact with ones, more specifically related to mathematics as a subject?

Multiple literacies are fundamental for the 21st century learning, but their role in mathematics education is yet to be clarified. For this Discussion Group, the facilitators invited several experts to briefly present different international perspectives on the topic and to formulate discussion questions. Each day started with a short introduction by the facilitator, and closed with a summary of the day activities.

On Day 1, we discussed, "What are the new types of literacies that are relevant to mathematics education?"

- *Financial literacy and math education* (Annie Savard, McGill University, Canada). A financial concept might be introduced in the sociocultural context; learning probability and developing critical thinking in regard to gambling activities can be done through simulators of games of chance.

D. Martinovic (✉)
University of Windsor, Windsor, Canada
e-mail: dragana@uwindsor.ca

- *Problem solving and math education* in the digital era (Eleonora Faggiano, Università degli Studi di Bari Aldo Moro, Italia). Continuous investigation of mathematical potential of new technologies in view of mathematizable life situations and use of the synergy between the traditional and digital artefacts are crucial.
- *Digital literacy and its connection to math teaching and learning: experience of the CompéTICA partnership network (ICT Competences in the Atlantic Canada)* (Viktor Freiman, Université de Moncton, Canada). There is a need for development of a life-long continuum of digital competences as combination of specific digital literacy skills and 'soft' skills. Connections of such competences to math education need to be investigated.

On Day 2, we discussed, "The 21st century learning skills." We inspected models, such as those developed by Thoughtful Learning, that define the 21st century learning skills as the 4 C's—critical and creative thinking, communicating, and collaborating and discussed the (new) inter- and trans-disciplinary connections and the next generation of mathematics standards in view of the Vision of the Framework for K-12 Science Education.

- *'Soft' skills and math education in the digital era* (Allen Leung, Hong Kong Baptist University, China, & Anna Baccaglini-Frank, University of Pisa, Italy). Learners put into action 'soft' skills to interact with and relate to a digital environment for the purpose of acquiring mathematical knowledge and developing new digitally-based conceptions of math notions and ways of thinking.
- *Creative and critical thinking* in technology-rich environments (Antonella Montone, Università degli Studi di Bari Aldo Moro, Italia). Connecting mathematical and digital literacies is not only necessary, but it should also become "indispensable." The children's fairy tales which involve mathematical objects provide a natural environment in which children develop their own thinking. Technological tools like Scratch and Lego allow children to create characters, the environment, and a situation.

Open Access Except where otherwise noted, this chapter is licensed under a Creative Commons Attribution 4.0 International License. To view a copy of this license, visit http://creativecommons.org/licenses/by/4.0/.

Scope of Standardized Tests

Raimundo Olfos, Ivan R. Vysotsky, Manuel Santos-Trigo,
Masami Isoda and Anita Rampal

The aim of the Discussion Group was to capture the sense of the community about standardized testing and provide implications to global policies. One of the regular views reads as follows: Standardized tests are needed because they can provide a high amount of information and evidence of validity. Of course there can be incorrect interpretations, but these can be reduced if the quality of the test fulfills the requirements that are associated with standardized assessments. Critical approaches highlight other issues: The limited scope of standardized tests in school math, because these tests undermine abilities to conjecture and to encourage open problems in class. Standardized testing devalues abilities to collaborate and to engage in real-world experience, thereby failing the mission of the pursuit of happiness and justice of all. Ethics issues are unsolved: policymakers do not know how to use test-based incentives. Some school systems are under great pressure to raise their scores. Tests create competition between schools. Standardized testing does not take into account diversity, test anxiety, home language of students and special needs. So they fail in democratic systems.

These ideas were shared in two sessions on Tuesday, 26 July, and Friday, 29 July 2016. In a plenary format, Raimundo Olfos introduced the aims and structure of the discussion, followed by a discussion amongst the participants. Masami Isoda presented some clarifications about standardized testing and Ivan Vysotsky problematized their political potential. Ethical consequences of standardized tests were commented on by Anita Rampal, and participants shared their vision and contributed with new ideas. Valeria Di Martino referred to multicultural classrooms and standardized tests in mathematics. Audrey Paradis commented on the use of

---

R. Olfos (✉)
Pontificia Universidad Católica de Valparaíso, Valparaíso, Chile
e-mail: raimundo.olfos@pucv.cl; raimundo.olfos@gmail.com

standardized tests in terms of teacher perception. Federica Ferretti, Alice Lemmo, and Francesca Martignone talked about the use of large-scale assessment in teacher education, and Johan Yebbou reflected about the introduction of international assessment studies in a country. Finally, the group shared their views in collaborative writing to provide implications for global policies.

Open Access Except where otherwise noted, this chapter is licensed under a Creative Commons Attribution 4.0 International License. To view a copy of this license, visit http://creativecommons.org/licenses/by/4.0/.

Mathematics for the 21st Century School: The Russian Experience and International Prospects

Sergei A. Polikarpov and Alexei L. Semenov

The sessions started with a lecture, "A personal experience in international employment of Russian pre-university math," which was given by Mark I. Bashmakov, Russian Academy of Education, St. Petersburg, RF, a major figure in the Kangaroo Olympiad and a creator of the productive education movement in Russia.

Trailers of the movies *Senses of Math* and *The Discrete Charm of Geometry* on mathematical research and education were presented by films' director, Ekaterina V. Eremenko, Berlin University, who graduated from Moscow School 91 and obtained MS and PhD degrees from the Lomonosov Moscow State University.

Alexei L. Semenov presented a perspective on the development of Russian mathematical education during the last 100 years, stressing the crucial role of the social context. He outlined: (1) industrial and authoritarian models for society, family, and school, Vygotskian theories of the 1930s; (2) changes that began in the 1960s: specialized schools, the Kolmogorov reform, programming as second literacy, constructionism and technologies in education; and (3) current important changes caused by the implementation of the Conceptual Framework for Development of Mathematical Education in the Russian Federation adopted by the Russian Government in 2013.

Alexander P. Karp discussed highlights of the Russian school mathematics and its differences from today's situation in U.S. schools.

Nikolai Konstantinov has been involved in math education reforms since the 1950s and elaborated a new system of learning math where pupils under the guidance of university students solve research tasks and discover new and unexpected facts. He presented the history of the Tournament of Towns.

In the second session, the following presentations were given:

"Math for 21st century school: An expert opinion" by Anatoly Kushnirenko, Scientific Research Institute for System Analysis of RAS; "1C: MathKit in fifth and

S.A. Polikarpov (✉)
Moscow State University of Education, Moscow, Russia
e-mail: sa.polikarpov@mpgu.edu

sixth grade students' adaptive testing taking into account the type and degree of their math giftedness" by Vladimir N. Dubrovsky, Kolmogorov Boarding School of Moscow State University; and "Online educational platform for primary school students" by Anna Shvarts, Uchi.ru, an internet company.

Also participating were Alexander Soifer, professor at University of Colorado at Colorado Springs and the President of WFNMC; G. Mikaelyan and S. Arutunyan from Armenian State Pedagogical Abovyan University, Erevan; and Sergei Dorichenko, the editor-in-chief of *Kvantik*, a mathematics magazine for primary school teachers and pupils.

Open Access Except where otherwise noted, this chapter is licensed under a Creative Commons Attribution 4.0 International License. To view a copy of this license, visit http://creativecommons.org/licenses/by/4.0/.

Lesson/Learning Studies and Mathematics Education

Marisa Quaresma and Carl Winsløw

Lesson Study (Shimizu, 2014) and Learning Study (Runesson, 2014) (LS) have a growing importance in pre- and in-service mathematics teacher education, and are increasingly objects and methods of research in mathematics education. Many social, cultural, cognitive, and affective issues influence the way Lesson Study/Learning Study develops, and its outcome (Ponte et al., 2014). Research into these issues generates a quest for more solid theorization of the lesson-study process (e.g., Clivaz, 2015; Miyakawa & Winsløw, 2009). Both practice and research were focused on in this Discussion Group.

Both sessions were opened by short interventions by invited members of a panel (different for the two sessions) followed by questions and a general discussion led by a discussant.

The first session focused on regional/national particularities of and approaches to Lesson Study/Learning Study in mathematics education around the world. The panel members were Takuya Baba (Japan), Lim Chap Sam (Malaysia), Aoibhinn Ní Shúilleabháin (Ireland), and João Pedro da Ponte (Portugal), and the discussant was Stéphane Clivaz (Switzerland). Each of the panelists presented observations and questions from their experience with LS in their respective countries. In Japan, the use of LS has more than 100 years of history and is common in primary and lower secondary schools. In some countries outside Japan, many schools have become involved in LS, but in all countries it remains a major challenge to make LS "sustainable" in the sense that it can continue as a teacher-led activity without special funds or aid from outside. The rationales and effects of LS were also discussed; it was emphasized that the ultimate goal of LS is to further students' learning, even if this passes through teachers' learning.

The second session had theoretical, methodological and epistemological issues as its theme. The panel members were Toshiakira Fujii (Japan), Stéphane Clivaz

---

M. Quaresma (✉)
Instituto de Educação, Universidade de Lisboa, Lisboa, Portugal
e-mail: mq@campus.ul.pt

(Switzerland), Klaus Rasmussen (Denmark), María Soledad Estrella (Chile), and Akihiko Takahashi (USA), and Takuya Baba led the discussion. The main topics raised were ethical aspects of "repeats of lessons," which seem to be favored by some proponents of LS; recent theoretical developments in didactical research on LS; the roles and interaction of teachers and researchers in LS; and the use of new technology (such as the Lesson Note app). The crosscutting issue was what the key elements of LS are and which of these can and cannot be adapted to different contexts.

References

Clivaz, S. (2015). French Didactique des Mathématiques and Lesson Study: A profitable dialogue? *International Journal for Lesson and Learning Studies, 4*(3), 245–260.

Miyakawa, T., & Winsløw, C. (2009). Didactical designs for students' proportional reasoning: An "open approach" lesson and a "fundamental situation". *Educational Studies in Mathematics, 72*(2), 199–218.

Ponte, J. P., Quaresma, M., Baptista, M., & Mata-Pereira, J. (2014). Teachers' involvement and learning in a lesson study. In S. Carreira, N. Amado, K. Jones, & H. Jacinto (Eds.), *Proceedings of the problem at web international conference: Technology, creativity and affect in mathematical problem solving* (pp. 321–333). Portugal: Faro.

Runesson, U. (2014). Learning study in mathematics education. In S. Lerman (Ed.), *Encyclopedia of mathematics education* (pp. 356–358). Netherlands: Springer.

Shimizu, Y. (2014). Lesson study in mathematics education. In S. Lerman (Ed.), *Encyclopedia of mathematics education* (pp. 358–360). Netherlands: Springer.

Open Access Except where otherwise noted, this chapter is licensed under a Creative Commons Attribution 4.0 International License. To view a copy of this license, visit http://creativecommons.org/licenses/by/4.0/.

Mathematics Houses and Their Impact on Mathematics Education

Ali Rejali, Peter Taylor, Yahya Tabesh, Jérôme Germoni and Abolfazl Rafiepour

Since 1999, teams of the Iranian high school teachers and university faculty developed and promoted the idea of **Mathematics Houses**. A "Math House" is a community center that aims to provide a learning environment and opportunities for students and teachers at all levels to experience deeper understanding of mathematical concepts and develop creativity through teamwork and cooperation to work on real-life problems through (Barbeau & Taylor, 2009). The first Mathematics House was established in Isfahan, Iran, as an NGO for the occasion of Mathematics Year 2000. Mathematics Houses have been developed in more than 30 cities in Iran under the Iranian Union of Math Houses.

Mathematics Houses have also been established in other places around the world, for example at the Maison des Mathématiques et de l'Informatique (Lyon, France), La Maison des Maths (Quaregnon, Belgium), Mathematicum (Giessen, Germany), and the Archimedes Premises (Belgrade, Serbia).

Goal of the Discussion Group: Introduce Mathematics Houses and similar institutions around the world, discuss their effects on mathematics education and their important impacts on promoting teamwork and popularizing mathematics, and look for some new ways of cooperation and exchange of experiences.

Structure of the Discussion Group: First day of the program started with an introduction by Peter Taylor, followed by a talk by Ali Rejali introducing Isfahan Mathematics House. Christian Mercat then introduced the House for Mathematics and Informatics in France and Abolfazl Rafiepour discussed on developing Mathematics Houses in Iran, followed by a discussion session chaired by Peter Taylor. The second day started with a talk on content development in Mathematics Houses by Yahya Tabesh, which was then followed by panel discussions on goals, strategies, and programs for Mathematics Houses chaired by Peter Taylor with Christian Mercat, Yahya Tabesh, and Abolfazl Rafiepour as the panel members.

A. Rejali (✉)
Isfahan University of Technology, Isfahan, Iran
e-mail: a_rejali@cc.iut.ac.ir; a.rejali@yahoo.com

The closing session consisted of discussions about conclusion and future works and was chaired by Ali Rejali.

Outcome: Exchanging experiences globally, establishment of an International Network of Mathematics Houses and promotion of establishing Mathematics Houses in other regions worldwide.

Reference

Barbeau, E. J., & Taylor, P. J. (Eds.). (2009). *Challenging mathematics in and beyond the classroom, the 16th ICMI study, new ICMI study series*. New York: Springer.

Open Access Except where otherwise noted, this chapter is licensed under a Creative Commons Attribution 4.0 International License. To view a copy of this license, visit http://creativecommons.org/licenses/by/4.0/.

An Act of Mathematisation: Familiarisation with Fractions

Ernesto Rottoli, Sabrina Alessandro, Petronilla Bonissoni,
Marina Cazzola, Paolo Longoni and Gianstefano Riva

The discussion in our group concerned activities that have been carried out in some third and fourth grade classes in primary schools aimed at familiarizing children with the concept of fractions. Our proposal seeks to give an effective answer to the long-standing problem of unsatisfactory results in teaching and learning fractions. It starts with a process of mathematization that identifies the comparison of two homogeneous quantities by a pair of natural numbers. The concept of fractions is then introduced as the comparison-measure: A fraction is the measure of the quantity Q with respect to the whole W; the second quantity in the comparison, W, is always referred to as the "whole." Studies have shown that children can identify the relationship between fractions and division themselves. Therefore, the next step is a mutual interaction between the teacher and the children that allows them to arrive at Euclidean division. Thanks to this approach, Euclidean division is experienced by the children not as a formula to be memorized but as the icon of their active process of learning. Euclidean division is covered in fourth grade: All the *subconstructs of the construct of rational numbers* have their roots in Euclidean division and are related to it. In this way a new universe of fractions is structured with Euclidean division as core; the didactic process, while exploring the various contexts, keeps Euclidean division in mind and comes back to it constantly.

The discussion within our group covered three main points.

Linguistic splitting. The measure is the comparison between a quantity and the "special" quantity called "the whole" (formalized as n/n). The term *unit* (formalised as $1/n$) is reserved to indicate the common unit. So the whole differs from *the unit*. However: Two Names → Two Substantives → Two "Substances." This splitting results in the unusual classroom activity of the construction of the whole.

Exercise books as tools for noticing. In our class activities, the children's work in finding answers was relatively light in that they were able to come up with

---

E. Rottoli (✉)
Università Milano Bicocca, Milan, Italy
e-mail: ernerott@tin.it

adequate results fairly easily. However, the teachers found many difficulties, e.g. the usage of common manipulatives. In our discussion group, the following question was addressed: Were teachers called on to change their paradigms for a revolution? The discussion has outlined the key role of children's exercise books as objects of noticing: they allow notice of the features that are fundamental in scaffolding the universe of fractions.

Mathematisation is first. Our approach is based on a process of mathematisation and it aims to construct a *new universe* of fractions. Which is the *true universe* of fractions? What is its relation with the *technical universe* of fractions? We have proposed that the universe of fractions is a plurality of *tuned universes*. What does *tuned* mean in this context? The discussion about these questions remains open. It certainly requires further analyses and reflections.

Open Access Except where otherwise noted, this chapter is licensed under a Creative Commons Attribution 4.0 International License. To view a copy of this license, visit http://creativecommons.org/licenses/by/4.0/.

The Role of Post-Conflict School Mathematics

Carlos Eduardo Leon Salinas and Jefer Camilo Sachica Castillo

We arrived at the following conclusions after discussing and thinking about the role of mathematics in school following an armed conflict:

First of all, although mathematics has been an important factor in human culture, it has not been present in schools; therefore, students' knowledge that shows mathematical practices is not taken into account. For example, measurement is a practice that responds to the procedures and forms of a social group, and the school seeks to homogenize this process, keeping in mind the management of units and instruments that sometimes do not coincide with those that are part of the daily life of the students.

In the multicultural classrooms that are occurring in the post-conflict period in Columbia, it is necessary to involve knowledge-building practices in the learning and procedures of the schools. These should be part of the policies and guidelines designed by the Colombian Ministry of Education, finding unified ways to address problems and planning designs that have student learning as a starting point. Returning to the previous example of measurement, the idea should not be to tell the student how to measure but to explore the students' ways and resources so that they can be contrasted with what the school proposes to the relationship be identified.

Secondly, it is essential to manage coexistence based on knowledge that demonstrates the necessity of comparing procedures performed by students. Post-conflict mathematics classes cannot remain individual workspaces where the communication of ideas and the discussion of points of view are not considered. In a period when eradication of the violence of Colombian society is desired, it is very

---

C.E. Leon Salinas (✉)
La Gran Colombia University, Bogotá, Colombia
e-mail: carlos.leon@ugc.edu.co

important in the classroom to form communities of practice that revolve around solving the common problems and interests of a particular group. The students are not the only actors in these groups; the idea is to create a more inclusive classroom for other actors in the academic community, such as parents and teachers in other areas of knowledge, so that the math class becomes an interdisciplinary space.

Finally, it is necessary to think of mathematics as an instrument to understand the world and not as only a purpose; that is, mathematics must be understood as knowledge that allows us to comprehend what happens in our daily lives. Mathematics classes cannot be thought of as merely places where students assimilate concepts; in the post-conflict period, it is necessary for students to understand that mathematics can help them to verify hypotheses raised in the solution of a problem or the analysis of a situation.

These are aspects of three fundamental elements in the educational proposal must have for the post-conflict mathematics class.

Open Access Except where otherwise noted, this chapter is licensed under a Creative Commons Attribution 4.0 International License. To view a copy of this license, visit http://creativecommons.org/licenses/by/4.0/.

Applying Contemporary Philosophy in Mathematics and Statistics Education: The Perspective of Inferentialism

Maike Schindler, Kate Mackrell, Dave Pratt and Arthur Bakker

The aim of this discussion group was to put contemporary philosophy to work (cf. Cobb, 2007). Inferentialism is an example of contemporary philosophy (Brandom, 2000) that increasingly receives interest in mathematics and statistics education. It can be considered an orienting framework that provides epistemological foundations for conceptualizing and analyzing knowledge, learning, communication, and reasoning in the fields of mathematics and statistics. Inferentialism avoids a representationalist perspective on knowledge and learning by focusing on reasoning and inferences (Bakker & Derry, 2011). The Discussion Group (DG) brought together researchers who are interested in the role and use of inferentialism or other contemporary philosophies in mathematics and statistics education. It gave the attendants the opportunity to share perspectives, to question, to discuss, and to make joint efforts in answering the posed key issues. The DG format at ICME provided the opportunity to discuss the significance and the restrictions of the perspective of inferentialism and other contemporary philosophies on the learning and teaching of mathematics and statistics. The discussion was initiated by several talks: Arthur Bakker (Utrecht) introduced inferentialism as a semantic theory and Maike Schindler (Örebro) gave an overview on researchers presently working with inferentialism in mathematics and statistics education. Paul Ernest (Exeter) talked about meaning in mathematics and mathematics education and anti-representationalism, and Dave Pratt (London) gave a talk on constructionism. Alexandra Thiel-Schneider (Dortmund) presented an empirical study using inferentialism and Luis Radford (Ontario) summarized the discussion elaborating on how inferentialism relates to existing theories in our domain. The participants experienced the discussion group as a fruitful gathering of researchers interested in philosophy in mathematics education; and of various perspectives on inferentialism and its possible use. The talks were welcomed as an input and promoter of discussion among all participants.

M. Schindler (✉)
University of Cologne, Cologne, Germany
e-mail: maike.schindler@uni-koeln.de

The discussion has helped authors of articles for a special issue to appear in Mathematics Education Research Journal (e.g., Bakker, Ben-Zvi, & Makar, 2017; Derry, 2017; Mackrell & Pratt, 2017; Noorloos, Taylor, Bakker, & Derry, 2017; Schindler, Hußmann, Nilsson, & Bakker, submitted).

References

Bakker, A., Ben-Zvi, D., & Makar, K. (2017). An inferentialist perspective on the coordination of actions and reasons involved in making a statistical inference. *Mathematics Education Research Journal.*

Bakker, A., & Derry, J. (2011). Lessons from inferentialism for statistics education. *Mathematical Thinking and Learning, 13*(1–2), 5–26.

Brandom, R. (2000). *Articulating reasons: An introduction to inferentialism.* Cambridge, MA: Harvard University Press.

Cobb, P. (2007). Putting philosophy to work: Coping with multiple theoretical perspectives. In F. Lester (Ed.), *Second handbook of research on mathematics teaching and learning* (pp. 3–38). Greenwich: Information Age.

Derry, J. (2017). An introduction to inferentialism in mathematics education. *Mathematics Education Research Journal.*

Mackrell, K., & Pratt, D. (2017). Constructionism and the space of reasons. *Mathematics Education Research Journal.*

Noorloos, R., Taylor, S., Bakker, A., & Derry, J. (2017). Inferentialism as an alternative to socioconstructivism in mathematics education. *Mathematics Education Research Journal.*

Schindler, M., Hußmann, S., Nilsson, P., & Bakker, A. (submitted). Sixth-grade students' reasoning on the order relation of integers as influenced by prior experience: An inferentialist analysis. *Mathematics Education Research Journal.*

Open Access Except where otherwise noted, this chapter is licensed under a Creative Commons Attribution 4.0 International License. To view a copy of this license, visit http://creativecommons.org/licenses/by/4.0/.

Teaching Linear Algebra

Sepideh Stewart, Avi Berman, Christine Andrews-Larson and Michelle Zandieh

In this discussion group, the members of the leadership team gave a brief overview of their research, and posed the following set of research questions for discussion:

(a) How can applications of Linear Algebra be used as motivation for studying the topic?
(b) What are the advantages of proving results in Linear Algebra in different ways?
(c) In what ways can a linear algebra course be adapted to meet the needs of students from other disciplines, such as engineering, physics, and computer science?
(d) How can challenging problems be used in teaching Linear Algebra?
(e) In what way should technology be used in teaching Linear Algebra?
(f) What is the role of visualization in learning Linear Algebra?
(g) In what order (picture, symbols, definitions and theorems) should we teach Linear Algebra concepts?
(h) How can we educate the students to appreciate the importance of deep understanding of the Linear Algebra concepts?

Members of the leadership team provided materials for at least 3–4 groups building on themes and resources from their own work which provided the basis for more discussions. Over the two sessions the groups worked on various tasks. These tasks ranged from drawing concept maps of some major linear algebra concepts, to activities that pressed participants to coordinate geometric and algebraic interpretations of solutions to systems of linear equations, to challenging linear algebra tasks.

In the context of these activities, participants offered insights and perspectives from their experiences related to the teaching and learning of linear algebra from their country and institution.

S. Stewart (✉)
University of Oklahoma, Norman Campus, Norman, USA
e-mail: sstewart@math.ou.edu

Moving Forward

After the conference we accepted an invite from ICME-13 and proposed an edited manuscript. The work will appear in:

Stewart, S., Andrews-Larson, C., Berman, A., & Zandieh, M. *Challenges and strategies in teaching linear algebra*. Berlin: Springer.

Open Access Except where otherwise noted, this chapter is licensed under a Creative Commons Attribution 4.0 International License. To view a copy of this license, visit http://creativecommons.org/licenses/by/4.0/.

Creativity, Aha!Moments and Teaching-Research

Hannes Stoppel and Bronislaw Czarnocha

There were four presentations of different approaches to the observation and analysis of Aha! Moment in the classroom. Czarnocha presented bisociation theory and argued for the affective/cognitive duality as an essential component of the phenomenon. Liljedahl presented the point of view developed on the basis of his dissertation and subsequent research proposing a linear process based on Wallace/Poincare/Hadamard four stage theory. As conclusions cognitive experience is unremarkable whereas affective experience is remarkable for illumination. Stoppel presented his research from the course on coding and cryptography, pointing to two different condition of occurrence and related views of students on creativity, and Palatnik presented expanded analysis of his dissertation, which documented an Aha! Moment in connection with the Problem of dividing a Pizza into a maximal number of pieces by a given number of cuts. He argued for the high complexity of the process, expressing the point of view that bisociation is too simple to account for it.

By Czarnocha the definition of bisociation suggests that the construction of mathematical schema takes place during the Aha!moment. An Aha!moment is characterized by both an essential cognitive process and by an affective experience leading to cognitive/affective duality to the Aha! moment, since construction of the schema is a unique mathematical process.

According to Liljedahl Aha! Moments are not unique for mathematics. What characterizes Aha! moments in distinction to other mathematical activities is only affective experience, although Perter was heard accepting the fact that in the light of Koestler definition of bisociation as connecting two frames which by themselves are not connected, there is an important mathematical cognitive process of constructing a schema of thinking taking place. Palatnik argued that bisociation is not enough to imagine what happened before. Furthermore it is too simple a concept to account for the complexity of mathematical issues arising during the Aha! moments as e.g. not only the solution makes sense. Here the question appeared whether sense is necessary for an Aha!moment.

There were altogether around 15 participants, although their composition had changed between the first and the second session.

Important questions and comments during the discussion were as follows:

1. Is the creation of a schema of thinking relatively to the problem at hand a central cognitive component of the Aha!Moment or is Aha!Moment primarily an affective experience?
2. How do we know the Aha!Moment has taken place? Do we need the student to explain the content of the Aha!Moment? Is every generalization obtained through Aha!Moment? These questions referred to one Czarnocha's example; the other example, "the Elephant" was accepted as an example exemplifying bisociation.
3. Does the level of complexity of analysis is intrinsic to the nature of Aha! Moment or is the result of the methodology used?
4. Are different conditions in which student see the occurrence of creativity related to different approaches to problem solving?
5. Interesting complementary connection was observed joining Palatnik problem and one of Liljedahl's problems.

Open Access Except where otherwise noted, this chapter is licensed under a Creative Commons Attribution 4.0 International License. To view a copy of this license, visit http://creativecommons.org/licenses/by/4.0/.

White Supremacy, Anti-Black Racism, and Mathematics Education: Local and Global Perspectives

Luz Valoyes-Chávez, Danny B. Martin, Joi Spencer and Paola Valero

Understanding the forms wherein racism operates in mathematics teaching and learning practices requires taking into account the local meanings of race and the composing features of racialized social systems. The mechanisms through which racism is reproduced within school mathematics across racialized social systems differ subtly, but manifest themselves in strikingly similar ways. So, while the meanings of race are malleable and locally situated, its production of privilege, power, exemption, and disenfranchisement is stable. Likewise, the still widespread view that mathematics and mathematics education are non-political practices has allowed little interrogation of how racialized social practices permeate mathematics education. The minimization of attention to racism in international mathematics education research is perplexing, particularly in contexts characterized by long histories of anti-Blackness and colonialism and in other contexts that are facing increased levels of xenophobia, anti-immigration, and contestations of national identity.

The Discussion Group was an opportunity to create awareness on these issues and to push for analyses of race and racism in local, global, and international contexts. We aimed at facilitating discussion among colleagues around the world on experiences of racisms and racialization of mathematics education. We also intended to find and imagine different ways of collaborative work that move the field forward and have an impact on research, practice, and policy. Around 30 participants mainly from the United States, England, Brazil, South Africa, Portugal, and Germany took part in the discussions. An initial recalling of experiences that could be seen as instantiations of racism in relation to math education was a starting point for recognizing the overt as well as the subtle ways in which racism is instantiated and becomes present. In some contexts, being the "black body" is associated with incapacity for mathematical thinking. In other contexts, the color of skin as a marker intersects with belonging to

---

L. Valoyes-Chávez (✉)
Universidad Santiago de Cali, Cali, Colombia
e-mail: luz.valoyes00@usc.edu.co

© The Author(s) 2017
G. Kaiser (ed.), *Proceedings of the 13th International Congress on Mathematical Education*, ICME-13 Monographs, DOI 10.1007/978-3-319-62597-3_109

other categories such as "minority," "immigrant," or "indigenous." An important realization was that participants who would consider themselves concerned with diversity and equity were surprised by not only how racism can operate inside well-intentioned and progressive views of mathematical participation but also how well-intentioned progressivism in mathematics education can allow racism to maintain its role in shaping the larger social order. The aim of questioning the conflation of notions of race with "black bodies" was achieved as participants' contexts allowed them to evidence the local meanings and practices of racism.

Group leaders presented some key contemporary notions of race and racialization in connection to the notion of white supremacy. The reflection on how assumed characteristics of universality and abstraction of mathematics embedded in notions of school mathematics are part of the narratives of mathematics as the invention of White cultures were challenged. A significant concluding point was the importance of questioning the White epistemologies on which widespread ideas of mathematics education build. Such exploration could become an area of international collaboration for practitioners and scholars interested in broadening participation in mathematics education practices for many students who have been opted out through the operation of racialized epistemologies of mathematics education.

Open Access Except where otherwise noted, this chapter is licensed under a Creative Commons Attribution 4.0 International License. To view a copy of this license, visit http://creativecommons.org/licenses/by/4.0/.

Research on Non-university Tertiary Mathematics

Claire Wladis, John Smith and Irene Duranczyk

Description of Activities and Presentations

This session focused on research being conducted at non-university tertiary institutions. Instructors in adult education as well as other non-university postsecondary institutions participated. The participants of the session shared concerns about wanting to know more about the non-university mathematics classroom through classroom-based research by practitioners. Examples of research presently being conducted were shared (Mesa, Wladis, & Watkins, 2014; Sitomer et al., 2012; Wladis, Conway, & Hachey, 2016) and are cited below. The goal for developing collaborative research and/or grant proposals among researchers of similar interest continues.

A presentation of the National Science Foundation grant *Algebra instruction at community colleges: An exploration of its relationship with student success* (Watkins, Strom, Mesa, Kohli, & Duranczyk, 2015) design was shared. The researchers are exploring the impact of students' and instructors' pre-existing and moderating variable on the relationship between student-instructor interaction with mathematics in the classroom and students' performance outcomes. Research on how individual and institutional characteristics factor into failure rates and performance measures exists, but there is little information about the fundamental work of teachers in the classroom, and the interaction that occur between instructors, students, and the mathematical content. A qualitative study (Smith, 2016), provided the student point of view. It explored students' reflections on their experiences (unsuccessful and successful) in mathematics at the community college. Student voices provided essential insights as to how postsecondary educators might foster positive learning transformations, and avoid being the source of needless obstacles to degree attainment. Then the City University of New York

---

C. Wladis (✉)
City University of New York, New York, USA
e-mail: cwladis@gmail.com

(CUNY) was presented as an example of a college system in the United States where research by community college faculty has been systematically supported, and the structures to support faculty research were described. A few examples of research projects coming out of this CUNY system were presented: An NSF-funded project exploring factors that predict which characteristics put students at higher risk of dropping out of online versus face-to-face STEM courses (Wladis, Conway, & Hachey, 2014); and a project, instigated by elementary algebra instructors, to create a concept inventory for elementary algebra at the tertiary level.

References

Mesa, V., Wladis, C., & Watkins, L. (2014). Research problems in community college mathematics education: Testing the boundaries of K—12 research. *Journal for Research in Mathematics Education, 45*(2), 173–192.

Sitomer, A., Strom, A., Mesa, V., Duranczyk, I. M., Nabb, K., Smith, J., et al. (2012). Moving from anecdote to evidence: A proposed research agenda in community college mathematics education. *MathAMATYC Educator, 4*(1), 35–40.

Smith, J. T. (2016). *Stories of success: A phenomenological study of positive transformative learning experiences of low-socioeconomic status community college mathematics students* (Doctoral dissertation, Knoxville, TN: The University of Tennessee).

Watkins, L., Strom, A., Mesa, V., Kohli, N., & Duranczyk, I. (2015). *Algebra instruction at community colleges: An exploration of its relationship with student success*, National Science Foundation Award # 1561436.

Wladis, C., Conway, K. M., & Hachey, A. C. (2014). *Can student characteristics be used to effectively identify students at-risk in the online STEM environment?*, National Science Foundation Award #1431649.

Wladis, C., Conway, K. M., & Hachey, A. C. (2016). Assessing readiness for online education—research models for identifying students at risk. *Online Learning [Special Section: Best Papers Presented at the OLC 21st International Conference on Online Learning and Innovate 2016], 20*(3), 97–109.

Open Access Except where otherwise noted, this chapter is licensed under a Creative Commons Attribution 4.0 International License. To view a copy of this license, visit http://creativecommons.org/licenses/by/4.0/.

Part VIII
Reports from the Workshops

Flipped Teaching Approach in College Algebra: Cognitive and Non-cognitive Gains

Maxima J. Acelajado

This study looked into the cognitive and non-cognitive gains from using the flipped teaching approach (FTA) by comparing it with the traditional classroom approach (TCA) in the delivery of the following topics in college algebra: factoring, rational expressions, radicals, and solving applied problems. A quasi-experimental design with switching replication was utilized with 55 freshman students from two comparable intact classes of the College of Education, De La Salle University-Manila, during the first term of the 2014–2015 school year.

The two classes, designated as the experimental group and the control group, were exposed to FTA and TCA, respectively. A validated teacher-made pre-test and post-test on the topics under consideration were administered to the respondents to gauge and compare their achievements in each topic. They were also asked to answer a perceptions inventory and to write a journal about their experiences with FTA after the experiment.

With FTA, the students learned new content, prior to class and online, by watching video lectures, websites, and PowerPoint presentations, which were provided by the teacher. During the face-to-face class time, they were asked to answer what used to be the homework/assignment and some assessment materials, either individually or in groups to test their understanding of the lesson. Although the teacher did not impart the initial lesson in person, she tutored the students in the classroom, whenever necessary. The TCA was done with the teacher handling the class discussion on exactly the same topics, after which the students were given the assignment/homework for submission and discussion during the next meeting.

Findings revealed cognitive gains such as better critical thinking ability, improved achievement, and significant learning gains in each topic as seen from the result of the t-test for dependent and independent samples, generally in favor of FTA. Moreover, the use of FTA produced non-cognitive gains such as the

M.J. Acelajado (✉)
De La Salle University, Manila, Philippines
e-mail: acelajadom@gmail.com; maxima.acelajado@dlsu.edu.ph

improvement of students' attitudes toward mathematics and greater cooperation among them. They seemed to be more motivated, confident, relaxed, responsible, and active in learning when exposed to the flipped teaching approach. The majority indicated that they were happy to have control of their own learning, as they were able to explore more mathematical concepts through various modes and resources outside the classroom at a time convenient for them. The students appreciated the personalized guidance of their teacher, which kept them engaged in learning while having their misconceptions regarding the day's lesson clarified by the teacher.

Open Access Except where otherwise noted, this chapter is licensed under a Creative Commons Attribution 4.0 International License. To view a copy of this license, visit http://creativecommons.org/licenses/by/4.0/.

A Knowledge Discovery Platform for Spatial Education: Applications to Spatial Decomposition and Packing

Sorin Alexe, Cristian Voica and Consuela Voica

This workshop introduces a novel kind of math manipulatives and activities aimed to develop spatial intelligence in middle school students. The participants are trained through a series of 2D and 3D geometric puzzles that involve hands-on activities. A short introductory presentation and several brief videos present the platform and the plan of the workshop. Participants receive the manipulatives and the instructions needed to accomplish two projects. They group in four teams and they work together within each team. Here we present one 2D puzzle and one 3D puzzle that were solved during the workshop:

The Shadow Puzzle

For each of the shadow regions in Fig. 1b find all feasible decompositions using only one element of the alphabet in Fig. 1a at a time. How many such decompositions do you find?

A 2D Packing Problem

Using folding and taping, one can transform the 2D element shown in Fig. 2a into the 3D module shown in Fig. 2b. You are given a 2D region shaped as an equilateral triangle such that on each side you can fit four identical 3D modules as depicted in Fig. 2c. Find the maximum number of 3D modules that could be packed

---

S. Alexe (✉)
XColony Project, Stamford, USA
e-mail: salexe@aol.com; sorin@x-colony.com

© The Author(s) 2017
G. Kaiser (ed.), *Proceedings of the 13th International Congress on Mathematical Education*, ICME-13 Monographs, DOI 10.1007/978-3-319-62597-3_112

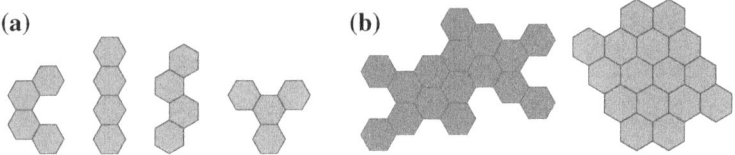

Fig. 1 **a** The alphabet of 2D elements, **b** the shadow regions

Fig. 2 Transformation from 2D to 3D

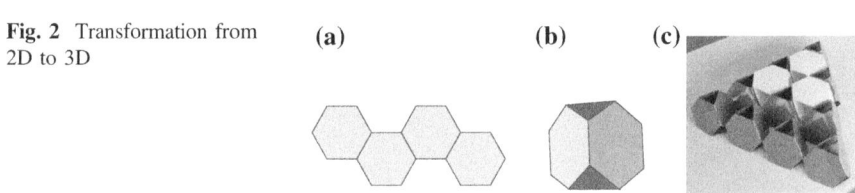

into this 2D region. The height of the packed region should not exceed the height of one 3D module. Do you find the answer is 16?

A Brief Brainstorming Session Concludes the Workshop

The participants are confident that this kind of manipulatives can be successfully used for increasing the effectiveness of teaching geometry to middle school students, for improving the level of communication in class and for increasing students'interest in learning geometry.

Further developments of the Knowledge Discovery Platform that expand the scope of this research are planned through ongoing joint international educational projects involving workshop's organizers and participants.

Open Access Except where otherwise noted, this chapter is licensed under a Creative Commons Attribution 4.0 International License. To view a copy of this license, visit http://creativecommons.org/licenses/by/4.0/.

Designing Mathematics Tasks for the Professional Development of Teachers Who Teach Mathematics Students Aged 11–16 Years

Debbie Barker and Craig Pournara

Theme

Our workshop focused on identifying and developing tasks appropriate for use in professional development with mathematics teachers of 11–16 year olds who self-identify as "non-specialists". We invited professional development practitioners, teachers and researchers to join us for a workshop that would involve active engagement with such professional development tasks and a reflection of these tasks from different international perspectives and practice.

Overview

We explained how our collaboration had started with a meeting in England July 2015 and how we had worked on two "seed activities" which are simple enough to support collaboration but rich enough to be developed for the needs of teachers in our different professional development contexts.

Seed 1—Area of polygons card set: Each card contains a triangle or quadrilateral drawn on a grid. Debbie demonstrated how these could be used as a classroom activity with students. The participants spent time ranking the cards in order of difficulty and discussing the different methods students might use to find the area of each polygon. We then discussed how the task is used during a professional development session. Participants shared their ideas about the activity and their

---

D. Barker (✉)
Mathematics in Education and Industry (MEI), Trowbridge, UK
e-mail: debbie.barker@mei.org.uk

experiences of using similar activities with students or in professional development. It was agreed that the task is useful as a starting activity for teachers developing their own subject knowledge.

Seed 2—Learning through variation: Following an initial discussion about the ideas underlying variation theory, we worked very briefly to the ideas of variation theory and then worked individually on a series of equations involving square roots, e.g. $\sqrt{x-3} = x - 5$. The choices within this example set were made explicit by analysing the dimensions of variation and the range of permissible change. This was followed by examples to explore the number of solutions for equations of the form $\sqrt{x-k} = x - k$, a question posed by teachers in a session with Craig. We acknowledged that while many present could recognise the pattern in the solutions by inspection, it was not likely that non-specialist mathematics teachers would do so. Consequently the task provides opportunity for developing fluency together with the larger purpose of investigating patterns in the solutions.

Finally, Debbie explained how she had used the same seed. At the MEI Conference 2016 she ran a session, *The Art of Choosing Examples*. This was very much a "beginners' guide" to variation theory and its application in the classroom of the everyday teacher. In contrast to the intended audience of our collaboration, the session was attended by people with a wide range of experience in mathematics education. It was a well-received session and feedback indicated that the ideas of variation are very accessible to teachers and useful in their daily practice. This resonates with Craig's experience in working with teachers in the SA context. There is likely much potential for further supporting teaching and learning using the principles of variation theory.

We were keen for the workshop to be of practical use to delegates, so the session ended with a sharing of a selection of tasks that have been used successfully in professional development with non-specialist mathematics teachers. They were well-received by the participants.

Open Access Except where otherwise noted, this chapter is licensed under a Creative Commons Attribution 4.0 International License. To view a copy of this license, visit http://creativecommons.org/licenses/by/4.0/.

Contributing to the Development of Grand Challenges in Maths Education

David Barnes, Trena Wilkerson and Michelle Stephan

David Hilbert (1902) presented a set of challenging problems before the International Conference of Mathematics and argued that mathematics was at a crossroads and a sense of direction was necessary to guide the field forward in the next century. He submitted a list of mathematics problems that had yet been solved and some questions that had not even been asked to spark creativity and discovery. He also understood the importance of a community of scholars being able to determine when a problem is solved, the need to develop new representations to express a problem, and the ability to explain its solution.

As others have undertaken defining Grand Challenges, Gould (2010), the following represent some guidelines:

1. Define complex and difficult questions that are solvable within a 10–20 year timeframe;
2. Positively impact the quality of life for potentially millions through educational, social, and economic outcomes;
3. Require multiple sub-disciplines to engage/collaborate on research;
4. Have a defined, measureable outcome to gauge progress and completion; and
5. Garner support publicly and within the field which understand, and appreciate the outcomes of the effort.

The Research Committee (Stephan et al., 2015) introduced the concept of Grand Challenges to our mathematics education research community. While there are significant differences in the local and national challenges when you consider

---

D. Barnes (✉)
Research, Learning and Development, National Council of Teachers of Mathematics, Reston, USA
e-mail: dbarnes@nctm.org

mathematics education globally, there are also likely to be significant overlap. The endeavor of ICME is founded on the desire to "promote the collaboration, exchange and dissemination of ideas and information on all aspects of the theory and practice of contemporary mathematical education." The voluntary participation in and adoption of a common set of Grand Challenges in Maths Education works to support the international collaboration toward common challenges.

The following key questions were discussed:

- As you consider the state of maths education in your community/country what would you see as research questions that could qualify as grand challenges?
- Which of a preliminary set of additional questions would be of interest to the maths education and research community for your country? What are the benefits and what are the risks?
- Are there opportunities to engage educators, researchers and policy makers in the development of Grand Challenges in Maths Education and collaborative efforts to solve these problems?

Small group discussion surfaced the universal nature of some challenges and the real and regional nature of other challenges. Consistent challenges include the access to education for all individuals no matter gender or status.

References

Gould, M. (2010). GIScience grand challenges: How can research and technology in this field address big-picture problems? *ArcUser, 13*(4), 64–65. Retrieved from http://www.esri.com/news/arcuser/1010/files/geochallenges.pdf

Hilbert, D. (1902). Mathematical problems. *Bulletin of the American Mathematical Society, 8*(10), 437–479. doi:10.1090/S0002-9904-1902-00923-3

Stephan, M. L., Chval, K. B., Wanko, J. J., Civil, M., Fish, M., Herbel-Eisenmann, B., Konold, C., & Wilkerson, T.L. (2015). Grand challenges and opportunities in Mathematics Education Research. *Journal for Research in Mathematics Education, 46*(2), 134–146. Retrieved from http://bit.ly/1Yd4q2V

Open Access Except where otherwise noted, this chapter is licensed under a Creative Commons Attribution 4.0 International License. To view a copy of this license, visit http://creativecommons.org/licenses/by/4.0/.

The Role of the Facilitator in Using Video for the Professional Learning of Teachers of Mathematics

Alf Coles, Aurelie Chesnais and Julie Horoks

Workshop Activities

The aim of this Workshop was to learn from comparing and contrasting two established ways of working with video in the professional learning of teachers of mathematics, one from a UK and one from a French background; we hoped to expand the possibilities our own practice, in using video.

We addressed the following key questions: (1) how can and do facilitators guide work with mathematics teachers on video in a particular context?; (2) what are the principles, based upon research on teacher practice and teacher education, that guide our choices for teacher education and in particular our use of the video?; (3) what are the implications, for mathematics teacher learning, of different choices made by facilitators?

In the workshop, we shared the detail of our practice and how wider principles are enacted when using video. The first way of working we offered was based on principles derived from Jaworski (1990) and Coles (2014). The second way of working was based on principles derived from Horoks and Robert (2007), Chesné, Pariès, and Robert (2009), Chappet-Pariès and Robert (2011). Interestingly, we found that it was not possible to use the same video excerpt. In discussion with participants, there was broad agreement that differences in ways of working could be characterized as follows: in the method demonstrated by Coles, participants were forced to suspend their usual ways of interpreting video—any evaluation or judgment was not allowed in the initial stages, in an attempt to allow new interpretations and possibilities for action to arise. In the method demonstrated by Chesnais and

---

A. Coles (✉)
University of Bristol, Bristol, UK
e-mail: alf.coles@bris.ac.uk; alf.coles@gmail.com

Horoks, there was a specific aim to draw on teacher knowledge and expertise and put that experience to use in interpreting events—the video clip is carefully selected in relation to specific curriculum items and research findings. The role of the facilitator is therefore markedly different in each case, either attending to the kind of thing being said (e.g., is it an evaluation or is it a description of detail?), with less attention on the content (Coles); or attending to the content of what is said (e.g., does it display awareness of the complexity of teaching and learning?), with less attention to issues around whether it is offered judgmentally or not (Chesnais and Horoks).

It was clear from the Workshop that both ways of working have affordances and constraints. What has been powerful is sharing the detail of what we do as this has emerged for us as the only way of beginning to understand how each of us interprets the words we use to describe what we do.

References

Chappet-Pariès, M., & Robert, A. (2011). Séances de formation d'enseignants de mathématiques (collège et lycée) utilisant les vidéos—exemples. *Petit x, 86,* 45–77.

Chesné, J.F., Pariès, M., & Robert A. (2009). «Partir des pratiques» en formation professionnelle des enseignants de mathématiques des lycées et collèges. *Petit x 80,* 25–46.

Coles, A. (2014). Mathematics teachers learning with video: The role, for the didactician, of a heightened listening. *ZDM—The International Journal on Mathematics Education, 46*(2), 267–278.

Horoks, J., & Robert, A. (2007). Tasks designed to highlight task-activity relationships. *Journal of Mathematics Teacher Education, 10*(4–6), 279–287.

Jaworski, B. (1990). Video as a tool for teachers' professional development. *Professional Development in Education, 16*(1), 60–65.

Open Access Except where otherwise noted, this chapter is licensed under a Creative Commons Attribution 4.0 International License. To view a copy of this license, visit http://creativecommons.org/licenses/by/4.0/.

Making Middle School Maths Real, Relevant and Fun

Kerry Cue

Over the last 20 years participation in STEM studies in senior school has steadily dropped in many western countries. In Australia participation rates in Year 12 Advanced Maths was 9% in 2012. In England 85% of all students drop maths at 16 years of age. While in the USA many students don't even have access to a full senior school maths program.

Claiming single issues from individual countries deliver successful maths education is simply 'cherry picking' results. For instance, 25% of Scottish students study maths until at least 18 years of age compared with only 15% in England. In Scotland a flexible curriculum helps promote maths participation. Moreover, the push by politicians to introduce back-to-basics curricula and methods of teaching in the UK, USA and Australia produces rigid, repetitive, pre-packaged, parrot-style learning when 'effortful, varied practice builds mastery'.

'Why Do so Many Students Hate Maths?' Because It Is Scary, Boring, Pointless and Everyone Hates It Anyway

Reasons why individual students drop out of mathematics before senior school cited across the research literature include peer group pressure, irrelevance, boredom, rigid curriculum, low quality text books, lack of innovation, cultural influences, over confidence of students (leading to less effort), low expectations of parents, teachers and the socio-economic group.

Some of the reasons are, however, systemic such as poorly trained teachers, a lack of funding and changes in university entrance requirements. While it is tempting to recommend that maths be made compulsory at senior school level,

---

K. Cue (✉)
MATHSPIG, Melbourne, Australia
e-mail: kerrycue@tpg.com.au; kerrycue@yahoo.com.au

there is a very strong argument against that move. Research by Professor Geoff Prince highlighted the Catch 22 of promoting maths education in schools. If maths is made compulsory at senior school level more students drop out of school. But if maths is not made compulsory in the senior years good students will drop out of higher level maths to boost their university entrance scores by pursuing a lower (and terminal) level maths. This will reduce the number of students available to study tertiary level maths.

Standardising maths curriculum coupled with the drive to meet national standards can restrict innovation and prevent classroom teachers from using a variety of methods to both motivate and inspire their students. Yet, it is vital to improve student attitudes to maths in middle school to increase numbers of students taking senior level and tertiary level maths. But how do you do that?

Make Middle School Maths Interesting, Fun, Exciting, Inspiring and Accessible to All Students

According to Cindy Moss, Global STEM Initiative, we should *'Empower our teachers to be able to show kids that STEM is fun'*. Middle school maths has to be real and relevant too. When I started my MATHSPIG BLOG (1,000,000+ hits) I had an epiphany. I realised that for all the years I had studied maths I never found an answer to a question that I wanted to know. I found answers I wanted to get right and then I moved on to the next question. The aim of Mathspig has been, therefore, to ask maths questions that prompt middle school students to be curious to know the answer. EG. How old is your hair? How long would it take a 14yo to bleed to death from an arrow wound? How many m&ms will kill a 14yo? And much more. https://mathspig.wordpress.com/

Open Access Except where otherwise noted, this chapter is licensed under a Creative Commons Attribution 4.0 International License. To view a copy of this license, visit http://creativecommons.org/licenses/by/4.0/.

"Oldies but Goodies": Providing Background to ICMI Mission and Activities from an Archival Perspective

Guillermo P. Curbera, Bernard R. Hodgson and Birgit Seeliger

Aim of the workshop

To draw attention to the importance and usefulness of archiving among the mathematics and mathematics education communities, present some of today's modern technical tools and focus on the particular case of the ICMI Archive—a subset of the Archive of the International Mathematical Union (IMU).

1. **The IMU Archive: Past and present**. Why is archiving necessary? Keeping the memory of the past is a joint social responsibility. Archiving is thus a way of honouring the efforts of our predecessors. Science is not excluded from this responsibility; institutions are responsible even to a larger extent. Mathematics and mathematics education will benefit from having an accurate image of their past.

    The IMU/ICMI Archive keeps records from 1950 on (the previous ones were lost). They were first stored at the Eidgenössische Technische Hochschule in Zurich. In the 1990s they were transferred to Helsinki, where Olli Lehto, former IMU Secretary, organized them. Currently they are at the permanent IMU office in Berlin. There are Regulations for the Archive, which include restrictions for certain materials (70 years for matters related to awards and prizes).

2. **Technical aspects of archiving today**. The IMU/ICMI Archive holds analogue and digital records. The criteria for professional archiving are authenticity, reliability, integrity, and usability of the material. Digital records cause new challenges, as they are in constant technological change. Therefore, their life

---

G.P. Curbera (✉)
Universidad de Sevilla, Sevilla, Spain
e-mail: curbera@us.es

cycle management requires long-term preservation storage and formats that may derivate from original formats to ensure the survival of their content. Special procedures for email archiving include a structured file migration into the established long-term format PDF/A, which guarantees record-keeping as well as search functions.

Images play an important role for the mathematical community. Hence the first online project of the IMU/ICMI Archive is a platform for photos. "The IMU Media Platform is offered by the IMU to help showcase and illustrate the history and activity of the Union and its commissions and associated persons or events, which supports the idea of the worldwide network of mathematicians. The IMU provides the platform so that platform members can install their own image objects in a database and offer these images to other members for their use." (Quoted from IMU website)

3. **The content of the ICMI archive.** Existing as a subset of the IMU Archive, the ICMI Archive comprises in particular five boxes of paper documents (mainly letters) collected during the last half of the 20th century. A large portion of these goes back to the early days of the Archive. Fewer documents have been filed since the early 1990s, when email was becoming the main channel of communication. The archived documents are mostly in English, with a few in French, German, or Italian. Of notable interest are, for instance, documents related to moments of turbulence in the life of the ICMI. The Archive also contains books produced in the context of ICMI activities (ICME proceedings, ICMI Study volumes, etc.).

Open Access Except where otherwise noted, this chapter is licensed under a Creative Commons Attribution 4.0 International License. To view a copy of this license, visit http://creativecommons.org/licenses/by/4.0/.

Using Braids to Introduce Groups: From an Informal to a Formal Approach

Ester Dalvit

The workshop was based on an outreach activity on braids given several times to high school students, from Grade 8 advanced students up to Grade 12, to give a taste of actively "doing" mathematics.

We reviewed and discussed some of the motivations to expose high school students to mathematics enrichment such as reinforcing motivation to learn, getting a broader view of mathematics, meeting professional mathematicians, and developing independent thinking and collaboration.

We then demonstrated a part of the activities have given to high school students: Starting from a concrete, physical object, turn it into an abstract and formal one. In particular, the participants, working in pairs, were asked to use strings to realize a braid given by a drawing, describe it in plain English "as one would on the phone", and finally to use symbols to shorten the description, "as one would in a text message". This process involves some notational choices and naturally leads to a description of a braid in levels, as in the picture. The symbols in the picture are just one of the many different notations proposed by students and by participants in the workshop.

We discussed the didactic importance of accepting different choices for the notation and stressing the motivations behind them. This approach gives an idea of mathematics as a creative discipline, in contrast with the usual perception of its nature as static, where conventions are imposed by teachers or books and motivations are considered too abstract and too difficult to be understood.

Ester Dalvit: INdAM-Marie Curie fellow.

E. Dalvit (✉)
University of Camerino, Camerino, Italy
e-mail: ester.dalvit@gmail.com

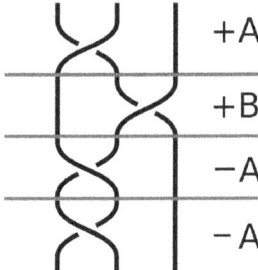

We then introduced the formal counterpart of the intuitive process: Dividing the braid into levels and assigning a symbol to each of them corresponds to the theoretical process of finding generators for the braid group and building braids via an operation called composition. We noted the power of this specific example, where students naturally construct a correct model. This is just a small example, but it is representative of the general mathematical activity. Relating the informal description to a formal one can help the students appreciate the importance of formalism.

Finally, we explored the axioms of groups using the specific cases of braids under the operation of composition and integer numbers under addition, drawing a parallel between the two examples and finding one difference. We discussed the importance of the idea of structure in mathematics, which is often not recognized at all. The idea that the subject of mathematics is computations has been widespread, but the justification as to why computations are possible has been overlooked: Operations, and ultimately structure, make computations and their rules correct and applicable. Braids are a simple example where this can be demonstrated: Some properties hold which are similar to those for numbers, yet there are some differences. Finding rules that hold for braids is an appealing task, because it is in the zone of proximal development for high school students.

Participants in the workshop had different backgrounds: Some were high school teachers, some researchers in mathematics education, and some scholars doing research in mathematics and teaching. The discussion was enriched by the differences in background, teaching experiences, and goals. The contributions of the participants also included considerations about teaching group theory at the college level and about braids in ethnomathematics.

Open Access Except where otherwise noted, this chapter is licensed under a Creative Commons Attribution 4.0 International License. To view a copy of this license, visit http://creativecommons.org/licenses/by/4.0/.

Curious Minds; Serious Play

Jan de Lange

Young children are great researchers. The natural amazement and curiosity of young children are the starting points for exploration of their world. Especially for parents with children aged 3–6 years the opportunity should not be missed.

The signals are clear and all point in the same direction: young children (3–6 years) can do much more than we think. As parents we just have to facilitate the development the talents. There are especially great opportunities to take the natural curiosity of children as a starting point for making talents visible such as logical reasoning and problem solving in a fun and challenging way.

Much research has been done under the name Curious Minds during the period starting in 2005. It's goal was to make visible the talents and insights of young children. The plan behind the *Young Parent Academy* is to use the results of this research: have parents look in a different way to the early development of their children and support parents with small suggestions in a practical way in order to offer children more challenges when they are playing. **The Curious Child, Existing.**

J. de Lange (✉)
Freudenthal Institute, Utrecht, The Netherlands
e-mail: j.delange@uu.nl

Literature and Our Challenge

Challenge: how do we get the two cars equally large at the photo? Core Mathematical Concept: Perspective

Can we keep the curiosity in the child? Can we further facilitate the development of these very valuable process skills?

If we go back to the start of the Curious minds project the original questions were:

1. What talents, possibilities, and qualities do children in the age-range 3 ton 6 have?
2. How can we optimally enhance the development of these qualities and talents?
3. How intertwined are these talents? Can they be attributed to disciplines as mathematics, science, technology and engineering, or are the more broad and connected?
4. How can scientific talent development be used for language development.

Seven Arguments (If You Need Them) to Get Started

The arguments to exploit these opportunities are manifold (apologies for deleting the sources).

1. Young children are natural scientists and researchers.
2. Young children can do much more in arithmetic than we think.

3. Investing in young children is necessary and promising.
4. Investing in young children provides a high return on investment.
5. Young brains give great opportunities to learn and develop.
6. The natural curiosity of young children is the basis for logical thinking.
7. (Serious) play is a way in which children learn. Curiosity, imagination and creativity are like muscles: if you don't use them, you lose them.

Therefore
DESIGN ACTIVITIES
We have designed many interesting play-activities with a solid scientific content that have led to new insights how young children reason and think.

The experiments have resulted in the development of more than a hundred activities and are recorded on more than 2000 interviews with 'playing' children.

Connections between formal and informal learning, and between play, reasoning and core scientific concepts are at the core.

Three of these activities were part of the workshop: how to 'classify' (play) animals?, can you copy a Lego building, using Duplo, and make a picture where two similar cars seem to have the same size.

The people attending actually did an excellent job, behaving as young kids. It was a challenge for them as well.

Open Access Except where otherwise noted, this chapter is licensed under a Creative Commons Attribution 4.0 International License. To view a copy of this license, visit http://creativecommons.org/licenses/by/4.0/.

International Similarities and Differences in the Experiences and Preparation of Post-Graduate Mathematics Students as Tertiary Instructors

Jessica Deshler and Jessica Ellis

In the United States, post-graduate students are used in the teaching of undergraduate mathematics in a variety of ways, from grading papers for experienced instructors to having full responsibility of their own class. There is an established network of scholars in the US whose members examine various aspects of how these instructors are utilized, how they develop as tertiary mathematics instructors and how they are professionally prepared. There are also nationally funded initiatives underway aimed at creating stronger networks among those that wish to provide professional preparation to these instructors, those that create resources for use in professional development, and scholars that conduct research in this area.

The goal of this workshop was to bring together scholars from mathematics, education and academic development from around the world to discuss how post-graduates are (1) involved in the instruction of undergraduate mathematics and (2) how they are prepared for their teaching roles. The organizers presented an overview of what is known about both of the topic questions based on data from a large national survey conducted as a joint effort of teams from two projects whose goals are to understand and support post-graduate mathematics student instructional development in the US.

There were eleven attendees from eight countries representing post-graduate students through administrators. Participants were asked to describe the following during the workshop:

- How post-graduate students are involved in the teaching of tertiary mathematics at their institution (and how typical this is of other institutions in their country, to the best of their knowledge).

J. Deshler (✉)
West Virginia University, Morgantown, WV, USA
e-mail: deshler@math.wvu.edu

- How post-graduate students are prepared for teaching tertiary mathematics at their institution (and how typical this is of other institutions in their country, to the best of their knowledge).

Workshop organizers collected written responses to these questions as the first step towards developing a written overview of the differences and similarities between nations in the practices surrounding post-graduate student instruction of tertiary mathematics. We plan to submit this overview for publication so that it can serve as the beginning of international conversations regarding the tertiary teaching preparation of post-graduate students, as well as other novice instructors (such as post-docs). Participants expressed interest in contributing to the development of an in-depth international understanding of practices related to the issues of the workshop through more formal and systematic data collection. Organizers plan to modify the US survey used to collect information on post-graduate involvement in and preparation for teaching based on the workshop discussion and comments from participants to make it applicable to the contexts of other nations. Through this ongoing work, we hope to provide international perspectives on these issues, to facilitate communication and collaboration between scholars working in similar fields in different contexts, and to identify issues for future research collaborations related to these issues.

Open Access Except where otherwise noted, this chapter is licensed under a Creative Commons Attribution 4.0 International License. To view a copy of this license, visit http://creativecommons.org/licenses/by/4.0/.

Using LISP as a Tool for Mathematical Experimentation

Hugo Alex Diniz

The following problem has been studied quite often (amongst others by Gardner, 1979):

> Two numbers (not necessarily different) are chosen from the range of positive integers greater than 1 and not greater than 20. Only the sum of the two numbers is given to Mathematician S. Only the product of the two is given to Mathematician P.
>
> On the telephone S says to P: "I see no way you can determine my sum."
>
> An hour later P calls him to say: "I know your sum."
>
> Later S calls P again to report: "Now I know your product."
>
> What are the two numbers?

This problem is known as the "Impossible Problem" because mathematician Martin Gardner (1914–2010) named it so in the article "A Pride of Problems, Including One That Is Virtually Impossible," published in his column Mathematical Games in *Scientific American* magazine in 1979. The interesting thing is that the problem really is "impossible" because it has no solution! But this was not the intention. Gardner wrote about this fact some months later (Gardner, 1980). Gardner tried to simplify a problem originally proposed by the Dutch mathematician Hans Freudenthal (1905–1990) (Freudenthal, 1969). It was published in German, but was translated by Davis Sprows in 1976:

Let x and y be two numbers with $1 < x < y$ and $x + y \leq 100$ Suppose S is given the value $x + y$ and P is given the value xy.

(1) P says: "I don't know the values of x and y".
(2) S replies: "I knew that you didn't know the values."
(3) P responds: "Oh, then I do know the values of x and y".
(4) S exclaims: "Oh, then so do I."

H.A. Diniz (✉)
Federal University of Western Pará, Santarém, Brazil
e-mail: hugo.diniz@ufopa.edu.br

© The Author(s) 2017
G. Kaiser (ed.), *Proceedings of the 13th International Congress on Mathematical Education*, ICME-13 Monographs, DOI 10.1007/978-3-319-62597-3_121

What are the values of x and y? (Sprows, 1976)

We call the above problem by Freudenthal's Problem with sum equal to 100, represented by P_{100}. The problem P_{100} is not impossible and has a unique solution!

We propose using the LISP language to approach these problems. For example, with the following code, we translate the Gardner's Problem to LISP and prove that it has no solution:

(defun domain (x y) (<= 2 x y 20))
(defun sums (p)

(loop for n from 2 to (isqrt p) when (and (domain n (/p n)) (zerop (mod p n))) collect (+ n (/p n)))))

(defun products (s)

(loop for n from 2 to (/s 2) when (domain n (- s n)) collect (* n (- s n)))))

(defun revealing (p) (= 1 (length (somas p)))))
(loop for sum from 4 to 40 when (notany #'revealing (products sum)) collect sum)

Modifying the above code, it is possible to study problem P_{100} and discover its unique solution.

References

Freudenthal, H. (1969). Formulering van het som-en-productprobleem. *Nieuw Archief voor Wiskunde, 17*(3), 152.

Gardner, M. (1979). Mathematical games. *Scientific American*, 241(6), 22–30, dez. 1979. doi:10.1038/scientificamerican1279-22

Gardner, M. (1980, March) Mathematical games. *Scientific American, 242*(3), 24–38. doi:10.1038/scientificamerican0380-24

Sprows, D. (1976, March). Problem 977. *Mathematics Magazine, 49*(2), 96.

Open Access Except whereotherwise noted, this chapter is licensed under a Creative Commons Attribution 4.0 International License. Toview a copy of this license, visit http://creativecommons.org/licenses/by/4.0/.

Mathematics Teachers' Circles as Professional Development Models Connecting Teachers and Academics

Nathan Borchelt and Axelle Faughn

Short Description of the Workshop: Aims and Underlying Ideas

Math Teachers' Circles (MTC) are professional development communities of mathematics teachers and professors who meet regularly to work on rich mathematics problems. Ongoing research has begun to demonstrate the benefits of MTC for teachers' confidence, knowledge, and teaching of mathematics. Mathematics professors gain an opportunity to share their enjoyment of mathematics with teachers, contribute to teacher education and enrichment, and become more involved in the local education community. During this workshop we introduced participants to MTC professional development models, engaged them in MTC-type mathematics, shared some results of MTC interventions, and opened the discussion to further ideas and/or questions on implementing MTC in various contexts.

Structure of the Meeting

Introductory comments provided ICME participants with information on the MTC model being used in the US. An engaging "Brownie Problem" was shared with participants in order to simulate a typical MTC meeting. The specifics of this activity and how it can be used is described in further detail at http://www.mathteacherscircle.org/resources/mathematical-materials/.

A. Faughn (✉)
Western Carolina University, Cullowhee, USA
e-mail: afaughn@email.wcu.edu

ICME participants reported enthusiasm for being challenged mathematically, genuine enjoyment of the group discussions following the initial challenge and frustrations of making little headway on their own. The workshop allowed us to make connections among existing MTCs and generate interest in starting new circles where they do not yet exist through guidance and experience-based advice.

Key questions and issues that participants were asked to consider include: (1) In what ways can Math Teachers' Circles contribute to the increase of mathematics content knowledge of teachers? (2) How can participation in Math Teachers' Circles impact the type of experiences or level of mathematics that teachers share with students? In other words, is there transfer to the classroom and what form does such transfer take? What resources used in professional development are particularly conducive to this transfer?

Unfortunately, workshop participants' demographics were hardly representative of ICME's varied and international flavor as the group consisted mostly of American mathematics educators. It would be very interesting to discuss this model and others similar to it from an international perspective. Still, we were able to share some initial research results with the group. Some categories have emerged from our data collection regarding what participants identify as areas of growth through participation in a MTC which somewhat overlap findings from prior studies. However, the role of resources used by the collective has not been studied fully, even though that is a stated goal of MTC interventions. We fill this gap in the research and emphasize how selected resources create a nature of mathematical practices. We are looking at the strategies that carry over into the classroom not only through instructional changes but also through teachers adapting their involvement with the Mathematics to engage students more meaningfully. In particular when participants were asked how MTC involvement can help them build upon existing resources, teachers' responses are varied, but consistently offer a picture of how they adjust their pedagogy after taking part in MTC sessions by letting students struggle more through inquiry-based activities and by adjusting the time they spend on problem solving. In other words, exposure and time to integrate new learning are key to sustained changes in practice. Participation in MTC does impact the type of experiences or level of mathematics that teachers share with students, especially when it comes to problem solving practices.

Open Access Except where otherwise noted, this chapter is licensed under a Creative Commons Attribution 4.0 International License. To view a copy of this license, visit http://creativecommons.org/licenses/by/4.0/.

Exploring and Making Online Creative Digital Math Books for Creative Mathematical Thinking

Pedro Lealdino Filho, Christian Bokhove, Jean-Francois Nicaud, Ulrich Kortenkamp, Mohamed El-Demerdash, Manolis Mavrikis and Eirini Geraniou

When we look at e-books designed for mathematics education, we can distinguish two streams. On the one hand, we see publishers of traditional mathematics textbooks have digital versions of their products, mostly static pdf documents that can be downloaded and used on different devices. In anticipation of new interactive possibilities, limited interactivity is sometimes built in. On the other hand, we see innovative groups of designers that have started to develop highly interactive tools and micro-worlds for mathematics education. Initially, many of these tools were implemented as standalone applications. These tools have been increasingly integrated with written tasks, producing interactive worksheets, dynamic web pages, and e-books for math. In some European countries, the M C Squared project has aimed at starting several so-called Communities of Interest (CoI) that work on digital, interactive, and creative mathematics textbooks called c-books. The c-books are authored in the M C Squared platform, where authors can construct books using various interactive "widgets." The workshop aimed to introduce the project and acquaint participants with the affordances and authoring process of the M C Squared platform.

A short overview of M C Squared and the architecture of the authoring tool platform was given, showing the possibilities of creating c-books individually or collectively. An example of a unit of a c-book, "Experimental Geometry" (Fig. 1) was given showing the different widget factories and how they perform inside the c-book and the creative mathematical thinking affordances present in the c-book.

P.L. Filho (✉)
Université Claude Bernard, Lyon, France
e-mail: pedrolealdino@gmail.com

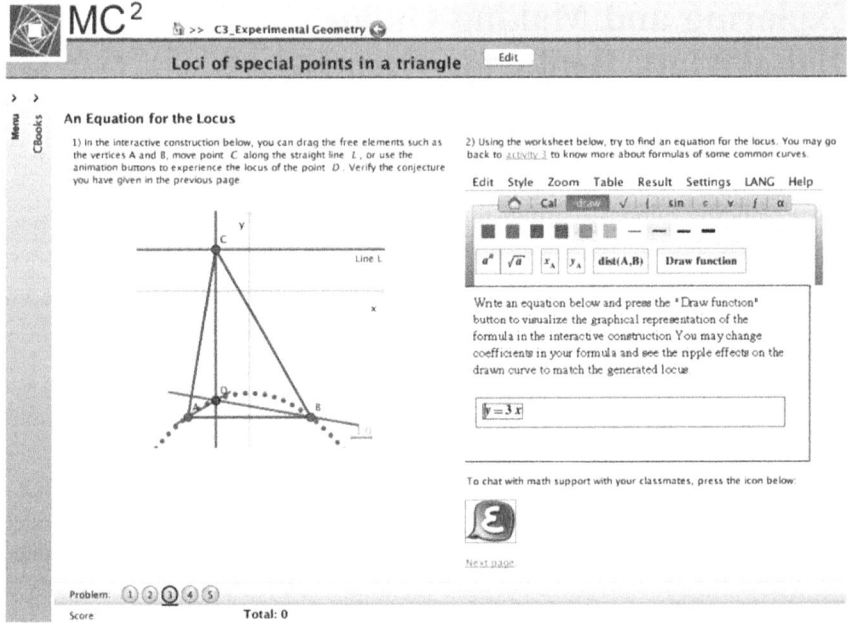

Fig. 1 Screenshot of a c-book page showing the widgets: Cinderella, EpsilonWriter and EpsilonChat

Another goal of the workshop was to teach participants how to author a simple interactive c-book and explore existing resources created along the duration of the project. The participants got acquainted with a selection of other M C Squared platform features such as student work storage and answer checking.

Open Access Except whereotherwise noted, this chapter is licensed under a Creative Commons Attribution 4.0 International License. Toview a copy of this license, visit http://creativecommons.org/licenses/by/4.0/.

The Shift of Contents in Prototypical Tasks Used in Education Reforms and Their Influence on Teacher Training Programs

Karl Fuchs, Christian Kraler and Simon Plangg

The aim of this workshop was to show the (new) importance of task-based learning in school in acquiring general and specific knowledge of our world. By focusing on this, we have had to face a significant shift in task-based philosophies of teaching mathematics. Traditional operation-based teaching has been increasingly replaced by other approaches, especially methods stressing (new) prototypical aspects. Thus activities such as transferring, interpreting, and reasoning have gained a much higher importance in mathematics classes.

Hence, one of the underlying ideas of the workshop was to focus participants' attention on these important parameters in ongoing national education reforms in the Austrian school system.

To elaborate this observation we presented a qualitative study that had focused on the development of teaching styles in Austrian secondary schools over the last 40 years: The shift to prototypical teaching approaches was fostered in at least two ways. In particular, young teachers and teachers with a thorough education in subject-specific didactics introduce a new understanding in learning mathematics that transcends the traditional analytical and algorithmic approach (transfer, interpretation, reasoning, etc.). Furthermore, new technologies have offered completely new possibilities in dealing with underlying principles of tasks. From the didactical point of view, the role of the genetic approach in its diversity, the concept of fundamental or universal ideas, and the multi-media approach change fundamentally. In brief, we referred to Geogebra as an example (interactivity, computer algebra, process orientation, and dynamical modelling possibilities).

K. Fuchs (✉)
Department of Mathematics, University of Salzburg, Salzburg, Austria
e-mail: KarlJosef.FUCHS@sbg.ac.at

All of these observations were discussed as significant tendencies and educational goals in an ongoing process in the context of a major educational reform in Austria's school system. In the final part of the presentation, we questioned whether the tasks designed for attaining these goals by the Federal Institute for Educational Research, Innovation, and Development of the Austrian School System (BIFIE) satisfy the aims mentioned. The detailed discussion of two tasks from final exams showed that the pattern change that is needed in the design of individual tasks has not yet been realized. Most exam tasks rather follow the traditional task approaches mentioned above.

Open Access Except where otherwise noted, this chapter is licensed under a Creative Commons Attribution 4.0 International License. To view a copy of this license, visit http://creativecommons.org/licenses/by/4.0/.

Analysis of Algebraic Reasoning and Its Different Levels in Primary and Secondary Education

Juan D. Godino, Teresa Neto and Miguel R. Wilhelmi

Description of the Workshop: Aims and Underlying Ideas

An important objective in various curricular guidelines (e.g., NCTM, 2000) has been the enhancement of algebraic reasoning beginning in the first educational levels. This objective implies that we assume a new view of school algebra as not being limited to handling algebraic expressions. The effective implementation of this new conception of school algebra poses a challenge for the training of mathematics teachers, because few current training programs include the development of such a new vision. The objective of the workshop was to implement some practical activities aimed at recognizing the main features of school algebraic reasoning that can be used to train teachers to promote algebraic thinking in primary (Aké, Godino, Gonzato, & Wilhelmi, 2013) and secondary education (Godino et al., 2015). The wider view of school algebra that was presented and discussed takes into account the processes of generalization and symbolization as well as structural and functional modelling and analytical calculation. It also created a meaningful link between algebraic thinking in primary and secondary education.

Planned Structure

Number and length of modules: 2; 45 min each.

Practical Activities: (i) Solving a selected set of tasks on school algebraic reasoning for primary (secondary) education, (ii) assigning levels of algebraic thinking to different solutions, (iii) enunciating related tasks whose solution involves

---

J.D. Godino (✉)
University of Granada, Granada, Spain
e-mail: jgodino@ugr.es

changes in the levels of algebraization, and (iv) presentation and discussion of results.

References

Aké, L., Godino, J. D., Gonzato, M., & Wilhelmi, M. R. (2013). Proto-algebraic levels of mathematical thinking. In A. M. Lindmeier & A. Heinze (Eds.), *Proceedings of the 37th Conference of the International Group for the Psychology of Mathematics Education* (Vol. 2, pp. 1–8). Kiel, Germany: PME.

Godino, J. D., Neto, T., Wilhelmi, M. R., Aké, L., Etchegaray, S. & Lasa, A. (2015). Levels of algebraic reasoning in primary and secondary education. In K. Krainer & N. Vondrová (Eds.), *Proceedings of the Ninth Conference of the European Society for Research in Mathematics Education* (CERME9, February 4–8, 2015) (pp. 426–432). Prague, Czech Republic: Charles University in Prague, Faculty of Education and ERME TWG 03: Algebraic Thinking.

National Council of Teachers of Mathematics (NCTM). (2000). *Principles and standards for school mathematics*. Reston, VA: Raupach.

Open Access Except where otherwise noted, this chapter is licensed under a Creative Commons Attribution 4.0 International License. To view a copy of this license, visit http://creativecommons.org/licenses/by/4.0/.

Designing and Evaluating Mathematical Learning by a Framework of Activities from History of Mathematics

Lenni Haapasalo, Harry Silfverberg and Bernd Zimmermann

Description of the Workshop: At first, on the basis of his long-term studies of the history of mathematics, BZ represented the eight sustainable activities that proved to lead frequently to new mathematical results at different times and in different cultures for more than 5000 years (Zimmermann 2003). After that, LH represented how this octagon may be used as an instrument to measure how the eight activities are supported within school mathematics, university mathematics, and the usage of ICT in everyday life, respectively. The results suggest that the support gained from all those areas is modest, and amazingly the support gained from the overall usage of ICT seems to have even a descending trend. The studies suggest that design of ICT-based learning environments orchestrated within the so-called pit-stop philosophy, promote a promising support for the Z-activities. This would mean a thorough shift in curriculum design, including dynamic assessment. HS represented a method for applying computer-aided analysis of the Finnish mathematics curriculum for the comprehensive school. Even though his computer-based datamining (Silfverberg, 2016) revealed that the curriculum expressions refer to many of the above-mentioned activities, they seem to be supported poorly in reality.

L. Haapasalo (✉)
University of Eastern Finland, Joensuu, Finland
e-mail: lenni.haapasalo@uef.fi

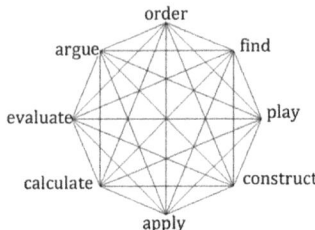

The Workshop activities included discussions about curricula in different countries and use of history of mathematics in mathematics instruction. There were participants from Africa, Asia, South-America, and Europe.

References

Haapasalo, L. (2007). Adapting mathematics education to the needs of ICT. *The Electronic Journal of Mathematics and Technology, 1*(1), 1–10. https://php.radford.edu/~ejmt/deliveryBoy.php?paper=eJMT_v1n1p1

Haapasalo, L. (2008). Perspectives on instrumental orchestration and assessment—From challenges to opportunities. In *Plenary speech in the 13th Asian technology conference in mathematics. (ATCM 2008)*. December 15–19, 2008, Bangkok, Thailand: Suan Sunandha Rajabhat University. Internet: http://atcm.mathandtech.org/EP2008/papers_invited/2412008_15968.pdf

Haapasalo, L. (2013). Adapting assessment to instrumental genesis. *The International Journal for Technology in Mathematics Education, 20*(3).

Haapasalo, L., & Zimmermann, B. (2011). Redefining school as pit stop: It is the free time that counts. In W.-C. Yang, M. Majewski, T. de Avis, & E. Karakirik (Eds.), *Integration of technology into mathematics education: Past, present and future. Proceedings of the sixteenth Asian technology conference in mathematics* (pp. 133–150). September 19–23, 2011. Bolu, Turkey.

Haapasalo, L., & Zimmermann, B. (2015). Investigating mathematical beliefs by using a framework from the history of mathematics. In C. Bernack-Schüler, R. Erens, T. Leuders, & A. Eichler (Eds.), *Views and beliefs in mathematics education. Results of the 19th MAVI conference*. Wiesbaden: Springer Spektrum.

Silfverberg, H. (2016). Using 'Zoctagon' as a frame for evaluating curricula. In L. Eronen & B. Zimmermann (Eds.), *Learning, technology, assessment. Festschrift in honour of Lenni Haapasalo* (pp. 133–141). Münster, Germany: WTM Verlag für wissenschaftliche Texte und Medien.

Zimmermann, B. (2003). On the genesis of mathematics and mathematical thinking—A network of motives and activities drawn from the history of mathematics. In L. Haapasalo & K Sormunen (Eds.), *Towards meaningful mathematics and science education. Bulletins of the faculty of education* (Vol. 86, pp. 29–47).

Open Access Except where otherwise noted, this chapter is licensed under a Creative Commons Attribution 4.0 International License. To view a copy of this license, visit http://creativecommons.org/licenses/by/4.0/.

Sounding Mathematics: How Integrating Mathematics and Music Inspires Creativity and Inclusion in Mathematics Education

Caroline Hilton and Markus Cslovjecsek

The workshop focused on the principles underpinning our integrated approach to the teaching of mathematics and music, within the context of "low threshold, high ceiling" tasks. Participants engaged with a number of activities from the Comenius Project "EMP-M—Sounding Ways into Mathematics" (2013–2016). By actively participating in the activities, participants were enabled to experience the mathematical relationships and patterns through the music, thus reinforcing the interconnectedness of the two disciplines. Following the ideas of Barnes (2015), we explored the notion of integrated teaching as "interdisciplinary", rather than, for example, hierarchical or opportunistic.

It is worth briefly reflecting on why this interconnectedness between music and mathematics is so attractive. According to Leone Burton, mathematical thinking "is mathematical not because it is thinking about mathematics, but because the operations on which it relies are mathematical operations" (Burton 1984, p. 36). Citing Hofstadter (1979), Burton suggests that mathematical thinking relies on pattern recognition, iteration and repetition; processes found not only in mathematics, but also in music and art, for example in the works of Bach and Escher, to name but two. Learning mathematics, according to this definition, would require children to explore mathematical questions and relationships in messy and often haphazard ways, in attempts to make sense of what they find. For this reason, exploring mathematics and music together, in an interconnected way, where it is hard to identify where the music ends and the mathematics begins, for instance, seems to make sense. What we have found is that while the language of mathematics and music and their representations are different, when we look for similarities we see that we can often make direct translations.

C. Hilton (✉)
UCL Institute of Education, London, UK
e-mail: c.hilton@ioe.ac.uk

Du Sautoy (2008) tried to capture the creative nature of mathematics in an interview in the Guardian newspaper:

> I think very often the exciting moments in mathematical history are moments when suddenly there's a leap of imagination - for example, the idea of negative numbers, or zero—I mean, that's almost as imaginary as a four-dimensional shape. What's a negative number? I can't show you minus three potatoes - but let's come up with the idea of a negative number and the way that it will behave and explore that. That's why it's a creative subject. It's a lot about creative intuition…in Einstein's view, the ultimate test for an equation was an aesthetic one. The highest praise for a good theory was not that it was correct or that it was exact, simply that it should be beautiful.

Thus, by integrating mathematics and music in a truly interdisciplinary way, we can support children to explore patterns and relationships in ways that develop their creativity, understanding and aesthetic pleasure.

References

Barnes, J. (2015). *Cross-curricular learning 3–14*. London: Sage.
Burton, L. (1984). Mathematical thinking: The struggle for meaning. *Journal for Research in Mathematics Education, 15*(1), 35–49, 36.
Du Sautoy M. (2008). *The Guardian Newspaper*. November 3, 2008 [Online]. Available at: http://www.theguardian.com/science/2008/nov/03/marcus-dusautoy
Hofstadter, D. R. (1979). *Gödel, Escher, Bach: An eternal golden braid*. London: Harvester.

Open Access Except where otherwise noted, this chapter is licensed under a Creative Commons Attribution 4.0 International License. To view a copy of this license, visit http://creativecommons.org/licenses/by/4.0/.

Adopting Maxima as an Open-Source Computer Algebra System into Mathematics Teaching and Learning

Natanael Karjanto and Husty Serviana Husain

The workshop introduced and explained the computer algebra system (CAS) Maxima for teaching and learning of calculus and linear algebra at the tertiary level. The didactic principle underlying this approach is a necessity to combine an element of technology into our classroom to enhance student understanding of calculus and linear algebra concepts. Maxima is an open-source computer software that can be used for the manipulation of symbolic and numerical expressions, including limit calculation, differentiation, integration, Taylor series, systems of linear equations, polynomials, matrices, and tensors. It can also sketch some graphical objects with excellent quality (http://maxima.sourceforge.net/).

The workshop started by providing information on getting help in Maxima, which can be done using the command `describe` or ?, for instance, `describe(diff)` and `describe(integrate)` to obtain information about the derivative and the integral, respectively. The symbol `%` refers to the most recent calculated result. The workshop continued with simple examples of calculus computation, as presented in the following table. The participants were also invited to try exercises related to the presented materials. Other examples presented were sketching curves in two and three dimensions, including several interesting parametric plots. Three examples of the plots are displayed in this article. The first is a cardioid, which comes when studying polar curve; the second is a Möbius band as an example of a non-orientable surface; and the third is a torus, which appears when discussing a solid of revolution.

N. Karjanto (✉)
Sungkyunkwan University, Seoul, Republic of Korea
e-mail: natanael@skku.edu

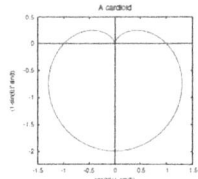
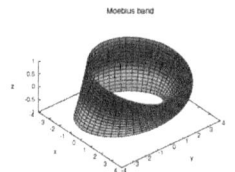
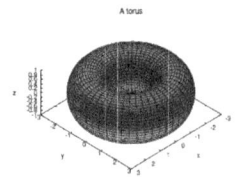

Description	Symbolic	Maxima code	Output
Finding a limit	$\lim_{x \to 1} \dfrac{3x^2 - 4x + 1}{x^2 + x - 2}$	`limit((3*x^2 - 4*x + 1)/` `(x^2 + x - 2),x,1);`	$\dfrac{2}{3}$
Evaluating first derivative	$\dfrac{d}{dx}(x^2 e^x)$	`diff(x^2*exp(x),x);` `factor(%);`	$x^2 e^x + 2xe^x$ $x(x+2)e^x$
Evaluating second derivative	$\dfrac{d^2}{dx^2}(x^2 e^x)$	`diff(x^2*exp(x),x,2);` `factor(%);`	$x^2 e^x + 4xe^x + 2e^x$ $(x^2 + 4x + 2)e^x$
Calculating indefinite integral	$\displaystyle\int \dfrac{1}{1+x^2}\,dx$	`integrate(1/(1 + x^2),x);`	$\arctan(x)$
Calculating definite integral	$\displaystyle\int_0^1 \dfrac{1}{1+x^2}\,dx$	`integrate(1/(1 + x^2),x,0,1);` `%,numer;`	$\pi/4$ 0.7853981634

Open Access Except where otherwise noted, this chapter is licensed under a Creative Commons Attribution 4.0 International License. To view a copy of this license, visit http://creativecommons.org/licenses/by/4.0/.

The Power of Geometry in the Concept of Proof

Damjan Kobal

To improve learning we need to challenge students' thoughts. Through intuitive geometric statements, the human brain instantly poses smart questions and offers hypotheses while engaging in self-challenging explorations. Using geometry, which is too often neglected in our schools, engaging teaching can sometimes be achieved in contemplative pantomime settings almost without words. The aim of the workshop was to show the power of geometry in the development of the concept of proof. Several relatively easy geometric ideas were presented through simple thought-provoking questions and by the use of technology. The aim of these questions is not solely to motivate an answer; rather, it is much deeper and educationally wider. Namely, the aim is to motivate the understanding and the beauty of the resolved uncertainty brought by the certainty of a proof. In a way, a proof should be as much an emotional experience as a rational achievement. Participants were challenged by several thought-provoking questions followed by individual engagements in the form of short problem-solving sessions and concluded by joint discussions. Geometry was used to show that to learn and appreciate mathematics, it is necessary to understand the concept of proof. In order to understand the concept of proof, one also needs to experience the challenge of uncertainty that precedes the certainty of a proof.

At the workshop, several motivational techniques and ideas were actively presented. Participants were challenged using concrete, thought-provoking puzzles. In the introduction, some tricks were presented and participants were challenged to observe their own reasoning and motivation powered by cognitive puzzlement. Using thought-provoking questions, participants were guided to understanding (and feeling) the need for proof. Within the main part of the workshop, participants were presented with several problems that were mostly visual. Particular problems were

---

D. Kobal (✉)
Department of Mathematics, Faculty of Mathematics and Physics,
University of Ljubljana, Ljubljana, Slovenia
e-mail: damjan.kobal@fmf.uni-lj.si

given in "pantomime" fashion, i.e., without words, with the first challenge for participants being to formulate the problem as briefly and as elegantly as possible. By that, participants were invited to increase their sensitivity to "meaning" and provoked to consider the inflation of content that occurs when too many words are used (in explaining). Some challenges included "hands-on" tasks. During the workshop, geometric ideas and teaching accents involving the following subjects were discussed: Missing angle of a triangle, midpoints of a quadrilateral, apparent regular octagon, line through centroid, intersection of two squares, triangle on top of a square, geometric series formula, constructing parabola geometrically, ellipse by folding paper, parabola-ellipse analogy, geometric paradoxes, and sound technology geometrically.

All of the problems were very easy to formulate. Most of the problems were illustrated and communicated by the use of technology. Participants were challenged to formulate problems "formally" and to explore several "upgrades of (their) understanding". Proofs and answers were obtained as cognitive solutions and conclusions to "provocative uncertainty." In most cases, we showed that a proof is as much a rational conclusion as it is also an "emotional experience" (that comes as a solution to felt problem). Because of this, it is very important how the problem is introduced.

In the "sum-it-up" conclusion, participants' feedback was discussed. Participants were also given access to the interactive dynamic presentations/visualisations (https://www.geogebra.org/m/mZpYbUmK, accessed 10 November 2016) that were used during the workshop.

Open Access Except where otherwise noted, this chapter is licensed under a Creative Commons Attribution 4.0 International License. To view a copy of this license, visit http://creativecommons.org/licenses/by/4.0/.

Workshop: Silent Screencast Videos and Their Use When Teaching Mathematics

Bjarnheiður Bea Kristinsdóttir

Silent screencast videos are animated short videos without any text, subtitles or verbal commentary that show mathematics dynamically, focusing on one mathematical concept. The silent videos can be used in mathematics classrooms cross-culture and cross-language, giving students the assignment to add their commentary to the video.

The workshop started with an introduction explaining the background of silent screencast videos, how they are made, what they look like, and what they are used for. Next, instructions on how to add a commentary to a silent video were given and examples of commentaries made by students in Iceland were presented. In addition, participants had the opportunity to view and discuss results from experiments using the silent videos in 5th–12th grade mathematics classrooms in Estonia, Iceland, Latvia and Lithuania. After these discussions, the participants worked in groups of two to add their own commentaries to ready-made silent videos, and received guidelines on how to make their own silent videos. All workshop materials were made available on Google Drive (goo.gl/DheZ42) and some of the commentaries and silent videos created by the participants were shared in a Padlet (https://padlet.com/wall/oa1xnbbepfy9).

The activities presented in this workshop were initiated by a group of researchers and mathematics teachers from the *Nordic GeoGebra Network* (NGGN). NGGN was formed in 2010 to promote the use of the open-source dynamic mathematics software GeoGebra in mathematics classrooms in the Nordic and Baltic countries. The network also hosts annual conferences for teachers and education researchers. A grant from the Nordic Council of Ministers' program *Nordplus* made a three-year collaboration project between mathematics education researchers and mathematics teachers from the Nordic and Baltic countries possible. Each of the three years had a special theme and to address these themes, key topic groups were formed. The first

---

B.B. Kristinsdóttir (✉)
School of Education, University of Iceland, Reykjavík, Iceland
e-mail: bjarnhek@hi.is

key topic group started in 2013 at the NGGN conference in Copenhagen working on the theme "Learning mathematics through screencast technology and video" and there the idea of the silent screencast videos came up. Several short silent videos showing one mathematical concept each were made, and teachers of 5th–12th grade in four different countries tested three of them in class in Autumn 2014; making their pupils write or record their own commentary to the videos. The participating teachers were surprised in a positive way: the tasks encouraged communication in the classroom and enabled teachers to "see what the pupils were thinking" (Hreinsdóttir & Kristinsdóttir, 2016). All the tested videos were made freely available online in a GeoGebra-book (https://goo.gl/66lNbX) and teachers in the different countries have continued working on new silent videos to add to the collection.

Reference

Hreinsdóttir, F., & Kristinsdóttir, B. (2016). Using silent videos in the teaching of mathematics. In S. Ceretkova (Ed.), *Staircase to even more interesting mathematics teaching* (pp. 157–164). Nitra, Slovakia: Constantine the Philosophers University in Nitra.

Open Access Except where otherwise noted, this chapter is licensed under a Creative Commons Attribution 4.0 International License. To view a copy of this license, visit http://creativecommons.org/licenses/by/4.0/.

Shout from the Most Silent Nation, North Korea: Can Mathematics Education Be Politically Neutral?

JungHang Lee

North Korea has been making CNN headlines on their nuclear missile tests and human rights issues. However, North Korea still remains as the most closed and shadowed country. This workshop addressed mathematics education in one of the most closed countries in the world—North Korea, as an extreme example of political influences on mathematics education. Questions on political influence on mathematic educations were proposed and discussed.

North Korean secondary school mathematics education is examined through the review of North Korea's social and educational structures as well as its political and ideological position.

Since it is almost impossible to obtain any type of information about North Korea, I conducted in-depth interviews with defectors, who are now in South Korea, former secondary school mathematics teachers and students, to understand their real life experiences in secondary school mathematics in North Korea. Interviewees responded to questions concerning typical ways their teaching and learning were carried out in mathematics classes; the Workers' Party's influence in every aspect of education, from teacher education to curriculum and textbooks issued; and the impact the March of Suffering (nine consecutive years of negative economic growth, which killed ten percent of North Korean population) had on the teaching and learning of mathematics as well as its lingering effects in secondary mathematics education.

There are two main focuses of this workshop. One is to introduce an extreme case study of mathematics education in North Korea influenced by political and ideological standpoint. This would broaden the participants' understanding of

---

J. Lee (✉)
Nyack College, New York, USA
e-mail: junghang.lee@nyack.edu

mathematics education as not only a self-regulating subject, but also as an interwoven matter shaping and shaped by the vessel and the people in it. This also proposes a chance to reassess the participant's own mathematics education system with possibly enhanced span. For example, U.S. school mathematics education is greatly influenced by the No Child Left Behind Act, which was signed by former President George W. Bush on January 8, 2002. It sounded very attractive to the general public, yet caused many problems over the years. On December 10, 2015 President Barack Obama signed legislation replacing No Child Left Behind Act with the Every Student Succeeds Act. Mathematics education in U.S. will again be modified and promoted by its society's political stance.

After my presentation on the subject matters, several small groups were formed and we discussed the following questions:

- Should mathematics be a politically neutral subject?
- Is there any political influence on mathematics education in your country?
- What are the benefits and detriments of politically motivated mathematics educations?

Open Access Except where otherwise noted, this chapter is licensed under a Creative Commons Attribution 4.0 International License. To view a copy of this license, visit http://creativecommons.org/licenses/by/4.0/.

Workshop Theme: "Use of Educational Large-Scale Assessment Data for Research on Mathematics Didactics"

Sabine Meinck, Oliver Neuschmidt and Milena Taneva

Workshop Description

As a leading organization in the field of educational research for nearly 60 years, IEA promotes capacity building and knowledge sharing to facilitate innovation and foster quality in education. IEA's manifold empirical studies inspire fruitful dialogue on critical educational issues, informing the development of evidence-based policies and practices across the globe.

Only within the past decade, IEA conducted nine international large-scale assessments (ILSA), each with up to 59 participating countries, studying various topics and target populations. IEA studies approach educational reality in all its complexity, collecting not only achievement data but also a wide range of information about the contexts within which teaching and learning occurs. In the context of mathematics, two IEA studies are of special interest. The Trends in International Mathematics and Science Study (TIMSS) investigates mathematics and science achievement of fourth and eighth grade students, and considers achievement within the context of the in- and outside school environment of students. This study has been conducted every four years since 1995. Secondly, the Teacher Education and Development Study in Mathematics (TEDS-M), conducted in 2008, examined how future teachers around the world are prepared to teach mathematics in primary and lower secondary schools. The key research questions focused on the relationships between teacher education policies, institutional practices, and future teacher's mathematics and pedagogy knowledge at the end of their pre-service education. The study gathered information on various characteristics of teacher education

S. Meinck (✉)
IEA Data Processing and Research Center, Hamburg, Germany
e-mail: sabine.meinck@iea-dpc.de

institutions, programs, and curricula; the opportunities to learn within these contexts; and future teacher's knowledge and beliefs about mathematics and learning mathematics.

All data arising from IEA studies are publicly available and extensively documented; they provide a valuable and rich source for in-depth analysis in many fields of educational research, including the didactics of mathematics. However, due to their complexity, thorough methodological skills are needed to analyse and interpret this data correctly which increases the threshold for researches to actually make use of it.

The primary objective of the workshop was to show how IEA study data can be used for research aiming at improving the teaching of mathematics. In the first part of the workshop, the structure of IEA data was introduced and then the access paths to data sources, technical documentation, analysis guides, and software tools were shown to participants. Sources and contacts available for support when working with IEA data were mentioned as well. In the second part, the possible uses of data for researchers who focus on the didactics of mathematics were discussed. The methodological challenges and limitations of ILSA data were explained along with solutions about how to address them for data analysis and interpretation, for example, by disclosing appropriate statistical analysis methods. Participants had the chance to study questionnaire and achievement test materials and to develop ideas for suitable research questions that can be answered through the use of TIMSS and TEDS-M databases.

Open Access Except where otherwise noted, this chapter is licensed under a Creative Commons Attribution 4.0 International License. To view a copy of this license, visit http://creativecommons.org/licenses/by/4.0/.

Curriculum Development in the Teaching of Mathematical Proof at the Secondary Schools in Japan

Tatsuya Mizoguchi, Hideki Iwasaki, Susumu Kunimune,
Hiroaki Hamanaka, Takeshi Miyakawa, Yusuke Shinno,
Yuki Suginomoto and Koji Otaki

Aim and Key Questions of the Workshop

In the workshop, we aimed to share a theoretical framework as well as some issues on the teaching of mathematical proofs through Grades 7–12 of secondary schooling in Japan. The difficulties faced by students in learning mathematical proofs are well known. The key questions in the workshop were as follows: (1) What kind of teaching content should be included in the secondary curriculum for the teaching of mathematical proofs? (2) What kind of evolution should be envisioned in the course of the curriculum? (3) How can we allow comparing different curriculums for teaching mathematical proofs with different countries in terms of our proposed framework?

Activities and Presentations During the Workshop

We first introduced our research project and theoretical perspective. In this introduction, the first and second questions were considered. Participants were then divided into three small topic groups based on their interests. The contributors who took part in each of the topic groups, theoretical framework, what is a proof in Japan, and teaching materials, are shown in the following table. Through these activities we intended to discuss the third question with international participants.

T. Mizoguchi (✉)
Tottori University, Tottori, Japan
e-mail: mizoguci@rs.tottori-u.ac.jp; tatsuya_ds@me.com

Topic group activities	Contributed team members and contents
Theoretical framework	T. Mizoguchi and Y. Shinno made a presentation about the framework and gave examples from upper secondary school textbooks in Japan
What is proof in Japan?	T. Miyakawa and S. Kunimune explained what we call "proof" in Japan by referring to geometrical proofs in lower secondary school textbooks
Teaching materials	H. Hamanaka introduced the mathematical proof related to parallelograms. Y. Suginomoto, K. Otaki, and H. Iwasaki introduced the operative proof related to Sylvester's theorem

After the group activities, we had an opportunity to engage participants in reporting the activity of each group. At the end of the workshop, Hideki Iwasaki made a closing remark about the importance for future research of sharing this research theme in mathematics education.

Open Access Except where otherwise noted, this chapter is licensed under a Creative Commons Attribution 4.0 International License. To view a copy of this license, visit http://creativecommons.org/licenses/by/4.0/.

Symmetry, Chirality, and Practical Origami Nanotube Construction Techniques

B. David Redman Jr.

A Short Description of Our Workshop: Aims and Underlying Ideas

Our workshop illustrated several educational and entertaining applications of origami in the classroom. The activities illustrated symmetry, chirality, and duality in simple modular origami as well as the flexibility of Pentagon-Hexagon Zig-Zag (PHiZZ) units in constructing more sophisticated models. Additional illustrations of counting and graph coloring were provided.

The workshop highlighted a variety of materials, including recycled papers and packaging material and their preparation, ordinary office sticky notes, and traditional origami paper. We considered how to illustrate interesting mathematical concepts with simple models, demonstrating several concepts and encouraging participants to develop further illustrations. We made a more detailed study of using PHiZZ units to construct models such as Buckyball, tori, and carbon nanotubes.

Participants practiced constructing a variety of units used in modular origami models, studied previously assembled models, and learned how to construct models themselves.

B.D. Redman Jr. (✉)
Delta College, Saginaw, MI, USA
e-mail: bdredman@delta.edu

ICME-13 Practical Origami Workshop Handouts (https://goo.gl/4mm8aY).

Open Access Except where otherwise noted, this chapter is licensed under a Creative Commons Attribution 4.0 International License. To view a copy of this license, visit http://creativecommons.org/licenses/by/4.0/.

Reflecting Upon Different Perspectives on Specialized Advanced Mathematical Knowledge for Teaching

Miguel Ribeiro, Arne Jakobsen, Alessandro Ribeiro,
Nick H. Wasserman, José Carrillo, Miguel Montes and Ami Mamolo

Teachers' knowledge assumes a major role in practice and in students' learning and achievement. In particular, the construct of horizon knowledge or what can be termed *specialized advanced mathematical knowledge for teaching* (in order to capture the overall perspectives we are dealing with) has been the focus of attention from a variety of researchers with different foci. From this perspective, and aiming to deepen our understanding of such a construct, the aim of this working group was to discuss (and reflect upon) different theoretical perspectives, methodological approaches, and analytic methods used when focusing on such specialized advanced mathematical knowledge for teaching. In particular, we consider the activities of analyzing and conceptualizing situations where access and development of such teachers' knowledge is of primary importance.

Following work previously developed (in a DG at PME 34; Wasserman, Mamolo, Ribeiro, & Jakobsen, 2014), this workshop aimed at continuing and deepening the discussions. We had three slots. In the first, a brief overview of the different perspectives of conceptualizing the specialized advanced knowledge for teaching was given by the proposed organizers, representing four different approaches and understandings of this construct. Afterwards, the participants were engaged in commenting, solving, reflecting on, and discussing two situations (one vignette and one episode) designed to access and develop teachers' advanced mathematical knowledge linked with the tasks of teaching. This discussion aimed at discussing the participants' interpretations of the different aspects of advanced mathematical knowledge involved that can (potentially) be explored having the provided situations as a starting point—and, in case of a need for changes in such situations, what would be the focus of such changes. A global discussion followed, aiming at both

---

M. Ribeiro (✉)
State University of Campinas—UNICAMP, Campinas, Brazil
e-mail: cmribas78@gmail.com

synthesizing and enhancing the participants' views and understanding of the construct at hand and fostering a deeper understanding of what such a construct comprises and the nature of the associated tasks for developing it (and its differences with other aspects of teachers' knowledge). Based on the different perspectives of the construct and the subsequent analyses of the two explored situations, we proposed to the WG participants the possibility of collaborating on papers/book chapters that would be an outcome of the group and would lead to a broader understanding of what comprises a construct of advanced mathematical knowledge for teaching as well as its potential implications for future research in this area.

Reference

Wasserman, N., Mamolo, A., Ribeiro, M., & Jakobsen, A. (2014). *Exploring horizons of knowledge for teaching*. Discussion Group at PME 38, July 15–20, 2014, Vancouver, Canada.

Open Access Except where otherwise noted, this chapter is licensed under a Creative Commons Attribution 4.0 International License. To view a copy of this license, visit http://creativecommons.org/licenses/by/4.0/.

Collaborative Projects in Geometry

José L. Rodríguez, David Crespo and Dolores Jiménez

Sierpinski Carpet Project (2014–16)

We started our workshop showing the largest Sierpinski carpet in the world, built by more than 40,000 children from 400 centers in 39 countries. The small pieces were assembled in the Palace of Mediterranean Games in Almería on May 13, 2016, with the help of 1000 people. All the information related to this project, including activities on fractals, can be found at http://topologia.wordpress.com/sierpinski-carpet-project.

J.L. Rodríguez (✉)
Department of Mathematics, University of Almería, Almería, Spain
e-mail: jlrodri@ual.es

© The Author(s) 2017
G. Kaiser (ed.), *Proceedings of the 13th International Congress on Mathematical Education*, ICME-13 Monographs, DOI 10.1007/978-3-319-62597-3_136

Let's Play to Classify Surfaces! (2016–17)

In the second part of the workshop, we presented a project based on the construction and analysis of surfaces from a topological viewpoint. It was awarded with the First Prize in Mathematics at the Spanish edition of Science on Stage held in Algeciras in October 2016.

Starting with manipulative activities, the project was developed using computational software such as Mathematica or virtual reality (VR).

We spent some time to build several polyhedral surfaces with pieces of our manipulative game, 3D Polyfelt. We then computed their Euler characteristics, among other topological properties.

Materials of this project are available at: https://sites.google.com/a/ual.es/surfaces.

Open Access Except where otherwise noted, this chapter is licensed under a Creative Commons Attribution 4.0 International License. To view a copy of this license, visit http://creativecommons.org/licenses/by/4.0/.

Workshop on Framing Non-routine Problems in Mathematics for Gifted Children of Age Group 11–15

Sundaram R. Santhanam

Consider the following problem. All the even numbers from 2 to 128 (both inclusive) excepting the numbers end with 0 are multiplied together. What is the unit's digit of the product?

Teacher's tool: If a problem is too big then consider a simple problem with the same conditions.

Solution: Consider the numbers 2, 4, 6, 8.

These are the numbers from 2 to 8 (both inclusive) and even, where is 0 is not there in the unit place.

The product is $2 \times 4 \times 6 \times 8 = 384$. The unit's digit is 4.

Consider the even numbers between 2 and 18 (both inclusive and unit digit non zero) 2, 4, 6, 8, 12, 14, 16, 18.

We observe that the unit's digit by multiplying 2, 4, 6, 8 is 4. Clearly the unit's digit of multiplying the numbers 12, 14, 16 and 18 also must be 4.

∴ The unit's digit in the product must be 6.

Take numbers between 2 and 28 with the same conditions.

(2, 4, 6, 8) (12, 14, 16, 18) (22, 24, 26, 28). The units digit in the final product must be 4.

So, there is a pattern formed.

Table:

Group	1	(1, 2)	(1, 2, 3)	(1, 2, 3, 4)	(1, 2, 3, 4, 5)
Unit digit in the product	4	6	4	6	6

S.R. Santhanam (✉)
Association of International Mathematics Education and Research, Bhubaneswar, India
e-mail: santhanam2015aimer@gmail.com

S.R. Santhanam
Sri Prakash Synergy School, Kakinada, India

Now the given question can be solved easily.

The unit's digit in the product must be 4 because it is the 11th group.

Note for the teachers: For a gifted child of higher grade a good question will be when all the even numbers between 2 and 198 (both inclusive-deleting the numbers with O in the unit digit) are multiplied, what is the tens digit in the product?

Of course, this is not as simple as the first one. But a teacher can follow the same advice as to consider simple problem along the same line.

2, 4, 6, 8 give the product 384,   12, 14, 16, 18 give the product 48, 384

22, 24, 26, 28 give the product 384, 384,   32, 34, 36, 38 give the product 1, 488, 384

What we find here is that the last two digits of the products of the groups is 84. When we multiply all the numbers in two groups, because we need the ten digit of the product only, a simple calculation gives

$$\frac{84 \times 84}{56} \quad \frac{56 \times 84}{04} \quad \frac{04 \times 84}{36} \quad \frac{36 \times 84}{24} \quad \frac{24 \times 84}{16}$$

$$\frac{16 \times 84}{44} \quad \frac{44 \times 84}{96} \quad \frac{96 \times 84}{64} \quad \frac{64 \times 84}{76} \quad \frac{76 \times 84}{84}$$

Thus the tens digit repeats at the 11th stage. The tens digit of the product $2 \times 4 \times 6 \times 8 \times 12 \times 14 \times 16 \times 18$ is 5.

(i.e.) if we take (2, 4, 6, 8) as the first group and

(12, 14, 16, 18), as the second group
(22, 24, 26, 28), as the third group

(92, 94, 96, 98) is the 10th group.

Let I denote the first group, II the first and second group etc.

Group	I	II	III	IV	V	VI	VII	VIII	IX	X	XI	XII	XIII
Ten's digit	8	5	0	3	2	1	4	9	6	7	**84**	5	0

Now when this analysis is done then the framing of the question by a teacher becomes easy.

Open Access Except where otherwise noted, this chapter is licensed under a Creative Commons Attribution 4.0 International License. To view a copy of this license, visit http://creativecommons.org/licenses/by/4.0/.

Enacted Multiple Representations of Calculus Concepts, Student Understanding and Gender

Ileana Vasu

Multiple representations of mathematical ideas can provide students with a better understanding of mathematical concepts (Janvier, 1987). Well-chosen representations are powerful at conveying mathematical concepts. They can be effective at the novice level or for students who perceive themselves as weak in math. They may also provide access to mathematical concepts for those students who lack operational expertise.

Despite interest in multiple representations as a trademark of mathematical success, Calculus instruction is mostly symbolic in nature and lacks consideration of gender specific issues. The gender make-up of the students who opt out of STEM majors after taking Calculus is disproportionately female (Bressoud, 2011; Rassmussen, 2012). Many studies in mathematics education report student difficulties with multiple representations in Calculus but do not explore connections with the enacted curriculum.

Workshop Activities: In this workshop participants shared ideas about mathematical representations in the curriculum as they explored their connections with student understanding and gender. Questions addressed by participants include:

1. How do representations appear in the enacted curriculum and in assessments in Calculus?
2. What is the link between student experience with multiple representations and student understanding?
3. If experience with multiple representations is necessary for a deeper understanding, how should they be incorporated in the curriculum?
4. How do female students relate their learning to multiple representations?

I. Vasu (✉)
Holyoke Community College and University of Massachusetts, Amherst, MA, USA
e-mail: ivasu@educ.umass.edu; ivasu@hcc.edu

To tighten the discussion, participants were asked to examine Calculus exams from various colleges for the presence and quality of representations. Then, snapshots of Calculus curricula, including the use of multiple representations, the classroom culture and discourse, and the pedagogical approaches were then presented to the participants, in conjunction with videos of students in these curricula as they were solving Calculus problems.

Trends: Participants then fleshed out common themes, patterns, and emergent ideas based on the videotapes and student artifacts. The trends noted were:

- The majority of the assessments were symbolic and procedural. Most exams contained one other representation. No exam contained all representations.
- Student use of multiple representations is aligned with their experience in the classroom, but this is especially true for female students
- Students in the active curricula we examined, made use of more representations in solving problems, were more able to take risks when thinking about strategies and to back track when they reached areas of conflict.
- Students in the lecture class were limited in the number of strategies they attempted and ignored areas of conflict
- Curricula that use contextual representations help students become more self sufficient and generate multiple methods in solving Calculus problems.

These trends were in agreement with quantitative results obtained by the organizer. The workshop concluded that mathematical rich tasks in active curricula may serve as a catalyst for student understanding of Calculus and discussed possible teaching and research directions.

Reference

Bressoud, D. (2011). The Calculus I instructor. From *Launchings*. Retrieved from https://www.maa.org/external_archive/columns/launchings/launchings_06_11.html

Janvier, C. (1987). Translation processes in mathematics education. In C. Janvier *Problems of Representations in the Learning and Teaching of Mathematics* [Papers derived from a symposium organized by CIRADE of Université du Québec à Montréal] (pp. 27–32). Hillsdale, NJ: L. Erlbaum Associates.

Rasmussen, C. (2012). A report on a national study of college calculus: Who is switching out of STEM and why. *Plenary address at the 15th Conference on Research in Undergraduate Mathematics Education*, Portland, OR.

Open Access Except where otherwise noted, this chapter is licensed under a Creative Commons Attribution 4.0 International License. To view a copy of this license, visit http://creativecommons.org/licenses/by/4.0/.

Using Inquiry to Teach Mathematics in Secondary and Post-secondary Education

Volker Ecke and Christine von Renesse

Both organizers are co-principal investigators of the project "Discovering the Art of Mathematics" (www.artofmathematics.org) which is dedicated to bringing inquiry-based learning into mathematics classrooms. While the project was originally designed to work with faculty at the college or university level, both organizers have taken the work into the schools (K-12) on an ongoing basis. Additionally to co-authoring 11 freely available books with inquiry activities appropriate for high school and college level they also wrote a freely available electronic book about pedagogy in the inquiry classroom, including lots of videos, see https://www.artofmathematics.org/classroom.

For K-12 professional development, the organizers have done year long support of several local school districts, including mentoring teachers, co-teaching in the classrooms, bringing college students into the classrooms on a regular basis, facilitating K-16 professional learning communities and helping teachers develop inquiry-based materials. The college level workshops are listed at https://www.artofmathematics.org/workshops-professional-development. Additionally, both organizers have extensive experience in teaching pre-service teachers at the primary and secondary level.

Workshop

Using active learning and inquiry approaches in the mathematics classroom has positive effects on students' beliefs, attitudes and learning outcomes, see for instance the study by Freeman et al. (2014). Yet it is difficult for teachers to make the shift from traditional lecture style to a more active classroom happen, partially

---

C. von Renesse (✉)
Westfield State University, Westfield, MA, USA
e-mail: cvonrenesse@westfield.ma.edu

because most of us only experienced traditional teaching ourselves. In this workshop the participants first experienced inquiry-based learning as students. We then used the shared experience to discuss inquiry-based teaching and learning: what does it feel like as a student, what gets in the way of faculty exploring this way of teaching, and what are some of the many tools helpful for teaching successfully using inquiry (see Ecke & von Renesse, 2015 or https://artofmathematics.org/classroom/mathematical-conversations). The 90-minute workshop allowed participants to engage deeply in thinking about inquiry—from a student and a teacher practice perspective.

References

Ecke, V., & von Renesse, C. (2015). Inquiry-based learning and the art of mathematical discourse, *Problems, Resources, and Issues in Mathematics Undergraduate Studies (PRIMUS), 25*(3), 221–237.

Freeman, S., et al. (2014). Active learning increases student performance in science, engineering, and mathematics. *Proceedings of the National Academy of Sciences.* http://www.pnas.org/content/111/23/8410.full

Open Access Except where otherwise noted, this chapter is licensed under a Creative Commons Attribution 4.0 International License. To view a copy of this license, visit http://creativecommons.org/licenses/by/4.0/.

Making of Cards as Teaching Material for Spatial Figures

Kazumi Yamada and Takaaki Kihara

A static figure is used in the learning of plane figures. In contrast, it is important to present three-dimensional shapes and dynamic movements, when a teacher teaches spatial figures. There are the following advantages in using pop-up card creation as teaching material. When making a card, a three-dimensional card is completed by trial and error, making a cut in a plane (card) plan, and opening and closing a card repeatedly. In this process, the instruction that connected the plane figures and the spatial figures is attained. In particular, a pop-up card called "origami architecture" is effective as teaching material from this respect. When you open a card that is folded in two to 90°, the three-dimensional spatial object appears. When you fold this card, this card is returned to its original state (Figs. 1–3).

When people see a pop-up card, they often wonder how a solid is made from one sheet of plane paper. Students can observe the spatial motion of the work first and understand that a three-dimensional work can be made from a card by opening and closing it repeatedly. They will want to make an original pop-up card. They will consider how to write a plan on a flat card while imagining the state a card will be in while opening and shutting it. They will infer how a line on a plane changes into a solid edge. They may come to observe opening and shutting of a card from the front, then view it from various directions and observe. In particular, they will notice that it is important that they observe the state of the transformation of the section of the card from the side. They can learn to understand projection view through this activity.

K. Yamada (✉)
Niigata University, Niigata, Japan
e-mail: mathexpnet@hotmail.com

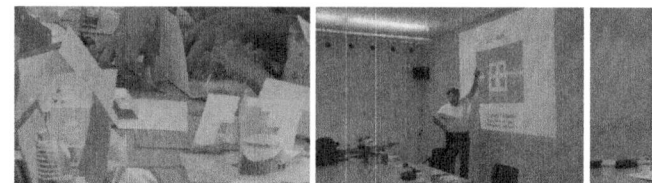

Figs. 1–3 Making of cards

Making a card stimulates the intellectual curiosity of students and they become excited and interested in solving the problem. Thus, pop-up cards are attractive teaching materials (http://www.maroon.dti.ne.jp/kihara/home.htm).

Open Access Except where otherwise noted, this chapter is licensed under a Creative Commons Attribution 4.0 International License. To view a copy of this license, visit http://creativecommons.org/licenses/by/4.0/.

Creative Mathematics Hands-on Activities in the Classroom

Janchai Yingprayoon

Many children find mathematics difficult and boring. But they are curious and they love to have fun with exciting things around them. Appropriate activities can be found to stimulate them to have fun and love learning mathematics. The workshop showed ways of developing creativity in mathematics and technology education to increase intellectual curiosity, develop problem-solving and thinking skills, promote discovery, and unleash creativity. There were five activities in the workshop.

Curves in Nature

A picture of a real bamboo stem and a sheet of graph paper were given to participants to find the relation between two variables using curves. The relation between the sectors of bamboo tree and their lengths were studied and discussed.

Reaction Time Test

From the given materials, participants studied the nature of a free-falling object by measuring reaction time. The plot between reaction time and the number of participants with that reaction time (frequency) was studied. The distribution plot tends to be a normal distribution.

J. Yingprayoon (✉)
International College, Suan Sunandha Rajabhat University, Bangkok, Thailand
e-mail: janchai@loxinfo.co.th

Simple Balance

The participants made their own simple balance from given materials. They learned about the principle of moment and how to calibrate their scales and calculate error in the measurement. The mathematic relation was discussed.

Mathematics of Robot Arms

A simple robot arm will be constructed using ice cream sticks. The learners worked on finding the mathematic relations of extended lengths of a robot arm.

Augmented Reality (AR) in Mathematics Education

This workshop described how to develop a simple AR system for learning mathematics. Sample AR materials used for mathematics education in high school and at the university level were discussed. The workshop showed how to generate and view geometrical objects in three dimensions using mobile phones or computer tablets in order to better understand their mathematical structures.

Open Access Except where otherwise noted, this chapter is licensed under a Creative Commons Attribution 4.0 International License. To view a copy of this license, visit http://creativecommons.org/licenses/by/4.0/.

Part IX
Additional Activities

Teachers Activities at ICME-13

Nils Buchholtz, Marianne Nolte and Gabriele Kaiser

The 13th International Congress on Mathematical Education in Hamburg was not only a scientific experience for researchers in mathematics education from all over the world. As Convenor and Local Chair of ICME-13, we are particularly pleased that also more than 250 teachers from all over Germany and five other countries have benefited from this congress. Despite the school holidays taking place in many federal states of Germany in July, mathematics teachers could take part in a special conference for teachers taking place parallel to ICME-13 on the 27th–29th of July 2017. On three days, the teachers were able to participate in a comprehensive lecture and workshop program, which included not only professional development courses in the area from primary to secondary mathematics education, but also thematically relevant excursions. A special highlight was already offered to the teachers before the official opening ceremony of the conference for teachers. They were allowed to participate at the ICME-13 Plenary Session by Günter Ziegler: "What is mathematics?—and why we should ask, where one should learn that, and who can teach it?" and thus take a look at the regular conference program.

The opening ceremony for the teachers at the Auditorium Maximum gave the expected program an appropriate setting. The opening speeches of Gabriele Kaiser (Convenor of ICME-13), Rudolf vom Hofe (President of the German Society of Didactics of Mathematics), State Councilor Michael Voges (Hamburg School Authority) and Marianne Nolte (Local Chair of ICME-13) were marked by the appreciation of the work of the colleagues from the school practice, but they also

---

Nils Buchholtz (Member of the Local Organizing Committee), Marianne Nolte (Local Chair of ICME-13), Gabriele Kaiser (Convenor of ICME-13).

N. Buchholtz (✉)
University of Hamburg, Hamburg, Germany
e-mail: nils.buchholtz@uni-hamburg.de; n.f.buchholtz@ils.uio.no

N. Buchholtz
University of Oslo, Oslo, Norway

discussed the importance of continuing professional development of teachers facing an increasing heterogeneity of students in schools. In this respect, the opportunities of professional development of ICME-13 were certainly unique. In his lecture on "Mathematical experiments—little effort, great impact" with many illustrations and mathematical experiments Albrecht Beutelspacher subsequently got the teachers in the right mood for the following sessions. Luckily, many members from the German Society of Didactics of Mathematics and the German Teacher Association for the Advancement of Mathematics and Natural Science Teaching (MNU) agreed to participate in the lecture and workshop program. With the offered excursions, the teachers also had some opportunities to get to know about teaching-relevant out-of-school learning opportunities, which provide mathematical and scientific learning content for students For example, the teachers were able to visit the logistics of the Hamburg harbor, learn experiments in a student laser laboratory, gain an insight into Hamburg's aluminum smelting and processing, explore the connections between mathematics and art in the Hamburg Art Gallery or take a mathematical city walk through the Hamburg city center. In the Auditorium Maximum, the teachers were able to visit the mathematical exhibition in guided tours and put their "hands on" many mathematical exhibits.

Open Access Except where otherwise noted, this chapter is licensed under a Creative Commons Attribution 4.0 International License. To view a copy of this license, visit http://creativecommons.org/licenses/by/4.0/.

Early Career Researcher Day at ICME-13

Gabriele Kaiser, Thorsten Scheiner and Armin Jentsch

Over 450 early career researchers from more than 50 different nations joined a special event at the 13th International Congress on Mathematical Education in Hamburg, Germany. On July 24th, the Early Career Researcher Day (ECRD) of ICME-13 took place comprising, an intensive one-day program of 18 workshops, and 6 sub-plenary survey presentations. The welcoming address by Gabriele Kaiser as convenor ICME-13 and the opening speech by Ferdinando Arzarello as ICMI president set the stage for multiple, rich learning opportunities for early career researchers advancing their understanding of scientific research and o allowing them to interact with more than 40 international experts in mathematics education. During the day, ECRD participants had the opportunity to attend workshops on empirical methods and lectures on important mathematics educational themes, both given by international experts in the field. Workshops included: Design Research, Mixed Methods, Video-based Research, Qualitative Text Analysis, Grounded Theory, Large Scale Assessments, Socio-Cultural Studies, Ethnographic Studies, Argumentation Analyses, Interaction Analyses, and Networking Theories. Lectures included: Theoretical Aspects of Mathematics Education Research, Frameworks and Principles for Task Design in Mathematics Education, False Choices in Research Paradigms, International Comparative Studies, Professional Education and Development of Teachers, and Thinking about Mathematics as Discourse. The afternoon was marked by a panel presentation and discussion on major journals in mathematics education given by journal representatives (Merrilyn Goos (Educational Studies in Mathematics); Jinfa Cai (Journal for Research in

---

Convenor ICME-13: Gabriele Kaiser.
Co-Organizers ECRD: Thorsten Scheiner, Armin Jentsch.

G. Kaiser (✉)
University of Hamburg, Hamburg, Germany
e-mail: gabriele.kaiser@uni-hamburg.de

Mathematics Education); Marcelo Borba (ZDM Mathematics Education); Carolyn Maher (Journal of Mathematical Behavior); Olive Chapman (Journal of Mathematics Teacher Education); Peter Liljedahl (International Journal of Science and Mathematics Education); Charalambos Charalambous (Mathematical Thinking and Learning), followed by workshops on academic writing and academic publishing (Richard Barwell, Helen Forgasz, Vince Geiger, Aiso Heinze, Jeremy Kilpatrick, Cynthia W. Langrall, Norma Presmeg). Alan H. Schoenfeld (University of California, Berkeley) provided a look ahead in his keynote presentation on 'What Makes for Powerful Classrooms, and How Can We Support Teachers in Creating Them? A Story of Research and Practice, Productively Intertwined'. The welcoming reception in the Congress Center following the ECRD provided further opportunities to intensify networking and socializing among early career researchers and international experts.

These activities will hopefully contribute to the establishment of firm promotional structures of early career researchers at the level of ICMI and the next ICMEs.

Open Access Except where otherwise noted, this chapter is licensed under a Creative Commons Attribution 4.0 International License. To view a copy of this license, visit http://creativecommons.org/licenses/by/4.0/.

The manufacturer's authorised representative in the EU is Springer Nature Customer Service Centre GmbH, Europaplatz 3, 69115 Heidelberg, Germany. If you have any concerns regarding our products, please contact ProductSafety@springernature.com

Printed and bound by CPI Group (UK) Ltd, Croydon, CR0 4YY

23/03/2026

02076671-0003